Blake's Design of Mechanical Joints

Mechanical Engineering

Series Editor:
L. L. Faulkner
Battelle Memorial Institute, and The Ohio State University
A Series of Textbooks and Reference Books

PUBLISHED TITLES

High-Vacuum Technology: A Practical Guide, Second Edition
Marsbed H. Hablanian

Shaft Alignment Handbook, Third Edition
John Piotrowski

Applied Combustion, Second Edition
Eugene L. Keating

Introduction to the Design and Behavior of Bolted Joints, Fourth Edition: Non-Gasketed Joints
John H. Bickford

Design and Optimization of Thermal Systems, Second Edition
Yogesh Jaluria

Mechanical Tolerance Stackup and Analysis, Second Edition
Bryan R. Fischer

Vehicle Dynamics, Stability, and Control, Second Edition
Dean Karnopp

Pump Characteristics and Applications, Third Edition
Michael Volk

Principles of Composite Material Mechanics, Fourth Edition
Ronald F. Gibson

Handbook of Hydraulic Fluid Technology, Second Edition
George E. Totten and Victor J. De Negri

Mechanical Vibration: Analysis, Uncertainties, and Control, Fourth Edition
Haym Benaroya, Mark Nagurka, and Seon Han

Blake's Design of Mechanical Joints, Second Edition
Harold Josephs and Ronald L. Huston

For more information about this series, please visit: www.crcpress.com

Blake's Design of Mechanical Joints

Second Edition

Harold Josephs and Ronald L. Huston

CRC Press is an imprint of the
Taylor & Francis Group, an **informa** business

CRC Press
Taylor & Francis Group
6000 Broken Sound Parkway NW, Suite 300
Boca Raton, FL 33487-2742

Printed on acid-free paper

International Standard Book Number-13: 978-1-4987-6629-6 (Hardback)

Library of Congress Cataloging-in-Publication Data

Names: Josephs, Harold, author. | Huston, Ronald L., 1937- author. | Blake, Alexander, author.
Title: Blake's design of mechanical joints / Harold Josephs and Ronald Huston.
Description: Second edition. | Boca Raton: CRC Press, Taylor & Francis, 2019. | Series: Mechanical engineering | Revised edition of: Design of mechanical joints / Alexander Blake. c1985. | Includes bibliographical references and index.
Identifiers: LCCN 2018022860| ISBN 9781498766296 (hardback: alk. paper) | ISBN 9781315153827 (e-book)
Subjects: LCSH: Joints (Engineering)
Classification: LCC TA660.J64 B43 2019 | DDC 621.8/2—dc23
LC record available at https://lccn.loc.gov/2018022860

Visit the Taylor & Francis Web site at
http://www.taylorandfrancis.com

and the CRC Press Web site at
http://www.crcpress.com

Contents

Preface

This book is a revision and update of Alexander Blake's very popular and highly regarded volume on the design of mechanical joints. The reason for this revision is to include recent developments and to bring the language and terminology up-to-date. At the same time, however, we attempt to retain Blake's lucid style by providing intuitive and "straightforward" analyses of complex, indeterminate, and multi-constrained joint and fastener problems.

The fasteners and joints of modern mechanical systems and structures are becoming increasingly sophisticated with requirements for lighter weight, improved quality and reliability, enhanced safety, longer life, and reduced cost. These are usually conflicting constraints. Therefore, another objective of this book is to enable the reader to make rational and optimal balances between these constraints.

Since there are mechanical joints in all machines and structures, whether moveable (articulating) or simply fixed fasteners, they present design and manufacturing issues for engineers, designers, and fabricators. Moreover, the kinds of joints and fasteners are now virtually unbounded. Therefore, in this book we attempt to expand and update each of Blake's original chapters. We also include a rather lengthy and self-contained chapter on the use of statistical methods for fastener design. Since statistical analysis may be new to many readers, we illustrate it with numerous worked out examples.

To enhance the utility of this book, we have also added worked out examples for several of the chapters that will further illustrate the use of Blake's intuitive and insightful style. In addition, we have developed a comprehensive glossary defining the terminology of the subject matter.

Blake's original volume was intended primarily for engineers and designers. We believe the present edition can well serve as a classroom text given the present inclusion of numerous examples and the expanded chapter of statistical methods applied to fastening systems. Additionally, the book will also be of interest as an independent study resource for students and as a supplement to mechanical design texts. Finally, and perhaps of greatest importance, the book is expected to serve as a working reference for anyone encountering difficult joint design problems and for those attempting to modify and/or optimize existing designs.

An underlying objective of the book is to provide readers with the ability to avoid the confusion often occurring with statistically independent problems requiring tradeoffs, and also to dispel incorrect notions and traditions which have led to both underdesigned and overdesigned joints and fasteners.

The book is intended to be readily accessible to those with a standard undergraduate background in engineering fundamentals (basic mechanics and strength of materials), calculus, and elementary physics. Each chapter provides the basic formulas for developing reliable, safe, low maintenance, and long-lived designs.

Although there is some "cross-over" between the various chapters, each chapter is intended to be self-contained, so that a designer/engineer with a specific problem may identify the appropriate chapter and thus readily find guidelines for solution.

The use of increasingly available computational mechanics hardware and software is encouraged for design refinement and for stress, strain, and graphical representation but only after the basic design is established using the methods provided herein.

The book itself is divided into 11 chapters with the first of these being simply an introduction to the overall subject matter. The next three chapters provide a basis for the detailed design and analysis in the balance of the book.

The second chapter provides a review of mechanics of materials, including fracture mechanics.

Then in Chapter 3, we provide a rather extensive overview of fasteners with an emphasis upon rivets and bolts.

Finally, in Chapter 4, we introduce the new chapter (not in Blake's original volume) on statistical methods. This is a self-contained chapter for readers with little or no prior knowledge of statistics. The objective is to enable readers to make statistical analyses so frequently needed and useful in modern design.

Beginning with our fifth chapter, we provide more detailed discussions. Specifically, we look in greater depth at riveted and bolted joints. The next two chapters discuss flanges, stiffeners, clamps, and piping joints.

In Chapter 8, we consider couplings and pin connections including the design of keys, splines, pins, and links. The last three chapters discuss welding, membranes, diaphragms, and adhesions. Each chapter has an extensive list of references and a list of symbols, and there is a glossary of terms used throughout the text.

We acknowledge and appreciate the patient work of Charlotte Better who typed the entire manuscript. We are also appreciative of the work of Jeremy Czanski in the initial planning of the book. We are grateful to the publisher for the continual encouragement and patience with the many delays in bringing the work to completion.

Authors

Harold Josephs is a Professor Emeritus of Engineering at the Lawrence Technological University in Southfield, Michigan. Dr. Josephs received his BS in Mechanical Engineering from the University of Pennsylvania, his MS in Engineering from Villanova University, and his PhD from the Union Institute. In 1983/1984, he was awarded the title of Outstanding Engineer in Education by the Michigan Society of Professional Engineers. In 2012, Dr. Josephs was a Visiting Professor in the Department of Mechanical Engineering at the Shanghai University of Engineering and Science. From 1986 to 1999, Dr. Josephs served as the Director of the Fastener Research Center at Lawrence Technological University. Dr. Josephs is the author of numerous peer-reviewed publications, co-author of two engineering texts, holds nine U.S. patents, and has presented numerous seminars to academe and industry in the fields of safety, fastening, bolting, and joining. He is a licensed professional engineer, a certified safety professional, a certified professional ergonomist, a certified quality engineer, a certified reliability engineer, a fellow of the Michigan Society of Engineers, and a fellow of the National Academy of Forensic Engineers. Dr. Josephs maintains an active consultant practice in safety, ergonomics, fastening and joining, and accident reconstruction.

Ronald L. Huston is Professor Emeritus and Distinguished Research Professor at the University of Cincinnati. He is also Director of the Institute for Applied Interdisciplinary Research. He is a member of the Department of Mechanical and Materials Engineering. Dr. Huston received a BS in Mechanical Engineering in 1959; an MS in Civil Engineering in 1961; and a PhD in Engineering Mechanics in 1962 with all degrees awarded by the University of Pennsylvania, PA. He is a licensed professional engineer (PE) in Ohio, Kentucky, Alabama, Louisiana, and Texas. He has been a faculty member at the University of Cincinnati since 1962; Head of the Department of Engineering Analysis 1969–1975; Acting Senior Vice President and Provost 1982; and Interim Head of Chemical and Materials Engineering 2002–2003.

During sabbatical leaves he served as a Visiting Professor of Applied Mechanics at Stanford University (1979) and as Visiting Lecturer of Mechanical Engineering at Tianjin and Chongqing Universities in China (1985). In 1979–1980, Dr. Huston served as Division Director of Civil and Mechanical Engineering at the National Science Foundation. His fields of research are: Biomechanics and Cell Mechanics. He is the author of 171 journal articles, 156 conference papers, 8 books, and 89 book reviews. He has served as a paper reviewer for over 35 journals. He has served as Technical Editor of Applied Mechanics Reviews, Associate Editor of Journal of Applied Mechanics, and regular reviewer for Zentralblatt Math. He is also a regular reviewer for the Hong Kong Research Council. He has been the recipient of 34 research grants and contracts. He is an active consultant in safety, accident reconstruction, and injury biomechanics.

Authors

1 Introduction

1.1 HISTORICAL REVIEW

We can trace the appearance of bolted joints to the very history of technology. There is evidence that threaded fasteners were first used in connecting a base (or stem) to the bowl of goblets. Surgical instruments with mechanical joints date back to about 79 AD.

In the 6th century, golden bracelets had small screws holding two halves together. Around the beginning of the 15th century, rivets began to appear—a milestone in the development of true fasteners.

Despite their obvious limitations, rivets were then used extensively in the construction of machines, ships, buildings, and machinery. But an initial drawback in the use of rivets in manufacturing assembly was that very few components were interchangeable. Finally, in the 18th century, Whitworth in England and Sellers in the United States developed interchangeable bolted joints based upon knowledge gained in rivet applications.

Nevertheless, it was not until the first quarter of the 20th century that engineers began to employ analysis for efficient design of mechanical joints. In 1927, Rotscher [1] began to question the relation of preload to external load in a bolted joint. In essence, he examined the contribution of an external loading of the bolt using springs as models for the bolt and the components to be held together.

Since then fully engineered fasteners have been recognized as the fundamental component of assembled members of products—particularly metal products. This in turn resulted in various rules to guide future bolt designs and loading limits. Indeed, a number of select committees of the engineering societies have developed detailed guidelines and standards for the design of both bolted and riveted connections.

In this book, we focus upon developments since World War II [2, 3]. These developments have occurred worldwide with contributions from the United States, the United Kingdom, Canada, Germany, Holland, Austria, Japan, France, Belgium, Switzerland, Italy, and others as well.

In view of these developments and advanced design, the task of evaluating and compiling the thousands of references fell largely upon universities and regulatory bodies with the onus of establishing standards largely replacing the long established "rules of thumb" so that reliable design manuals could be written.

For example, prior to World War II, the focus of engineers was to determine the optimal tightness of bolts in frame assembly. The concern was balancing the strength of the bolt versus the tendency of yielding at the joint. This was commonly referred to at that time as the "balanced design concept." From tests at the time it was concluded that the minimum yield strength of a steel bolt should not be less than about 50,000 psi. It was also concluded that the fatigue strength of a bolt should be as strong as a well-driven tight rivet. This implied that the bolt preload should be in the plastic range. But then the issue surfaced—how to accurately attain this condition. Faced with a number of fatigue failures in floor beams, and similarly intermittently loaded structures, the Research Council of the American Society of Civil Engineers (ASCE) and the American Railway Engineering Association were in the forefront of high-strength joint development. Their studies concluded that excessive rivet-bearing loads could be substituted by a high clamping force provided by bolting.

The pioneering work of these technical societies culminated eventually in material specifications for high-strength bolt applications—and for the first time it was officially recognized that rivets could be replaced by bolts on a one-to-one basis. This recognition by the ASCE Research Council was published in 1951—a rather recent development when viewed in the context of the overall history of bolted joints.

A number of tests and field investigations followed, which essentially confirmed the quality and even superiority of high-strength bolts over rivets.

There were also early indications that the fatigue strength of bolted joints could exceed the fatigue strength of riveted joints. Additional tests were made at the same time on flat plate-joints to determine the basis for joint efficiency and a number of empirical rules were developed, including the strength of hot-driven rivets subjected to combined shear and tension. This appeared to be a natural reaction to the progress made in the area of a bolted-joint design and it also helped with the studies of riveted connections. Extensive experimental and theoretical investigations of riveted double-shear aluminum joints addressed a number of questions concerning the behavior of various types of double-shear splices.

The work on bolted and riveted joints and the relevant U.S. specification had an additional impact on the research in Germany and the subsequent issue of the German Code of Practice in 1956. The design methodology and experience with nonslip, high-strength bolted connections, according to German records, was finally reviewed in 1959 [4].

About the same time, considerable attention was paid to the theory, methods, and equipment of bolt pretension control leading to the development of calibrated torque wrenches. This particular era of investigations resulted in a pretension procedure still known as one-full-turn-of-the-nut pretention method starting from the finger-tight (or snug) torqueing preload of the bolt.

In the usual cases of joint design, when rivets or bolts are used, the questions of realistic bearing and shear strength criteria must be addressed. A number of investigations between 1952 and 1958 were concerned with the strength of the fasteners in shear and the effect of bearing stresses on the integrity of a mechanical joint. The effects of punched holes, misalignment, slip, and tension on the rivet and bolt assemblies were also studied. These studies and other tests led, finally, to the modified turn-of-nut procedure and bolt installation techniques for bridge construction in the United States.

The general practice of bolted joint design in the United Kingdom was similar to the American procedures and resulted in the British Standard 3294, dealing with the bolted joint design methodology and field applications [5]. Different schemes for bolting interior beam-to-column connections were analyzed and tested to prove that the bolted joint could be designed to develop stress levels with the connected material.

As the reliance on bolted connections developed, it was officially recognized that bolt overall strength properties were superior to those of rivets. However, this analysis considered the shear carried by the bolts as designed to go through the body and not through the threads. At the same time, the allowable stresses in tension were increased significantly to be equal to the bolt-proof stresses because the tests demonstrated no adverse effects on fatigue resistance.

It was generally realized that relatively high stress concentrations would develop under the head or nut of the heavy bolt and several studies were made concerning the practice of applying hardened washers as the key components of the bolted assemblies. The original results from such studies, however, indicated that a heavy structural bolt with or without such washers had the same performance under static and fatigue conditions. These findings led to the modification of the original specifications so that the washers were not mandatory after 1962.

With time, research on bolted joints and individual fasteners became more sophisticated. Experiments on large bolted and riveted lap joints showed that the end bolts sheared prior to the development of the full strength of the joint. Particular attention was then paid to the slip behavior (where slip is defined as lateral relative motion of two adjoining bolted parts) and the ultimate strength of the joint, and it was concluded that the allowable stresses should be based on the shear strength of the individual fastener rather than on the original concept of balanced design. Gradually, the studies were extended to the problem of friction in a bolted connection and mathematical models were developed for the relationship between deformation and loading throughout the elastic and plastic regions of bolted plates. The emphasis of these studies focused on the effect of a joint length, pitch, and variation in fastener diameter and bolt shear area. The studies ranged from the treatment of a single fastener under combined tension and shear, up to the experimental analysis on a full-size bolted railway bridge. It was learned that the bolted bridge connections did not slip under the full working loads. Given the different evolving design criteria in various countries, wide differences still existed in the slip coefficients and the factors of safety in various country standards. The resistance to slip was often modified by the galvanizing process or through the use of epoxy adhesives. The turn-of-nut method, at the same time, was considered to be more reliable than that of the conventional torque control utilizing wrenches. Beeswax was found to be the best for lubrication purposes, yielding an average slip coefficient of approximately half the normal value for that of clean mill scale. The resultant specifications also characterized a number of joint and bolt materials.

The time interval during the 1950s and 1960s may be termed the "golden era" of developments in the field of high-strength bolted connections [7]. This is again a remarkably short period of time when considering the early beginnings with the bowl goblets of 79 AD or bracelets in the 6th century.

1.2 AN OVERVIEW OF MECHANICAL JOINTS

Traditionally, the design of joints and fasteners has focused upon individual structural components and/or requirements for a particular component or subassembly. Nevertheless, numerous models and design formulas have been developed from fundamental principles which can shortcut and improve upon the traditional "rule of thumb" approach. Moreover, with continued studies and use, these formulas have become refined and increasingly reliable in terms of safety, reliability, and economy.

In recent years, these formulas and their refinements have been enhanced by finite element methods and associated numerical analyses.

As opposed to other structural elements, the design of mechanical joints presents a special challenge. The behavior of a joint is often so complex that a single design formula cannot usually be used to make a reliable design. Emphasis on a single formula, while useful and convenient, may result in neglect of important aspects of material characteristics of the various joint components, leading to defects in the overall design. Consequently, a poorly designed joint can become this weak link in the overall behavior of a structure or machine.

The principal objective of a mechanical joint is to hold parts together and thereby transmit a loading force from a given structural component to a corresponding adjoining component. Since an external load creates a stress field, the mission of a mechanical joint is to transfer the stress across a mechanical boundary—and hopefully, without a deleterious gradient.

Figure 1.1 depicts an example of a joint as an adhesive section (a "scarf" joint). The inclined dark line represents a glue joint, fastening together the left and right ends of wood components.

Figure 1.2 provides a similar connection for a metal structure in the form of a "butt" joint. Here the adhesive is a weld holding the left and right ends together.

Among the obvious disadvantages of a welded joint is the possibility of a tensile failure in the case of crack propagation across a heat affected zone (HAZ). Nevertheless, the stress concentration per se is avoided because of a relatively smooth transition from one side of the joint to the other.

Although there are similarities between scarf and butt joints we consider various types of welds in Chapter 8 and then design requirements for adhesive scarf joints in Chapter 10.

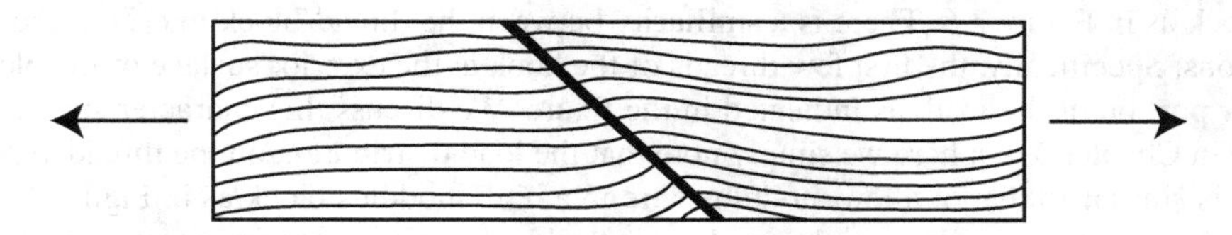

FIGURE 1.1 A scarf joint.

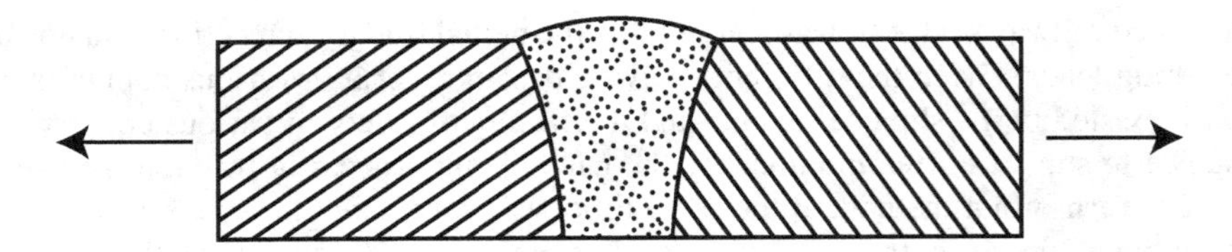

FIGURE 1.2 A butt joint.

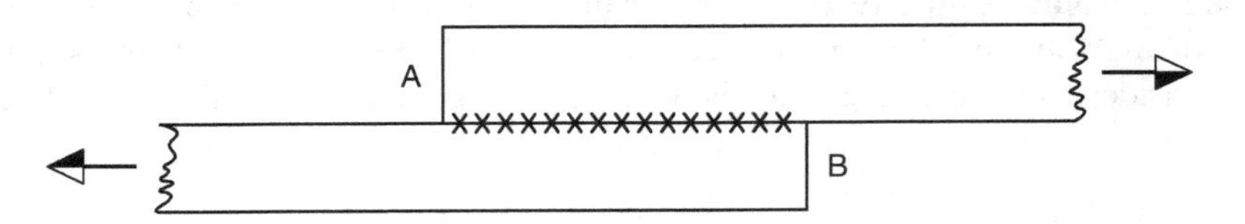

FIGURE 1.3 A lap joint with stress concentration.

When a mechanical joint involves a lap configuration such as that shown in Figure 1.3, some stress concentration cannot be avoided. The actual bonded length is AB with the maximum stresses likely to develop at A and B.

As long as a lap joint is relatively rigid and secure (and not requiring replacement), it makes little difference whether the joint is riveted, bolted, welded, screwed, nailed, or glued.

Figure 1.4 shows yet a different kind of joint where a nylon thread is cast into a hardened plastic block. Depending upon the relative modulus (or rigidity) of the thread and block it can be shown that the higher stress (loading) is either at the interior end or exterior beginning of the thread.

This last example illustrates an important principle of joint mechanics: When a low-stiffness thread is being pulled out of a relatively rigid block, the resistance of the thread to pull-out is essentially independent of its length; and conversely, when a stiff thread is being pulled out of a relatively soft block, the thread length imbedded in the block increases the pull-out force. Thus, as far as a bird is concerned, a short worm is just as hard to extract from the same soil sample as a long one.

This interesting principle has important practical application where it may be necessary to make a long and complicated cavity in a relatively large block of a softer material. Suppose, for example, that we want to place several steel wires in a frayed out configuration within the body of a plastic (polymer) block as represented in Figure 1.5. It might appear that the design of the attachment between the wires and the block might be secure due to the frayed-out configuration of the wires. But with the metal (say steel) wires being stiff relative to the polymer (say soft plastic) block the stress distribution along the wires is similar to that of the lower example of Figure 1.4. Then an issue which arises is that not all wire stresses are likely to be identical. Then, as one wire approaches the adhesive strength and begins to pull out of the block, the remaining wires must carry an additional load. In practice then, all wires eventually lose their grip, one by one. The conclusion then is that the joint of Figure 1.5 can at times carry only a small fraction of a possibly anticipated design load.

Joints relying on adhesion forces between metals and polymers (plastics) must be designed carefully. The reason is that as soon as the plastic reaches its yield strength, the adhesion will begin to fail. We discuss various aspects of adhesive bonding in Chapter 10.

For another illustration, consider a threaded connection between a threaded metal hook and a metal block as in Figure 1.6. There is a similarity between the thread/block interface and adhesive connections: Specifically, the first few threads of the hook at the exterior surface of the block carry the major portion of the load, as indicated in the figure. We discuss this characteristic of threaded fasteners in Chapter 3, but here we simply note that the load distribution on the threaded portion of the hook is similar to that of a low modulus wire in a high modulus block as in Figure 1.4. From a practical perspective, with the initial threads near the block surface carrying most of the load, there is very little to be gained in making the threaded connector longer.

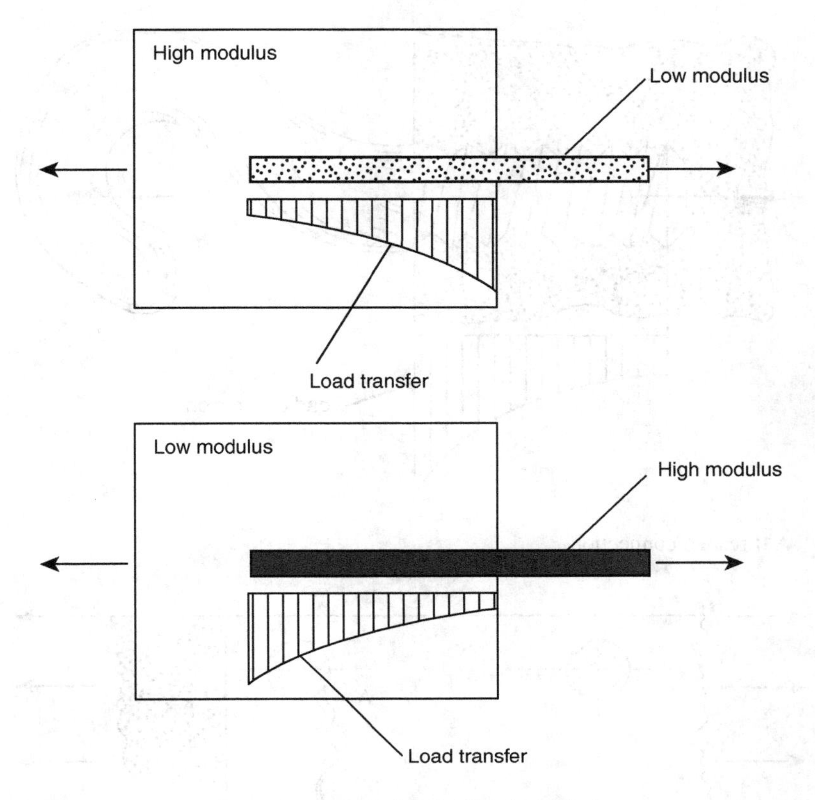

FIGURE 1.4 Dependence of load transfer on material rigidity (modulus).

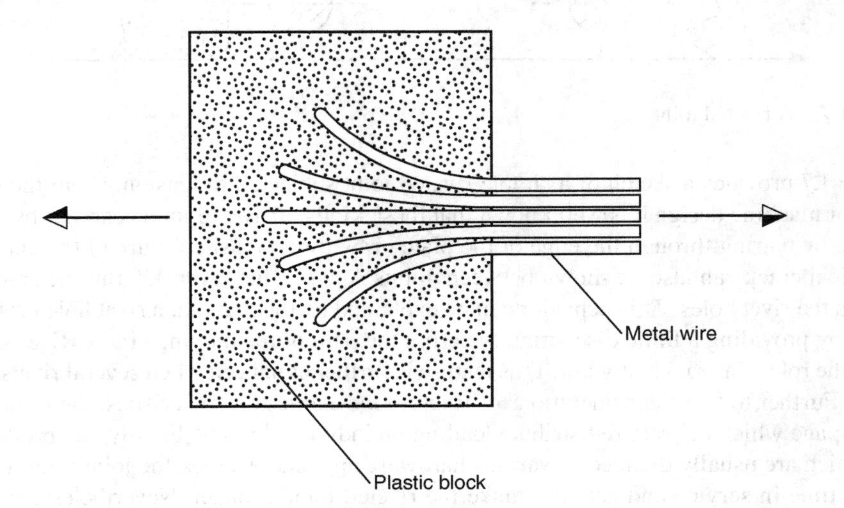

FIGURE 1.5 Frayed-out wire joint.

In modern manufacture of machine structures traditional riveted fasteners are increasingly being replaced by welded connections. The reason is that welded connections are lighter and less expensive than rivet fasteners. This is an example where economic considerations determine the engineering design. But, riveted fasteners are very reliable and easy to inspect. Moreover, design procedures for rivets are well established and rivets provide a number of advantages in complex joints found in many modern machines and structures.

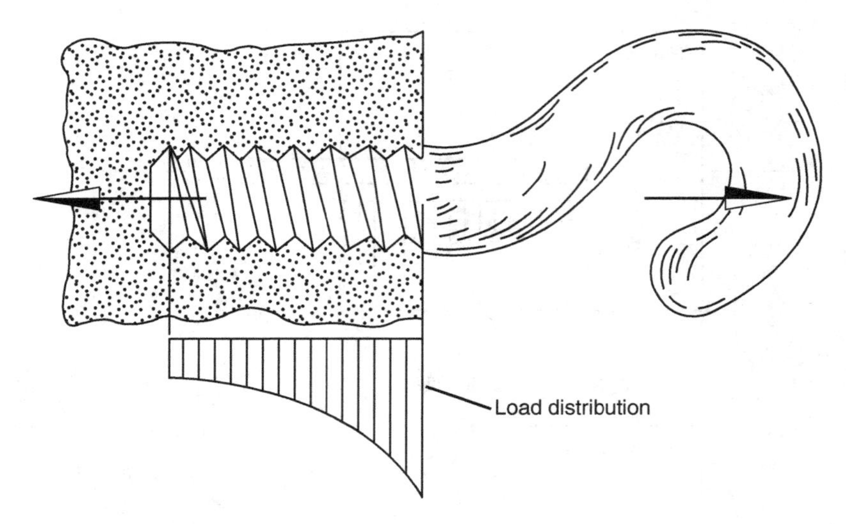

FIGURE 1.6 A threaded connection.

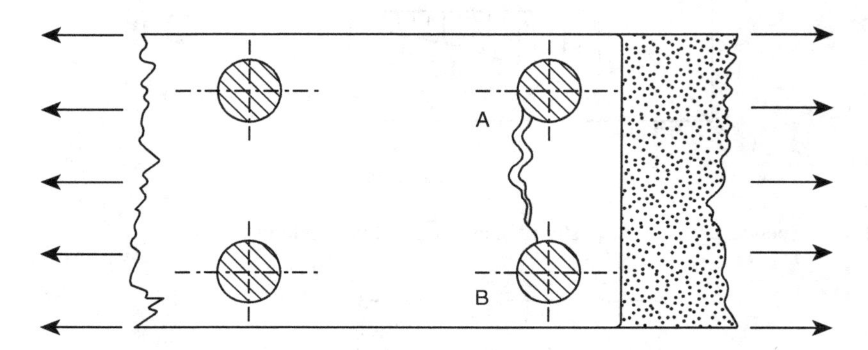

FIGURE 1.7 A riveted joint.

Figure 1.7 provides a sketch of a simple riveted joint subjected to tension. From the theory and practice of machine design it is well known that these kinds of riveted joints can fail by shearing of the rivets, by tearing through the edge of the plate or by direct tensile failure of the entire plate. A hypothetical crack can also be shown between points A and B in Figure 1.7. But the crack does not get across the rivet holes. This behavior can be explained by the fact that a rivet hole acts as a crack inhibitor by providing a finite discontinuity across the path of the moving crack. Riveted joint then can play the role of a crack mitigator. This feature is further enhanced when several rivets are placed in series. Further to these considerations of fracture mechanics characteristics, the joint undergoes some slippage which helps to redistribute loading on individual rivets. Finally, the products of corrosion which are usually dreaded in various hardware applications lock the joint after a reasonable length of time in service and actually make the riveted joint stronger. Nevertheless, care must be taken during the fabrication of the riveted joints. Specifically, rivet holes should be reamed after a punching operation to remove brittle layers of material caused by the plastic working of the plate. Chapter 4 provides additional details of a riveted joint design and practice.

In general, the analysis and design of mechanical joints in tension is rather difficult and often require special attention to subtleties involving both design and configuration. The difficulty is in the variety of material properties and the complexities of the geometry and the mechanics of joint interaction. For example, masonry joints are very weak in tension. That is, mortar in brick structures primarily transmits compressive loads with adhesion of the bricks being a comparatively weak feature.

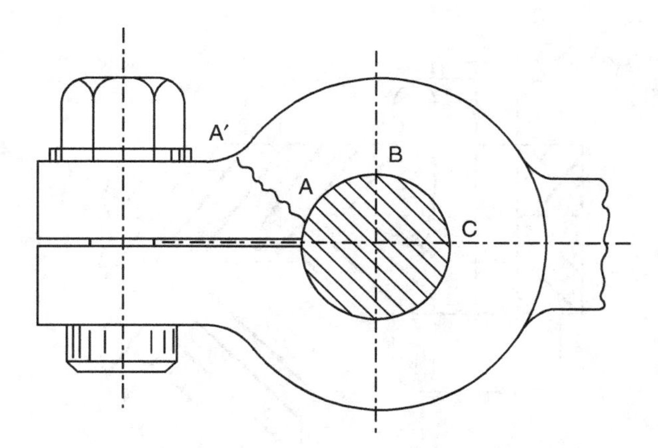

FIGURE 1.8 A split hub joint.

In wooden roof-truss applications, the joints are often made by using wooden pegs or steel straps.

It is remarkable how many different approaches have been tried to design adequate connections at the end of a tension member. These approaches range from a conventional knot used in ancient times to today's high tensile capacity eye bar joint. A thorough understanding of joint design, configuration, and analysis is necessary to prevent joint failure and the potential for catastrophic or even fatal results.

From a practical viewpoint a knot or a splice cannot be very economical or light. Alternatively the concept of an eye bar, no matter how attractive, also has limitations. The stress distribution is complex and the materials may not be sufficiently crack resistant.

In building suspension bridges, for example, short links are often used as mechanical joints. These links need to be made of tough and ductile material to safeguard against failure.

Chapter 8 presents stress and design analyses for eye bars and links in some detail. It may also be noted that in biological systems the overall connection at a tensile joint is provided by the attachment of ligaments to the bones.

We frequently learn with dismay that an assembly of relatively simple components in a mechanical joint is seldom easy to analyze. For example, consider the case of a split hub held by a system of clamps and bolts about a circular shaft as in Figure 1.8 wherein we assume all surfaces to be in perfect contact. We must first address the distribution of contact stress around the circular contour ABC taking into account the effect of friction [6]. Assuming that we can calculate the distribution of pressure, we would then attempt to determine the preload needed on the bolt to hold the right contact pressure and not to overstress the joint section such as that shown by A'A in Figure 1.8. If we next add considerations for the frictional torque on the circular shaft, and/or the axial force needed to keep the entire joint in equilibrium, the number of equations and assumptions needed to solve the design problem is significant. It then becomes clear that knowledge of strength and rigidity of individual components is a necessary but not sufficient condition of joint mechanics because of the often subtle effects of interfaces. It may be stated with little oversimplification that joint mechanics is in fact the mechanics of interfaces.

In Chapter 7, we consider various forms of clamping devices and similar joints.

Flanged connections also constitute a large and complex set of mechanical joints. In Chapter 5, we discuss flanged connections involving plane flanges used in piping including high pressure joints with gaskets and heavy flanges intended to provide both rigidity at the connection and also leakage control. Again, we begin with a relatively simple assemblage of components such as shown in Figure 1.9. In considering the mechanics of this joint, the relative rigidity of the various components such as the bolts, the flanges, and the gaskets will affect the outcome of the analyses. The extent of gasket compression depends on the bolt preload, which in turn is a function of the torque designed to

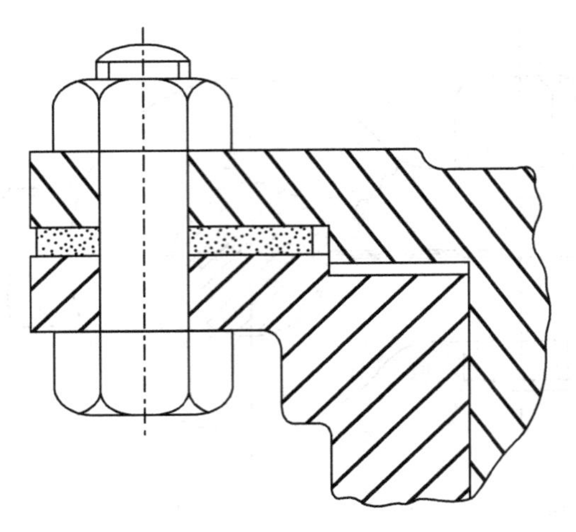

FIGURE 1.9 Another example of bolted joint design.

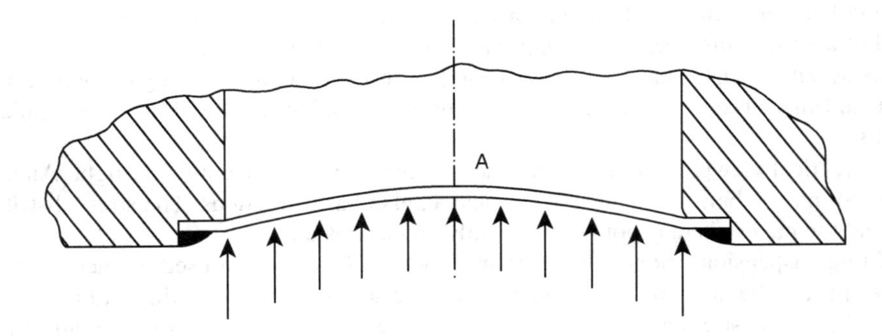

FIGURE 1.10 A window joint.

overcome the frictional resistance in the bolted assembly. The parameters of friction, compressibility of gasket material, and the overall stress analysis of the joint can quickly complicate the global joint analysis, design, and quality control.

There are a number of mechanical interfaces in the design of machines which perform the functions of separating liquids or gasses under high pressure differences. These interfaces need to be included in pressure vessel design. To bring out the important features of these interfaces we devote a portion of a chapter to consider windows, diaphragms, and expulsion devices. For example, a window panel made of a nonmetallic material may be designed to contain a vacuum in a diagnostic device. Figure 1.10 provides a simple illustration, where A denotes a vacuum region. The window must, therefore, be tested to withstand an atmosphere of pressure (14.7 psi or 100 kPA). The window, together with its clamp or adhesive seal at the periphery must be able to withstand bending and membrane forces occurring as the window is deflected. The design may thus be required to accommodate large window deflection and creep effects particularly if the device is destined to withstand a pressure/vacuum environment over a long time period. The design and analysis of these kinds of joints requires skill in stress analysis, knowledge of material properties, familiarity with fabrication procedures, and awareness of experimental results.

Chapters 7 and 9 of this book, however, include some of the design formulas and evaluations of more commonly needed mechanical joints, as well as those cases of joint technology which are not always readily documented in the open literature. However, we will not include discussions of all the types of mechanical interfaces involving friction drives, splines, step joints, shrink-fit assemblies, and other systems discussed in many conventional books on machine design.

The foregoing examples of fastening and joint configurations are intended to illustrate some of the typical problems of the mechanics and design discussed throughout the book. Gordon [6] provides a good review of the many factors which have influenced the development of mechanical joints throughout their history including their successes and failures going back to the design of ancient ships to that of today's automobiles, aircraft, and space vehicles. Structural optimization to provide a robust joint has consistently been the goal in these designs. Unfortunately, however, in the manufacture of many products and machines with widespread application, lack of a proper analysis and economic constraints often prevent the desired optimization.

Another inhibition to obtaining optimal design is our limited knowledge of the strength and rigidity of available materials. Tables of strength and elastic moduli are general and for materials without internal defects. This in turn requires designers to employ factors of safety.

When a given machine is to be designed for rigidity, the external or applied stresses are often very low and in such cases it is unlikely that the overall strength of the joints in the machine will be critical. Nevertheless, it is probably useful to design for joint flexibility to redistribute the external loading. This may be important, for example, when using wood and other construction materials sensitive to environmental conditions such as changes and extremes in temperature and humidity.

The standard objective of weight optimization in mechanical systems has been continually plagued by poor quality control and poor reliability or by the limitations of the mechanical joints. This is especially the case with ships, aircraft, and windmills. With ships and aircraft the problem of obtaining reliable, but yet lightweight joints has persisted throughout the 20th and 21st century. But in recent years there have been significant advances in materials (for example, carbon fiber composites) and in adhesives.

It has been said that a good engineering design philosophy is nothing more than our ability to predict the behavior of existing structures or selecting the configurations and materials for new designs. Current machines and structures have evolved over a long period of time in a competitive market and ever improving numerical analysis technology. But even today the progress in design of tension structures and machine elements is marred by the difficulties in devising end fittings and connectors to efficiently and reliably support the applied loads. There are still many theoretical and practical issues to be resolved. The elementary problems of weight reduction and material costs are very challenging—particularly in the manufacture of large quantities of identical products.

The stress distribution in all end fittings is always complex and there is always a potential for stress concentration, crack development, and loss of toughness. It is widely accepted that toughness is likely to diminish as tensile strength increases. It appears, therefore, that we are faced with incompatible material requirements. In the case of a tension structure, we would like to use a high-strength material for the tie-bar and a tough material for the end joint. This implies that the joint material must be of low strength to assure a sufficient toughness. A compromise, however, is not difficult to obtain in the case of end fittings for the cables on modern suspension bridges. There is perhaps a mile of wire between the two anchorages of the cable, balancing the control weight and cost losses for even the most complicated design of the end fittings. Such a compromise does not exist in the case of chain links, plate links, riveted brackets, and similar connectors where a weaker but more ductile material is recommended. The application of ropes, wires, and tension rods is on an increase in various mechanical systems, making the problem of mechanical joints and fittings of special importance.

REFERENCES

1. Rötscher, F. 1927. *Die Maschinenelements.* Berlin: Decker.
2. DeJonge, A. E. R. 1945 *Riveted Joints: A Critical Review of the Literature Covering Their Development.* New York: American Society of Mechanical Engineers (ASME).
3. American Society of Civil Engineers (ASCE). 1967. *Bibliography of Bolted and Riveted Joints.* New York: American Society of Civil Engineers (ASCE).

4. Appendix. A Translation of the Preliminary Directives for the Calculation, Design and Assembly of Non-Slip Bolted Connections (H.S. Connections) for Steel Structures, Bridges and Cranes. German Committee for Structural Steelwork, Jubilee Symposium on High Strength Bolts, Institution of Structural Engineers, London, 1959.
5. General Requirements for the Use of High Strength Friction Grip Bolts in Structural Engineering, Part 1. *General Grade Bolts*, British Standards Institution, No. 3294, London, 1960.
6. Gordon, J. E. 1978. *Structures or Why Things Don't Fall Down.* New York: Plenum Press.
7. Dobrovolsky, V. et al. 1968. *Machine Elements.* Moscow: Mir Publishers.

2 Basic Concepts of Fasteners

2.1 INTRODUCTION

As we noted in Chapter 1, almost any system and/or subsystem contains parts that are joined together to form a joint. When considering the various approaches to fastening, the designing engineer is faced with an enormous number of choices, types, and materials of fastener types and approaches that can be utilized. Obviously, the engineer's choices are constrained by the design requirement and/or specifications defining the system/subsystem applications. The initial consideration to be analyzed by the engineer is to determine if the subject joint is to be either a permanent or a detachable joint. A detachable joint is defined herein as a joint that can be assembled and disassembled without any structural or geometric change to the fastener. A list of typical permanent/detachable fastener category examples is presented in Table 2.1.

A brief overview of the primary joining approaches will be presented.

2.1.1 Adhesive Bonding

A large spectrum of various materials can be joined using adhesive technology. Materials are joined adhesively by intermolecular or interatomic bonding. This permits the joining of many dissimilar materials using adhesive bonding, e.g. ceramic to metal, plastics to metal, metal to metal, etc. The aerospace industry has paved the way for the increasing use and application of adhesive bonding in industry.

An overview of the advantages and disadvantages of adhesive bonding is presented in Table 2.2.

2.1.2 Brazing and Soldering

The applications of brazing and/or soldering joints are fairly limited with the major applications now being oil cooler and radiator seams. Being cheaper than welding, brazing, and soldering joints can be utilized where there are relatively low stress and low temperature service loads and environments. Solder filler materials melt below approximately 840E F and brazing filler materials melt above approximately 840E F. Typically, soldered joints utilize metal formed edges to transfer loads across the joint. Advantages/disadvantages of soldered/welded joints are shown in Table 2.3.

2.1.3 Rivets

The use and applications of rivets in industry has been declining in recent years. However, there are a number of niche uses which still champion the use of rivets. An overview of the advantages and disadvantages of rivet bonding is presented in Table 2.4.

2.1.4 Threaded Fasteners

As with other joining processes, the primary function of most (but certainly not all) threaded fastener joints is to hold two or more things together. Some bolting systems are utilized to pull things together as part of assembly operations, while other bolting systems are utilized to counteract shear forces. As we shall see in the coming chapters a threaded fastener joint can be

TABLE 2.1
Broad Category Examples of Permanent/Detachable Joints

Permanent Joints	Detachable Joints
Adhesive	Couplings and clamps
[a]Braze/Solder	[b]Interference fit
Button joints	Keys/Keyways
(Modified captured rivet)	Pins
Corrosion	Retaining rings
[b]Interference fit	Set screws
Rivet	Snap fasteners
Weld	Splines
	Threaded fasteners (bolts and screws)

[a] Can at times be disassembled with the judicial application of heat.
[b] Interference fit joints can often be disassembled by heat application.

TABLE 2.2
Advantages/Disadvantages of Adhesive Joints

Advantages	Disadvantages
Can join thick or thin materials	May have limited shelf life (especially two-part adhesive systems)
Can join dissimilar materials (as noted above metallic and nonmetallic)	May require extended set time
Provides uniform stress distribution over joint	Possible need for heat application and/or surface preparation
Spreads load uniformly over a large area	Adhesive joints often have temperature limitation—typically at about 500E F
Lack of required perforation, minimizes stress concentration	Sustained loads could cause adhesive joints to creep
Provides leak-proof joints	Potential for outgassing
Minimizes joint weight	Disassembly often requires total joint destruction
Minimization of stress risers, enhances fatigue resistance	

TABLE 2.3
Advantages/Disadvantages of Brazed and Soldered Joints

Advantages	Disadvantages
Provides good electrical and thermal conductivity	Higher service temperatures limit application of brazing and welding
Less expensive than welding	Joint strength is usually lower than strength of joint materials
Liquid tight joints can be formed	Pre-operation surface cleaning is typically required
Joints can be formed from dissimilar metals	Use of toxic fluxes will require ventilation
Complex components can be built (in stages)	
Some joints can be disassembled by the judicious application of heat	

modeled as a series of springs. However, caution is in order in that the spring energy and therefore the stress distribution in the fastener spring system is non-uniform. This results in the fastener joint being an unstable system wherein multiple long- and short-term effects can (and often do) modify the distribution of stress and stored energy in the system. The terms bolt and screw

TABLE 2.4
Advantages/Disadvantages of Riveted Joints

Advantages	Disadvantages
Can join dissimilar materials	Little control over clamp load
Can join sandwich materials	Difficult (but often possible) disassembly of riveted joints
Some rivet operations can provide an almost flush surface	Installation often requires access to both sides of the joint
Provides high shear strength	Comparable bolts and nuts have higher tensile strength
Can provide vibration resistance	
Low per rivet cost	
Automated assembly possible in some applications	

TABLE 2.5
Advantages/Disadvantages of Threaded Fastener Joints

Advantages	Disadvantages
Provides for disassembly and reassembly	Unintended loosening (vibration loosening) can result from improper clamp load.
Provides for customer/consumer application using common and inexpensive equipment	Operator/technician may introduce errors or defects, e.g., cross-threading, missing bolts, wrong bolts, improper torqueing and preload, etc.
Minimum operator skills required	Bolted joint is typically weaker than the joined object.
Provides high tensile force (clamping force) in a small area	Stress concentrations are introduced.
Provides for controlled and measurable clamp load	Sealing is required to provide a fluid-tight bolted joint.
Some applications use threaded fasteners to provide adjustments	Often bolted joints have poor electrical conductivity and uncertain thermal conductivity.
Dissimilar materials (metallic and/or nonmetallic) can be joined	Exposed joint members can be subject to corrosion, e.g., nut end or bolt head.
Different material thicknesses can be joined – limited only by bolt length	Relatively high piece part and assembly costs.
Thermal stresses or thermal warping are not introduced	Thermal changes can weaken clamp load or cause joint loosening.
	No backing plates and/or overlap is required whenever plates are joined.

are often used interchangeably. We, however, will take the term bolt to include those threaded fasteners that are typically associated a nut or tapped hole and the term screw to include those threaded fasteners with a sharp end as in self-drilling screws for metals or as typically found in wood screws. An overview of the advantages and disadvantages of bolted joints is presented in Table 2.5.

2.1.5 Welded Joints

Although welded joints are increasingly being replaced by adhesive joints, there are nevertheless many applications for which welding is the preferred and sometimes the only joining approach available. There are many different types of welding approaches available, e.g. electron beam, mig, tia, friction welding, etc. However, an overview of the welding process advantages/disadvantages will provide insight into the applications of all welding processes, as show in Table 2.6.

TABLE 2.6
Advantages/Disadvantages of Welded Joints

Advantages	Disadvantages
When properly formed, welds have strength equal to materials being joined.	Special equipment is necessary to detect weld defects, e.g., voids or occlusions and this approach is often not successful.
Temperature changes do not affect strength and rigidity within joint service range.	Residual stresses may be introduced.
Provides good thermal and electrical conductivity.	Welding may cause joined objects to warp.
When properly formed, joints are fluid tight.	Heat treatment of the metals being joined may change especially in the "heat affected zone."
Typically lighter than bolted joints.	Number of dissimilar metals that can be joined is limited.
May be utilized to build-up or construct configurations other processes cannot form.	Disassembly of welded joints is very difficult if at all possible.
Can be designed to introduce no additional stress concentrations by properly grinding weld beads.	Welding requires: skilled (certified) operator and relatively expensive equipment.
	Pre-welding surface treatment may be required.

2.2 DESIGN WITH RIVETS

The use of rivets has steadily declined during the past 30 years, largely due to the development of the high-strength bolts for numerous applications. Further competition comes, of course, from welding technology and now increasingly from adhesive technology which can produce reliable joints even under rough field conditions. Even so, it should not be forgotten that rivets performed admirably in the past and still have unique applications. The famous Eiffel Tower, now close to 100 years old, is a monument to structural integrity and reliability due to riveted joints. Indeed, 15,000 steel members of the tower are still held in place by 2,500,000 rivets. Nevertheless, there are still sufficient applications of rivet technology being applied to warrant its review.

The function of a rivet is to fasten two plates (or panels) permanently. This can be accomplished by a cold or hot process; although, in more recent times, the cold method is preferred in industry because of the availability of efficient and fast riveting tools. Rivets are generally less expensive than threaded fasteners; although their strength in tension and shear may be lower than that of threaded fasteners.

The common practice of riveting involves holding the rivet in an assembly by holding the rivet outward and then forcing the rivet end downward into contact with the sides of the upper plate (or panel). This process may be accomplished by any one of three operations:

1. *Impact forming*: This is used when the thickness and/or hardness of the plates vary.
2. *Spin forming*: This case is recommended for accuracy of the connection. The rivet is stationary while the plate is turned and the clamping pressure is controlled by the spin rate. This method offers a design advantage where one of the assembled parts is required to rotate freely.
3. *Squeeze forming*: This is a process where the rivet is compressed beyond its elastic limit and thereby also causes it to bulge into the rivet hole. This method is less sensitive to mismatch of the rivet holes.

For design purposes the rivet material should be as strong as that of the plates with a minimum rivet spacing of three times the rivet diameter. When the pitch is too large, plate buckling between rivets may occur. To prevent local tearing or shearing of the material, the edge distance should be at least 1.5 times the hole diameter.

The selection of the design clearance between the hole and the rivet is of special importance. In general, this parameter together with the rivet size, length of the rivet and the type of rivet head are recommended by the manufacturers. There are, however, certain practical rules which should be kept in mind for design purposes. These are:

- The rivet diameter should be within one to three times the plate thickness. Below the lower limit the rivet head may break away. Alternatively, a rivet diameter too large may damage the plate.
- The rivet length should be determined by the plate thickness. When in doubt, select a rivet long instead of too short.

The basic mode of a rivet failure is in shear, since rivets are not generally designed to transmit direct tension loads. While cylindrical pins are often put in the same category as rivets in terms of their performance and strength, additional bending considerations may be imposed on pins as described in Chapter 7.

The shear strength of rivets in a particular joint is based on experiments which can simulate the actual application very closely. It is important to note that the actual strength properties are derived from a double-shear test such as that show in Figure 2.1. The shearing load F is divided by 2 in order to estimate an average ultimate shear stress. Although this is an elementary case, the actual stress distribution over the cross-section of the rivet is complex because of the interaction of shear with the load compressive stresses. The presence of these stresses produces a bending moment: although this effect can be mitigated by limiting the clearances and employing harder materials for the test fixture.

The fundamental relation for the calculation of the average shear stress of the device of Figure 2.1 is:

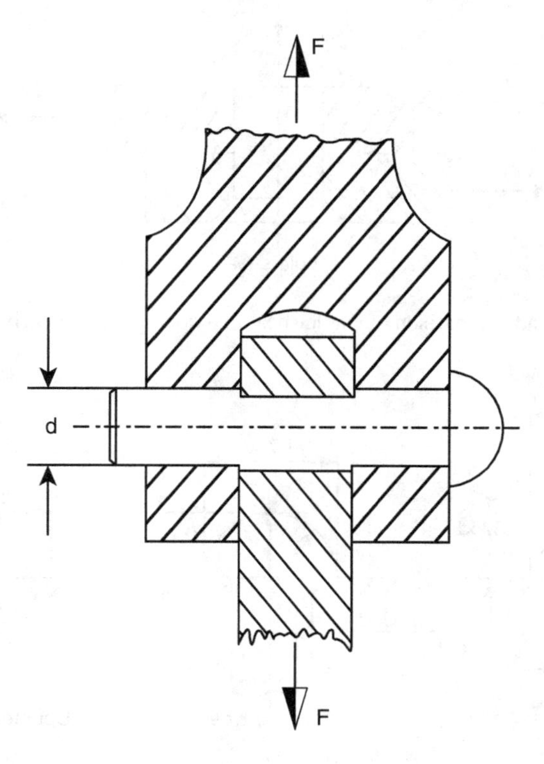

FIGURE 2.1 A double shear rivet test.

$$\tau = 0.64 \frac{F}{d^2} \tag{2.1}$$

The maximum theoretical shear stress for the case of a solid round bar is higher so that Eq. (2.1) becomes:

$$\tau_{\max} = 0.85 \frac{F}{d^2} \tag{2.2}$$

Figure 2.3 shows the common design geometry and typical proportions of rivet heads. A particular approach taken by a designer depends, of course, on the function and needs of an intended application. For example, the countersunk-type rivet, although more expensive, should be used with surfaces exposed to a fluid stream.

When aluminum is used as a rivet material, the rivets must be of good quality to assure a balanced amount of ductility and strength. Heat-treated alloys often do not have sufficient ductility to

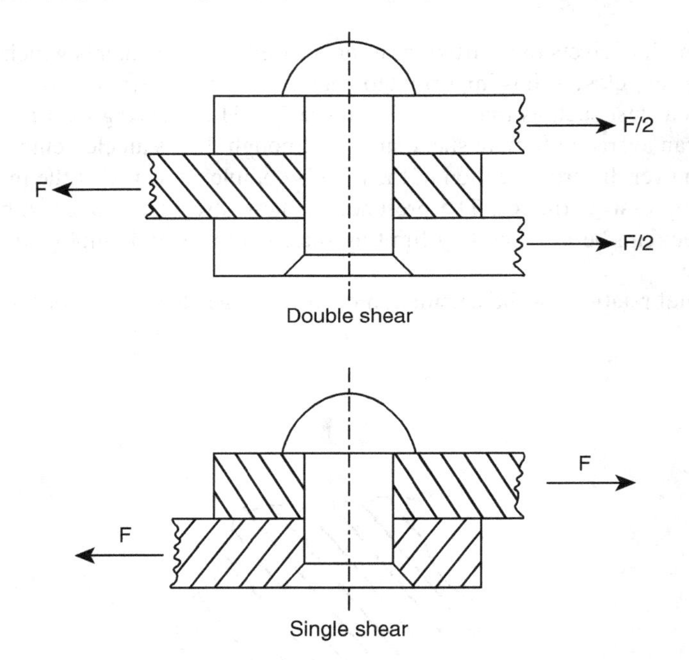

FIGURE 2.2 Illustration and comparison of single shear and double shear of a rivet in a lap joint.

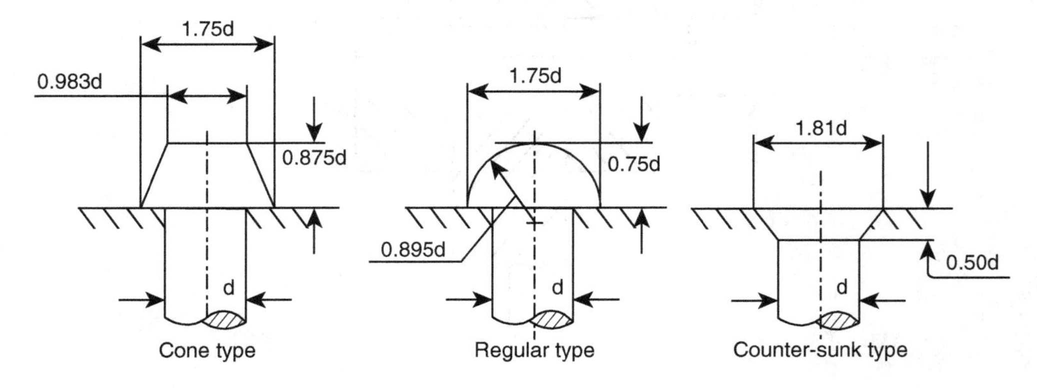

FIGURE 2.3 Typical rivet head designs.

permit a head to be formed without the potential of cracking. Alternatively, nonheat-treated aluminum alloys may not be strong enough after driving. Another problem with aluminum is that even a good heat-treated alloy can age at room temperature. The common remedy is to store the rivets in a refrigerating atmosphere until immediately prior to driving. Although this procedure is usually not very practical, it can be employed, in the case of larger rivets, exceeding say 0.25 in. in diameter.

Current construction practice recognizes at least two steels especially suitable for rivet fabrication. These are ASTM A502 grades 1 and 2 for the low carbon and manganese steels, respectively. The strength of the rivets can be specified in terms of the hardness of the rivet bar stock.

The majority of rivets made of steel are heated to about 1800° F prior to their installation. The process results in the development of a clamping force approaching the yield condition, followed by a residual tension. This effect is much smaller in cold formed rivets. In general, the process of driving hot rivets depends on a number of variables such as initial temperature, final temperature, driving time, and the riveting method.

The comments which follow are based on an excellent account of rivet technology by Fisher and Struik [1]: For example Figure 2.4 shows an approximate stress–strain relation for A502 rivet steels.

Field experience and studies of A502 rivets have shown that driving temperatures between 1800 and 2300° F have only a limited influence on the ultimate strength of the rivets. However, the degree of dynamic working of the metal increases the tensile strength by about 10% and 20% for pneumatic and machine-type driving techniques, respectively. The increase of strength, however, is accompanied by a reduction in elongation for hot as well as cold working of the metal. These effects, observed in practice, are essentially independent of the length/diameter ratios of the rivets.

Tests on many rivets have shown that the average shear to tensile strength is about 0.75 and the grade of the rivet material had a limited effect on this ratio. The scatter in the data defining the strength ratio appeared to be relatively small and dependent on the variation in testing procedures and the preparation of the test specimens. The load-deformation curves obtained in double-shear tests are influenced to some degree by the length of the rivets. Longer rivets, for example, indicated larger deformations due to bending. Similar effects could be postulated when single-shear tests are

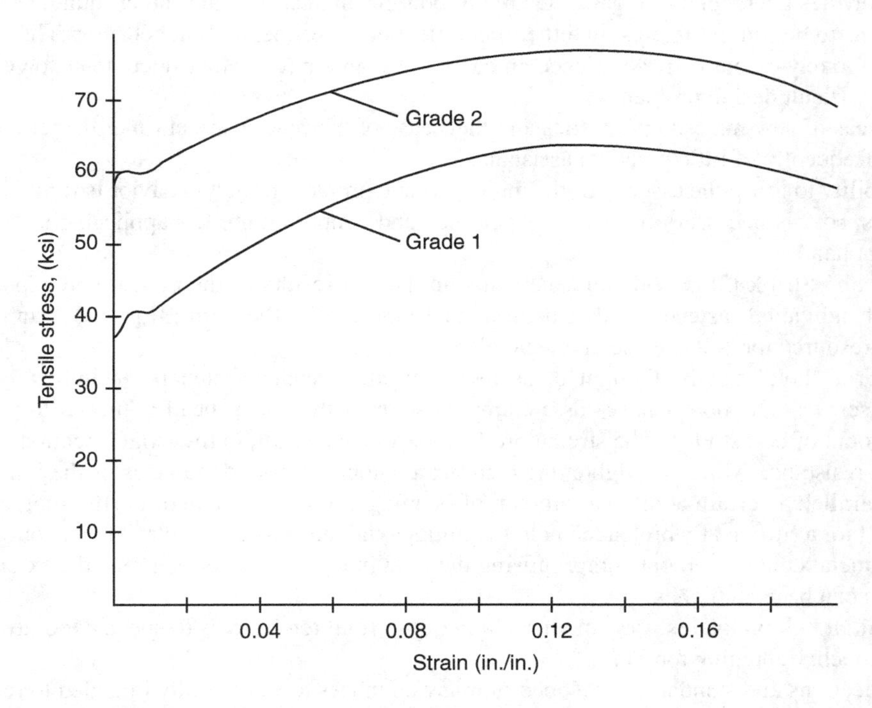

FIGURE 2.4 Stress strain relations for A502 steel.

compared with the double-shear test results. It appears that single-shear joints are subject to inherent eccentricity effects.

The problem of the combined tension and shear strength of individual rivets can be defined in terms of an elliptical interaction curve. Based on the interaction relation obtained from the experiments, the following working equations can be used in the calculations:

$$S = \left(S_u^2 - 1.78\tau^2\right)^{0.5} \tag{2.3}$$

and

$$\tau = 0.75\left(S_u^2 - S^2\right)^{0.5} \tag{2.4}$$

In the foregoing equations, τ denotes the shear stress applied to the rivet while S and S_u define the applied tensile stress and the ultimate tensile strength of the rivet material. The tensile strength capacity F of the rivet is:

$$F = AS_u \tag{2.5}$$

In Eq. (2.5), A is calculated as the undriven cross-section area of the rivet. A lower bound estimate of the tensile strength of the materials can be taken as 60,000 and 80,000 psi for A502 steel in grades 1 and 2, respectively.

2.3 DESIGN WITH BOLTS

Although many tests have been conducted for individual fasteners and joints and considerable progress has been made in numerical analysis, our knowledge of bolt behavior continues to be incomplete. Many designers tend to regard the area of bolt mechanics as being rather mundane and thus there seems to be limited interest in either theoretical or experimental bolt behavior. Therefore, we often have to rely on our own experience, in handbooks, and in regulatory documents to guide us in the more difficult design problems.

The areas of new material properties and their ease of fabrication are at times difficult to digest and thus frequently of little practical assistance.

Our ability to grasp the essence of the multi-variable problem of bolt behavior is limited by various issues, so it is necessary to rely on experience and simplified models applicable to the design problem at hand.

Fisher and Struik [2] provide an excellent summary of results of theoretical and experimental studies of individual fasteners and structural steel joints. Also Bickford [3] provides an excellent practical resource for bolt science and technology.

The term "bolt" can be thought of as including all threaded fasteners, including studs and machine screws. The bolt behaves as a clamp as soon as the nut or head is turned to produce a small amount of bolt stretch. The stretch produces stress and strain in the axial direction while the frictional resistance while bolt tightening causes a torsional stress. If the faces of the joint are not exactly parallel, a certain additional amount of bending cannot be avoided. Furthermore, practice shows that for a sufficiently preloaded bolt, the initial axial tension can be relaxed by about 15% in a metal-to-metal contact without compromising the joint integrity. For soft gaskets, the extent of this relaxation can be even higher.

The initial bolt preload is very important and our normal tendency is to specify too little, rather than too much, tightening torque.

Specifications and standards developed in many countries are essentially intended to reduce the cost and to promote reliability. These two characteristics are not necessarily compatible because of the many variables involved. Fastener standards have their origin in engineering societies, trade associations, and manufacturing experience. Much of the information is still of empirical nature because of the theoretical uncertainties in design.

The number of different fasteners defined by the various standards in the world is truly enormous. This situation, combined with the long and tedious process of conversion of all specifications to one international standard, creates additional difficulties in correlating the information on the theory and practice of bolt behavior.

The methods for dealing with uncertainties of bolt behavior are primarily statistical which in itself is a major branch of mathematics. We provide a summary of statistical methods for bolt analysis in Chapter 4.

2.4 SELECTION OF BOLT MATERIALS

The design criteria for threaded fasteners are defined by SAE International and the American Society of Testing and Materials (ASTM). They characterize the materials not only by tensile strength, yield, and hardness, but also to corrosion resistance, magnetic properties, electrical conductivity, and thermal conductivity. Although the majority of bolts are made of steel, the societal (SAE and ASTM) specifications also include nonferrous materials.

The characteristics of most common structural bolts are detailed in the following three ASTM standards: ASTM A307, A325, and A490. ASTM A307 establishes that low carbon steel can be used for all bolt sizes intended for light structures subjected to static conditions. The corresponding minimum ultimate strength of A307 bolts is about 60 ksi. When ASTM A325 steel is used, the bolts are made by heat-treating, quenching, and tempering with the minimum strength of 120 and 105 ksi for the smaller and larger bolt diameters, respectively. The ASTM A490 standard defines a quenched and tempered alloy steel bolt having a minimum required ultimate strength of 150 ksi. This particular bolt, in the diameter size of 0.5–1.5 in., was especially developed for use with high-strength steel members.

Unlike rivets, the strength of bolts can be defined simply by using a tensile test of the threaded portion. The load-elongation curve of a bolt's threaded section is more significant than the conventional stress–strain characteristics of the bolt shaft because the ultimate overall bolt capacity is governed by the reduced strength of the threaded section.

Figure 2.5 shows a comparison of test results for the ASTM A490 with a bolt having a 0.875 in. diameter wherein the bolt-threaded section exhibits the greatest strain. In the case of rivets, in contradistinction to bolts, the three curves in Figure 3.5 would be expected to be much closer due to the absence of bolt-type stress concentrations. As noted above, the weakest part of a bolt is the threaded portion where the relevant bolt "stress area" can be given by the expression:

$$A_S = 0.7854\left(D - \frac{0.9743}{n}\right)^2 \tag{2.6}$$

As in the bolt stress area, D defines the nominal bolt diameter and n is the number of threads per in. This equation was used in converting the nominal load deflection curve into stress–strain diagram shown in Figure 2.5. The basic material properties were obtained from a test on a standard cylindrical specimen of 2 in. gage length, turned from the bolt having nominal diameter of $D = 0.875$ in. According to the SAE standard, the relevant number of threads for a coarse thread series is: 9. Table 2.7 lists several mechanical properties for a number of SAE steels suitable for bolts and cap screws. Special stripping tests can be used here to determine proof strength, which is usually a little less than the conventional yield strength indicated in Table 2.7.

Table 2.8 provides a summary of properties for nonferrous materials, suitable for fastener (bolt/rivet) fabrication. As in any mechanical or strength standard, these values are provided for general guidance only. The reader is advised to consult more detailed relevant specifications for critical designs, or even better, to obtain strength data from results of sample testing.

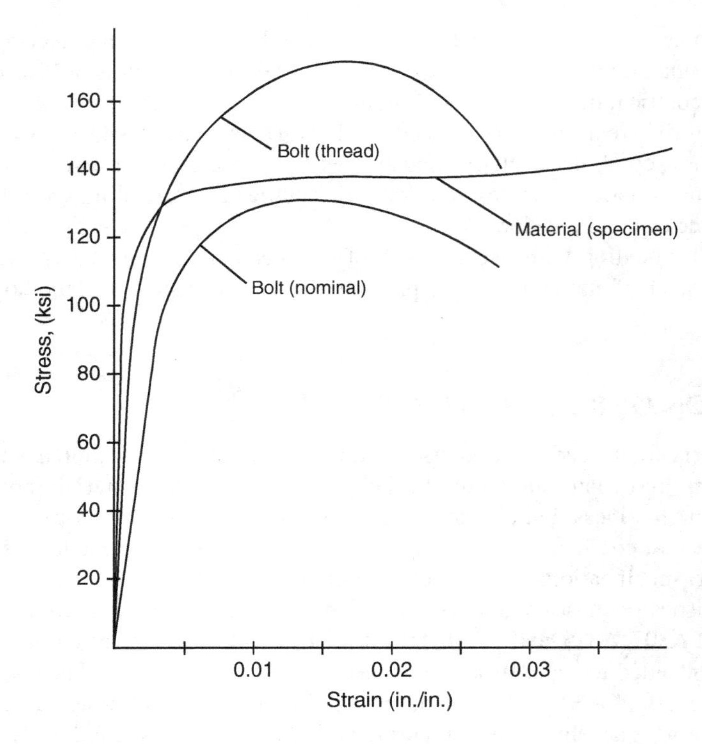

FIGURE 2.5 Comparison of bolt/bolt material for strength using the ASTM A490 test standard.

TABLE 2.7
Mechanical Properties of Bolt Materials

SAE Grade	Description of Steel	Nominal Bolt Diameter, in.	Yield Strength, psi	Ultimate Strength, psi	Brinell Hardness	Remarks
0	Low carbon	All sizes	–	–	–	General use
1	ASTM A307 low carbon	All sizes	–	55,000	207 max	Cold or hot heading
2	Low carbon bright finish	Up to 0.5	55,000	69,000	241 max	Automotive
	0.28C, 0.04P 0.05S	0.5 to 0.75	52,000	64,000		applications
3	Med. Carbon, 0.28 to	Up to 0.5	85,000	110,000	207 to	High fatigue
	0.55C, 0.04P, 0.05S	0.5 to 0.75	80,000	100,000	269	resistance
	Aged to 700E F					
4	Commercial	0.75 to 1.50	28,000	55,000	207 max	General use
5	ASTM A325	<0.75	85,000	120,000	241–302	High-preload
	0.28 to 0.55C	0.75 to 1.0	78,000	115,000	235–302	applications
	0.04P, 0.05S	1.00 to 1.50	74,000	105,000	223–285	
	Q&T at 800E F					
6	Medium carbon	<0.625	110,000	140,000	285–331	Higher strength
	Q&T at 800E F	0.625 to.75	105,000	133,000	269–331	
7	Fine grain,	Up to 1.50	105,000	133,000	269–331	Improved fatigue
	0.28 to 0.55C					strength
	0.04P, 0.05S					
	Oil Q&T at 800E F					
8	As above	Up to 1.50	120,000	150,000	302–352	For higher strength

TABLE 2.8
Selected Nonferrous Materials for Bolts and Other Fasteners

Material	Tensile strength, psi	Yield strength, psi	Rockwell hardness
2024–74 Aluminum	55,000 min	50,000	54 B
Yellow brass	60,000 min	43,000	57 B
Commercial bronze	45,000 min	40,000	40 B
Copper	40,000 avg	20,000 avg	40 B
Silicon bronze	70,000 min	38,000	74 B
Silicon-aluminum bronze	80,000 min	42,000	76 B
Monel	82,000 min	60,000 min	90 B
302 stainless	90,000 min	35,000 min	89 B
304 stainless	85,000 min	30,000 min	87 B
316 stainless	90,000 min	30,000 min	89B
347 stainless	90,000 min	35,000 min	89 B

Stainless steels are generally characterized by a considerable spread between the minimum and maximum values of the yield strength. The corresponding ratios of the minimum to maximum can be 0.30–0.35. Only the copper ratio is likely to be lower than 0.30. All the yield values given in Table 2.8 correspond to 0.5% elongation.

Where the temperature effect is a problem, such as in a relaxation process occurring in a locknut and bolt assembly, the following materials may be recommended:

60–400E F	7075-T6 aluminum alloy, SAE 4140 steel and SAE 8740 steel
400–800E F	300–400 alloy series and titanium
Up to 900E F	Martensitic chromium steels and 17-7 PH steel
900E F	Inconel and superalloys

The proof load and the tensile strength of threaded fasteners are subdivided into two broad categories, depending on the type of the thread required; either course or fine threads. The so-called UNC specification refers to the coarse thread series, while UNF denotes the fine thread category. Tables 2.9 and 2.10 summarize some of the mechanical requirements for the UNC and UNF series, respectively. The relevant proof load can be defined as that which stresses the bolt, screw, or stud up to a maximum value, without a measurable permanent set [4].

The tables show 0.125 in. steps in the nominal bolt diameter and the selected three grades of the best quality bolts according to the SAE J429 standard. Fastener manufacturers are generally required to conduct periodic tests to assure the quality of the product. The table gives a reliable guide to the bolt strength for all practical purposes, since it is based on the accepted industrial and regulatory standards. It is to be noted that the values indicated above make no reference to the actual stresses in the bolts. Therefore, in critical applications, the bolt or joint engineer may wish to estimate the actual stresses in the bolt by actual testing and determine the appropriate safety margins for the case at hand. The appropriate design equations will be considered later in this chapter.

Table 2.10 highlights the fine thread category of the ultimate strength for bolts.

In the majority of cases, the ultimate strength of the bolt material is below 180,000 psi level. For higher strength levels, the ANSI (American National Standards Institute) standard recommends a modified formula for the calculation of the stress area:

$$A_S = \pi\left(\frac{D_{\min}}{2} - \frac{0.1624}{n}\right)^2 \quad \text{for strength levels} > 180{,}000\ \text{psi} \tag{2.7}$$

TABLE 2.9
Proof Load and Tensile Strength Capacity for Coarse Threads

		Grade 5		Grade 7		Grade 8	
Nominal Diameter Inch	Threads per Inch	Proof Load lb	Tensile Capacity lb	Proof Load lb	Tensile Capacity lb	Proof Load lb	Tensile Capacity lb
0.250	20	2,700	3,800	3,350	4,250	3,800	4,750
0.375	16	6,600	9,300	8,150	10,300	9,300	11,600
0.500	13	12,100	17,000	14,900	18,900	17,000	21,300
0.625	11	19,200	27,100	23,700	30,100	27,100	33,900
0.750	10	28,400	40,100	35,100	44,400	40,100	50,100
0.875	9	39,300	55,400	48,500	61,400	55,400	69,300
1.000	8	51,500	72,700	63,600	80,600	72,700	90,900
1.125	7	56,500	80,100	80,100	101,500	91,600	114,400
1.250	7	71,700	101,700	101,700	127,700	116,300	145,400
1.375	6	85,500	121,300	121,300	153,600	138,600	173,200
1.500	6	104,000	147,500	147,500	186,900	168,600	210,800

TABLE 2.10
Proof Load and Tensile Strength for Fine Threads

		Grade 5		Grade 7		Grade 8	
Nominal Diameter Inch lb	Threads per Inch lb	Proof Load lb	Tensile Capacity lb	Proof Load lb	Tensile Capacity lb	Proof Load lb	Tensile Capacity lb
0.250	28	3,100	4,350	3,800	4,850	4,350	5,450
0.375	24	7,450	10,500	9,200	11,700	10,500	13,200
0.500	20	13,600	19,200	16,800	21,300	19,200	24,000
0.625	18	21,800	30,700	26,900	34,000	30,700	38,400
0.750	16	31,700	44,800	39,200	49,600	44,800	56,000
0.875	14	43,300	61,100	53,400	67,700	61,100	76,400
1.000	12	56,400	79,600	69,600	88,200	79,600	99,400
1.125	12	63,300	89,900	89,900	113,800	102,700	128,400
1.250	12	79,400	112,700	112,700	142,700	128,800	161,000
1.375	12	97,300	138,100	138,100	174,900	157,800	197,200
1.500	12	117,000	166,000	166,000	210,300	189,700	237,200

Here, D_{min} stands for the minimum pitch diameter and the stress is based on the average of D_{min} and the nominal root diameter. Other symbols are the same as those in Eq. (2.6).

Although very little has been published on this topic, it is known that a bolt preload can relax over a period of time due to such effects as tension loading vibration or temperature. If a steel rod is supporting a heavy load in direct tension and the temperature of the environment is raised, the rod will gradually lengthen and, over a period of time accompanied by elevated temperature, the rod may eventually break. This long-term effect is often called stress relaxation of the material. For a given load, the part will continue to lengthen with temperature, since the elongation is inversely proportional to the modulus of elasticity $\left(\delta = \frac{PL}{AE}\right)$. Table 2.11 presents several values of the modulus of elasticity for typical bolting materials.

TABLE 2.11
Modulus of Elasticity of Several Bolting Materials

Material	Modulus Psi × 10^6	Modulus MPa × 10^6
Carbon Steel	29.5	0.200
Stainless steel	28.0	0.193
High strength, A286	29.1	0.201
AISI 4340	30.0	0.207
Inconel X	31.0	0.214
Aluminum	10.8	0.075
Monel	27.0	0.186
Brass	15.0	0.103
Titanium alloy	16.5	0.114

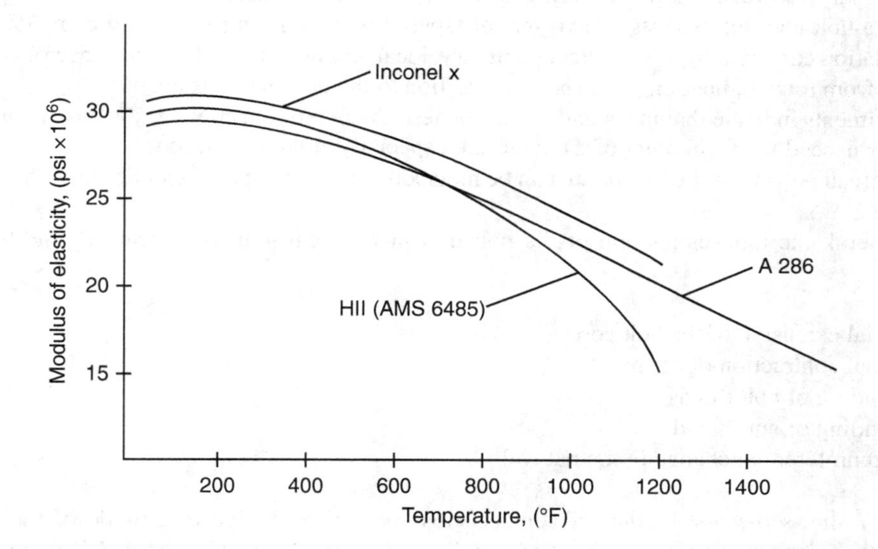

FIGURE 2.6 Effect of temperature on elastic modulus.

Increasing the temperature has the net effect of decreasing the modulus of elasticity and strength of metallic materials. The influence of a large change in temperature is to cause the material to behave in a different manner. An example of temperature effect on the modulus of three bolting materials is given in Figure 2.6.

In the case of high-temperature bolts, thread dimensions should be reduced by 0.003 in. from standard dimensions. This is done to minimize the potential galling of the bolt thread in the nut due to temperature expansion. As a rule, bolts are not reusable above 500E F.

2.5 STRESS AND STRAIN OF BOLTED JOINTS

Extensive experience with differing materials shows that lower strength bolt materials can often be deformed beyond their yield points. Whereas higher strength materials can be found to work-harden, due to cyclic loading, which in turn promotes brittleness. Both these situations can lead to failure of the joint. As a general guideline, to avoid joint failure it is necessary to maintain a reserve of bolt strength within desired design limits, given the joint's exposure to all foreseeable service loads.

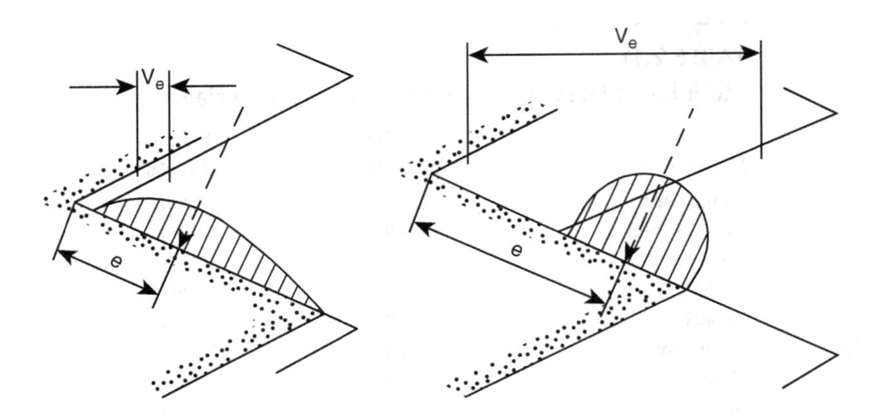

FIGURE 2.7 Model of thread loading.

There is a wide variation in stress throughout the bolt/nut connection. Given that the geometry between a bolt and nut consists of two sets of tapered threads in imperfect and complex contact. This variation can arise from imperfect geometry, localized defects, and the presence of combined loadings from torsion, bending, and shear in addition to the nominal tension.

Experiments indicate that in a standard tensile test, the maximum stress at some point of the bolt can easily exceed the yield point of the material, especially at the thread root.

The thread portion of the bolt or nut can be modeled as a short, tapered cantilever beam as shown in Figure 2.7.

In general, the total elongation of the bolt in a mechanical joint is affected by the following strains:

- Axial extension of the bolt core
- Axial contraction of the nut
- Bending of bolt thread
- Bending of nut thread
- Circumferential change in the nut wall

Figure 2.7 shows the distribution of contact stress on a thread. The magnitude of the bending moment then depends upon the quantities e and V_e shown in Figure 2.7, where V_e is defined as the thread recession caused by the clearance and radial deformation due to the lateral expansion of the nut in compression. The distance e is measured from the center of contact pressure to the bottom of the thread. This region of the thread, therefore, is subjected to a finite stress gradient with highly localized high stresses.

The classical case of a bolt under tensile load discussed by Bickford [5] assumes perfect symmetry, perpendicularity of the interfaces in a joint and some localized stress gradients as represented in Figure 2.8. Such high stress areas can precipitate a failure mode within the fastener, unless there is a sufficient plastic flow to redistribute the loading. It is clear that even this idealistic case of stress distribution is sufficiently complex to engage finite element analysts and design practitioners in all areas of industry and research. Furthermore, the lines of principal tension and compression point to a number of locations where the stress gradients may be above the conventional values derived from the calculations and thereby cause bolt failure. The more obvious areas of significant gradients are the fillets, the thread run-out, and the first thread to engage, which are difficult to treat analytically and experimentally. Figure 2.9 presents a simplistic representation of the stress distribution in a bolt along its central axis.

Essentially, the stress pattern of Figure 2.9 is based on the nominal stress value $\sigma = F/A$ where F is the axial load and A is the cross-section of the bolt body at the point considered. This simple

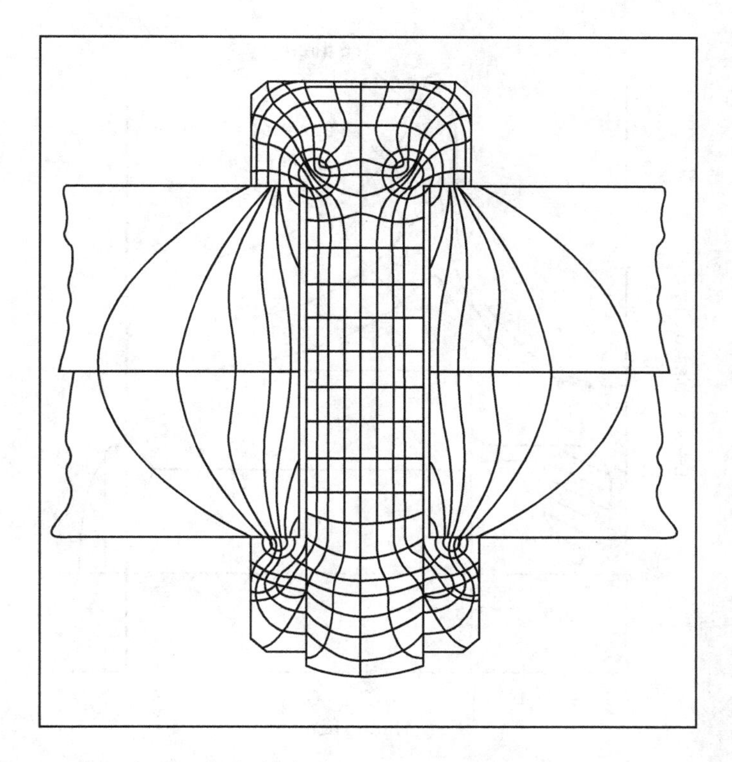

FIGURE 2.8 Stress trajectories in a bolted joint.

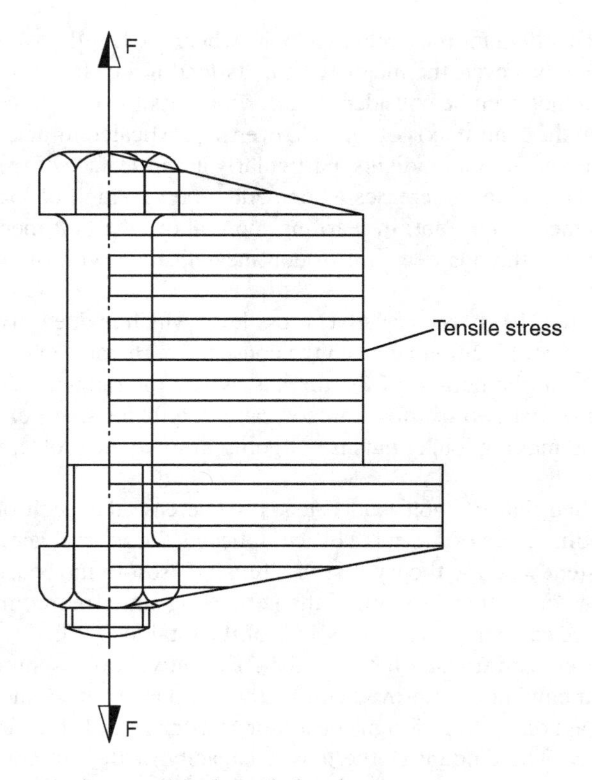

FIGURE 2.9 Simple stress pattern produced on a bolt axis.

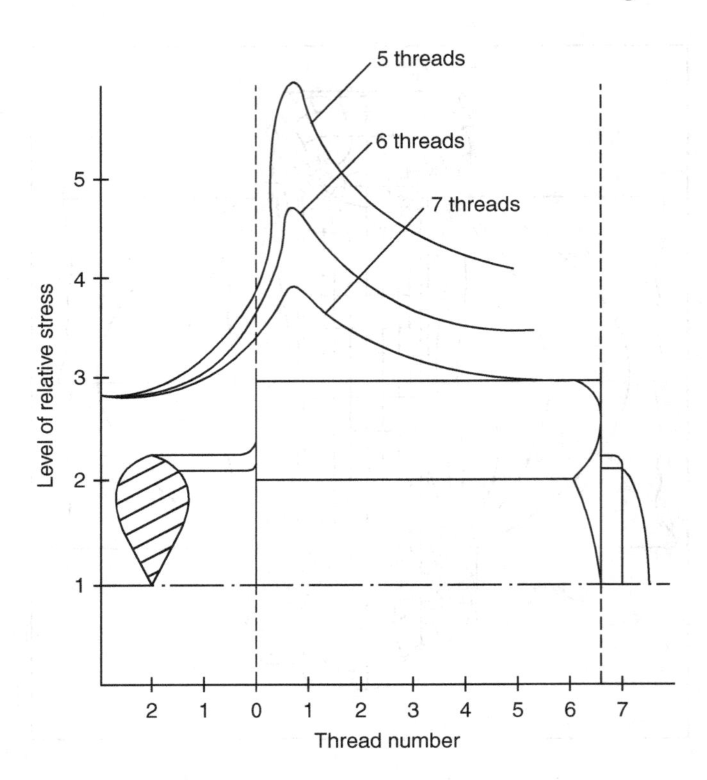

FIGURE 2.10 Variation of peak stresses with the number of threads in a conventional design.

modeling is reasonably justified for relatively long bolts where the length to diameter ratio is greater than about 4. In practice, however, the majority of bolts tend to be short and stubby. Under these conditions F/A stress cannot even be considered uniform across the bolt diameter. As to the actual stress picture away from the central axis of the bolt, even sophisticated finite element modeling falls short in predicting the actual stress gradients, particularly at the surface of the fastener.

The complexity of the problem increases as one adds the variation of load transfer across the thread array. It is extremely important, in learning more about the bolt mechanics, to know that peak stresses in nut or bolt threads decrease in nonlinear fashion with the number of threads as indicated in Figure 2.10.

The diagram in Figure 2.10 shows a relative stress level which is due to two main factors. One is the nonlinear distribution of loading between the consecutive threads, and the other is concerned with stress concentration at the bottom of the threads, where the radius of curvature is extremely small. It appears that the first two or three threads carry the major share of the bolt load, so that adding more threads and making longer nuts is not going to solve the problem of peak stresses and the nonlinear response.

It should also be noted that the bolt load F tends to increase the pitch of the bolt thread and decrease the corresponding pitch of the nut. This complicates the general geometry of load transfer further, and it is consistent with the theory that the thread closest to the bearing surface of the nut will be overloaded. In a conventional design of the nut having six or more threads, the thread next to the bearing nut surface can carry as much as 15% of the total load [6].

The problem of uneven load transfer between the consecutive threads stimulated designers and researchers to develop a number of improved configurations such as those shown in Figure 2.11.

The most common approach to redesign for a more uniform load transfer is to use a nut as a partially tensile member. The amount of the tensile capacity of the nut configurations shown in Figure 2.11 is assumed to be increasing when viewed from left to right. The far right version may

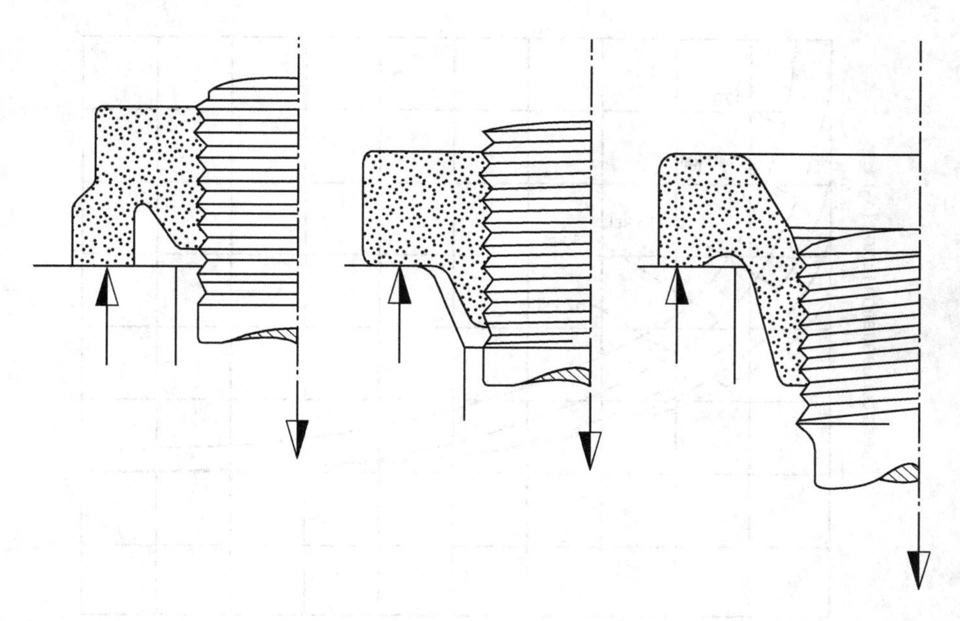

FIGURE 2.11 Examples of improved nut designs for more uniform load transfer.

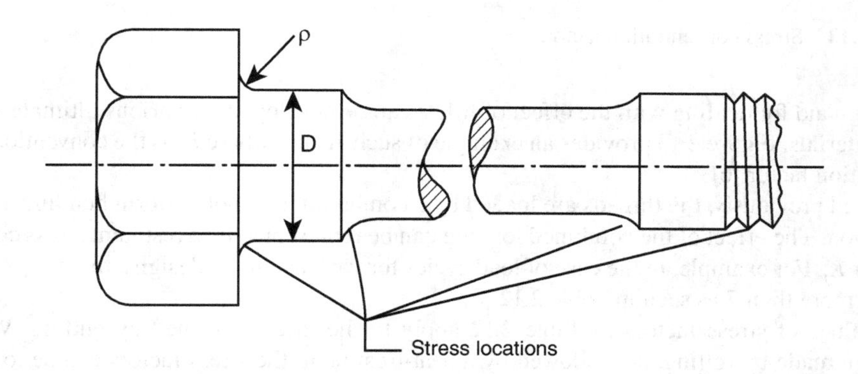

FIGURE 2.12 Sites of stress concentration.

be considered to have a nearly even load distribution among the threads. In more conventional nut designs, with nonuniform thread loading, the nut expands as a result of compression which tends to distribute the load more uniformly. However, this mitigating effect is usually rather small. Also, when a nut is made out of a relatively soft material compared with that of the bolt, the load on individual threads tends to be more uniform as they proceed to deform plastically.

The mechanics of load interaction is a threaded joint, such as a bolt and nut combination, is strain dependent and as such it relates to the important and complex area of load transmission in riveted, spliced, and adhesive joints of great variety. There is a definite similarity between the response of a number of consecutive threads, rivets, bolts, pins, and adhesives which tend to deform in proportion to the forces to which they have been exposed.

Stress concentrations in a threaded joint also occur at the junction between the shank and the head, as well as at all other locations where the diameter changes. Figure 2.12 shows the typical areas of stress concentration in bolts. The stress concentration factor in the fillet transitions from the shaft to the head or end is a function of ρ/D where D is the shaft diameter and ρ is the fillet radius (See Figure 2.12).

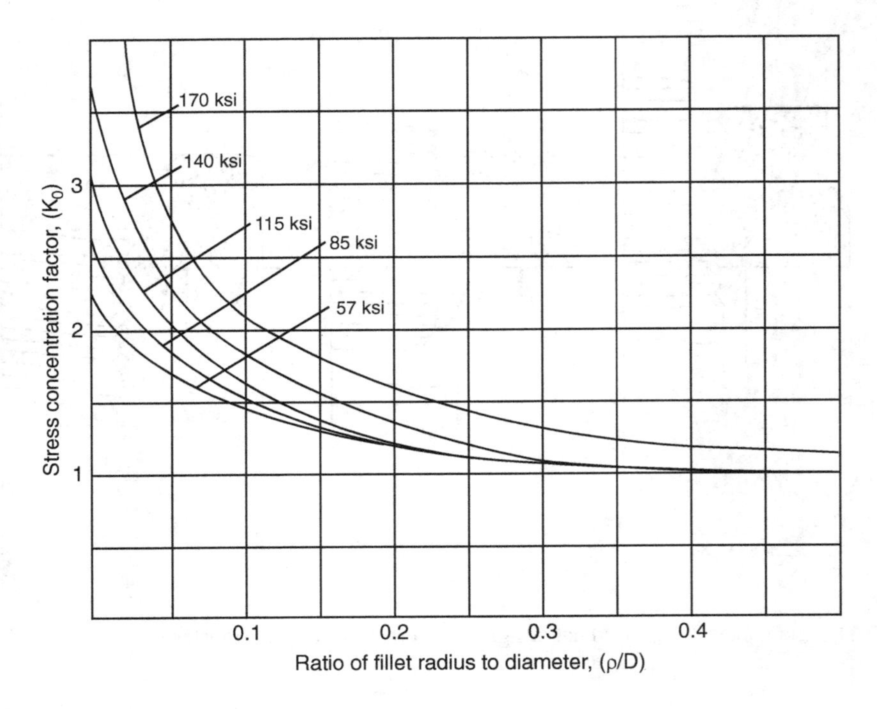

FIGURE 2.13 Stress concentration factors.

A design aid for dealing with the effect of fillets can be developed for various ultimate strengths of bolt materials. Figure 2.13 provides an example of such an aid, where K_0 is the conventional stress concentration factor [6].

As stated previously, the threads are loaded by a combination of nonuniform bending, shear and compression. The effect of the combined loading can be represented by a resultant stress concentration factor K_f. For example, in the case of load cycles for various thread designs, this factor can vary from 3 to more than 7 as seen in Table 2.12 [6].

The values of stress factors in Table 2.12 apply to the threads formed by cutting. When the threads are made by rolling, not followed by a heat-treatment, the stress factors can be lowered by 20%–35% for the medium-carbon and heat-treated steels, respectively. The lower values of K_f, from Table 2.12, are reserved for softer and more ductile materials.

The condition of shear loading on a single bolt depends on the type of joint and the number of the effective shear planes involved. Essentially, we can have a cross-section through the body of the shank, across the threaded portion only, or involving both the non-threaded shank and the threaded sections. To estimate the shear strength of the bolt, we can multiply the allowable shear stress of the bolt material by the total cross-sectional area of the shear planes.

TABLE 2.12
Effective Stress Factor K_f for Cycling Loading of Threads

Thread form	Medium carbon steels	Heat-treated alloy steels
Unified thread with rounded root, Whitworth thread	3.2 to 3.8	5.4 to 6.0
Metrical standards	4.4 to 5.0	5.6 to 6.4
American National thread with flat root, Sellers	5.0 to 5.8	6.4 to 7.2

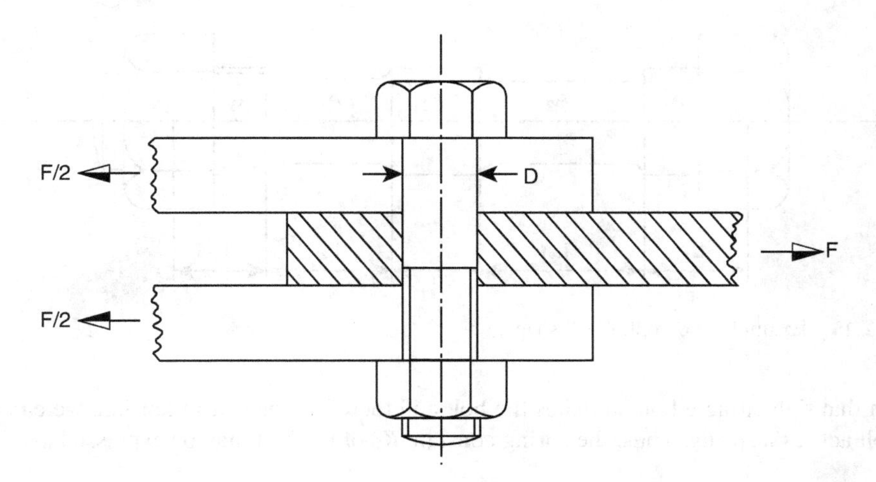

FIGURE 2.14 Example of a bolt in double shear.

For example, consider the double shear joint, involving one shank, and one threaded cross-section as illustrated in Figure 2.14. The maximum shear stress may be expressed as:

$$Q_{max} = (A + A_S)\tau_{max} \tag{2.8}$$

Where A denotes the cross-section based on the nominal bolt diameter D, while A_S, the relevant bolt stress area, can be given by Eq. (2.6) or (2.7) depending on the ultimate strength level of the bolt material. If the fastener material has an ultimate strength lower than say 180,000 psi, then the foregoing formula can be expressed in terms of the nominal bolt diameter D and the number of threads per inch, n. Therefore, substituting Eq. (2.6) into Eq. (2.8) we find:

$$Q_{max} = \left(\frac{\pi D^2}{2} - \frac{1.5D}{n} + \frac{0.75}{n^2}\right)\tau_{max} \quad \text{for strength levels} < 180{,}000 \text{ psi} \tag{2.9}$$

There are numerous configurations where either one or several shear planes through the bolt are involved in a clamped condition. Such planes could pass through either the nominal cross-section (with area $\pi D^2/4$), or through the cross-section at the bottom of the thread as governed by Eqs. (2.6) or (2.7).

We can also obtain the factor of safety pertinent to Figure 2.14 by dividing Q_{max} by F.

The experimental load-elongation characteristics obtained for bolts during tightening configurations show that if, instead, the bolt were loaded in pure tension it would be able to support a higher load. The lowered strength for tightening bolts is due to the simultaneous loading of service bolts by both tension and torsion. However, after the torqueing wrench is removed, the torsional stress will gradually decrease due to the relaxation of the applied torque.

The issue of strain in a bolted joint should be considered in conjunction with the analysis of stiffness and deflection of a bolt. These parameters are necessary for determining preloads and clamping forces. Since the tensile loads are not applied to bolts "from end to end," an effective length must be assumed which will occur somewhere between the actual length and the grip length.

For a bolt with a more complex shape as depicted in Figure 2.15, the overall change in length ΔL due to axial tension is:

$$\Delta L = \frac{F}{E}\left(\frac{L_1}{A_1} + \frac{L_2}{A_2} + \frac{L_3}{A_3} + \frac{L_4}{A_4} + \frac{L_5}{A_5}\right) \tag{2.10}$$

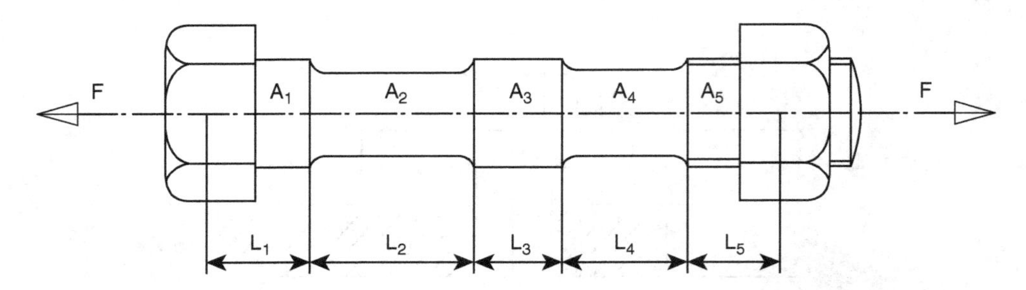

FIGURE 2.15 Example of complex bolt shape.

Given that tightening a bolt stretches the bolt and therefore loads it in tension, we can consider that a bolt acts as a spring. Thus, the spring constant K_B of the bolt may be expressed as:

$$K_B = \frac{F}{\Delta L} \tag{2.11}$$

where K_B is obtained directly by inspection of Eqs. (2.10), and (2.11).

It is clear that Eqs. (2.10) and (2.11) are easy to use in the calculations, although the accuracy of predictions must be questioned because of the unavoidable variations in dimensions, mechanical properties, and fabrication techniques. Such effects can be of special importance if the measurement of bolt stretch is used as a means of preload control.

It may be instructive here to touch on the topics of torsion and bending of individual fasteners prior to evaluating the design criteria for torque and preload control of a bolt: The torsional moment M_{to} applied to the bolt overcomes frictional, geometrical, and tensile constraints. The tensile constraint arises due to stretching of the bolt. The geometrical constraint relates to the form of the thread. The frictional resistance is essentially based on the interaction of the thread surfaces, while a certain amount of effort is also required to overcome the friction between the nut and the clamped surface. Assuming that the effective radius of contact between the threads is not substantially different from $D/2$, then the torque M_{to} is approximately

$$M_{to} = \frac{F}{2}\left(\frac{P}{\pi} + \frac{\mu D}{\cos\beta}\right) \tag{2.12}$$

Where P is the pitch, μ is the coefficient of friction between the threads and β is the half-angle of the thread profile.

For international thread design, $\beta = 30°$. If we assume that the bolt shank behaves as a cylinder in pure torsion, then the maximum torsional stress produced at the outermost fiber of the shank is:

$$S_{to} = 0.81\frac{F(P\cos\beta + \pi\mu D)}{D^3\cos\beta} \tag{2.13}$$

Normally, the bolt shank is assumed to carry only F/A type of tensile stresses unless there is a question related to flange alignment and uniformity preventing a purely axial response of the bolt. If the deformed bolt axis has a radius of curvature R, then the external bending moment causing the curvature is:

$$M_b = \frac{\pi E D^4}{64R} \tag{2.14}$$

Where, as before, E is the modulus of elasticity of the bolt material, and D is the nominal bolt diameter. The corresponding maximum bending stress S_b is then:

$$S_b = \frac{ED}{2R} \tag{2.15}$$

In the majority of applications, the bolt will be subjected to a combined stress system involving tension, torsion, shear, and bending. The right choice of load combination, geometry, and physical properties is usually not known to be able to develop a reliable approach to the determination of the final design formula.

Bickford [5] derives an energy equation, however, to indicate the degree of complexity involved in a simple process of applying the torque to the nut. It may be instructive to keep this conclusion in mind when selecting the appropriate design equations for the particular problem at hand.

While our primary interest here is to determine the tensile rather than the shear strength of the bolt, it is important to recognize that the tensile strength is reduced when torsion or shear loads are present. Typically, such a condition exists only during the actual process of applying torque. As soon as we remove the torque wrench from the nut, the induced torsional stress should decrease due to the relaxation. If we now reapply an external tensile load to the fastener, a higher level of tension than previous will be required to yield the material.

2.6 TORQUE EQUATIONS AND FRICTION

Design theory attempts to model important bolt parameters and, over the years, these attempts have been influenced by experimental work aimed at defining yield, reliability, and failure causes of a single bolt. This is important because any criteria of torque, axial load, and stress in the bolt must be realistic when compared with empirical parameters. The main task here is to estimate the tensile strength of the threaded portion of the bolt. Specifications have been developed, based on a large number of fasteners made from well-characterized materials being tested over the years. Manufacturers are now required to repeat such tests periodically to prove to the customer that their products meet the specifications.

While a great many test results have been published, our need to compute the design limits for the specific cases at hand has not gone away. At the same time, researchers have found that the yield load divided by the cross-sectional area, based on the mean thread diameter, resulted in a theoretical stress at yield. The problem here, of course, is that the actual stress levels vary within the fastener and that the particular fastener seldom sees the yield stress in practice.

In reviewing the simplest theoretical models, we find that a number of variables, such as several types of friction, are difficult to control and determine. The design formulas, Eqs. (2.12) and (2.13), given so far indicate the effect of the torsional action on the bolt based on the frictional resistance of the threads. The required initial torque to overcome all the frictional resistance for a given preload F_i can be obtained from the following expression:

$$T = F_i r\left(\frac{\cos\beta\tan\alpha + \mu}{\cos\beta - \mu\tan\alpha} + \frac{\mu D}{2r}\right) \tag{2.16}$$

where F_i is the initial bolt load, α defines the helix angle of the thread at mean radius r with β and D denoting the same symbols as before.

The well-known approximate "rule of thumb" formula for the tightening torque is:

$$T = 0.2F_iD \tag{2.17}$$

The stress determined for the combined tension and torsion may be calculated from the following equation [7].

$$S = \frac{0.024F_i}{\mu^2 D^2}\left[\left(1 + 36\mu^2\right)^{3/2} - 1\right] \tag{2.18}$$

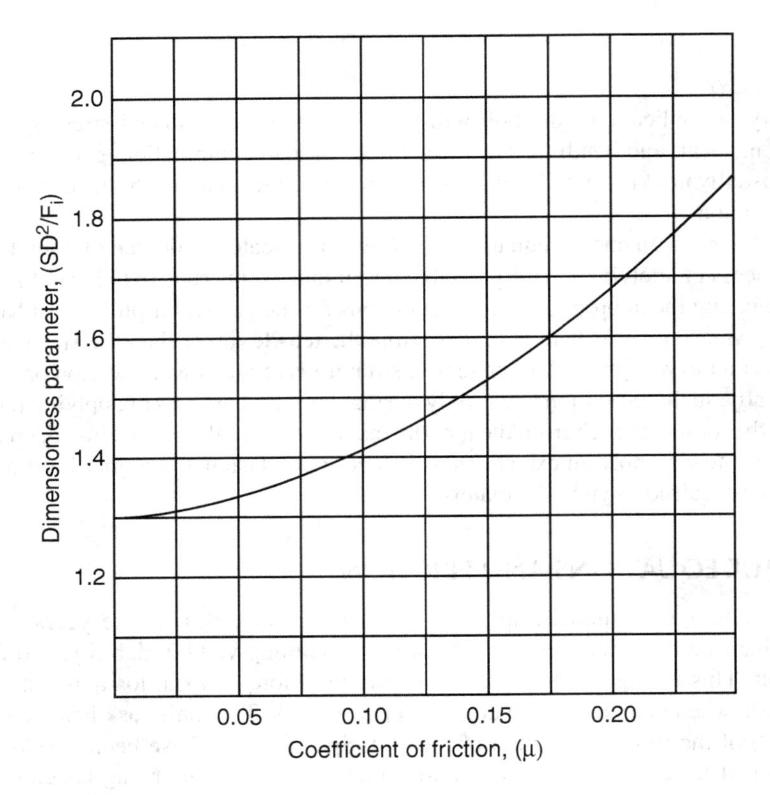

FIGURE 2.16 Effect of contact friction on stress intensity.

The parameter SD^2/F_i can now be graphed for the common range of friction coefficients corresponding to typical industrial lubricants used in bolted joints. Figure 2.16 illustrates a typical graph between the dimensionless parameter SD^2/F_i and the friction coefficient μ.

The more general form of the design equation defining the torque typically referred to as the nut factor equation is

$$T = KDF_i \tag{2.19}$$

where if D is expressed in inches, K is known as the "nut factor" and F_i is given in pounds, the initial tightening torque T will be expressed in inch-pounds. Table 2.13 provides typical nut factors K applicable with steel fasteners [3].

TABLE 2.13
Average *K* Factors for Steel Fasteners

Industrial Lubricant	*K*
As received steel	0.20
As received cad plate	0.19
Fel-Pro 65A	0.13
Moly grease	0.14
Parkerized and oiled	0.18
Petroleum, light oils	0.12
Phosphate and oil	0.19

TABLE 2.14
Friction Factors, μ

Lubricant	Range
Red lead, graphite, mineral oil	0.078–0.110
Red lead, graphite, machine oil	0.055–0.065
Graphite, mineral oil	0.035–0.060
Graphite, machine oil	0.035–0.055
Molokote G	0.030–0.075
Fel Pro C5-A	0.045–0.080
Crane Compound 425A	0.070–0.090
Grafo 360	0.100–0.115
Neolube	0.030–0.090

While the general form of Eq. (2.19) is convenient and easy to interpret, there are some problems with the assignment of the nut factor or torque coefficient K. The average K values summarized in Table 2.13 appear to be fairly consistent. Unfortunately, when the extreme values of K are noted as reported by industry, the range may be between 0.08 and 0.27 instead of 0.12–0.20 as shown in Table 2.13 the value 0.20, often recommended in textbooks, should be considered for first-order, rough, preliminary estimates only. The value $K = 0.20$ should be further restricted to nuts and bolts having a plain, unplated finish of the contact surfaces without lubrication. Furthermore, analysis also shows that about 50% of the wrenching torque is used to overcome friction between the nut and the bearing surface, with some additional 40% representing the torque used to overcome the sliding resistance of the threads. Typically, then, only 10% of the wrenching torque goes into useful work to develop the required preload in the bolt.

Since the type of lubricant used on a bolt or nut has a significant effect on the preload, it is of interest to define the design rationale for the lowest and highest K values in use. For example, the presence of the lowest K value assures that we will have the highest preload and stress intensity in a bolt of stud. Alternatively, the highest torque coefficient K will result in the most conservative design by producing the lowest preload. These rules follow directly from Eq. (2.19), indicating that the preload is inversely proportional to K.

It should be pointed out that the K factors given in Table 2.13 represent an integrated effect of geometry and friction as applicable to the simplified formula: Eq. (2.19). The actual kinetic coefficient of friction μ shown in Eq. (2.16) is generally lower than the corresponding value of torque coefficient K. Table 2.14 presents some typical values of friction coefficients providing some comparisons of values [1].

In a more detailed analysis, the torque coefficient K can be expressed in terms of geometry and the specific coefficients of friction acting on contact surfaces of the torqued bolt in the following way:

$$K = K_1 + K_2 + K_3 \tag{2.20}$$

Where K_1 relates to the bearing friction under the nut, K_2 defines the frictional resistance on the thread surfaces and K_3 is a geometrical parameter involving the helix angle of the thread and other bolt dimensions. K_1 and K_2 represent the loss while K_3 defines the useful portion of the applied torque. As mentioned previously, the useful portion amounts to about 10% of the total torque. The magnitude of K_3 is slightly larger for coarse threads. The above K theory applies equally well to the three basic configurations shown in Figure 2.17.

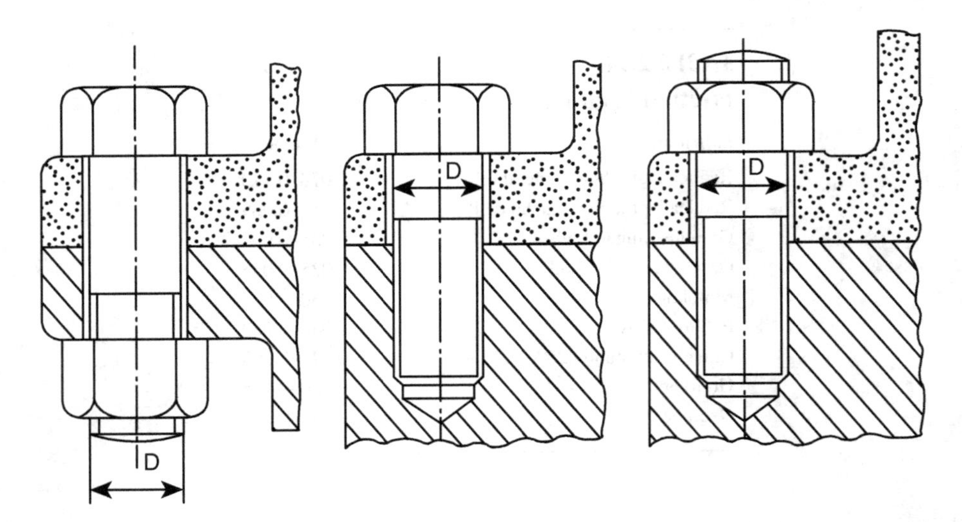

FIGURE 2.17 Standard bolt configurations.

2.7 STRENGTH AND DESIGN CRITERIA

In most joint designs, fasteners are subjected to tension, shear, bending, or a combination of forces requiring sophisticated techniques to make design choices. However, regardless of bolt application, certain procedures are common to all conditions and can be outlined briefly as follows:

- Determine the external force or service conditions for the most highly loaded fastener.
- Select a factor of safety based on the type of application and best engineering judgment.
- Select a fastener material for a given environment and economic constraints.
- Select fastener dimensions.
- Calculate the required tightening torque.

Table 2.15 presents dimensions and mechanical properties in relation to the clamping loads for a pair of SAE grade bolts. Handbooks and trade publications provide more detailed tables—many for specialized design applications. The ultimate tensile strength S_u and the corresponding proof strength S_p are considered to be minimum values. The S_p/S_u ratios for the Grade 5 and 8 are 0.7 and 0.8, respectively. The preload is often known as the initial load; where the load remaining after initial tightening (preload) and relaxation, and application of service loads is known as clamp load. The table provides load values for coarse and fine threads. The fasteners with the coarse threads are used more often because they are easier to assemble, they have more thread clearance for plating and they are easier to thread into materials of a lower tensile strength. Also, for bolt sizes 1 in. and higher, coarse threads are stronger than the corresponding fine threads.

The best feature of a higher strength fastener is its ability to maintain a higher clamping load. This is evident from Table 2.15 and this approach is compatible with good design practice. In some instances, the preload tension may extend beyond the yield range of the fastener without any damaging effects because the bolt load carrying capacity has a significant safety margin. For example, the slope of the curve for an SAE Grade 5 bolt material tapers off rather gradually above the elastic limit to an ultimate strength value 35% greater than the minimum yield. This behavior represents a 35% margin and it is essentially a reserve of strength. The result is that the bolts are not weakened by a preload to yield and should perform well under static and dynamic conditions.

TABLE 2.15
Design Data for Selected SAE Bolts

Threads Per in.	Diameter in.	Stress Area in.^{2}	SAE Grade 5 Bolts S_u, psi	S_p, psi	Preload, lb.	SAE Grade 8 Bolts S_u, psi	S_p, psi	Preload, lb.
20	0.2500	0.0318	120,000	85,000	2,020	150,000	120,000	2,860
28	0.2500	0.0364	120,000	85,000	2,320	150,000	120,000	3,280
16	0.3750	0.0775	120,000	85,000	4,940	150,000	120,000	7,000
24	0.3750	0.0878	120,000	85,000	5,600	150,000	120,000	7,900
13	0.5000	0.1419	120,000	85,000	9,050	150,000	120,000	12,750
20	0.5000	0.1599	120,000	85,000	10,700	150,000	120,000	14,400
11	0.6250	0.2260	120,000	85,000	14,400	150,000	120,000	20,350
18	0.6250	0.2560	120,000	85,000	16,300	150,000	120,000	23,000
10	0.7500	0.3340	120,000	85,000	21,300	150,000	120,000	30,100
16	0.7500	0.3730	120,000	85,000	23,800	150,000	120,000	33,600
9	0.8750	0.4620	120,000	85,000	29,400	150,000	120,000	41,600
14	0.8750	0.5090	120,000	85,000	32,400	150,000	120,000	45,800
8	1.0000	0.6060	120,000	85,000	38,600	150,000	120,000	54,500
12	1.0000	0.6630	120,000	85,000	42,200	150,000	120,000	59,700
7	1.1250	0.7630	105,000	74,000	42,300	150,000	120,000	68,700
12	1.1250	0.8560	105,000	74,000	47,500	150,000	120,000	77,000
7	1.2500	0.9690	105,000	74,000	53,800	150,000	120,000	87,200
12	1.2500	1.0730	105,000	74,000	59,600	150,000	120,000	96,600
6	1.3750	1.1550	105,000	74,000	64,100	150,000	120,000	104,000
12	1.3750	1.1350	105,000	74,000	73,000	150,000	120,000	118,400
6	1.5000	1.4050	105,000	74,000	78,000	150,000	120,000	126,500
12	1.5000	1.5800	105,000	74,000	87,700	150,000	120,000	142,200

2.8 CONTROL OF TORQUE AND PRELOAD

The problem of preload and torque control is not trivial. The designer needs a correct preload since too much or too little may be detrimental. Furthermore, when several fasteners are involved, it is also necessary to have a uniform preload to avoid local warping of the part and overloading some of the fasteners.

Considerations of stress and strain lead directly to the theory and practice of torque and preload control. Any failure to estimate and to apply the correct preload can result in a number of problems and at times, very serious incidents. Examples of such problems can be summarized as follows:

- Breaking of bolt shank or stripping of threads
- Crushing or deformation of joint members
- Vibration loosening of the nut
- Low preload fatigue failures
- Joint separation and leakage
- Joint slip, misalignment, and danger of bolt shear
- Excessive weight for lower preload
- High number of fasteners at insufficient preload

Prior to getting involved in a study of torque and preload control, Bickford suggests [3] a number of influencing factors which are likely to change our estimate of preload:

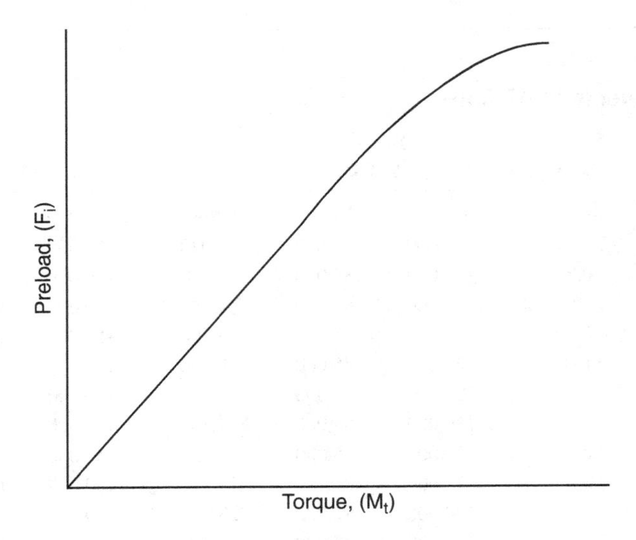

FIGURE 2.18 Example of preload-torque curve.

- Accuracy of torque wrench
- Operator error
- Accuracy of input variables
- Performance of gaskets
- Prediction of external loads
- Long-term relaxation phenomena
- Quality of component parts

The design formulas, Eqs. (2.17) and (2.19) show that there is a linear relationship between the applied torque and the developed preload. Thus, the general theory appears to be rather elementary and experience shows that such a linear relationship exists as in Figure 2.18.

Observe that the straight line in Figure 2.18 eventually becomes curved when either the fastener or the joint begin to yield. It was noted previously that the initial input torque T can be subdivided into three components on the basis of three torque coefficients K_1, K_2, and K_3. While such components may not reflect the correct input energy, it is often argued that only a small portion of the entire torque is transformed into a useful torqueing effort producing the preload. Whether the more simplified formula, Eq. 2.19 or the more complex equations of energy is used, the argument appears to be overshadowed by the complexities of the frictional effects which are difficult to control. It does not take long to list a dozen or more variables affecting the friction.

The parameter K called "torque coefficient" or "nut factor" is extremely useful because it apparently integrates all the geometrical and frictional effects. The problem with K, however, is that it can only be determined experimentally for each new application. At best, we can hope to apply a standard deviation to the test results and calculate a confidence level for a particular design. This practice is well known, although the evaluation of a particular confidence level and its interpretation in relation to economic and safety constraints is not a simple matter.

If one applies the same torque to a large number of bolts, it is possible to obtain a histogram giving a probability versus preload. Experiments show that we can find either a normal (Gaussian) or a skewed distribution. Experience also shows that it is particularly difficult to predict the range of K values for a small number of tests, even for the case of using the same operator, the same joint, tools, and other possible variables under our control.

One of the basic questions in this research is: What can be the effect of thread lubricant on the scatter of preload results for a fixed value of torque? The practice indicates that lubricants tend to reduce scatter in preload and the torque required for a given preload appears to have a Gaussian distribution. The drier the contact surface, the greater the scatter seems to be. In summary, we can say that thread lubrication helps to improve repeatability of test results. Also, whenever we use a torque wrench, such factors as friction, operator, geometry, tool accuracy, and relaxation must affect the final outcome in addition to quality of design, workmanship, human error, and similar constraints. The latter effects can be very serious and they will tend to overshadow our ability to correctly predict a value of K or μ.

The problem with torque control, even with the aid of a variety of modern tools, poses a question of other means of control such as the commonly known relationship between the amount of bolt turn and the preload. What happens when we rotate the nut a well-defined number of degrees? Will the bolt stretch be directly proportional to the amount of nut rotation so that the amount of preload can be related to the bolt stretch? Unfortunately, the answer is not simple, since the proper analog of bolt joint behavior requires knowledge of the spring constants of the fastener and the body of the joint. Further review of the problem of spring constants is deferred to Chapter 5, which will discuss the mechanics of a bolted joint. The presence of washers, gaskets, and geometrical variations within the fasteners will not make an easy task out of defining the relation between the amount of turn and the preload. The geometric variations need to include the perpendicularity of surfaces and flatness of joint members, as well as thread deformation. On this basis, it appears that turn measurement used alone would not be any more accurate than the measurement of torque. However, a combination of torque and turn control can have certain advantages:

- Torque and turn until yielding provides useful information.
- The method helps to identify the presence of blind holes, wrong parts, and crossed threads.
- The torque and turn information can be used as an input to computer-control systems for a more sophisticated study.

It should be pointed out that the so-called turn-of-nut method is still used in structural steel applications. Historically, it was the first control method of bolt preload. This procedure is illustrated graphically by the solid line in Figure 2.19.

In the turn-of-nut technique of torque control shown in Figure 2.19, the nut is first tightened with an approximate torque A, snugging the bolt and providing bolt stretch well within the yield strength. Such an initial preload is accurate because of the effects of friction and geometry. The subsequent turn, however, takes the torque to point B in Figure 2.19 which corresponds to the bolt stretch well

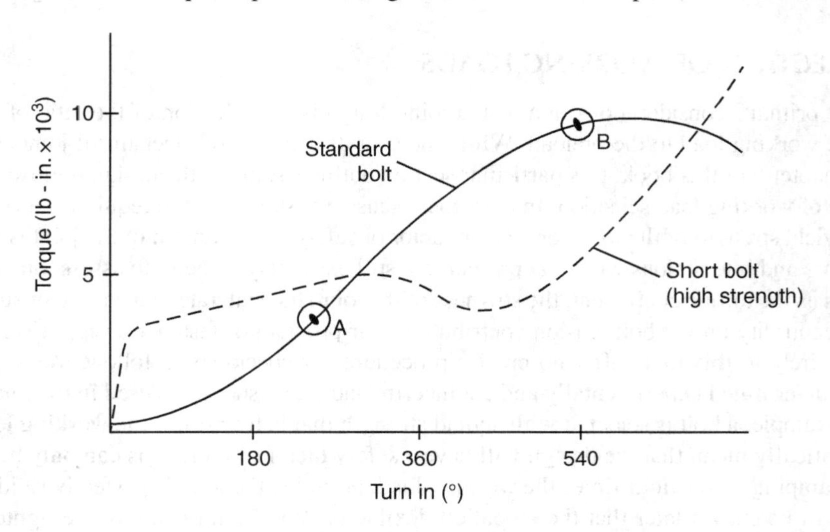

FIGURE 2.19 Example of turn-of-nut characteristic.

past its yield point. This usually happens when we turn the nut about 180E past the linear response of the torque-turn curve. While this technique is quite admissible for ductile bolts, it is a poor means of defining the yield point and it is not recommended for brittle materials.

A number of industrial organizations have developed guides for torque/turn procedures. In the case of aerospace applications, the tightening of the fasteners is assumed to be below the yield point of bolt materials. This procedure is called "turn of nut" and it shows better accuracy than using the torque or turn alone. The accuracy is improved through the cycling and retightening which reduces the degree of relaxation. It should be mentioned, however, that "turn of nut" technique in aerospace industry is used on high-quality materials and assembly only.

It is unfortunate that the problem of fastener control, under the usual conditions of production, has very little to do with the theoretical behavior of an ideal fastener. Bickford lists a number of problem areas [3] which can have serious implications:

- Blind holes
- Insufficient depth of tapped hole
- Wrong size of bolt or a hole
- Inclusions of particles
- Partially stripped or crossed threads
- Soft materials
- Gross misalignment of parts
- Galling of threads
- Warpage of assembled components
- Machine burrs
- Tool malfunction

Nevertheless, there are a number of measurement systems which can help with the control decisions. Some of the preload-monitoring devices are computer controlled and work well unless we are using short bolts made from very high-strength materials. Under these conditions, the basic S-curve of torque versus turn can become grossly distorted as indicated very roughly by a broken line in Figure 2.19. For a more complete review of the various control systems employed in tightening of the bolted joint, the reader is referred to the book by Bickford [3]. The extensometer, when used correctly, is still a remarkably reliable instrument of control under static or dynamic conditions. However, fasteners and joints, in general, still constitute a very complex area of design.

2.9 SELECTION OF WORKING LOADS

One of the primary considerations in a bolted joint design is the selection of the ratio of the service load or the working load to the preload. While the overall behavior of mechanical joints is the main topic of Chapter 5 of this book, this particular section outlines some of the design philosophy and the basic rules of working load selection. In a practical sense, we start with the required service load and select the yield strength of the joint for a given factor of safety. The strength of the joint is affected by many factors and interactions which, at present, are still extremely difficult to assess. Such important parameters as the torque coefficient, the stiffness of the joint, the leak rate of a gasket, or subtle effects of load eccentricity on the bolt tension contribute to our problem of fastener design. Even the books devoted entirely to this topic offer no precise procedures or checklists to follow. Many parameters have to be determined experimentally and engineering judgment should be used in the final analysis.

If, for example, a bolt is selected with a yield strength that is four times the working load, it does not automatically mean that the design will have a safety factor of four. This can only be true if the relevant clamping load is four times the working load, provided the joint is perfectly rigid or a "hard joint." It will be shown later that the so-called flexible or "soft" joint can only be tightened to the working load.

In many practical configurations, the ratios of the working loads to the preloads range between 0.3 and 0.8. It appears that, as this ratio increases, the margin between the working load and the preload decreases. It also follows that use of the more efficiently designed bolts requires more reliable information which is not easy to obtain. From a safety perspective, a wide margin between the external load and the preload is desirable which, however, implies a relatively high cost.

A cardinal rule says that the working load should not be permitted to exceed the bolt preload. This is especially important in the case of dynamic or fatigue applications. As long as the bolt does not experience any appreciable stress variation, there can be no failure in fatigue regardless of the number of load cycles. Again, this rule loses its meaning when the joint has considerable flexibility and the variable stress present may be high enough to cause failure by fatigue regardless of the initial preload. As shown in Figure 2.10, the highest stress is found at the first thread inside the nut. This is the most critical location for fastener failure due to fatigue. Other critical points will be found at the junction of the head and shank or at the thread runout. In general, it appears that the best fatigue resistance is obtained when the rigid joint members are held together by elastic bolts. Furthermore, it is good to emphasize again that a high preload provides assurance that the joint can withstand higher static and cyclic loads trying to shear, bend, or pull the joint apart.

While a reasonably high preload is beneficial, it is well to keep in mind the fact that plastic elongation of the bolt can occur. Unfortunately, however, joint loosening is often unavoidable. It can, however, be minimized by tightening larger bolts to somewhat higher values of initial tension. It appears that the ultimate balance between the high preload and the minimum joint loosening is a matter of a rather delicate compromise.

Knowledge of the factors of safety, yield strength of the bolt material and joint stiffness enters the deliberations of the working loads. If the assumption is made that the joint components do not separate as a result of the application of the external working load, then the decrease in the deformation of the connected members must be equal to the increase in the deformation of the bolt. This basic relation permits expressions for the calculation of the working load in terms of the spring constants and the bolt preload.

2.10 CURRENT VIEW OF BOLT PRELOAD

It is advisable, even at the risk of being repetitious, to comment on modern views of initial preload and joint behavior as a function of time and other variables [5]. The correct preload, even with modern controls, is still difficult to determine. The apparently simple problem is complicated by a multitude of variables: Bickford quotes a number as high as 76. This obviously includes design, materials, fabrication, and environment variables during and after the bolted joint is constructed. Many changes occur to the joint after we discontinue preload control for economic reasons and the question then is: What can be done to monitor preloads in service?

In some applications, the preloads can be checked utilizing ultrasonic techniques which have been improved in recent years. This can be achieved by recording the initial length of each fastener acoustically and monitoring this parameter as a function of time. As the fasteners loosen, for example, their length should decrease. This decrease represents a good measure of the amount of preload lost in the fastener during the initial torqueing operation.

During the initial preload the joint contact surfaces must have rested on the microscopically small high spots. Because of very high local stresses, such spots will creep and break down until larger contact spots develop and the process is stabilized. The development of contact stresses is made much more complex if the face of the nut is not perfectly perpendicular to the axis of the threads. Generally, the relaxation of the preload is on the order of 5%–10%, although higher numbers have also been observed. Some correction of this problem can be made by retightening.

It is known that significant short-time relaxation is caused by gasket creep, particularly under high contact pressures. However, with good design involving spiral-wound, oval ring, and similar gaskets, the initial loss of preload can be reduced to about 5%. For most configurations, the gasket should creep a little to perform its intended function.

Another cause of short-term relaxation is elastic interaction. This phenomenon depends on the stiffness of the joint members, whether or not a gasket is present. The joint, in the vicinity of the first bolt, is compressed during the preload. When the next bolt is tightened, the joint is further compressed near the first bolt, allowing this bolt to undergo a slight relaxation. As the subsequent bolts are progressively tightened the losses due to the elastic interaction can be quite substantial. There appears to be two methods of compensating for these effects. One is based on the idea that more torque should be applied to the first bolts than to the latter ones. The problem is that we may be limited by the amount of preload that can be applied. The selection of torques for the "so called" multipass procedures can be recommended as the second control method, provided ultrasonic measurements are made in monitoring the preloads. Under these conditions, we must end up with unequal torque and bolt tension in order to minimize the loss of preload.

As mentioned in other sections of this book, the mechanism of vibration loosening can eliminate practically all initial preload. Although this effect is severe, it can be prevented by a proper joint design, use of special thread forms, application of spring nuts, and anaerobic adhesives. Other methods involve pinning or bonding of the joint members together in order to prevent the transverse slip, which is found to be most damaging.

When temperature or nuclear radiation is involved, the fasteners are subject to long-term stress relaxation. This phenomenon usually develops under a high, constant load. The only method to mitigate this effect is to use high-strength materials such as A286 or Nimonic 80A. These and other effects continue to be of modern concern despite the significant progress in design and materials technology. Better understanding of the 76, or a similar number of the variables and their interactions affecting our bolted joint design judgment, may eventually make the difference between the proper use or misuse of some of the six billion bolts manufactured every year.

2.11 SUMMARY OF DESIGN CONSIDERATIONS

Recent development trends point to the increased applications of high-preload bolts. Rivets are being replaced by bolts on almost a one-to-one basis. Turn-of-the-nut procedures of torque control have been used in bridge construction. Heavy structural bolts could be employed without the use of washers. The performance of a joint depends on the strength of individual fasteners and the control of slip.

The remnants of riveting practice still refer to such operations as impact forming and spin, as well as squeeze forming. The rivet material should be as strong as that of the plates. Rivet spacing should be at least equal to three rivet diameters. The edge distance should be at least 1.5 times the rivet diameter. This diameter should be within one to three times the plate thickness. The rivet is designed to work in shear. The process of driving hot rivets depends upon temperature and time. The average ratio of shear to tensile strength for the rivets should be about 0.75. Combination of tension and shear in rivets is analyzed according to an elliptical interaction curve. The Eiffel Tower is an example of extensive use of rivets.

Development of bolted joints continues to be hampered by limited data and thus it requires considerable engineering judgment. The problem is worsened in that in a given structure we can usually only afford to test a few samples.

The materials for externally threaded fasteners are normally characterized by the tensile yield, ultimate strength, and hardness. These are covered by SAE and ASTM specifications. The weakest section of the bolt is typically in the threaded area. The maximum shear and tensile strengths of bolts currently available are 130 and 230 ksi, respectively. The thread designs include UNC and UNF standards for the coarse and fine thread categories. The proof load is that at which the bolt carries the maximum load without a measurable, permanent set. Bolt preload can be relaxed due to vibration and temperature effects. Bolts are generally not reusable above the service temperature of 500E F.

Lower strength bolts can be strained beyond yield without becoming brittle due to work hardening. The combined state of stress in a bolt can be due to tension, shear, and bending. Stress concentrations in a bolt designed by elastic theory can be very high. The peak stresses in a threaded portion vary in a nonlinear fashion. The first two or three threads carry the major share of the bolt load. Improved design configurations for both bolts and nuts are available for assuring more even load transfer. Effective stress factors for cycling and fatigue loading of bolts depend, to a significant degree, on the thread form. The response to shear loads is a function of bolt material, the shear surface, and the number of threads.

In considering the problem of bolt strain, it is first necessary to define the effective length of the bolt. The bolt is designed primarily to act as a spring in tension. The torque required to give a specified preload depends on the bolt diameter and a complex friction parameter. Only about 10% of the wrenching torque constitutes a useful portion of the work required to give the preload. The most conservative bolt design is attained with the maximum frictional resistance under static conditions.

Useful load tables are available for selecting maximum bolt preloads. The ratios of proof to maximum strength are on the order of 0.7–0.8. Fasteners with coarse threads are typically used more often than fine threads, where the fine thread series is preferred in automotive and aerospace industries. The threaded lengths for all bolts typically vary between one to 2.5 bolt diameters.

Poor control of bolt preload is the primary source of operational problems and incidents. It is fortunate that despite the great number of influencing factors, there is a linear relationship between the applied initial torque and the preload over a useful range of these parameters. This statement applies to a given bolted joint with a known value of a frictional constraint. A number of industrial organizations still practice "turn of nut" procedures as their means of bolt preload control. There are also a number of sophisticated devices on the market suitable for monitoring bolt preload.

One of the important decisions in bolted joint design is the selection of the ratio of the service or working load to the preload. Here the torque coefficient, stiffness of the joint and potential leak rate past a gasket, as well as load eccentricity, can affect our technical decisions. As a practical guide for designers, the foregoing ratio can be in the range of 0.3–0.8. In the case of fatigue, the working load to preload ratio should be as low as economically acceptable. Plastic elongation of the bolt should be controlled in order to avoid any loosening of the joint. Ideally, the decrease in the compression of the connected member should not be numerically higher than the deformation of the bolt.

REFERENCES

1. Rochrich, R. L. 1967. "Torquing stresses in lubricated bolts," *Machine Design*.
2. Fisher, J. W. and J. H. A. Struik. 1974. *Guide to Design Criteria of Bolted and Riveted Joints*, New York: John Wiley & Sons.
3. Bickford, J. H. 1981. *An Introduction to the Design and Behavior of Bolted Joints*. New York and Basel: Marcel Dekker, Inc.
4. "Mechanical and quality requirements for externally threaded fasteners," SAE J 429d, Society of Automotive Engineers, Warrendale, PA, 1967.
5. Bickford, J. H. 1985. "That initial preload—What happens to it?" *Mechanical Engineering*.
6. Rothbart, H. A. 1964. *Mechanical Design and Systems Handbook*. New York: McGraw Hill.
7. Huston, R. L. and H. Josephs. 2009. *Practical Stress Analysis in Engineering Design*, 3rd Ed., Boca Raton, London, and New York: CRC Press.

SYMBOLS

A Nominal bolt cross-section
$A_1,\ldots,A_5$ Arbitrary areas
A_S Stress area
D Nominal bolt diameter
$D_{\min}$ Minimum thread diameter

d	*Rivet diameter*
E	Modulus of elasticity
e	Distance from root of thread
F	General symbol for force
F_i	Initial bolt load
K	Torque coefficient or nut factor
K_1, K_2, K_3	Components of torque coefficient
K_b	Spring constant of bolt
K_f	Combined stress factor
K_0	Conventional factor of stress concentration
$L_1, \ldots, L_5$	Arbitrary lengths
ΔL	Change in length
M_b	Bending moment
M_{to}	Applied torsional moment
n	Number of threads per inch
P	Pitch of thread
Q_{max}	Maximum shear load
R	Radius of bolt curvature
r	Mean radius of thread
S	General symbol of stress
S_b	Bending stress
S_{to}	Torsional stress in bolt shank
S_u	Tensile strength
T	Initial tightening torque
V_e	Thread recession
α	Helix angle, deg
β	Half-angle of thread profile, deg
ρ	Fillet radius
μ	Coefficient of friction
τ	Shear stress
τ_{max}	Maximum shear stress

3 Concepts of Strength and Failure

3.1 PRELIMINARY CONSIDERATIONS

The topics discussed in this chapter will undoubtedly be familiar to most readers with an undergraduate engineering background. What is different here is a practical, philosophical discussion of the elementary principals, and the ways that relatively simple equations can be utilized to analyze large/complex mechanical systems. We hope to provide tips and insights for using the simple textbook illustrations applied to large systems having non-simple geometries. Our focus is upon applying practical stress analysis to the design of mechanical joints and fasteners.

Stress values form the basis for the analysis and design of "robust" mechanical systems. That is, stresses govern (or should govern) the sizes and shapes of the components of mechanical systems, in anticipation of probable and possible external loads.

Stress considerations blend the values computed by relatively simple equations determined by mechanics and types of applied forces or load with known values of the properties of the materials involved. Surprisingly, the number of basic formulas appears to be rather small and their use is generally straightforward. What then, in real design situations, makes the application of strength and fracture equations and analysis so challenging?

There are several reasons for this: First, textbook examples generally have relatively simple shapes—often different and far simpler in complexity than components typically found in "real-world" mechanical systems. Secondly, the issues of safety and economic concerns are seldom considered in text problems. Finally, tables of material values typically represent optimal values at a single measured point, assuming no defects in the materials, nor any difference for other values in the spectrum beyond the specific point measured.

The literature on structural analysis and mechanics of materials contains a number of stress definitions and descriptive terms such as: "direct stress," "bending," "flexure," "torsion," "twist," "transverse shear," "ultimate stress," "principal stress," "primary," or "secondary stresses" and others. In reality, however, there can only be two basic kinds of stress: normal and shear when we consider a state of stress at a given point of a mechanical system. At a given cross-sectional plane of a structure the normal stress is, of course, that which is perpendicular to that plane, while the shear stress acts along the given plane. In a simple beam, loaded in the transverse direction (perpendicular to the long axis of the beam), the normal stress is that which is perpendicular to the transverse cross-section or, in other words, the normal stress is directed along the axis of the beam. We recall that a simple beam is considered to be a 2-D component.

The familiar bending stress of $\sigma = MC/I$, known to design engineers, is simply a normal stress at the section where the bending stress is considered. The transverse stress at this location, on the other hand, is the shear stress which acts at 90° to the normal stress just described. These two stresses, of course, can be combined according to a suitable strength of materials theory depending on the type of stresses and the materials from which the component is made.

For example, if the material is brittle, we will initiate our analysis by first examining the maximum normal stress in our failure considerations. Alternatively, if the material is ductile, we will focus upon the shear stresses. We will discuss failure in greater detail later in this chapter.

Where published information makes a reference to the "bending stress" or a "flexural stress" treating it as a specific material's property, some confusion may arise unless we can keep in mind that both terms are describing the same stress and that, therefore, they both produce the same kind of normal stress as indicated above.

As shown by the elementary theory of bending of beams given in the normal stress can be either tensile or compressive as discussed in detail in many elementary books on strength of materials. These stresses are proportional to the distance C, measured from the neutral axis. The parameter C then enters the following standard equation for the bending stress:

$$\sigma = \frac{MC}{I} \tag{3.1}$$

In Eq. (3.1), M is the bending moment and I is the second moment of area, or as it is popularly described, the "moment of inertia" which is a well-defined property of a cross-section. The term σ is referred to as the "bending stress," "flexural stress," or "normal stress" caused by the external bending moment M.

Eq. (3.1) indicates that for stress calculation, we need to know the external loading on the beam (M) and the geometry of the beam (c/I).

3.2 EFFECT OF LATERAL FORCES

Since the most frequent problem in engineering design is concerned with bending of beams, it is helpful to distinguish between pure bending and the response of a beam to lateral forces. The effect of pure bending is to form the beam into a circular arc of radius equal to (EI/M) where E is the elastic modulus. Under these conditions normal stresses act at each transverse cross-section with no shearing stresses present.

When a beam is loaded with lateral (or shear) forces, however, bending and lateral forces exist simultaneously. The normal stresses, as before, are given by the bending moment as given by Eq. (3.1). For all practical purposes the normal stresses are unaffected by the shearing forces at a given cross-section. The effect of a shearing force on beam deflection is also rather small for most beam configurations encountered in practice.

Therefore, for a beam supporting lateral forces, it is necessary to know the magnitudes of the bending moment and the shearing force at a particular beam cross-section before stress analysis can be made. Suppose we have a cantilever beam with an inclined end load W, such as that shown in Figure 3.1. If we denote the angle at which the external load is inclined by θ, simple resolution of W into the two components gives the lateral load equal to $W \sin\theta$ and the axial compressive load equal to $W \cos\theta$.

At an arbitrary point on the beam, such as that defined by distance x measured from the built-in end of the beam (see Figure 3.1), we can compute the bending moment and the shearing force as:

$$M = W(L - x)\sin\theta \tag{3.2}$$

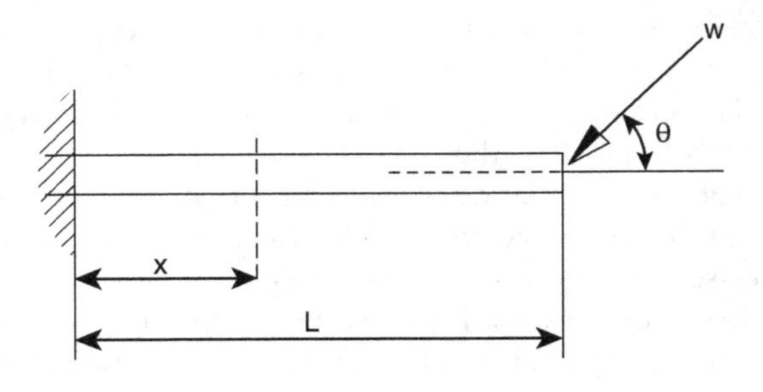

FIGURE 3.1 A cantilever beam with an inclined end load.

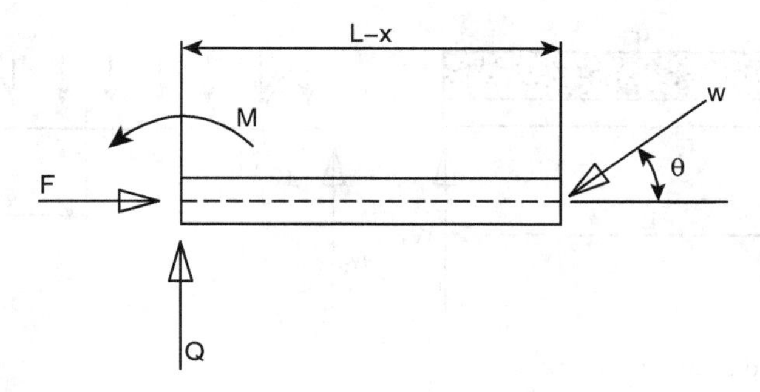

FIGURE 3.2 Equilibrium diagram for a portion of the cantilever beam.

and

$$Q = W \sin\theta \tag{3.3}$$

When $\theta = 0$, the external load W becomes equal to the maximum axial thrust while, at the same time, the bending moment and the shearing force vanish as per Eqs. (3.2) and (3.3). Alternatively, for $\theta = \mu/2$ the bending moment and the shearing force attain maximum values while the axial thrust becomes equal to zero.

It should be noted that the choice of x is quite arbitrary and the relevant distance can be measured from either the built-in or the supported end. What we are doing in effect is considering the state of static equilibrium of the right-hand portion of the cantilever as shown in Figure 3.2.

The rules of static equilibrium dictate that since we have three unknowns we have to establish two equations involving the forces and one equation defining the bending moments for problem solution. In terms of our diagram in Figure 3.2, the equations for the axial force F, shearing force Q and bending moment M can be stated as:

$$F - W\cos\theta = 0 \tag{3.4}$$

$$Q - W\sin\theta = 0 \tag{3.5}$$

$$M - W(L - x)\sin\theta = 0 \tag{3.6}$$

Eqs. (3.4) through (3.6) constitute three equations for three unknowns: F, Q, and M. In actual practice, when we have *statically determinant* beams,* we can evaluate the bending moment and the shearing force using Eqs. (3.2) and (3.3) which are equivalent to Eqs. (3.5) and (3.6).

3.3 STATICALLY INDETERMINATE BEAMS

Consider the beam configuration shown in Figure 3.3 where a beam with length L has cantilever (or "built-in") support at its left end and a vertical (or "pin," "roller," or "simple") support at its right end, herein called a "propped cantilever beam."

When we construct a free-body of the beam, we discover that we have more unknown support forces than the number of equations that we can obtain from the free-body diagram. That is, if we set moments about an end of the beam and add forces vertically, we obtain *two* equations. [Adding forces horizontally produces only a null (0 = 0) expression.] There are, however, *three* unknown reactions M_f, W_1, and W_2, as seen in Figure 3.3.

* Beams where a free-body diagram produces the same number of equations as there are unknowns.

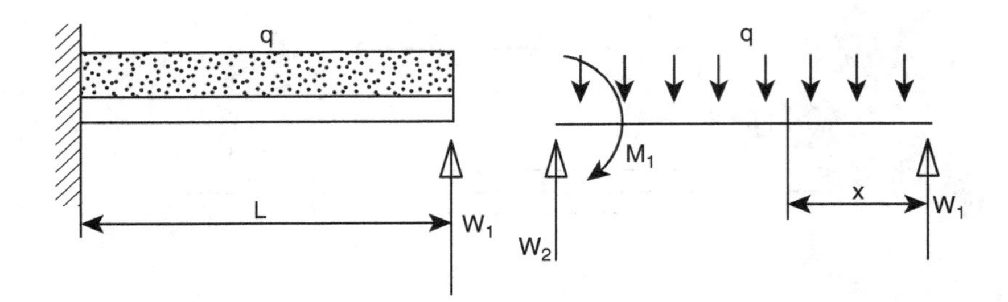

FIGURE 3.3 Propped cantilever beam.

With the number of unknowns (3) being fewer than the number of equations (2), we have a "statically indeterminant" beam. This means that we need to consider the deformation of the beam to obtain all of the unknown reactions.

These simple illustrations of statically determinant and statically indeterminant structures illustrate the importance of supports in our structural analyses. This means that the supports generally affect the difficulty in the analysis. This is one of the inherent complexities of stress analysis which influence the choice of the methods of design.

For example, in Figure 3.3, the bending moment M at a distance x from the simple support is:

$$M = \frac{gx2}{2} - W_1 x \tag{3.7}$$

To determine the reaction W_1, we know that the displacement at the W_1 support is zero. By referring to Castigilano's principle [1] we have the expression:

$$\int_0^L M \frac{\partial M}{\partial W_1} dx = 0 \tag{3.8}$$

From Eq. (3.7), we have:

$$\frac{\partial M}{\partial W_1} = -x \tag{3.9}$$

Substituting Eqs. (3.7) and (3.9) in (3.8) and integrating gives:

$$\frac{W_1 L^3}{3} - \frac{gL^4}{8} = 0 \tag{3.10}$$

By solving Eq. (3.10), we find W_1 to be: $3qL/8$, the known value of the end supported cantilever beam subjected to a uniform load.

3.4 CROSS-SECTION GEOMETRY

Once the external loads and the support (or "reaction") forces and moments are known, we need to know the geometry and cross-section of properties of the beam, to complete the analysis. In Eq. (3.1) we find the term: *I/C*, also known as the "section modulus" where I is the section moment of inertia. This term has a linear dimension raised to the third power. To use the section modulus, we need to determine the location of the centroid of the cross-section. For a symmetrical section or a section with a common shape, the centroid location can usually be found by inspection. For more complex cross-sections, however, the centroid location may be less evident and may require referring to one of the numerous figures and formulas enabling location of the centroid. Due to its frequency of use and its significance, a brief review of the operational concepts of the moment of inertia is in order here. Every rigid body has a moment of inertia, which is always a positive real number and whose value is dependent upon the axis from which the value is determined. Hence in its more formal sense, the moment of inertia is denoted by I_{xx}, I_{yy}, I_{zz} to denote the moment of inertia taken with respect to the X, Y, and Z axes, respectively. This is in contradistinction to the product of inertia, which is denoted by I_{xy}, I_{yz}, and I_{zx} whose values for the same rigid body can be either positive, negative, or zero dependent upon the axes chosen. In dealing with problems involving the products of inertia, it is often advantageous to choose those axes that cause the products of inertia to equal zero. Typically, these axes which cause the products of inertia to go to zero are the centroidal principal axes.

Often, to provide for each of calculation, it is found desirous to shift the inertial axis to a different but parallel axis. This is accomplished by utilizing the parallel axis theorem, which states:

$$I = I_G + MD^2 \tag{3.11}$$

where:

I = moment of inertia through the "desired-new" axis
I_G = moment of inertia of the mass center parallel to the axis taken with respect to I
M = mass of body or body section being transferred
D = distance between the two parallel axes.

It is to be noted from Eq. 3.11 that the moment of inertia through the centroidal axis is the minimum value of moment of inertia for any specific rigid body. Additionally, the moment of inertia can be written as:

$$I = MK^2 \tag{3.12}$$

or

$$K = \sqrt{I / M} \tag{3.13}$$

where K is defined as the radius of gyration.

In a manner similar to above, we can define the parallel axis theorem in terms of the radius of gyration by the equation:

$$K^2 = (K_G)^2 + D^2 \tag{3.14}$$

where:

K = Radius of gyration of a "desired-new" axis
K_G = Radius of gyration of the mass center parallel to the axis with respect to K
D = Distance between the two parallel axes.

Often we are able to compute the moments of inertia of complex bodies by the judicious application of the parallel axis theorem to a combination of simple body shapes—either line, area, or solid body shapes. Tables of common body shapes and their inertial properties are shown in reference [7].

3.4.1 Example

Determine the moment of inertia of the area element shown with respect to its centroidal axis.

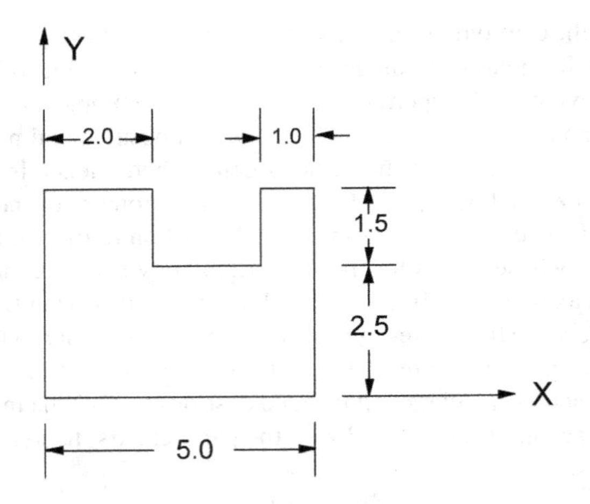

3.4.2 Approach

Prior to determining the moment of inertia of the given element with respect to its centroidal axis—we must first determine the location of the centroid.

To approach this problem, we will first devolve our given composite area element into three simple rectangles:

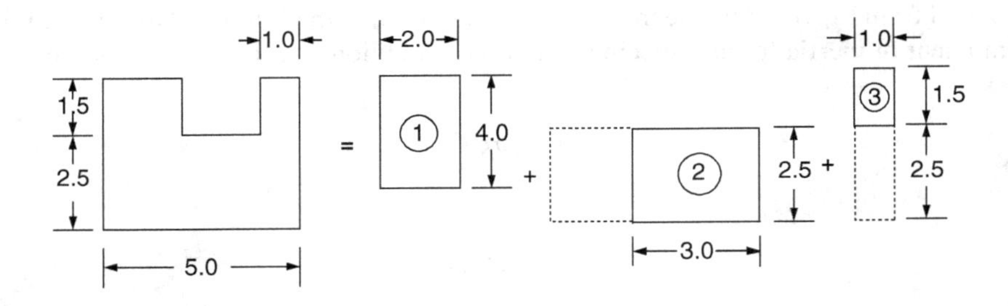

To find the location of the centroid, we will find the location of each of the three simple area elements and combine the three.

Part	Area	$\bar{x}$ (in.)	$\bar{y}$ (in.)	$\bar{x}A$ (in³)	$\bar{y}A$ (in³)
1	$4 \times 2 = 8$	1.0	2.0	8.0	16
2	$3 \times 2.5 = 7.5$	$2.0 + 1.5 = 3.5$	1.25	26.25	9.375
3	$1.0 \times 1.5 = 1.5$	$4.0 + 0.5 = 4.5$	$2.5 + 0.75 = 3.25$	6.75	4.875
Σ	17			41	30.25

$$\bar{x} = \frac{\Sigma \bar{x}A}{\Sigma A} = \frac{41}{17} = 2.41$$

$$\bar{y} = \frac{\Sigma \bar{y}A}{\Sigma A} = \frac{30.25}{17} = 1.78$$

The location of the centroid on our composite body is shown below.

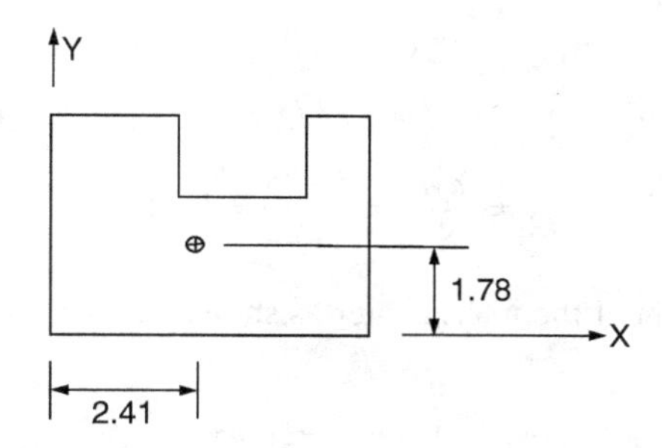

Now we can use the parallel axis theorem and the value of the moment of inertia for a rectangular area element $\left(I = \dfrac{bh^3}{12}\right)$ to find the moment of inertia of the composite body.

Part	Area	I_{xx}	$+MD^2$ (parallel axis)	
1	8	$\frac{1}{12}(2)(4)^3$	$8(2-1.78)^2$	$= 10.67 + 0.39 = 11.06$
2	7.5	$\frac{1}{12}(3)(2.6)^3$	$7.5(1.78-1.25)^2$	$= 3.91 + 2.11 = 6.02$
3	1.5	$\frac{1}{12}(1.0)(1.5)^3$	$1.5(3.25-1.78)^2$	$= 0.28 + 3.24 = 3.52$
Σ				$I_{xx} = 20.6$

Part	Area	I_{yy}	$+MD^2$ (parallel axis)	
1	8	$\frac{1}{12}(4)(2)^3$	$8(2.412-1.0)^2$	$= 2.67 + 15.95 = 18.62$
2	7.5	$\frac{1}{12}(2.5)(3)^3$	$7.5(3.5-2.41)^2$	$= 5.63 + 8.88 = 14.51$
3	1.5	$\frac{1}{12}(1.5)(1.0)^3$	$4.5(4.5-2.41)^2$	$= 0.13 + 6.54 = 6.67$
Σ				$I_{yy} = 39.80$

It would be instructive to address the same problem by using negative areas. Here, we will then have the configuration as such.

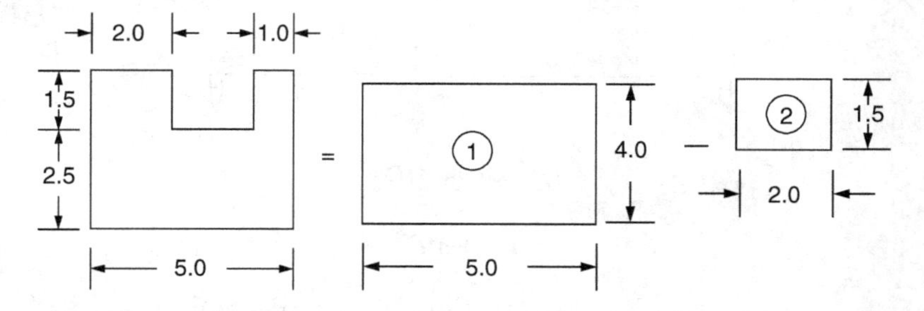

Here again we will proceed by first finding the location of the mass center.

Part	A (in²)	$\bar{x}$ (in)	$\bar{y}$ (in)	$\bar{x}A$ (in²)	$\bar{y}A$ (in²)
1	$5 \times 4 = 20$	2.5	2.0	50	40
2	$-2 \times 1.5 = -3$	3.0	3.25	−9	−9.75
Σ	17			41	30.25

$$\bar{x} = \frac{\Sigma \bar{x} A}{\Sigma A} = \frac{41}{17} = 2.41$$

$$\bar{y} = \frac{\Sigma \bar{y} A}{\Sigma A} = \frac{30.25}{17} = 1.78$$

And again, we find the location of the mass center as shown below:

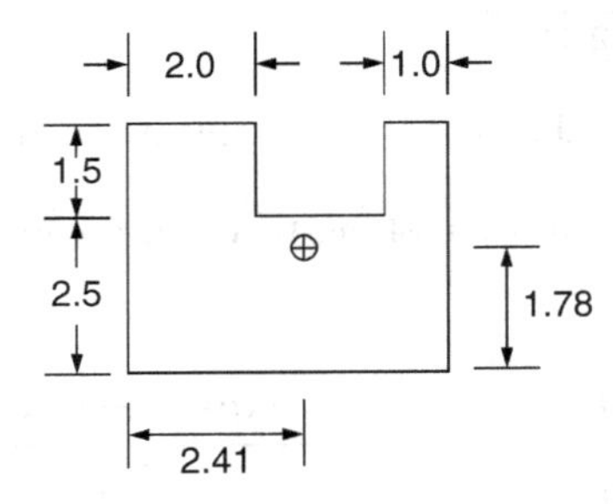

Again using the parallel axis theorem, we find:

Part	Area	I_x	$+MD^2$ (Parallel Axis)	
1	20	$\frac{1}{12}(4)(5)^3$	$+20(2.5-2.41)^2$	$= 41.67 + 0.16 = 41.83$
2–3		$\frac{1}{12}(1.5)(2)^3$	$+3(3-2.41)^2$	$= -(1.0 + 1.04) = -2.04$
Σ				$I_{yy} = 39.8$

We will further demonstrate application of the parallel axis theorem by moving the centroidal moment of inertia I_{xx} to the basis axis H-H.

1.78

H H

$$I_{HH} = I_G + MD^2$$

$$= I_{XX} + MD^2$$

$$= 20.6 + [(4 \times 2) + (2 \times 2.5) + (1.0 \times 4.0)](1.78)^2$$

$$= 20.6 + 17(1.78)^2$$

$$I_{HH} = 74.46 \text{ in}^4$$

As we noted above, we observe here that the moment of inertia about the centroidal axis is the minimum inertia value.

Note that, if the most remote fiber from the neutral axis, as given by the parameter C, happens to be in compression while the material is much weaker in tension, the relevant part could fail on the tension side before full strength in compression is attained. This situation is typical of concrete behavior where the tensile strength is only a small percentage of the strength in compression. For this reason, introduction of steel reinforcement into the concrete mix is essential for the protection of concrete structures against a premature failure in bending or tension.

In many cases, we can simulate the structural behavior using loading and support characteristics of beams. However, when considering such structures as a machine housing, support base, bridge flooring, and similar configurations, the effect of beams of relatively great width should be accounted for. Since lateral deformation in a wide beam is accordingly constrained since its rigidity is increased in proportion of the reciprocal of $(1-\upsilon^2)$ where υ denotes Poisson's ratio. A typical value of υ for common metals is approximately: 0.3. The theoretical limits for this ratio are 0 and 0.5 for the perfectly brittle and ductile materials, respectively. The problems of very short and wide beams are more complicated for which only a limited number of handbook equations are available.

3.5 COMBINED STRESS CRITERIA

As we progress in our thinking through the concepts of slender beams, wide beams, slabs, and plates, the methods of analysis become increasingly involved. In general, however, all types of loading on a structural part produce only three basic types of response such as axial, bending, or torsional. These, in turn, can be expressed in terms of the normal and shear stresses. A strictly uniaxial stress in a tensile specimen, for instance, can cause this part to fail in shear. The strength of a given structural material is usually defined by uniaxial tests. The basic question then is how can we use uniaxial strength, when in actual application the problem can be compounded by a normal stress or a shearing stress acting in some other direction? It is obvious that for good economic and technical reasons, we cannot use material's tests under all possible combinations of loading and support. One would desire, therefore, to have a relatively simple and practical method of predicting the maximum normal and shearing stresses in a component to be able to relate such values to a conventional, uniaxial strength of the material. Our frame of reference can generally be the conventional yield strength of the material, σ_y, which is a normal stress criterion. The allowable yield strength in shear, τ, can then be taken as follows:

$$\tau_a = \frac{\sigma_y}{\sqrt{3}} \tag{3.15}$$

Geometrical interpretation of two-dimensional stress systems, such as those which can be simulated by Mohr's circle, leads to relatively simple design formulas. The maximum and minimum normal stresses based on the Mohr's circle of stress are:

$$\sigma_{\max} = \frac{S_x + S_y}{2} + \frac{1}{2}\left[\left(S_x - S_y\right)^2 + 4\tau^2\right]^{0.5} \tag{3.16}$$

$$\sigma_{\min} = \frac{S_x + S_y}{2} - \frac{1}{2}\left[\left(S_x - S_y\right)^2 + 4\tau^2\right]^{0.5} \tag{3.17}$$

In Eqs. (3.16) and (3.17), S_x and S_y denote uniaxial stresses oriented at 90° to each other while τ is the applied shearing stress. The normal stress given by either of the above formulas is referred to in the technical literature as the principal stress acting normal to the principal plane on which the resultant shearing stresses are equal to zero. The maximum principal shearing stress, using the above notation becomes:

$$\tau_{\max} = \frac{1}{2}\left[\left(S_x - S_y\right)^2 + 4\tau^2\right]^{0.5} \tag{3.15}$$

The iconic Mohr's Circle of stress is shown graphically in Figure 3.4.

$$2\phi = \tan^{-1}\frac{2\tau_{xy}}{S_x - S_y}$$ ⇒ angle between principal axes and the X & Y axes

⇒ angle between principal plane and the X & Y plane

The design factors of safety in a more general sense can be defined as follows:

$$\text{Factor of safety on normal stresses} = \frac{\sigma_y}{\sigma_{max}}$$

$$\text{Factor of safety on shear stress} = \frac{\sigma_y}{\sqrt{3}\tau_{max}}$$

The above statements imply that the component part is likely to fail due to a normal stress at some location where the normal stress reaches the value σ_y.

As mentioned previously, this may well be the case for a sufficiently brittle material. However, there is always the issue of a more general failure criterion of more ductile materials and the main problem here is how to select the appropriate allowable design.

Over the years, engineers studied this problem and a number of strength theories have been developed. Essentially, the purpose of this development was to predict the material's failure under a combined system of stresses. Such failure can be defined as yielding or a sudden rupture, whichever occurs first. While Eq. (3.16), for example, is directly applicable to the analysis of brittle materials, the failure of ductile materials using symbols from Eq. (3.16) should occur when the maximum stress in a two-dimensional system reaches the following value of the von Mises-Hencky criterion [1]:

$$\sigma_y = \left(\sigma_{max}^2 + \sigma_{min}^2 - \sigma_{max}\sigma_{min}\right)^{0.5} \tag{3.19}$$

Therefore, for a cylindrical shaft subjected to a bending stress S_x and a torsional stress τ, the maximum stress at yield can be calculated from Eq. (3.19) by the appropriate substitution of Eqs. (3.16) and (3.17). This yields:

$$\sigma_y = \left(S_x^2 + 3\tau^2\right)^{0.5} \tag{3.20}$$

If our allowable normal stress happens to be equal to the uniaxial yield strength σ_y, the condition of ductile failure can be established by Eq. (3.20) when S_x and τ are known. The failure of a shaft made out of a brittle material, using the same symbols, gives the following criterion:

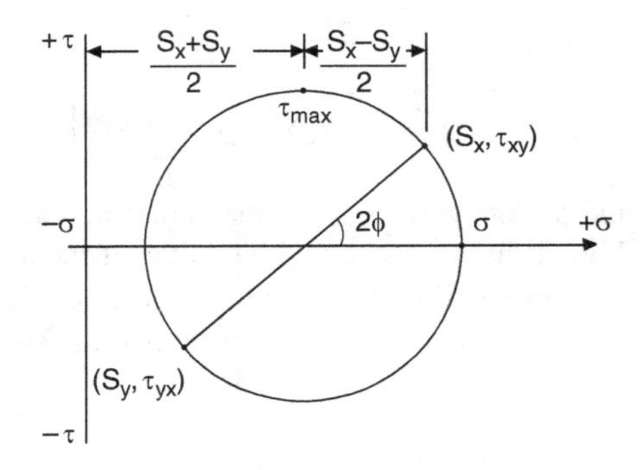

FIGURE 3.4 General depiction of Mohr's circle.

$$\sigma_y = \frac{S_x}{2} + \frac{1}{2}\left(S_x^2 + 4\tau^2\right)^{0.5} \tag{3.21}$$

In this discussion, we have used S_x as a symbol for simple uniaxial or bending stress, in accordance with the reasoning that they both essentially represent the same type of unit stress, normal to a particular plane. The stress S_x in a pure tensile test denotes uniaxial stress perpendicular to the transverse cross-section. But $S_x = \sigma$ can also be defined as the familiar bending stress, which according to Eq. (3.1) has also been interpreted as a normal stress.

In a simplistic practical way, engineers like to think about an allowable design stress which may or may not be equal to the uniaxial yield strength σ_y for a particular material. In using Eqs. (3.16), (3.17) or (3.19), we tend to consider them as specific examples of failure criteria in a two-dimensional stress system. The approach we typically use is based on the selection of the method we utilize to determine the limit of structural performance of a particular component or joint we have set out to design. It is advisable to remember, however, that the stress formulas, such as Eqs. (3.16), (3.17), or (3.19) may provide only a partial answer, if the design-life of a part happens to be subjected to fatigue, structural impact, or permanent deformation.

3.6 RESPONSE IN SHEAR

When considering shear stress, either due to a transverse load or due to a torsional moment, we describe it as a stress *tangent* to a surface. When we do this we find that the corresponding shearing force is *not* distributed uniformly over the cross-section.

Rivets are especially subjected to shearing action. Similarly, bolts, thrust collars, keys, welds, shafts, and pins frequently experience shear. Most of these components are found in mechanical joints.

The mechanics of shear transfer for a rectangular element of a beam, such as illustrated in Figure 3.5 can be explained by the tendency of all faces of the element to slide against one another—just as a scissors (or "shears") does it cutting.

Suppose we assume that the element shown in Figure 3.5 is very small and it is subjected to a pair of vertical shearing stresses which constitute a clockwise shearing couple. For the element to remain in static equilibrium, a shearing couple equal in magnitude is required to act in a counterclockwise direction. This counterclockwise couple can be represented by the two horizontal shear components τ. Since all sides of the element are equal in length, the equilibrium of the shearing couples leads to the conclusion that all the components must be identical as shown in the left portion of Figure 3.5. The counterclockwise couple in this instance is referred to as the "complementary shear."

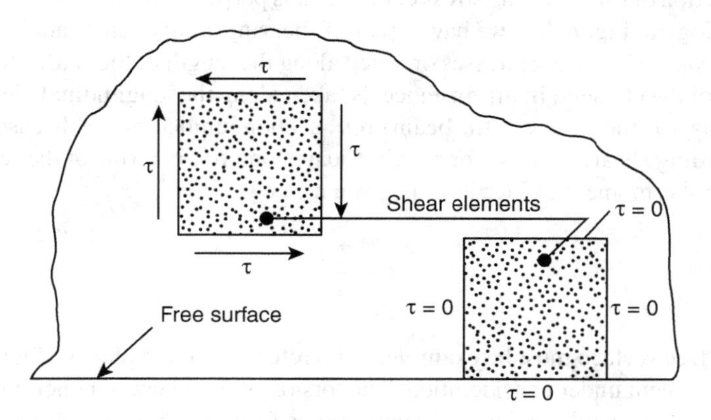

FIGURE 3.5 Shearing stress equilibrium.

If we now select an identical element of the structure, in such a location that the lower side of the element coincides with the free surface of the beam as shown in the right portion of the sketch, the original state of static equilibrium of the shearing couples cannot be maintained. If τ then vanishes at the free surface, the complementary shear couple must also vanish. For the state of the shearing stress to exist, the original clockwise shearing couple acting alone would have to be able to rotate the element in the clockwise direction and out of the free surface. Since such a rotation is obviously physically impossible, we must conclude that the element located right on the edge of the part must be free of any shearing stresses. This conclusions, although elementary in nature, is of special interest in stress analysis and experimental mechanics. From this, we can learn that in a structural beam subjected to transverse loading, the shearing stress at the outer fibers must be equal to zero for all types of cross-sectional geometry. Whereas, as we recall from Eq. (3.1) the normal stresses at the extreme fibers must attain their highest values. The nature of stress distribution on the various cross-sections of mechanical joints will be discussed in other sections of this book.

3.7 TORSION

At first glance it would appear that the mechanism of shearing should be the same for all types of external loading. Certainly this is true in the case of the equilibrium of a small element of a beam. This was helpful, for example, in explaining the concept of zero shearing stress on the free surface illustrated in Figure 3.5.

But what about the state of pure shear on the surface of a solid cylinder subjected to a torsional moment? The elementary theory of strength of materials tells us that for a given torsional moment, the shear stress on a shaft cross-section is directly proportional to the radius. The basic form of a torsional equation is then similar to that for simple bending.

$$\tau = \frac{G\theta r}{L} \tag{3.22}$$

Where L is the length of the component, θ is the angle of twist measured in the transverse plane of the axis of a cylindrical member, and G is the modulus of rigidity (or "shear modulus"). The shear modulus G can be obtained by dividing the shear stress into the shear strain which is analogous to the concept of the modulus of elasticity E, which is defined as the ratio of the normal stress to the axial strain.

The equilibrium of a small element of a shaft in torsion can be described as follows: According to Eq. (3.22), the shear stress is proportional to the distance from the center of the transverse cross-section. The direction of this shearing stress component is perpendicular to the radius drawn through the point. By analogy to Figure 3.5, we have a pair of shearing components caused by the torque and a complementary pair of shearing stresses oriented along the length of the shaft shown in Figure 3.6. The equilibrium of the element, in this instance, is taken along the longitudinal plane rather than the transverse plane as was the case with the beam cross-section implied by the discussion of Figure 3.5.

The corresponding shearing stress for a cylindrical geometry, in terms of the resisting torque T_m, radius r and the polar moment of inertia I_p, is given as:

$$\tau = \frac{T_m r}{I_p} \tag{3.23}$$

The concept of a free surface must be examined with reference to the planes of loading and support of the particular element under consideration. The torsion of members of noncircular cross-sections generally involves very complex solutions. Approximate formulas, however, are available for a number of noncircular geometries [2]. It is also well to note that when a cylindrical bar is subjected

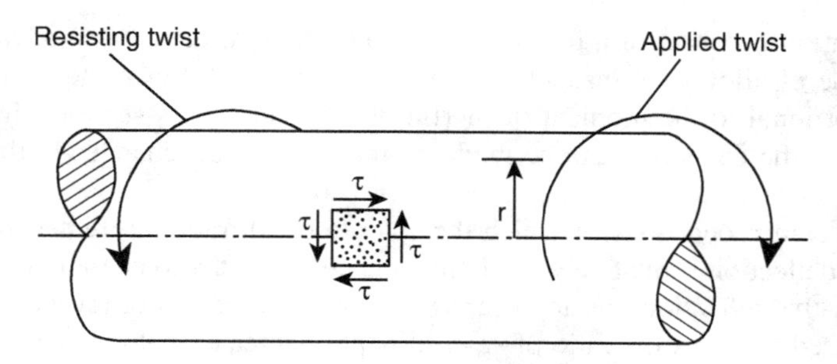

FIGURE 3.6 Shearing stress equilibrium for a shaft under twist.

to uniform twisting moment, tensile and compressive stresses can develop on a plane inclined at roughly 45° to the bar axis. These are the principal stresses of equal magnitude. For ductile materials, the failure of such components will be in shear. On the other hand, a shaft made of a brittle material, weak in tension, can be expected to fail by a tensile fracture along the 45° plane.

3.8 DEFLECTION ANALYSES

In addition to the stress criteria, stiffness and deflection characteristics of structural members often need to be considered in the course of a particular design. Such calculations are made rather frequently for a variety of beam and plate-like members, using the theory of small deflections of beams subjected to bending. If ρ is the radius of curvature to which the beam is bent, then the curvature which is defined as the inverse of ρ can be expressed as:

$$\frac{1}{\rho} = \frac{M}{ET} \tag{3.24}$$

Recall that M denotes the bending moment at a given section of an elastic beam and I is the moment of inertia or second moment of area about the neutral axis. It is important to note that Eq. (3.24) reflects only the elastic behavior and that deformations due to shearing forces are neglected. The curvature given in texts on analytical geometry is expressed in complicated terms, unless we can assume that the slope of the deflected beam is very small. This simplification was first suggested by Bernoulli almost three centuries ago and we are still using the differential equation of the elastic curve, based on the approximate expression for curvature as:

$$\frac{1}{\rho} = \frac{d^2y}{dx^2} \tag{3.25}$$

Textbooks on strength of materials, solid mechanics, and stress analysis use Eqs. (3.24) and (3.25) rather extensively in showing how to derive working deflection formulas. But only practical experience can teach us why and where we should calculate the deflections and what are the inherent limitations of such analyses. In many practical cases, the need for the deflection analysis is quite obvious. For example, old plastered ceilings could not stand excessive deflections of the supporting beams. Delicate machinery in a glass factory must have a rigid support system to avoid misalignment of the component parts. Special parts in the field of spectroscopy must be designed to have an absolute minimum of a geometrical distortion. Analysis of vibration of machinery and evaluation of statically indeterminate structures in general, require a considerable amount of knowledge of deflection characteristics.

One of the more fundamental practical questions in deflection analysis may be concerned with the accuracy of the calculations and the effect of the major variables involved. Interestingly, the only material property which influences beam deflection is the modulus of elasticity E. This is an important observation because, in a deflection-governed design, there is no advantage in using a sophisticated grade of alloy steel instead of a plain, carbon steel. Since the deflection of a beam is inversely proportional to the moment of inertia of the beam cross-section, significant changes in the magnitude of the deflection can be made by relatively small changes in the cross-sectional dimensions.

For the usual beam proportions in which the span to depth ratio is greater than about 10, the effect of shearing deflection is relatively small and it is not expected to be more than about 2%–3%. In wooden beams the deflection due to shear is somewhat more important because of the small value of G compared to E. As the ratio of span-to-depth decreases below about 3 the distribution of stresses and strains across the beam cross-section becomes complicated and simple beam theory and beam formulas are no longer valid.

Any discussion of small effects and finer points of the analysis of beam strength or deformation must be related not only to the variability of the material properties but also to the support conditions. These are usually referred to as "boundary conditions." In design situations involving mechanical joints, it is very difficult to define such boundaries with any reasonable degree of confidence. Hence, any great effort extended to obtaining greater precision in calculations of stresses and deflections does not appear to be justified. The apparent complexity of formulas and numerical techniques often encountered in modern practice, bears little relation to the correctness of the mathematical model or formula which allegedly and often mistakenly describes the loading and support conditions under consideration.

When in doubt regarding the choice of the formula or the numerical substitutions, it is essential to establish some sort of a "bracketing" assumption which could then be used as an independent check.

3.9 BUCKLING OF COLUMNS AND PLATES

So far our discussion about stresses and deformations has focused upon relatively simple structures (i.e., beams and rods). Our objective has been to alert our readers, and specifically those confronting actual design problems, with some basic concepts, some insights, and to identify some popular misconceptions. In our emphasis upon fundamentals we have attempted to also provide some practical procedures for making stress calculations.

Even with the basic concepts of stress, strain, and deformation being of primary importance in a mechanical system, a particular part can also fail in a still different mode such as a column buckling or a vessel collapse, or burst. These failures can occur at stresses far below the yield strength of the material, and they occur often enough that they merit special attention. We consider some of them in the following paragraphs.

Consider, for example, a pin-end supported beam in compression along its axis. The beam will become unstable and tend to buckle under the load F_c given by:

$$F_c = \frac{\pi^2 EI}{L^2} \tag{3.26}$$

Where L is the length of the beam (or column-like member) while E and I, as before, are the modulus of elasticity and centroidal second moment of area.

For a column having a solid circular cross-section with diameter d, the theoretical axial load corresponding to the yield strength of the material is:

$$F_y = \frac{\pi d^2 \sigma_y}{4} \tag{3.27}$$

Since for a circular cross-section, the second moment of area I is: $\pi d^4/64$, by substituting in Eq. (3.26) and dividing Eq. (3.27) by Eq. (3.26), we obtain:

$$\frac{F_y}{F_c} = 1.62\frac{\sigma_y L^2}{ED^2} \tag{3.28}$$

For $F_y = F_c$ and specified material properties, the ratio of column length to diameter becomes:

$$\frac{L}{d} = \frac{\pi}{4}\left(\frac{E}{\sigma_y}\right)^{0.5} \tag{3.29}$$

For all (L/d) ratios greater than that specified by the above formula, the pin-ended column member will buckle before the yield strength in compression is reached. For example, using ratio $E/\sigma_y = 300$, Eq. (3.29) gives $L/d = 13.7$. It should be emphasized, however, that for other end conditions and cross-sectional geometry, the relevant characteristic ratios will be numerically different, so that each specific case must be examined independently. In a more general sense, the critical buckling load of a slender column can be expressed as:

$$F_c = \frac{EIK_o}{L^2} \tag{3.30}$$

The column factor K_o depends on the manner of loading and support. Once its value is selected for the particular case at hand, other critical loads can be determined according to the numerical value of EI/L^2. That is, K_o represents the effect of boundary conditions while EI/L^2, independent of the boundaries, provides the complete numerical solution. The choice of appropriate boundary can be made on the basis of the following guidelines:

One end clamped – other free $K_0 = 2.47$
Both ends pin-jointed $K_0 = 9.87$
One end hinged – other clamped $K_0 = 20.19$
Both ends clamped $K_0 = 39.48$

The above conditions show how sensitive the buckling load is to the support conditions.

A number of intermediate boundary conditions can also be postulated using the above values of K_0. Such conditions would generally correspond to the theoretical upper limits within the elastic range. When it is expected that the column behavior is essentially inelastic the critical load can be found directly from Eq. (3.30) by substituting E_t for E. In this instance, E_t defines the tangent modulus measured as the local slope of the stress–strain diagram at the particular stress level considered. Although this concept is not new, it is still popular in preliminary calculations because it assures the practical upper limit of column strength. Any further improvement in load bearing capacity of the column will come from improvement in the base metal by appropriate alloying and heat treatment. The concept of tangent modulus is credited to Engesser [3]. Predictions based on Eq. (3.30), attributed to Euler, are generally conservative.

When flat plate sections are analyzed instead of conventional columns, local instability can occur without necessarily losing the significant portion of the overall buckling resistance of the plate.

The general equation for the critical buckling stress of a plate can be expressed as:

$$S_{CR} = K_p E\left(\frac{t}{b}\right)^2 \tag{3.31}$$

whereas: t = plate thickness; b = plate width

Here, the buckling stress factor K_p is similar to the column factor K_o found in Eq. (3.30), and depends to a large extent on edge constraints. It is a nondimensional quantity and it is sometimes referred to as the "plate coefficient."

Figure 3.7 illustrates the notation used in Eq. (3.31).

When the critical stress calculated from Eq. (3.31) is less than the yield strength of the material, the buckling is elastic and the modulus of elasticity E remains constant. The value obtained from Eq. (3.31) is an upper limit because the stresses actually measured are found to be smaller by an appreciable margin. The discrepancies between the theory and practice are mainly due to geometrical irregularities and become more significant with the decrease in plate thickness. When a long plate of width b is supported at the two long sides, the plate coefficients K_p listed in Table 3.1 can be used for predicting the elastic buckling stresses.

It is customary, in a number of design applications, to make the dimension b equal to the distance between the rivet lines or weld seams. In rolled sections and components made of cold forming or pressing, the distance b can be measured from the edges of the fillets. In the case of most actual design situations, the flat member is considered to be a plate when the ration L/b is greater than 5. It should also be noted that complete fixity (i.e., rigid support) of both edges is practically never realized. Hence, the most reasonable approach is to assume either both simple supports or the simple support-free conditions given as cases 3 and 1 in Table 3.1, respectively. Although some simply supported plates may have a tendency to behave as columns, there is always sufficient restraint along the two principal axes so that the plates develop double curvature and the elements located further from the central axis are subject to twist as well as bending. This provides an intuitive explanation of why the buckling load of a plate is considerably higher than the corresponding buckling load of a wide column.

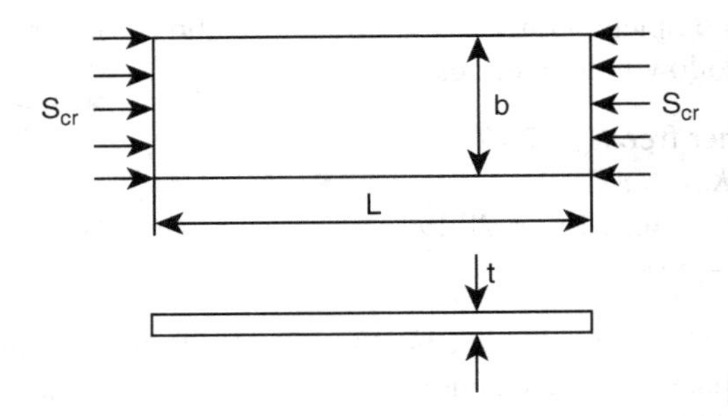

FIGURE 3.7 Notation for a rectangular plate in compression.

TABLE 3.1
Buckling Stress Factors for Edge-Loaded Flat Plates

Edge	Case	Edge	K_p
Simple support	1.	Free	$K_p = 0.38$
Fixed support	2.	Free	$K_p = 1.15$
Simple support	3.	Simple support	$K_p = 3.62$
Simple support	4.	Fixed support	$K_p = 4.90$
Fixed support	5.	Fixed support	$K_p = 6.30$

3.10 BENDING OF PLATES

The fundamental concepts of plate analysis and design have developed over the past 125 years from the rigorous solutions of the governing differential equations. While this general topic is of interest to the mathematicians, the number of simplifying assumptions required in finding an elegant solution remains rather staggering. As far as engineers in this country are concerned, the first classical treatment appeared not long before 1940 when Timoshenko published his monumental contribution to the theory of plates and shells [4]. Certainly, the names of Neuber, Bubnov, Galerkin, Vlasov, and others have also been associated with the development of plate fundamentals. The problem with the applications of plate theory was always difficult because of the complex mathematical procedures and laborious computations involved. These factors obstructed proper utilization of the theory at the design level and required the use of auxiliary approximate methods for the determination of internal forces and deformations. Both elastic and limit design theories of plate components are required in obtaining economy and reliability of the individual structures and of the overall mechanical system.

Bare [5] has prepared a most extensive compilation of practical formulas and tables for the analysis of plates in the plastic range. Metal and concrete plates of various geometry can be assumed to be isotropic, having essentially the same mechanical properties in all directions. Alternatively, wooden panels or plates involving stiffeners fall into the category of anisotropic bodies.

Plates, in general, fall into a number of classifications depending on the character of stress, loading, and geometry. When the ratio of the thickness to the least plate dimension exceeds 0.2, we are working with so-called slabs for which the three-dimensional theory of elasticity is required to correctly describe the state of stress. For ratios smaller than 0.2, the plates can be considered to be "thin." For the case of uniformly loaded plates and panels of rectangular geometry, the classification parameter sometimes used can be described as:

$$\Omega = \frac{q}{E}\left(\frac{a}{t}\right)^4 \tag{3.32}$$

Where:

Ω = plate parameter
a = plate length
t = plate thickness

The above criterion depends on the particular boundary conditions and it appears to be highest for the maximum restraint at the plate support. As a conservative number for this case, Bares [5] suggests: 55. Hence, plates having a value Ω higher than 55 can be classed as thin.

Very thin plates with relatively large deflections behave as diaphragms and can be used in design and construction of vacuum barriers, windows, and similar members. Such plates carry predominantly tensile forces in the middle plane and bending contribution is minimal. Examples of this type of plate design are given in Chapter 10, focusing upon special mechanical joints and interfaces.

The problem of bending of isotropic plates in the regime of small deflections must be considered with reference to the numerous simplifying assumptions: These assumptions impose certain constraints on the validity and accuracy of design calculations. The relevant constraints can be summarized as follows:

- The plate material is elastic, homogeneous, and isotropic.
- The plate thickness is constant and small compared with other dimensions.
- The plane sections remain plane after bending.
- The effect of transverse shear stress is negligible.
- The plate curvature under load is small so that the second derivative of deflection is applicable.

- The tensile stresses in the plane of the plate have minimal effect on the transverse bending deflections.
- The corners of the plate are secured against lifting.
- The deflection is smaller than the plate thickness.

In line with the above constraints and the general theory of plate bending, the differential equation for the deflection curve of a strip cut out of the plate is essentially the same as that for a beam: That is,

$$M_2 = EI\frac{d^2w}{dy^2} \tag{3.33}$$

Where w stands for the transverse displacement caused by the bending moment M_2 acting on the strip along Y-axis.

The corresponding bending stress is then:

$$\sigma = \frac{M_2}{Z} \tag{3.34}$$

where the term Z denotes the section modulus defined as I/C, where the symbols I and C have the same meaning as that given in Eq. (3.1).

Since the analysis considers a strip of unit width, Eq. (3.34) can be restated as follows:

$$\sigma = \frac{6^M 2}{t^2} \tag{3.35}$$

Despite its obvious simplicity, Eq. (3.35) is quite important in design of plates for bending. When a complete plate is bent under transverse loading, lateral restriction due to the presence of adjacent strips provides additional stiffness. The differential equation of the deflection curve is modified here to give:

$$D\frac{d^2w}{dy^2} = M_2 \tag{3.36}$$

Where D is

$$D = \frac{Et^3}{12(1-\nu^2)} \tag{3.37}$$

The parameter D defined by Eq. (3.37) denotes the plate flexural rigidity which differs from the rigidity of the conventional beam by the factor $1/(1-v^2)$, where v is the Poisson's ratio of the plate material. In a more general case, the plate can be bent in the two perpendicular directions simultaneously so that the resultant curvature of the entire plate can be expressed in terms of bending moments M_1 and M_2 acting along x and y axes as:

$$D\left(\frac{a^2w}{ax^2} + \nu\frac{a^2w}{ay^2}\right) = M_1 \tag{3.38}$$

and

$$D\left(\frac{a^2w}{ay^2} + \nu\frac{a^2w}{ax^2}\right) = M_2 \tag{3.39}$$

When the two bending moments are equal, the plate develops a spherical shape with the radius of curvature equal to the value ρ given by:

$$\rho = \frac{D(1+\nu)}{M} \tag{3.40}$$

For a linear thermal gradient ΔT acting across the plate thickness, the corresponding radius of curvature becomes:

$$\rho = \frac{t}{\alpha(\Delta T)} \tag{3.41}$$

If a mechanical restraint is provided to reduce the edge rotation to zero, the bending moment per unit length of the plate edge can be estimated from the expression:

$$M = \frac{\alpha(\Delta T)(1+\nu)D}{t} \tag{3.42}$$

The corresponding bending stress then becomes

$$\sigma = \frac{\alpha E(\Delta T)}{2(1-\nu)} \tag{3.43}$$

The curvature of the free plate, as given by Eq. (3.41), produces essentially no stresses as long as the restraint of the plate edge is not present. This is in agreement with the generally acceptable principle that expansion or contraction due to temperature is caused by thermal strains which control the shape of the body. By forcing the particular structural element back to its original shape, we must add mechanical forces that cause stresses. The free shape does not experience such forces, unless there are areas within the body where a natural expansion or contraction due to the thermal gradient is physically restrained.

It may be deduced from the foregoing considerations that when a plate can be analyzed as a conventional beam, the results will be quite acceptable for practical purposes. For example, taking a rectangular plate under uniform load q, supported only at two opposite edges, the deflection of the center of the plate varies from $0.0129qa^4/D$ to $0.0130qa^4/D$ when the length ratios a/b range from 0.5 to 2.0, respectively.

Many practical design charts for plates available in the literature are given for specific values of Poisson's ratio. For reinforced concrete, values of ν from 0.10 to 0.15 can be selected: while for steels, the value of 0.25 to 0.33 is quite common. Since the deflection is inversely proportional to plate rigidity D, given by Eq. (3.37), setting $\nu = 0$ will yield approximately 2% greater deflection than $\nu = 0.15$. The corresponding error, however, in predicting the bending moment can be 15%. In this case, Poisson's ratio appears to affect the stress distribution under statically indeterminate conditions. Increasing values of Poisson's ratio make the plate stiffer which will cause smaller deflections but higher moments. Only the shear forces can be assumed to be independent of Poisson's ratio. It also follows that the supports with rigid restraint contain reaction forces which are unaffected by the Poisson's ratio.

3.11 DESIGN OF CURVED ELEMENTS

Curvature affects the way a structural element resists external loading. For example, a simply supported beam under uniform load is subjected to shear and bending with essentially no axial force present, as long as the transverse displacement is small. Should the beam be formed into a parabolic arch, one would expect to be able to carry the entire vertical load, uniformly distributed over the span of the arch, by means of axial forces. Theoretically, the shear and moment components should vanish if the form is truly parabolic. In active situations involving various curvatures, however, the lateral loading can induce bending, shear, direct and torsional stresses.

When the ratio of radius to thickness of a curved element is greater than 10, there is close agreement between the calculated bending stresses for the curved element and those for a straight beam. This ratio can be relaxed to about 5 when the comparison is based on deflection.

When we are working with conventional curved elements, we can reasonably make the following simplifying assumptions:

- The cross-section is compact and uniform.
- The deflections are relatively small.
- The strains are elastic.
- The loads are applied gradually.

In view of the foregoing assumptions, when we are working with flanged, thin-walled, and hollow cross-sections, of curved members, we should consider these components separately, to make appropriate corrections. In the following paragraphs of this introductory chapter, we will restrict ourselves to simpler problems.

Consider, for example, the arched circular cantilever supported member as represented in Figure 3.8.

The bending moment at a section defined by the angle θ (measured from the horizontal), at bending moment M is:

$$M = WR(1-\cos\theta) \tag{3.44}$$

From Castigliano's principle [6] the partial derivative of the total elastic strain energy stored in a structure, with respect to one of the forces, gives the displacement at the point of application of the force in the direction of the force. In applying this principle to a curved configuration and using polar coordinates [7], the relevant expression for determining the deflection formula becomes:

$$Y = \frac{1}{EI}\int_0^\theta M\left(\frac{\partial M}{\partial W}\right)R\,d\theta \tag{3.45}$$

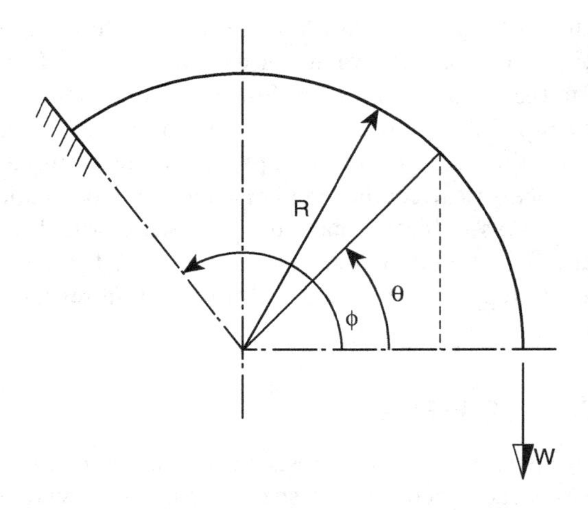

FIGURE 3.8 Circular cantilever subjected to an axial (vertical) end load.

From Eq. (3.44), the required partial derivative is:

$$\frac{\partial M}{\partial W} = r(1-\cos\theta) \tag{3.46}$$

Combining Eqs. (3.44), (3.45), and (3.46), and extending the integration over the entire curved member, gives:

$$Y = \frac{WR^3}{4EI}(6\phi + \sin 2\phi - 8\sin\phi) \tag{3.47}$$

By inserting the appropriate values of ϕ into Eq. (3.47), a number of design formulas can be obtained provided that the flexural rigidity remains constant. Where the second moment of area does not remain fixed, its variation with ϕ should be defined prior to integration. This process may require special numerical techniques, since only a handful of closed-form integrable functions is at our disposal.

The problem of out-of-plane bending of a curved element is much more complex, because of the existence of a twisting moment in addition to bending. It is fortunate, however, that the general Castigliano principles still apply. For the case of the transverse displacement, the general equation can be stated as:

$$Y = \frac{1}{EI}\int_0^\theta M\left(\frac{\partial M}{\partial W}\right)Rd\theta + \frac{1}{GK}\int_0^\theta T_M\left(\frac{\partial T_M}{\partial W}\right)Rd\theta \tag{3.48}$$

It may be noted that the first part of the above expression is identical with Eq. (3.45), consistent with the general principle that all components of the stored elastic energy are directly additive. The second part of Eq. (3.48) contains the term *GK*, known as the "torsional rigidity." It is important to realize that the parameter *K* represents the so-called torsional "shape factor." For a circular cross-section, I_p is equal to *K*, but for all other cross-sectional shapes, the *K* values cannot be deduced from the conventional formulas for the second moments of area. The torsional shape factor *K* has been defined in mathematical language for rectangular and other simpler cross-sections, but for other geometries the *K* values can only be determined experimentally.

Unfortunately, experimental values can involve errors of 10%—or more. It should also be noted that there is little uniformity in reference to the *K* factor in the literature. It is referred to as "torsional resistance," "torsional constant," "torsional rigidity factor," and other similar expressions. But in any event *K* should have the same dimensions as those of the second moment of area *I*.

Figure 3.9 presents an example of an arched cantilever subjected to out-of-plane loading. In this case the bending and twisting moments *M* and T_m, are:

$$M = WR\sin\theta \tag{3.49}$$

$$T_m = WR(1-\cos\theta) \tag{3.50}$$

In Eqs. (3.49) and (3.50) the relevant partial derivatives are:

$$\frac{\partial M}{\partial W} = R\sin\theta \tag{3.51}$$

and

$$\frac{\partial M}{\partial T_M} = R(1-\cos\theta) \tag{3.52}$$

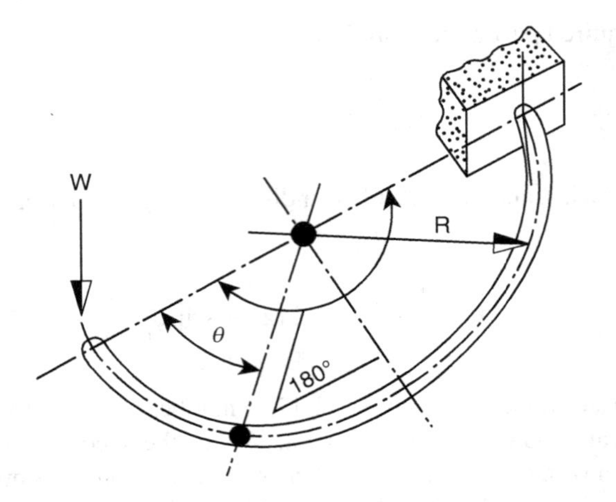

FIGURE 3.9 An arched cantilever curved bar subjected to out-of-plane load.

The results of Eqs. (3.49) through (3.52) can now be inserted into the integrals of Eq. (3.48) to evaluate the displacement *Y*. The result is:

$$Y = \frac{\pi W R^3}{2EI} + \frac{3\pi W R^3}{2GK} \tag{3.53}$$

This particular design case is relatively simple. However, when the displacement is required at a point other than that at which the load is applied, the derivation involves additional trigonometric terms. Examples of such calculations and a number of practical design aids for out-of-plane loading are available in references [7,8].

This brief reference to curved members would be rather incomplete without mentioning the classical problem of a closed circular ring which has many applications. Despite its striking appearance of symmetry and simplicity, the design analysis can become quite complex. The reason for this is that a circular ring is a statically indeterminate structure, where the conventional conditions of equilibrium of forces are insufficient for the problem solution.

Consider, for example, the simple case of a ring in diametral compression as in Figure 3.10. The equilibrium of forces for one quarter of the ring yields the following relations in terms of the external load *W* and geometry:

$$M = \frac{WR(1-\cos\theta)}{2} - M_f \tag{3.54}$$

$$N = -\frac{W\cos\theta}{2} \tag{3.55}$$

and

$$Q = \frac{W\sin\theta}{2} \tag{3.56}$$

where the moment *M* and the forces *N* and *Q* are defined by Figure 3.10.

It is clear that the evaluation of the redundant quantity M_f and ring deflections can involve rather complex derivations if we assume that all moments and forces, such as *M*, *Q*, and *N*, have to be included. For example, the expression for finding the fixing (redundant) moment M_f can be stated as follows:

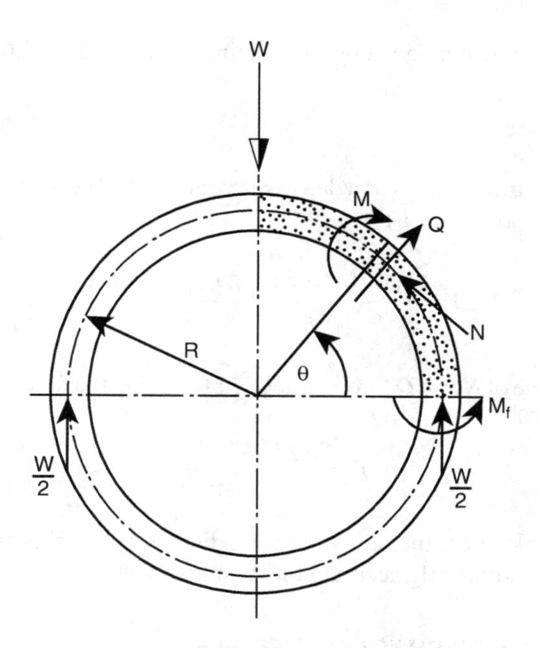

FIGURE 3.10 Circular ring in diametral compression.

$$\int_0^{\pi/2} (M - N\delta)\frac{\partial M}{\partial M_f} d\theta = 0 \quad (3.57)$$

The foregoing expression implies the condition that the rotation of the ring section at $\theta = 0$ is zero. The term δ defines the amount of radial displacement of the neutral axis from the geometrical, centroidal axis. The value of δ increases with the ring thickness, and it has been calculated for a number of typical cross-sections. It is customary to use this parameter with suitable correction factors for curved beams published by Wilson and Quereau over 60 years ago [9].

The term Q denotes shear force. In practice, the inclusion of the effects of N and Q on deflection and stresses of conventional rings is seldom justified.

In the particular case described by Eqs. (3.54) through (3.57), we obtain the following working formulas for the fixing couple M_f at $\theta = 0$ and the bending moment at any angle θ between 0 and $\pi/2$.

$$M_f = \frac{W(\pi R - 2R + 2\delta)}{2\pi} \quad (3.58)$$

and

$$M = \frac{W(2R - 2\delta - \pi R\cos\theta)}{2\pi} \quad (3.59)$$

When $\delta = 0$, that is the ring is thin, the above formulas become:

$$M_f = WR\left(\frac{\pi - 2}{2\pi}\right) \quad (3.60)$$

and

$$M = \frac{WR}{2\pi}(2 - \pi\cos\theta) \quad (3.61)$$

The maximum bending moment is found by substituting $\theta = \pi/2$ in Eq. (3.61). This gives:

$$M = \frac{WR}{\pi} \tag{3.62}$$

The total amount of diametral compression due to the combined effect of M, N and Q, but neglecting the displacement of the neutral axis, is:

$$Y = \frac{WR^3}{EI}\frac{\left(n^2 - 8\right)}{4\pi} + \frac{WR_{\pi}}{4A}\left(\frac{1}{E} + \frac{1}{G}\right) \tag{3.63}$$

For a thin ring the influence of N and Q forces is negligible, giving the following standard expression:

$$Y = \frac{WR^3}{EI}\frac{\left(\pi^2 - 8\right)}{4\pi} \tag{3.64}$$

In preliminary design work where the ratio of mean radius to ring thickness is greater than about 5, Eqs. (3.62) and (3.64) are sufficiently accurate for most practical applications.

3.12 STRENGTH AND STABILITY OF VESSELS

Vessels and vessel applications occur extensively in industrial design. Consequently, vessels with cylindrical and spherical features have been under investigation for a long time and the literature on the subject is truly enormous. Our objective in this section is simply intended to briefly review the fundamentals and the elementary formulas. No single section, chapter, or even an entire book can do justice to all the advancements in pressure and vacuum vessels.

Nevertheless, the material included here will hopefully be of some utility in those areas of design where vessels play a role in mechanical joint design and interfaces.

The majority of vessels, containers, and piping are designed for internal pressure. There are, however, important cases where the systems have to carry external pressure. In such situations, the decision should be made as to the most likely mode of structural response and the governing criteria. The choice can be either stress limit, elastic stability, or structural collapse.

This leads directly to the formula defining the hoop stress in a long cylinder on the premise that end closure does not provide any support.

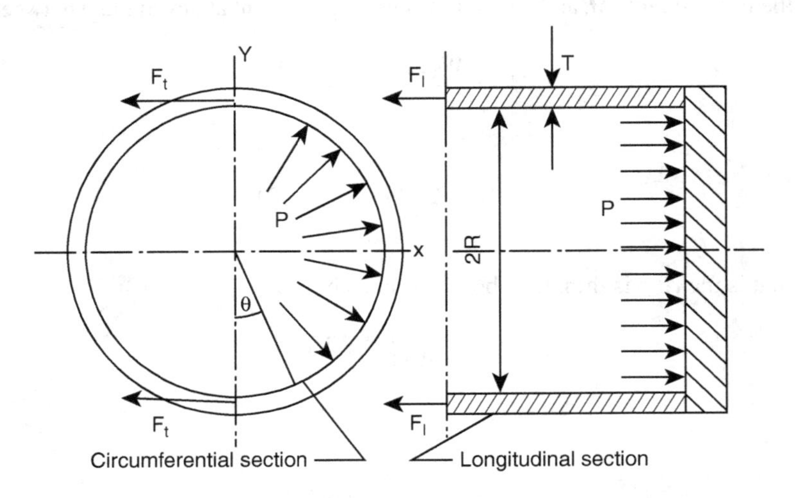

FIGURE 3.11 Loading on a thin cylinder.

$$\sigma_t = \frac{PR_i}{T} \tag{3.65}$$

Where:

R_i = Vessel inner radius
T = Pressure vessel thickness

Equilibrium considerations in the axial direction lead to the expression:

$$P_\pi R_i^2 = 2\pi\left(R_i + \frac{T}{2}\right)\sigma_\ell \tag{3.66}$$

Thus, we have the stress formula:

$$\sigma_\ell = \frac{PR_i^2}{2R_i + T} \tag{3.67}$$

It is evident that the cylinder radius R_i can be replaced by its mean value R without any significant error, provided the vessel is relatively thin. Hence, by introducing the parameter m for R/T we have

$$\sigma_t = Pm \tag{3.68}$$

and

$$\sigma_\ell = Pm / 2 \tag{3.69}$$

The foregoing equations show that the efficiency of the circumferential joints need only be half that of the longitudinal joints. It also follows from Eq. (3.69) that this formula applies to a spherical vessel, pointing to the fact that the relevant thickness requirement is only one half that for a cylinder. Hence, the sphere represents the most efficient pressure vessel configuration.

For hoop and longitudinal stresses in a thin conical vessel subjected to internal pressure, the relevant formulas are:

$$\sigma_t = \frac{Pm}{\cos\alpha_C} \tag{3.70}$$

and

$$\sigma_\ell = \frac{Pm}{2\cos\alpha_C} \tag{3.71}$$

The foregoing expressions pertain to the cone configuration subtending angle $2\alpha_C$.

The pressure vessel design equations quoted above are intended for elastic behavior. For the case of a double-riveted butt joint used in cover plates for boilers, the elementary pressure vessel equations can be helpful in establishing the forces acting on the rivets. Consider, for example, a section through the boiler plate joint such as that shown in Figure 3.12. If we assume that n rivets acting in double-shear are designed to resist circumferential tension σ_t, then the required rivet diameter d can be calculated with the aid of Eq. (3.65) and the equilibrium of forces.

The double-shear case is illustrated in more detail in the right corner of Figure 3.12, in which the force corresponding to hoop stress σ_t is balanced by a pair of cover plate forces acting along two parallel planes. The relevant shear areas are $\pi/d^2/4$ so that the required rivet diameter d becomes:

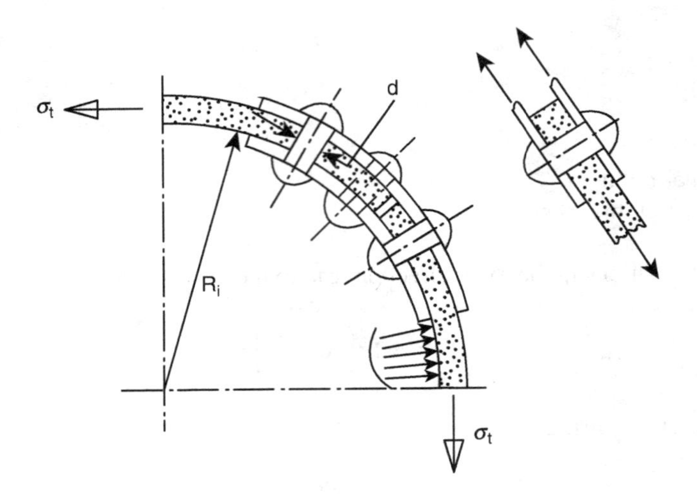

FIGURE 3.12 Example of a boiler plate joint.

$$d = 0.8\left(\frac{bR_iP}{n\tau_a}\right)^{0.5} \tag{3.72}$$

Where b is the width of the cover plate and τ_a denotes the allowable shear stress of the rivet material.

Eq. (3.72) shows that the result is independent of plate thicknesses, as long as the criterion is shear. The plate thickness, however, comes into play when other structural modes are analyzed, such as unit compression against the rivet shank or the shear-out (or tear-out) of the plate edge.

The majority of pressure vessel applications related to simpler mechanical interfaces and joints, involve cylindrical and spherical shells containing internal pressure. However, special formulas are also available for ellipsoidal and toroidal configurations [7], as long as the vessels can be classed as thin-walled. For the case of typical vessels and the conditions of ultimate strength of cylinders and spheres subjected to internal pressure, the relevant formulas and design charts can be summarized as:

$$P_C = \sigma_y \psi B_1 \tag{3.73}$$

$$P_S = \sigma_y \psi B_2 \tag{3.74}$$

$$P_C = \frac{\sigma_y}{m} B_3 \tag{3.75}$$

and

$$P_S = \frac{\sigma_y}{m} B_4 \tag{3.76}$$

where:

B_1 through B_4 = Strength factors for vessels
Ψ = Design factor

The foregoing design formulas can be used with working charts as given in Figures 3.13–3.15. In these charts $m = R/T$ and β denotes the materials strength ratio defined as σ_y/σ_u which relates to the

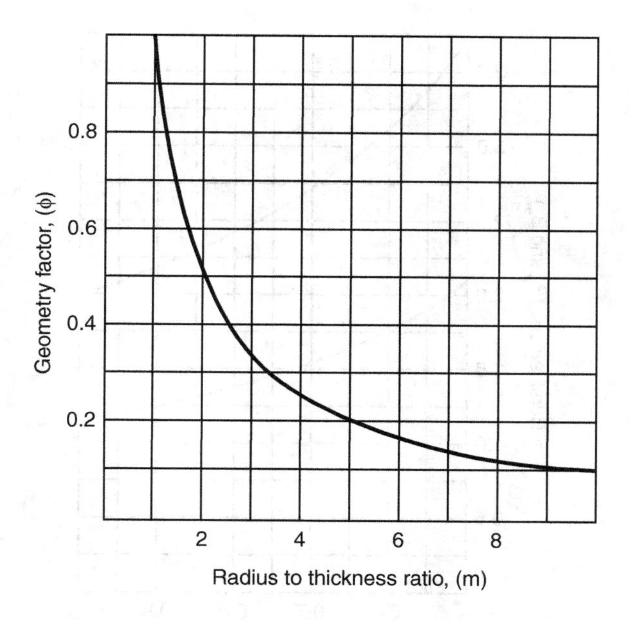

FIGURE 3.13 Geometry factor for thick cylinders and spheres.

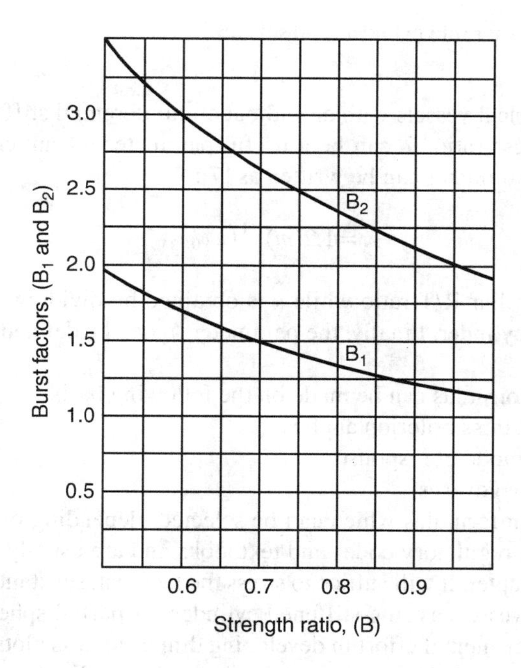

FIGURE 3.14 Burst factors for thick cylinders and spheres.

strain hardening of materials such as low and intermediate strength low-alloy and carbon steels. Values for ψ are shown in Figure 3.13.

The problem of stability of even the simplest shells, such as cylinders and spheres, is extremely complex because of the presence of manufacturing imperfections and the variations in material properties. Here we consider only the simplest rules and formulas, sufficient to make a preliminary assessment of the external pressure which can cause structural collapse.

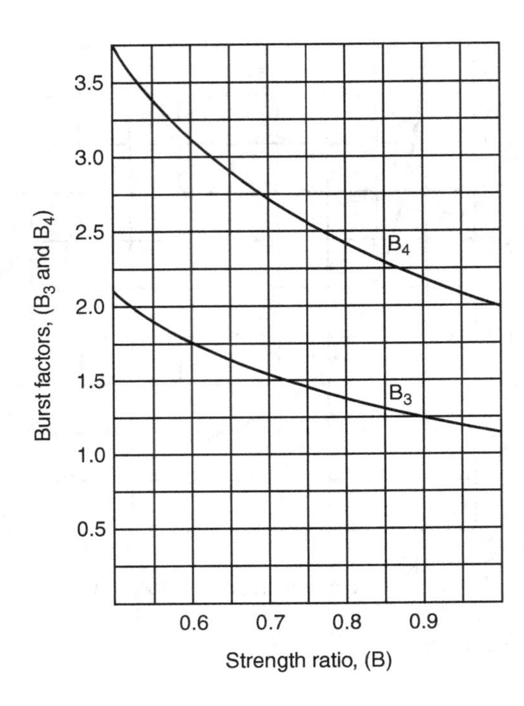

FIGURE 3.15 Burst factors for thin cylinders and spheres.

For the case of cylindrical vessels with or without circumferential stiffeners, experimental studies show that the "thinness ratio" λ can be a useful parameter for anticipating the behavior of a cylindrical vessel. This parameter can be written as [7]:

$$\lambda = 1.2(m)^{1/4}\left(k\phi_0\right)^{-1/2} \tag{3.77}$$

where m denotes the familiar R/T ratio while k is obtained by dividing vessel thickness T by the length L of the effective cylinder. Finally, the parameter ϕ_0 can be defined as the inverse strain calculated as the ratio E/σ_y.

The choice of design formulas can be made on the following basis:

$\lambda \leq 0.35$ Simple hoop stress criterion applies

$0.35 \leq \lambda \leq 2.5$ Mixed mode of response

$\lambda \geq 2.5$ Elastic stability governs

There are many design formulas which can be selected, depending on the above criteria. Such formulas can be found in regulatory codes and textbooks and are usually too involved for the brief summary given in this chapter. It will suffice to stress the fact that, substantial theoretical difficulties in the treatment of such vessels as ring-stiffened cylinders or partial spherical configurations have led to a considerable experimental effort in developing dimensionless plots for the available empirical data. Denoting, for example, the experimentally determined collapse pressures for ring-stiffened cylinders by P_e, the results can be represented as a lower-bound curve, as given in Figure 3.16 by Gill [10].

Similarly, for the case of hemispherical vessels subjected to external pressure, the lower-bound design curve can be illustrated as in Figure 3.17.

The design curve given in Figure 3.17 has been derived from the tests of hemispherical vessels in the stress-relieved and as-welded condition without, however, specifying the extent of geometrical imperfections [10]. The accuracy with which the collapse pressure can be predicted on the basis of experimental evidence must be influenced by the maximum scatter band involved. Since such

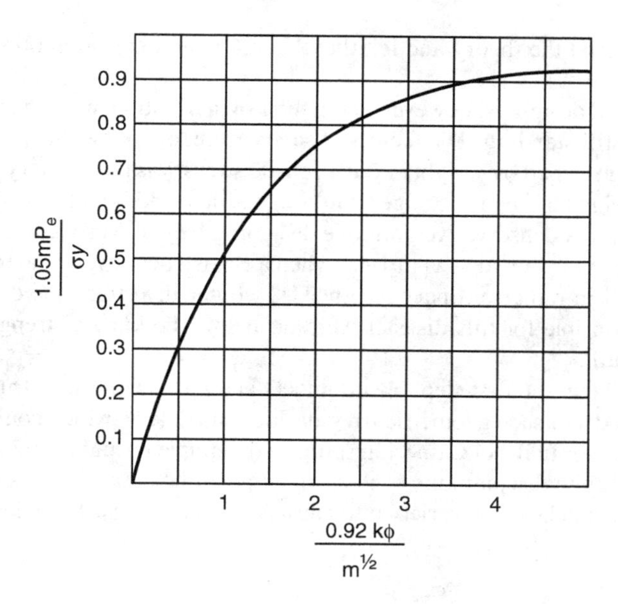

FIGURE 3.16 Design chart for ring-stiffened cylinders under external pressure.

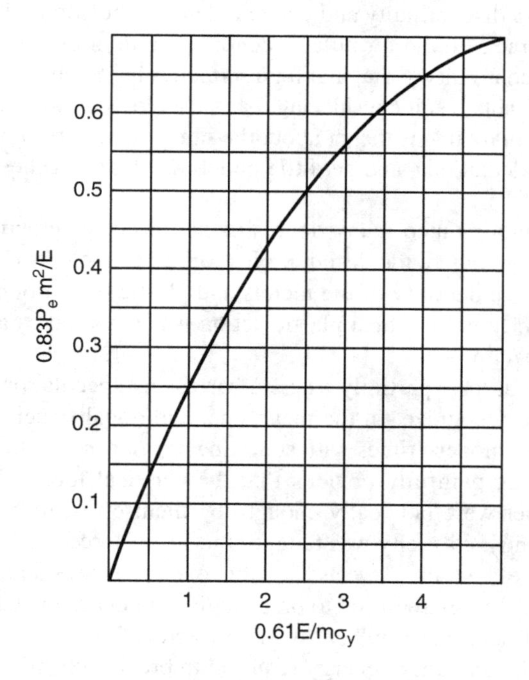

FIGURE 3.17 Design chart for hemispherical vessels under external pressure.

a scatter is sensitive to material and geometry variations, their probable extent should be known before a more reliable, lower-bound design curve can be developed.

3.13 EARLY EXPERIENCE WITH STRESS AND FRACTURE

Much of the earlier work in strength and stiffness of structures was somewhat abstract and mathematical in nature—particularly during the first half of the 19th century. Practical designers of that period did not understand the theory and felt that theoreticians were often blinded by the elegance of their own methods.

These same practical designers, however, using their own intuition, built magnificent bridges and structures which are still standing. The factors of safety were, of course, very high, and the designs were seldom, if ever, governed by weight or cost. Tensile stress was the primary governing criterion, particularly in the designs of beams, bridges, buildings, ships, and boilers.

But, unfortunately, this conservative, intuitive designing began to erode to the point where catastrophic failures began to occur such as in 1901 when a British destroyer suddenly broke in two and sank in the North Sea in rather ordinary weather [11]. Later investigations disclosed surprisingly that the stresses responsible for this disaster were far below the known strength of the steel from which the ship was built.

It was quite natural that the first step toward understanding this unusual structural failure was to develop a practical model of such geometric irregularities and defects which could help to explain the mystery. In 1913, Professor Inglis of Cambridge proposed a simple formula for calculating the increase in stress [12] due to a finite discontinuity such as an elliptical hole, porthole, door, or a hatchway, as well as cracks and/or scratches, in a variety of materials and structures. The relevant formula was:

$$\sigma_{\max} = \sigma\left[1 + 2(L/r)^{1/2}\right] \tag{3.78}$$

where σ_{max} is the stress at a discontinuity and where σ denotes the nominal stress existing at a point away from a notch or a crack. In the formula, L denotes length or depth of the notch, and r is the radius at the tip of the discontinuity. Note, that for a semicircular notch or a round hole, $L = r$, so that $\sigma_{max} = 3\sigma$. This is a remarkable result considering that the expression was proposed nearly 100 years ago. What is even more remarkable is that at about the same time, Kirsch in Germany and Kolosoff in Russia made similar calculations, and yet little notice was taken of these results in shipbuilding and other industries.

While the Inglis formula planted many doubts in the minds of practical engineers and some startling results could be predicted, the design profession, as a whole, was eager to dismiss Inglis's implications by invoking the ductility of the metals and plastic flow around the tip of a crack or a geometrical discontinuity. In effect, local plastic action was regarded as a "rounding off" mechanism for blunting the sharp tip.

While this theory was at least partially true, embarrassing speculations surrounding the Inglis formula were also easy to recognize. In the meantime, additional structural failures continued to crop up and persisted until modern times with some spectacular incidents involving ships, bridges, and oil rigs. It has become painfully obvious that the classical tools of elasticity developed by Hooke, Young, and Navier were not really enough, by themselves, to predict structural failures. After all, until quite recently, elasticity was taught in terms of forces and deformations. Even now, we seldom think of the stress–strain curve as a symbol of energy or as a measure of conservation of energy. Yet the quantity of energy required to break a given material or structure defines the toughness, sometimes called "fracture energy" or "work of fracture."

It is now known that the amount of energy required to break a chemical bond is not widely different among the majority of ductile structural materials.

Brittle materials, however, require very little energy to break because the work done during the fracture is pretty much confined to that needed to break the chemical bond. In ductile materials, the fracture is accompanied by significant deformations such as those observed in the necking and elongation process caused by the so-called "mechanism of dislocation."

Despite technological progress, structures contain cracks, scratches, holes, and other defects that, according to Inglis, may cause unacceptable levels of stresses. How and why are we able to live with these stress situations without widely spread catastrophic results in the structural shapes and mechanical systems? We consider this provocative question in the following sections.

3.14 GRIFFITH THEORY OF FRACTURE

In 1920, Griffith put forward a theory of fracture by energy [13]. With his remarkable foresight, he defined Inglis's stress concentration as simply a mechanism for converting strain energy into fracture energy, just as a can opener is a mechanism for using your muscular energy to cut through a tin can. Hence, stress concentration alone will not work unless it is continually supplied with enough energy to produce a fracture.

Since this energy must come from within, it can only be obtained from the accumulated strain energy within the system. And, of course, the internal strain energy is equal to the work done represented by the familiar stress–strain characteristics of the material. In simple terms, strain energy is equal to the area under the stress–strain curve, a familiar diagram in engineering studies.

In a rather brilliant move, Griffith proceeded to define a critical crack length that permeated all phases of linear elastic fracture mechanics (known as LEFM). The consequence of Griffith's theory is that even if the local stress of the crack tip is very high, the structure will not break as long as no crack or other opening is longer than the critical length. This design parameter depends on the ratio of the value of the work of fracture to that of the strain energy stored in the material.

The consequence of Griffith theory is the well-known inverse square root dependency of strength upon crack length as given by the following equation:

$$\sigma = \frac{K_{Ic}}{(\pi a_c)^{1/2}} \tag{3.79}$$

where K_{Ic} is often called "the plane strain fracture toughness." It describes the resistance of an existing crack to the driving force which tends to propagate the crack in an unstable manner, and the characteristic dimension, a_c, denotes the length or depth of the crack.

K_{Ic} is known to be a function of strain and temperature, and its proper value can only be determined by tests. For all practical purposes K_{Ic} represents an intrinsic materials property which should be independent of specimen design and dimensions. In this respect, K_{Ic} is as unique as the conventional yield strength of the material. However, its value cannot be derived from standard mechanical properties such as yield strength, ultimate strength, elongation, or modulus of elasticity. Furthermore, K_{Ic} is expressed in units of $(stress) = (length)^{1/2}$, unlike any other standard dimensional quantity in stress analysis.

It should be noted that a brittle fracture in general is of a catastrophic type and occurs with very limited or no plastic deformation at very high speeds or crack propagation. The corresponding fractured surface appears as a flat cleavage with virtually no shear lips.

As the working stress increases, the critical crack length becomes shorter. However, in weighing the merits and shortcomings of different steels, it is helpful to remember that the work of fracture of plain carbon steels decreases rather drastically with the tensile strength. Better combinations of strength and toughness can improve through the use of the appropriate alloying elements.

3.15 BEHAVIOR OF CRACKS

Application of fracture mechanics as a design tool is constrained by a number of variables: the theoretical stress, even using modern numerical methods, is unlikely to provide complete answers and there should be more attention paid to experimental results. Even the ingenious approach of Griffith, providing the well-known inverse square root dependency of strength upon crack length, is an oversimplification. Geometric and manufacturing effects are often overpowering and it is fortunate, indeed, that at least the basic concept of toughness has remained relatively invariant, helping us to distinguish between ductile and brittle behavior.

The response of metals, however, often tolerant of plastic deformation, has shown some dual behavior. Here the presence of a crack and the velocity of crack propagation pose a real design problem. Although a stationary crack stress field has been investigated with some success, our understanding of moving cracks has been much less satisfactory. In addition, the calculation of dynamic stress distribution with fast-moving cracks has always been classified as an intractable theoretical problem for structural members with finite boundaries. Fortunately, experimental analysis using fringe multiplication and interferometric techniques has established that dynamic stress distributions around cracks resemble their static counterparts.

This crack behavior observed on transparent plastic models, however, does not establish a similar mechanism in steel. Nevertheless, strain gage investigations [14] indicate that plastic deformations in steel could be transmitted with the high velocities of brittle crack propagation. This is indeed surprising and incompatible with the theory of stress wave translation in the plastic region of low-strength steel.

From a practical perspective, it is not the absolute velocity of propagation but, rather, the onset of crack extension that determines the most critical conditions. This is not to say that a rapid crack propagation is not a variable in the complex equation of brittle fracture associated with spectacular crack lengths. Such failures have been observed in pipelines in the field and in full-scale tests.

These experiments also confirm the classical Irwin theory that all the elastic strain energy available is released at the instant of fracture [15]. It is also of interest to note that while fracture toughness of materials is susceptible to fabrication variables, the rates of crack propagation change only with the changes in the modulus of elasticity.

The basic design consideration in fracture control is that fracture takes place by crack extension, and there is a much greater refinement of understanding studying test sample behavior, rather than that of the actual structural member. The stress analysis of a cracked component in the presence of a plastic flow represents a complicated problem even in plane components and simple structures due to the three-dimensional character of the model.

It is well to reemphasize that the traditional concepts of gradual yielding do not apply where the uninhibited growth of existing flaws or cracks leads to fracture stresses below the gross yield strength of the material. Rapid crack propagation usually occurs when either the operating temperature is lowered or the rate of loading is increased. The damaging loading is that which opens the crack, and it is observed that the element of material just ahead of a symmetrically loaded crack is always in a state of two-dimensional tension. The crack, however, remains harmless when the parameter $\sigma\sqrt{a}$ can be kept below the value of K_{Ic}.

Figure 3.18 shows the magnitude of variation of the plane strain toughness K_{Ic} with the specimen thickness unless of course the condition of plane strain prevails.

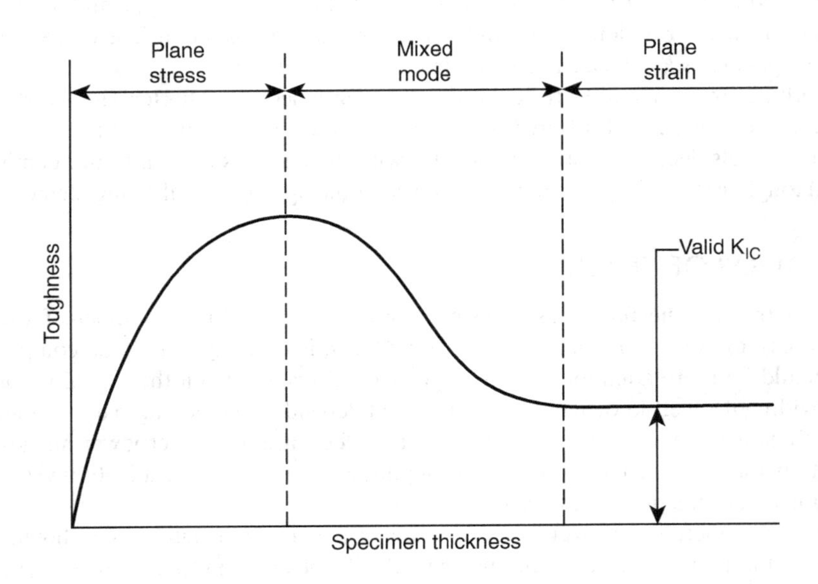

FIGURE 3.18 Effect of thickness on toughness.

Plane strain may be defined by the expression:

$$s_z = \nu(s_x + s_y) \tag{3.80}$$

Where S_z denotes the transverse normal stress on an element, while S_x and S_y are the in-plane stresses whose effect is controlled by Poisson's ratio ν.

As far as the designer is concerned, the knowledge of K_{Ic} and the working stress allows certain parametric freedom in computing the maximum allowable load a machine or a structural element can tolerate as a function of flaw size. This statement, however, is somewhat deceptive because in addition to knowledge of the flaw size, one has to know how rapidly this flaw can grow to a critical size. Hence, in a preventative mode of plane-strain fracture, it becomes necessary to rely on nondestructive testing.

K_{Ic} is essentially a materials property and, as such, if depends upon the metallurgical and mechanical aspects of the fabrication process. The quantitative approach to predicting the stresses governing crack initiation, however, must also be a function of crack geometry. This complexity has been resolved, to some degree, by providing tables of formulas. Nevertheless, work in fracture mechanics should be directed toward the translation of laboratory K_{Ic} data into the design of full-size structural systems under complex loading conditions. New theories of failure should also be explored such as that defined as the critical strain energy density factor [16].

3.16 DETECTION OF CRACK SIZE

The practical limitation of our ability to predict realistic stresses and to assure the prevention of structural failure depends on a blend of fracture mechanics knowledge with experimental ability to conduct nondestructive testing. The main problem is to be able to apply flaw detection probabilities to design and production control.

The object of a fracture-control program is to minimize the probability of a catastrophic failure due to the presence of undetected flaws. It may be of interest to observe that various nondestructive testing techniques have different detection sensitivities and accuracies. For example, Table 3.2 shows testing results for 4330 Vanadium modified steel in terms of flaw size (length).

Table 3.2 illustrates the critical lengths of surface flaws to cause failure at uniaxial yield strength for a number of typical aerospace materials. These examples have been taken at random and are thus not necessarily statistically valid. Nevertheless, it may be helpful to speculate on the probability of detection of surface flaws at some given confidence level. It appears that acceptable characteristics of probability versus flaw size can be of the type illustrated in Figure 3.19. The graph in Figure 3.19 determines the minimum flaw size that can be detected using standard techniques of nondestructive testing. Examples of material processing flaws are voids, inclusions or forging cracks that may have occurred during welding, bonding, or brazing. Service-induced cracks may be due to fatigue, overload, or stress corrosion (Table 3.3).

TABLE 3.2
Example of Detection Sensitivities for 4330 Vanadium-Modified Steel

Technique	Observable Crack Size (in.)
Visual	0.03
Ultrasonics	0.20
Penetrant	0.35
Magnetic particle	0.30

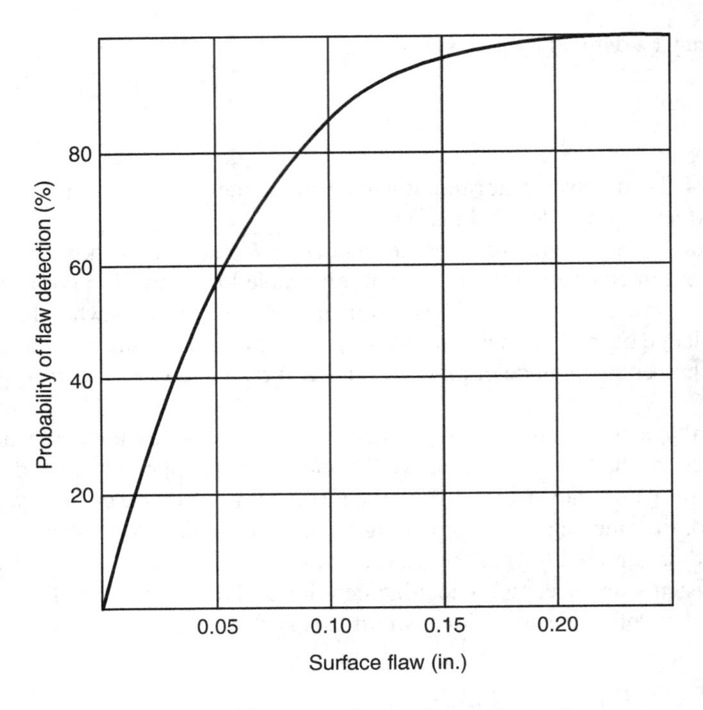

FIGURE 3.19 Characteristics of acceptable process for nondestructive testing.

TABLE 3.3
Examples of Critical Surface Flaws

Material	Yield Strength (ksi)	Toughness (ksi (in.)$^{1/2}$)	Flaw Size (in.)
2014-T-6 aluminum	57	22	0.17
2014-T3 aluminum	50	46	0.99
4330 vanadium-modified steel	190	55	0.10
4340 steel	230	54	0.06
Marage 300 steel	260	68	0.08
6 aluminum-4V	137	55	0.19

3.17 DUCTILE TO BRITTLE TRANSITION

The problem of "ductile-to-brittle transition" in metals was painfully recognized some 65 years ago when the welded plates of World War II ships were plagued by catastrophic failures. As a result of this experience, substantial effort at the Naval Research Laboratory was directed toward the development of various remedial measures and provided an excellent background for the evolution of fracture-safe design criteria in low and medium strength steels below the yield strength level of about 120 ksi [17].

The crack propagation and arrest characteristics in brittle fracture of steels were originally studied by Robertson [18] in the early 1950s. The investigation resulted in stress-temperature curves for crack-arrest behavior indicating that there was a specific nominal stress below which a crack would tend to arrest independently of crack size and geometry. The actual mechanism of structural failure may be opened to some interpretation in terms of what constitutes a critical level of damage in a particular situation.

For example, reduction of the net cross-sectional area in a tensile mode can be such that the structure can no longer support the desired loads, whether one considers elastic or plastic response of the crack tip. While the important elements of crack propagation and arrest have long been well-recognized and investigated, the engineering solutions to fracture control in design were slow in coming because of the theoretical complexity of the problem, changes in manufacturing techniques and economic constraints.

The key parameter of fracture mechanics K_{Ic} became essential to fracture control. It relates to yield strength, in accordance with Eq. 3.7, on the assumption of plane strain behavior. Figure 3.20 shows the classical crack arrest curve pertinent to fracture control.

In Figure 3.20, CAT represents Crack Arrest Transition while NDT represents Nil Ductility Transition temperature. The points A and A' denote the allowable stress S_a and the working stress S_w levels at a particular temperature, respectively. All the combinations of the nominal stress and temperature coordinates failing below the CAT curve indicate the conditions suitable for crack arrest. The area above the CAT curve represents the region of variables conductive to crack propagation. As the size of the crack increases, the nominal stress needed to propagate the crack is expected to be smaller.

For a conservative average estimate of stress on the lower portion of the CAT curve based on Robertson's work [18], 6 ksi may be considered. This criterion implies that at all stresses below 6 ksi, the crack will not propagate catastrophically under the usual conditions of loading and geometry, if the material is steel. When the appropriate edge-notch criteria of linear elastic fracture mechanics are invoked [19], the lower-bound nominal stress becomes a function of the material's thickness and fracture toughness. Figure 3.21 shows, the relevant functional relationship.

The design curve given in Figure 3.21 is based on a conservative, minimum value of K_{Ic} = 25ksi–in$^{1/2}$ which corresponds to the so-called garden variety steel such as that designated as A36.

The general trend of fracture-mechanics characteristics of aluminum alloys is the same as that for steels. The metallurgical transformations from high to low toughness are strength-dependent and can be modified by metal quality. Fracture resistance generally decreases with increase in yield strength of the aluminum. In addition, however, aluminum alloys can be described as metallurgically "dirty." These materials contain significant amounts of brittle intermetallic compounds that act as origins

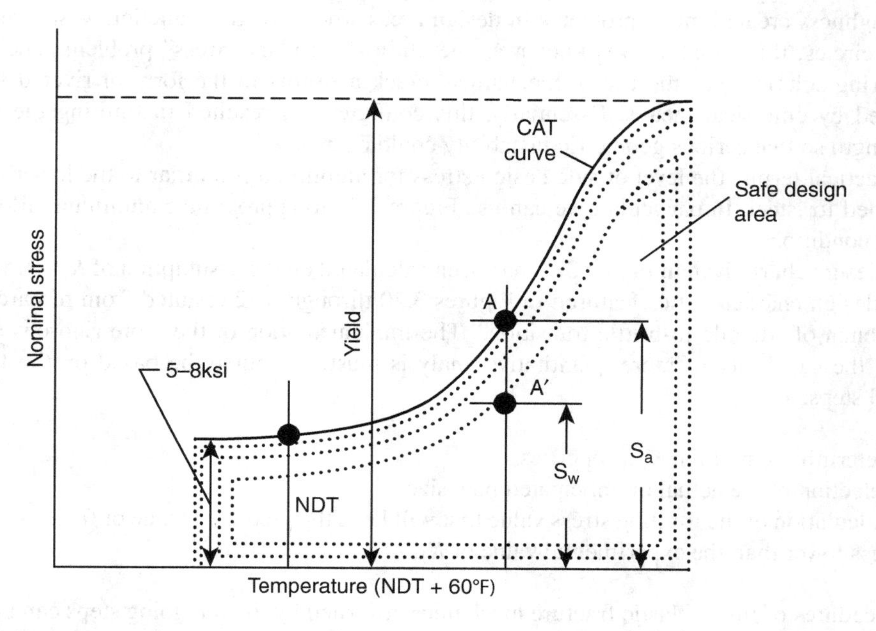

FIGURE 3.20 Typical crack arrest curve (qualitative).

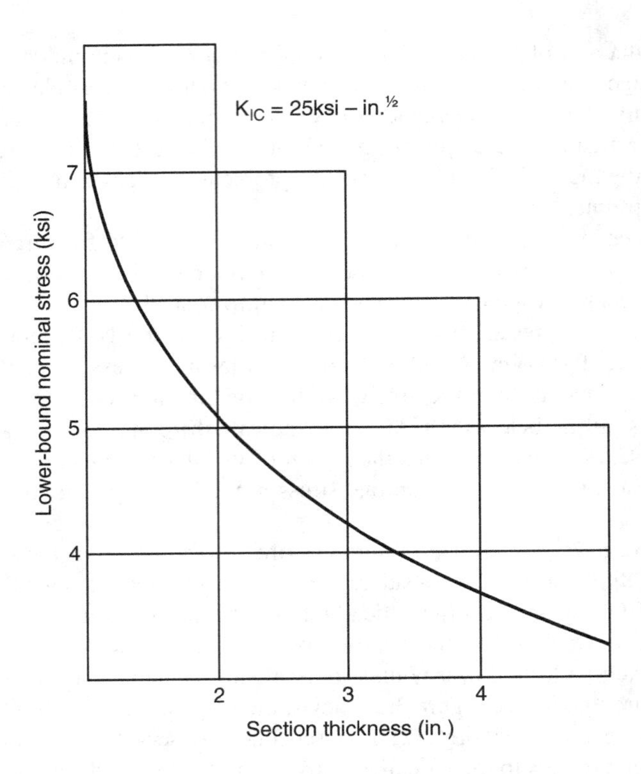

FIGURE 3.21 Lower bound design stress for fracture control for steel.

of microcracking and void formation. Production of metallurgically clean aluminum in the yield strength ranging from 45 to 65 ksi can markedly improve the fracture resistance of this material.

Aluminum catastrophic fractures under conditions of geometric instability have been experienced in the aerospace industry during the past 50 years. The combination of low elongation and poor toughness created many problems in design areas where weight reduction was necessary. In aircraft circles, this situation was known as the "thin-sheet plane-stress" problem. The ultimate engineering solution was the use of mechanical crack arrestors in the form of riveted stiffeners developed by empirical means. Essentially, this configuration resulted in limiting the potential crack length so that serious geometric instability could be avoided.

In practical terms, the level of safe design stress for aluminum is similar to the lower level recommended for steel. In particular, we can use Figure 3.22 to approximate aluminum alloys under ambient conditions.

The design chart given in Figure 3.22 has been calculated on the assumption of $K_{Ic} = 15\text{ksi–in}^{1/2}$.

The design characteristics featured in Figures 3.20 through 3.22 resulted from research on the phenomenon of "ductile-to-brittle transition." The implementation of the more rigorous approach offering the capability of more quantitative analysis must, of course, be based on the following essential steps:

- Determination of the K_{Ic} properties.
- Selection of the actual or anticipated flaw size.
- Calculation of the limiting stress value that will keep the predicted value of fracture toughness lower than the experimental value of K_{Ic}.

The procedures of linear elastic fracture mechanics governed by the foregoing steps can be used as a method of protection against brittle fracture when the elastic stress conditions prevail; although

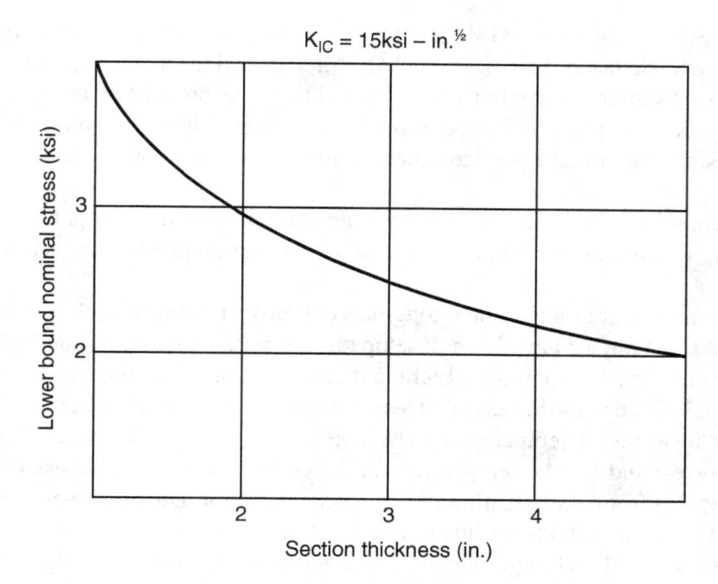

FIGURE 3.22 Lower bound design stress for aluminum.

it is known that some hardware cases are subject to inelastic behavior. The inelastic response technology, however, is continually in the process of development and is not the topic of this section. Unfortunately, we have to use approximations since the elastic and plastic stress regimes are often mixed when we try to correlate the stress with the critical size of the crack.

3.18 LIMITATIONS AND IMPLICATIONS OF FRACTURE CONTROL

Fracture mechanics, although well-defined, is not without important practical limitations. For example, the term "sharp crack" implies a *zero* tip radius. But also on the term "linear elastic fracture mechanics" (LEEM) implies fracture mechanics can be understood by an analysis of elastic behavior. A zero tip radius, however, must evolve from a plastic zone. This realization has led to the concept of a "plastic zone size correction," provided the failure is not due to a general yielding process.

Further limitations in current fracture mechanics are realized when we consider the various conditions under which K_{Ic} can be found, namely: static loading, dynamic initiation, or crack arrest fracture. In the case of structural steels, the numerical values of K_{Ic} based on the static initiation of fracture are usually the highest. For carbon and alloy steels of lower and intermediate yield strength, the K_{Ic} values are, of course, highly temperature-dependent. The choice is, therefore, between fracture mechanics and the transition temperature approaches to design.

While the theoretical equations of fracture mechanics are relatively simple, the tests to determine fracture-toughness parameters are expensive and time-consuming. It is not surprising that numerous efforts have been made to develop models and equations representing fracture toughness as a function of temperature.

The emphasis on knowledge of fracture control criteria given in this chapter stems from the fact that the traditional methods of design of mechanical fasteners and joints should be supplemented with the criteria of fracture as well as the strength. This trend is driven by the growing awareness of safety and liability in all sectors of manufacturing industry and research.

The role of crack characterization in design control is of the utmost importance. The critical crack size decreases with the increase in stress and it depends on the dimensions of the structural component as well as fracture toughness of the material. The plane-strain fracture toughness K_{Ic}

generally increases with the material's ductility and decreases with yield strength. However, the value of K_{Ic} cannot be deduced from the knowledge of standard mechanical properties.

Ideally, a material should behave in a generally yielding fashion in order to avoid various ramifications of brittle behavior. The brittle type of response is linear-elastic and requires a more explicit design approach since the actual flow size becomes an indispensable, but difficult to predict, design parameter.

In summing up, various fracture control considerations, it should be emphasized that the problem of developing or finding the data on brittle behavior and toughness characterizations is not a trivial one.

The effect of temperature on the notch toughness of most structural steels has been well-recognized. Various test techniques have been developed such as the Charpy V-notch (CVN), dynamic tear (DT), or the classical K_{Ic} approach. Technical and economical considerations have influenced the choice of the CVN index as the desirable test method for the development of fracture-toughness transition curves in terms of temperature. The familiar acronym NDT stands for Nil-Ductility-Transition and corresponds to the temperature at which the fracture toughness of steel begins to increase rather rapidly from plane-strain to fully ductile behavior. During the process of this transition, the material goes through the regime of elastic-plastic behavior.

The choice of the CVN technique for the characterization of material toughness is sometimes justified because rational correlation methods are available for analyzing CVN, DT, and K_{Ic} data. However, the CVN test should not be regarded as an invariant source for defining NDT for broad families of steels. The inherent difficulty is that a number of steels are sensitive to the shallow notch and the limited constraint effects provided by a relatively small CVN test section. The net effect of insufficient constraint caused by the design of the CVN test piece is the observed shift of the transition curve toward the lower temperature. This effect does not apply to the conventional DT data.

Nevertheless, the level of energy corresponding to the upper portion of the customary S-type shape of the transition curve may not be too different from that obtainable from the DT tests. In considering the importance of NDT, we must conclude that the use of the CVN method requires the development of specific correlations for each grade of steel, and the real practical solution is then to use the DT test method which is known to give the correct NDT indexing with the corresponding levels of energy to fracture.

Since most procurement specifications are still written in terms of the CVN criteria it clearly indicates that economic rather than technical considerations prevail for material purchase.

Since fracture mechanics and resulting methods for fracture control is a relatively new field of solid mechanics (developed largely during the past 50 years), it may be helpful to describe test piece geometry for the DT test.

Figure 3.23 depicts a typical specimen used in a dynamic tear test. It involves a deep notch, the root of which can be sharpened by pressing a knife edge into it. The test is conducted in a typical pendulum-type machine similar to that used in the CVN experiments. The maximum pressed-tip notch radius used, for example, in dealing with high strength steels, can be 0.001 in.

In other applications the deep, sharp crack can be introduced by an electron-beam weld, embrittled metallurgically. For example, Pellini quotes an example where a titanium wire added to the weld results in a brittle Fe-Ti alloy [17]. The resulting narrow weld can easily be preloaded to produce a sharp crack of very small but yet unspecified radius. The DT specimens are tested over a range of temperatures using the pendulum device, where the upswing of the pendulum after the fracture can be interpreted as a measure of energy absorbed during the process of fracture. For reasons of consistency in this type of an experiment, the size and geometry of a test piece need to be consistent.

The basic elements of fracture control in a conservative design can conceivably be questioned with respect to the influence of section size on the NDT. The amount of potential constraint provided by a relatively thick section also implies the tendency to brittle behavior regardless of metallurgical considerations. Until quite recently these particular features suggested that perhaps classical fracture mechanics negated the principles of physical metallurgy and that there was a real danger of

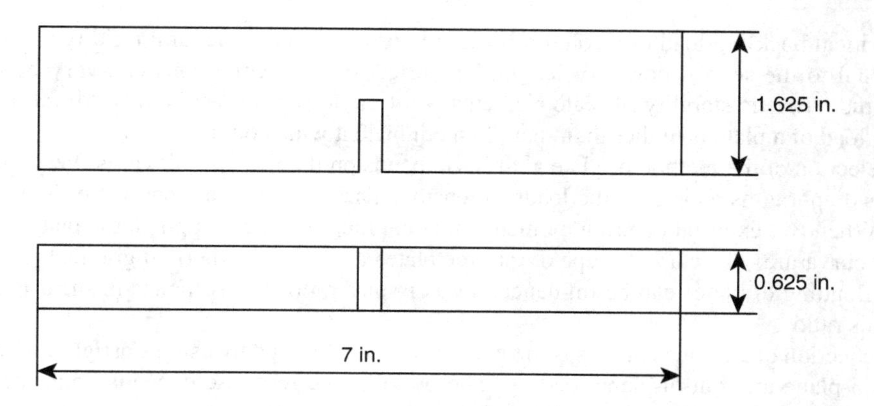

FIGURE 3.23 Specimen for dynamic tear test.

sudden failure, say of a thick-walled vessel or other structure, even at moderately elevated service temperatures.

Fortunately, these issues were settled some 40 years ago after a rather extensive series of tests [17]. One of the major findings from the application of unusually large DT specimens in those tests was that the increase in section size expanded but did not shift the transition temperature range. This may be an important consideration in fracture-safe design because it reduces the number of decision-making variables essentially to the NDT and to a lower stress.

3.19 SUMMARY

The fundamentals of practical stress analysis form the basis for the design of mechanical fasteners and joints. The number of working formulas is small but sufficient for sizing of individual components. A state of stress at a point can only be of normal or shear type. The stresses can be combined according to brittle or ductile theory of strength.

Bending moment and shearing force are the essential elements of beam analysis governed by the rules of static equilibrium. For this condition, the number of equations must be equal to the number of unknowns.

When the number of equilibrium equations is smaller than the number of unknowns, the structural system is considered to be statically indeterminate. The solution then requires use of deformation equations.

The second moment of area is a key cross-sectional value with each cross-section having two such values. The important value is dependent upon the direction of bending.

The choice of strength theory depends on the material used. The maximum principal stress formula applies to brittle materials. Prediction of ductile failure requires the calculation of the von Mises–Hencky stresses.

Response in shear is important in the analysis of mechanical fasteners and joints. In a two-dimensional element of a structure, two shearing couples are required for equilibrium. Transverse shear stress vanishes at the free surface and attains a maximum value at the center of the section. The torsional shear is greatest at the surface and zero at the center of twist. The mechanism of twist of noncircular cross-sections is very complex. The torsional moment can cause tensile and compressive effects.

A simplified formula for a beam curvature is useful in many deflection calculations. Deflection depends on the modulus of elasticity and geometry, but it is independent of the material's strength. Bending deflection of conventional beams is predominant. Deflection due to shear becomes significant when the ratio of span-to-depth decreases below three. The precision of deflection calculations depends strongly on the knowledge of the support conditions.

The critical buckling load of a column is directly proportional to flexural rigidity and inversely proportional to the second power of length. The effect of edge conditions can vary by an order of magnitude. Elastic stability of plate elements is a function of thickness to width ratio squared. Buckling load of a plate is higher than that of an equivalent wide column.

The selection of the method of plate analysis depends on the relative thickness. Very thin plates behave as diaphragms and carry the loads in tension. The classical plate parameter is the flexural rigidity. When two external bending moments of equal magnitude are applied, the plate develops a spherical curvature. The curved shape of the free plate caused by the thermal gradient produces no stresses. Bending of plates can be influenced by Poisson's ratio. The response in shear is independent of this ratio.

The deflection of curved elements is most conveniently obtained by using Castigliano's principle both for in-plane and out-of-plane loading. The out-of-plane response depends upon the flexural rigidity and the torsional shape factor. For a closed circular ring, the analysis is statically indeterminate. The neutral axis of the ring cross-section is not the same as the centroidal axis and the relevant discrepancy increases with the increase in radial thickness of the ring.

The design criteria and formulas for the vessels under internal and external pressure can be governed by either stress or stability. Formulas for thick vessels under either of the two pressure loads can be of the same type. Efficiency of longitudinal joints in cylindrical vessels is more critical.

Conical vessels can be designed by standard formulas intended for cylinders with only slight modifications. The ultimate strength of cylindrical and spherical vessels can be predicted once the geometrical parameters and the material's strength ratio (yield to ultimate) are established. The response of cylindrical shells can be anticipated with the aid of a thinness ratio, defined in the text.

Geometric irregularities and defects can cause serious stress concentrations and fracture. Local plastic action and the rounding-off mechanism of crack blunting are of practical interest in fracture control. Stress concentration is a mechanism for converting strain energy into fracture energy of a particular component. The key mechanical property is plane strain fracture toughness (K_{Ic}) but its value cannot be derived from standard mechanical properties. The parameter K_{Ic} is susceptible to fabrication techniques.

Knowledge of flaw size and K_{Ic} parameter for design is indispensable, although somewhat deceptive unless we can determine how rapidly a given crack can grow to a critical size. The probability of crack detection, even with the aid of modern techniques, decreases rapidly with the decrease in crack size.

The phenomenon of ductile to brittle transition with temperature remains a serious problem. The solution is to develop a crack arrest design curve featuring stress versus temperature. For a conservative design approach, the lower bound stress from this curve can be used. The allowable stress decreases with an increase in the material's thickness. The lower-bound method, normally recommended for steel, can be extended to nonferrous materials.

Linear elastic fracture mechanics, although applied to design with increasing frequency, has a number of practical limitations. The technique depends on a successful detection and characterization of the cracks. This process is continually under development.

REFERENCES

1. Timoshenko, S. 1956. *Strength of Materials*. New York: D. Van Nostrand.
2. Roark, R. J. 1965. *Formulas for Stress and Strain*. New York: McGraw-Hill.
3. Bleich, F. 1952. *Buckling Strength of Metal Structures*. New York: McGraw-Hill.
4. Timoshenko, S. 1940. *Theory of Plates and Shells*. New York: McGraw-Hill.
5. Bares, R. 1971. *Berechnung fur Platten und Wandscheiben*. Wiesbaden und Berlin: Bauverlag Gmb H.
6. Andrews, E. S. 1919. *Elastic Stresses in Structures*. London: Scott, Greenwood.
7. Huston, R. L. and H. Josephs. 2009. *Practical Stress Analysis in Engineering Design*. Boca Raton, FL: CRC Press.

8. Blake, A. 1979. *Design of Curved Members for Machines.* Huntington: Robert E. Krieger.
9. Wilson, B. J. and J. F. Quereau. 1927. *A Simple Method of Determining Stress in Curved Flexural Members.* Urbana: Circular 16, Engineering Experimental Station, University of Illinois.
10. Gill, S.S. 1970. *The Stress Analysis of Pressure Vessels and Pressure Vessel Components.* Oxford: Pergamon Press.
11. Gordon, J. E. 1978. *Structures or Why Things Don't Fall Down.* New York: Plenum Press.
12. Inglis, C.E. 1913. *Stresses in a Plate Due to the Presence of Cracks and Sharp Corners.* London: Trans. Institution of Naval Architects 60.
13. Griffith, A. A. 1920. *The Phenomena of Rupture and Flow in Solids.* London: Trans. Roy. Soc. A221.
14. Videon, F. F., F. W. Barton and W. J. Hall. 1963. *Ship Structure Committee.* Report 148, Arlington, VA: Office of Naval Research.
15. Paris, P. and G. C. Sih. 1965. *Fracture Toughness Testing*, ASTM STP 381, Philadelphia, PA: American Society for Testing Materials.
16. Sih, G. C. 1977. *Mechanics of Fracture.* Leyden: Noordhoff International Publishing.
17. Pellini, W. S. 1976. *Principles of Structural Integrity Technology.* Arlington: Office of Naval Research.
18. Robertson, T. S. 1953. Propagation of brittle fracture in steel. *Journal of Iron and Steel Institute*, 175.
19. Streit, R. D. 1982. *Design Guidance for Fracture-Critical Components at Lawrence Livermore National Laboratory*, Internal Report, UCRL-53254. Livermore, CA: Lawrence Livermore National Laboratory.

SYMBOLS AND ACRONYMS

a	Plate length
a_C	Characteristic dimension of crack
A	Area of cross-section
b	Width
B_1 through B_4	Strength factors for vessels
C	Distance to extreme fiber
CAT	Crack Arrest Transition
CVN	Charpy V-Notch Test
d	Diameter of solid cylinder
D	Plate flexural rigidity
DT	Dynamic Tear
E	Modulus of elasticity
E_t	Tangent modulus
F	Force
F_C	Critical column load
F_t	Tangential force
F_y	Yield load
F_1	Axial load
G	Modulus of rigidity
I	General symbol for second moment of area
I_b	Second moment of area
I_g	Moment of inertia about central axis
I_p	Polar moment of inertia
J	First moment of area
K	Torsional shape factor
K_0	Buckling factor for columns
K_p	Buckling factor for plates
K_{Ic}	Plane strain fracture toughness
$k = T/L$	Thickness to length ratio
L	Length
LEFM	Linear Elastic Fracture Mechanics

$m = R/T$ Characteristic ratio
M, M_1, M_2 Bending moments
M_f Fixing couple
n Arbitrary number
N Normal force
P Internal pressure
P_C Burst pressure for cylinder
P_e Experimental collapse pressure
P_s Burst pressure for sphere
Q Shear force
q Transverse unit load
r General symbol for radius
R Mean radius of curvature
R_i Inner radius of vessel
S_a Stress allowable
S_{cr} Plate buckling stress
S_w Working stress
S_x, S_y, S_z Stress components
t Thickness of plate
T Thickness of pressure vessel
T_m Twisting moment
W, W_1, W_2 Concentrated loads
w Transverse deflection
x Coordinate
Y Vertical deflection
y Coordinate
Z Section modulus
$\forall$ Linear coefficient of thermal expansion
$\forall_C$ Cone half-angle
$\exists$ Ratio of yield to ultimate strength
$*$ Displacement of neutral axis
ΔT Temperature gradient
θ Arbitrary angle
λ Thinness ratio
ν Poisson's ratio
Δ General symbol for radius of curvature
Φ General symbol for stress
Φ_{max} Maximum normal stress
Φ_{min} Minimum normal stress
σ_ℓ Longitudinal stress
σ_t Hoop stress
σ_y Yield strength of material
τ General symbol for shear stress
τ_a Allowable shear stress
τ_{max} Maximum shear stress
ϕ Subtended angle
ϕ_0 Inverse strain parameter
ψ Design factor
Ω Plate parameter

4 Quality Measurement and Statistics

4.1 INTRODUCTION

As noted earlier, the focus of this book is to provide the tools and approaches to produce a reliable bolted joint. Other terms that are utilized to describe a successful joint include "quality joint" or "robust joint." Yet, we know that "nothing lasts forever." This dictum is further complicated by the fact that bolted joints are designed to be easily assembled and disassembled when desired, which indicates that appropriate engineering analysis and design must be applied to a bolted joint so that it will stay assembled when required and not disassemble when not desired. How then can we be assured that we have designed a quality joint? The question arises then as to how to define a quality joint—or in other terms, we can ask, what is a quality bolted connection? Although the term quality can have a large subjective component, i.e., "I know quality when I see it," it also has a number of technical definitions that define a product's fitness for use. Other aspects of quality include:

- Perceived Quality
- Serviceability
- Conformance to Design
- Product Performance
- Product Features
- Durability
- Reliability
- Aesthetics

The above stated general criteria are often dependent upon numerous individual or combination of various inputs and/or parameters as outlined below.

Equipment:
- Different Machines
- Aging Machines
- Worn Tools
- Machines Out of Adjustment
- Etc.

Process:
- Heat Treatment
- Machining Approach
- Process Modification
- Etc.

Material:
- Different Material Properties
- Different Suppliers
- Different Material Shipments
- Different Lots

- Material Chemistry
- Etc.

Environmental:
- Plant Temperature
- Plant Humidity
- Plant Air Quality (Affecting Workers)
- Etc.

Testing:
- Test Type
- Test Accuracy
- Test Tool Accuracy and Repeatability

Operator:
- Different Operator Skill Levels
- Operator Fatigue (Beginning of shift to end of shift)
- Operator Discomfort
- Operator Age

Service Conditions:
- Static Conditions
- Dynamic Conditions
- Environmental Conditions

All of the above individually or in combination can contribute to a nonquality or defective threaded fastener joint. Given this abbreviated list, we note that many of the elements that can contribute to a defective bolted joint occur randomly without any particular trend or pattern—these are called chance variations, which permit the application of the rules of statistics to assist us in predicting bolted joint quality levels.

The objective of this chapter is to apply a number of elementary statistical concepts, models, and tools to assist in the design and development of quality threaded fastener joints. This presentation is by no means intended to be a thorough or complete description of statistics but rather a review of the theory of measurements and elementary statistical techniques as they can be applied to the joining process.

Consequently, the formulas and equations will be presented without proof or derivation. Additionally, it will be assumed that the reader has access to an inexpensive calculator, which can readily provide the statistical mean and standard deviation which forms the basis of much of this chapter.

4.2 THEORY OF MEASUREMENTS

All physical measurements are subject to errors. There are two general types of errors: systematic and random. Systematic errors are those due to inaccuracies in the measuring instruments and other failures of the instrument or instruments utilized for the measurements intended. Random errors are all others. (We neglect here blunders, which are incorrect manipulations by the observer, such as reading a scale incorrectly.)

A single observation does not give any indication of the size of the random error which may be present in the observations. If, however, a large number of observations are made and the numbers obtained are not all the same, we have a "spread" of the results which indicates how large the random error may be. There are several possible reasons for obtaining a spread of results in a series of supposedly identical measurements. One is that the scale of the measuring instrument is not sharply enough defined, so that we must make an estimate of the number corresponding to the desired location of the scale. An example of this would be an attempt to read a yard stick, calibrated in inches, from the other side of the room.

A more common reason for obtaining a spread of readings in a series of observations is that the item being measured is not sharply defined. A simple example would be an attempt to measure the length of a broken stick. The irregular shape of the ends would cause different answers to be obtained depending on just where the measurements were taken.

A third reason for obtaining a spread of readings is that the item being measured is not the same every time the measurement is made. There are two cases of this. The first is where the quantity being measured is changing always (or mostly) in one direction, this is a systematic error. The second is where the quantity being measured changes but not necessarily in one direction. An example of a quantity which is not constant, but which does not change systematically, would be the weight of a given size of bolts falling through a hopper in a given time (assuming the hopper starts full every time). Because of irregularities of the bolts, their motions through the neck of the hopper will not always be quite the same, so that on successive tries different weights will fall through in a given time.

As far as the spread of results are concerned, the three cases where:

a. The measuring instrument is not sharply defined;
b. The item being measured is not sharply defined;
c. The item being measured is not the same each time an observation is made (but does not change continuously in one direction);

are essentially the same. All three cases will create a spread of results. Typically it happens that, if a very large number of observations are made, the numbers will cluster in some small region with very few being far from the main cluster.

The accuracy of an experimental result is defined as the discrepancy between the result and the true value of the quantity measured. Since the latter is usually unknown, we seek to obtain by estimation or computation a measure of the quality of the result. Such a measure is called the precision or precision measure of the result.

The distinction between accuracy and precision may be illustrated by the results obtained by marksmen firing at a target, as shown in Figure 4.1 [1].

Case A illustrates high precision and a good accuracy. Case B illustrates high precision (as measured by the spread of the shots in the absence of information about the location of the center of the target), but poor accuracy (due perhaps to an erroneous "sight" adjustment or neglect of wind effects). Case C illustrates low precision but perhaps with erroneously claimed high accuracy if the "average" of all the shots cluster around the center of the target.

In a physical experiment, it is important to obtain some measure of the precision of the results, so we can interpret and use them properly. Also, an estimate of the precision of the individual process steps involved in an experiment may save time by showing the futility of expending time and energy to improve one step in the procedure when much greater sources of error are present in other procedural steps.

The following are a number of ways of improving the reliability of experimental results:

1. Care in performing the experiment is obviously a vital factor.
2. The theory should be considered carefully to make sure that no important item is overlooked.

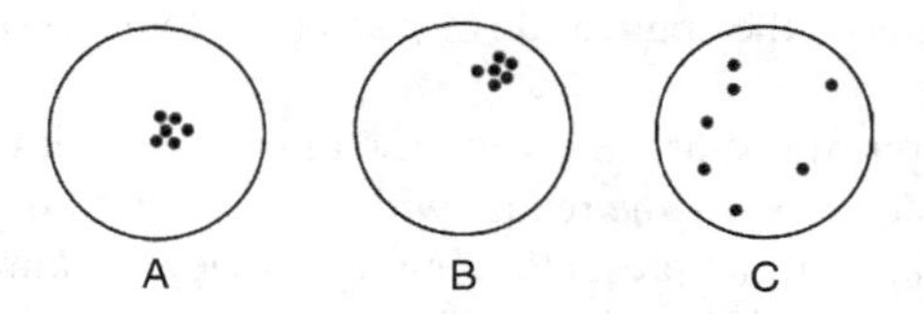

FIGURE 4.1 Marksman's targets illustrating the differences between accuracy and precision.

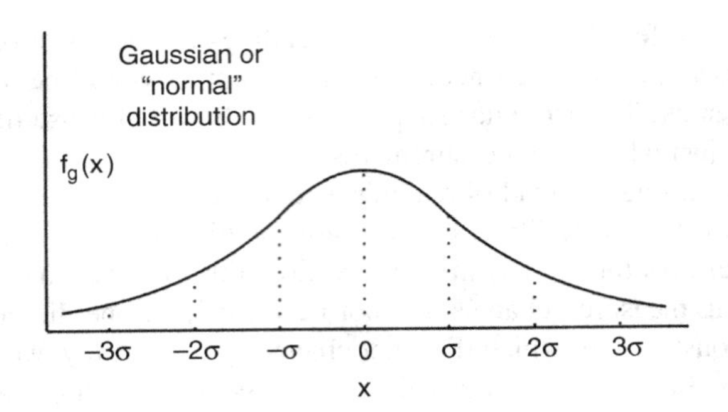

FIGURE 4.2 A normal curve showing symmetry about its central axis or mean [2].

3. All possible sources of error should be considered, and steps should be taken to eliminate them where/when possible or minimize them (if total elimination is not possible).
4. If it is possible, measurements should be repeated under a large variety of conditions, using as many different techniques as possible. This may give clues regarding errors previously not foreseen, which can then be eliminated. Also, it may reveal certain constant errors which were "averaged out," which were not foreseen, or which were foreseen but could not be corrected.
5. A large number of observations of the same quantities should be taken and averaged, to minimize the "random errors."

Random errors are primarily the errors "about which we can do nothing." They are called random because they are attributed to their independence and the chance of their occurrence. They include such quantities as errors of estimation between scale divisions, errors in perception, small and uncontrollable fluctuations in temperature, humidity, or other atmospheric conditions. The uncontrolled and independent existence of such factors leads to errors which are randomly distributed. In the target example listed in Figure 4.1, they are the source of the "spread" of the shots in the target.

It happens that in any series of observations of a quantity whose variations are random, the measurements cluster about a certain mean point in a pattern which is surprisingly universal. For example, if a thousand careful measurements of the weight W of a certain object were manually performed, the readings would not all be identical. If the number N of identical values obtained for W were graphically plotted against the value W, a set of points would be obtained through which a smooth bell-shaped curve would be drawn.

A curve of this form is called a "Gauss Error Curve," or "Normal Curve" or "Bell Shaped Curve" as shown in Figure 4.2. Since the derivation of the normal curve is based on the concept of random errors, hence "randomness" is fundamental to its creation and cannot be sufficiently overemphasized.

4.3 STATISTICAL ANALYSIS

As previously noted, this section on statistical analysis is meant to be a review of material previously learned in a course in elementary statistics. Hence, this section will not include any derivations or proofs but rather focus on the application of just a few elements of statistical analysis as they are applied to joint design.

Given the elementary applications and the wide use of examples, it is believed that the material in this section should be understood by the reader without any prior knowledge of statistics.

The material in this section will be presented first as a review of some of the fundamental principles and definitions of the applicable statistics. The definitions/review section will be followed by a series of examples and design studies which will be utilized to present the material.

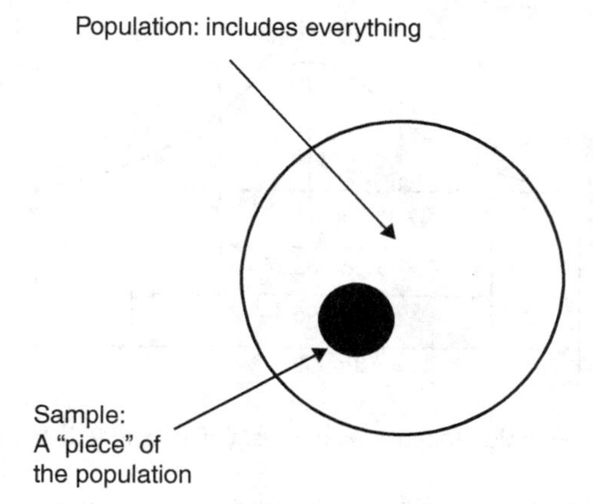

FIGURE 4.3 Graphical description of sample versus population.

1. **Population**
   - A population constitutes any finite or infinite collection of individuals, things, objects, or events which is "complete."
   - The term population connotes completeness or totality of observations of which we are concerned.
   - Population data is denoted by letters in the Greek alphabet (e.g., μ for the population mean and Φ for the population standard deviation).
2. **Sample**
   - A sample is a subset of a population.
   - The term sample connotes incompleteness.
   - Sample data is denoted by letters in the English alphabet (e.g., $\bar{x}$ for the sample mean and S for the sample standard deviation) (Figure 4.3).
3. **Statistical Inferences**
   - Statistical inferences involve reaching conclusions about population characteristics from a study of the sample or samples which as noted are only portions of the population.
   - Statistical inferences are basically predictions of what would be found to be the case if the parent populations could be fully analyzed with respect to the desired relevant characteristics.
   - There are two forms of statistical inferences:
     - Estimates of the magnitude(s) of population characteristics.
     - Tests of hypotheses regarding population characteristics.
4. **Random Sample**
   - The derivation of useful and valid generalizations from samples describing characteristics of the populations from which the samples originated, require that the individual data points included in the sample are chosen by random selection or random sampling.
   - Random sampling is defined by the requirement that each individual data point in the population has an equal chance of being chosen first from the sample.
   - Randomness of a sample is inherent in the sampling scheme utilized to obtain the sample and it is not an intrinsic property of the sample itself.
5. **Normal Distribution**
   - Many random variables are assumed to follow a "normal distribution." It is important to note that the normal distribution is a "continuous distribution," yet much (if not most) data analyzed as "normal" is obtained discretely, not continuously, violating the above stated "continuously" requirement.

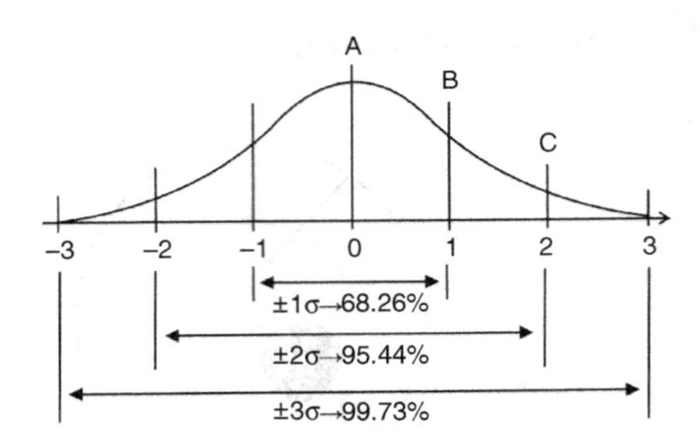

FIGURE 4.4 Percentage of population for various intervals of a normal distribution [1].

- A normal distribution is completely determined by two parameters:
  - The Arithmetic Mean (or simply mean)
  - The Standard Deviation.
- The normal curve is both unimodal and symmetrical with the area under the curve equal to 1.00.
- Properties of the normal distribution have been long established, are well known, and readily available in tabular form. Hence numerous statistical tests can be performed if the normal distribution assumption is valid. Values of the standard normal distribution are given in Table 4.1.
- The normal distribution is "normalized" and as noted above, the area under this normalized curve is 1.00. Hence for ALL normal distributions, given the population mean (μ) and standard deviations (σ), we find that 68.26% of all data will be within $\pm 1\sigma$ of the mean; 95.44% of all data will be within $\pm 2\sigma$ of the mean; 99.73% will be within $\pm 3\sigma$ of the mean; 99.994% will be within $\pm 4\sigma$ of the mean and 99.99994% will be within $\pm 5\sigma$ of the mean. See Figure 4.4 for a graphical depiction of some of these values.

Note that a normal distribution with a mean of zero and a standard deviation of 1 is also known as the "standard normal distribution."

Tables 4.1 and 4.2 present a more detailed list of normal distribution population percentages (P).

Table 4.2 presents a number of data values which are essentially an inverse of Table 4.1. In this table, the area under the normal curve is provided as the direct variable and the value of Z is the indirect variable. Additionally, special values of this table are presented which will prove to be helpful later, when we review the topic of levels of significance.

6. **Measures of Central Tendency**

   There are three measures of Central tendency:

   - Mean or Average—is the parameter defining the "centrality" of a normal distribution. A change in the mean will slide the entire normal curve to the right or left accordingly without changing the curve profile. The terms "mean" and "average" have identical meanings and will be used interchangeably.

$$\text{Sample mean: } \bar{x} = \frac{\left(\sum_{i=1}^{n} x_i\right)}{n} \tag{4.1}$$

   - Median—is the middle of the data values if the data is arranged in ascending or descending magnitude. If there are an even number of data values, then the average (or mean) of the two middle data values is utilized to define the median.
   - Mode—the mode is that data value which occurs most frequently in the data set.

TABLE 4.1
Values of the Standard Normal Distribution Function—Values of *P*

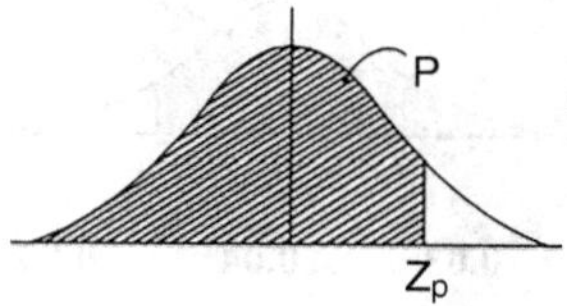

Z_P	0.00	0.01	0.02	0.03	0.04	0.05	0.06	0.07	0.08	0.09
0.0	.5000	.5040	.5080	.5120	.5160	.5199	.5239	.5279	.5319	.5359
0.1	.5398	.5438	.5478	.5517	.5557	.5596	.5636	.5675	.5714	.5753
0.2	.5793	.5832	.5871	.5910	.5948	.5987	.6026	.6064	.6103	.6141
0.3	.6179	.6217	.6255	.6293	.6331	.6368	.6406	.6443	.6480	.6517
0.4	.6554	.6591	.6628	.6664	.6700	.6736	.6772	.6808	.6844	.6879
0.5	.6915	.6950	.6985	.7019	.7054	.7088	.7123	.7157	.7190	.7224
0.6	.7257	.7291	.7324	.7357	.7389	.7422	.7454	.7486	.7517	.7569
0.7	.7580	.7611	.7642	.7673	.7704	.7734	.7764	.7794	.7823	.7852
0.8	.7881	.7910	.7939	.7967	.7995	.8023	.8051	.8078	.8106	.8133
0.9	.8159	.8186	.8212	.8238	.8264	.8289	.8315	.8340	.8365	.8389
1.0	.8413	.8438	.8461	.8485	.8508	.8531	.8554	.8577	.8599	.8621
1.1	.8643	.8665	.8686	.8708	.8729	.8749	.8770	.8790	.8810	.8830
1.2	.8849	.8869	.8888	.8907	.8925	.8944	.8962	.8980	.8997	.9015
1.3	.9032	.9049	.9066	.9082	.9099	.9115	.9131	.9147	.9162	.9177
1.4	.9192	.9207	.9222	.9236	.9251	.9265	.9279	.9292	.9306	.9319
1.5	.9332	.9345	.9357	.9370	.9382	.9394	.9406	.9418	.9429	.9441
1.6	.9452	.9463	.9474	.9484	.9495	.9505	.9515	.9525	.9535	.9545
1.7	.9554	.9564	.9573	.9582	.9591	.9599	.9608	.9616	.9625	.9633
1.8	.9641	.9649	.9656	.9664	.9671	.9678	.9686	.9693	.9699	.9706
1.9	.9713	.9719	.9726	.9732	.9738	.9744	.9750	.9756	.9761	.9767
2.0	.9772	.9778	.9783	.9788	.9793	.9798	.9803	.9808	.9812	.9817
2.1	.9821	.9826	.9830	.9834	.9838	.9842	.9846	.9850	.9854	.9857
2.2	.9861	.9864	.9868	.9871	.9875	.9878	.9881	.9884	.9887	.9890
2.3	.9893	.9896	.9898	.9901	.9904	.9906	.9909	.9911	.9913	.9916
2.4	.9918	.9920	.9922	.9925	.9927	.9929	.9931	.9932	.9934	.9936
2.5	.9938	.9940	.9941	.9943	.9945	.9946	.9948	.9949	.9951	.9952
2.6	.9953	.9955	.9956	.9957	.9959	.9960	.9961	.9962	.9963	.9964
2.7	.9965	.9966	.9967	.9968	.9969	.9970	.9971	.9972	.9973	.9974
2.8	.9974	.9975	.9976	.9977	.9977	.9978	.9979	.9979	.9980	.9981
2.9	.9981	.9982	.9982	.9983	.9984	.9984	.9985	.9985	.9986	.9986
3.0	.9987	.9987	.9988	.9988	.9988	.9989	.9989	.9989	.9990	.9990
3.1	.9990	.9991	.9991	.9991	.9992	.9992	.9992	.9992	.9993	.9993
3.2	.9993	.9993	.9994	.9994	.9994	.9994	.9994	.9995	.9995	.9995
3.3	.9995	.9995	.9996	.9996	.9996	.9996	.9996	.9996	.9996	.9997
3.4	.9997	.9997	.9997	.9997	.9997	.9997	.9997	.9997	.9997	.9998

Values of P corresponding to z_P for the normal curve.
z is the standard normal variable.
The value for P for $-z_P$ equals one minus the value of P for $+z_P$,
e.g., the P for −1.62 equals 1 − .9474 = .0526.

TABLE 4.2
z_P Corresponding to P for the Normal Curve, z is the Standard Normal Variable [3]

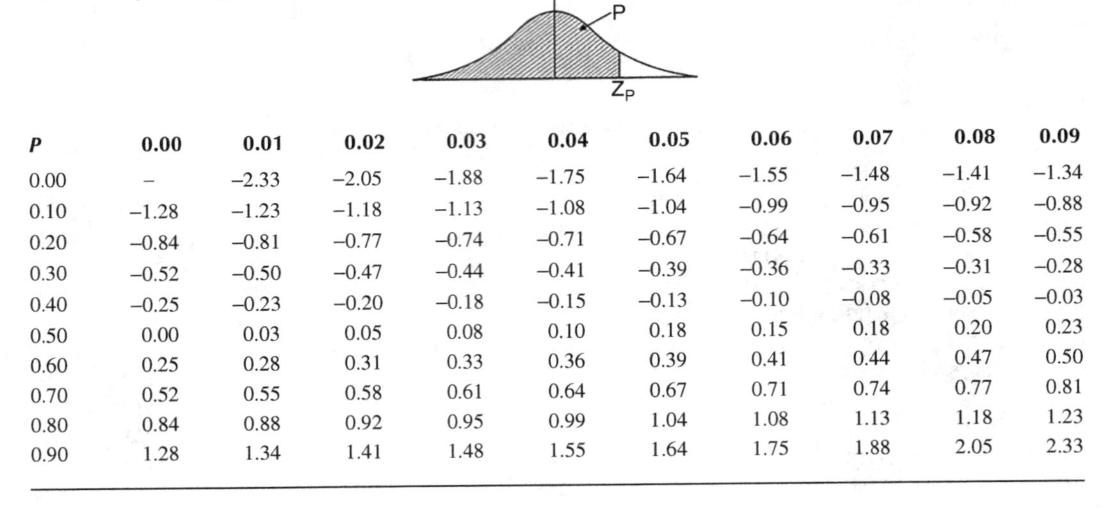

P	0.00	0.01	0.02	0.03	0.04	0.05	0.06	0.07	0.08	0.09
0.00	–	–2.33	–2.05	–1.88	–1.75	–1.64	–1.55	–1.48	–1.41	–1.34
0.10	–1.28	–1.23	–1.18	–1.13	–1.08	–1.04	–0.99	–0.95	–0.92	–0.88
0.20	–0.84	–0.81	–0.77	–0.74	–0.71	–0.67	–0.64	–0.61	–0.58	–0.55
0.30	–0.52	–0.50	–0.47	–0.44	–0.41	–0.39	–0.36	–0.33	–0.31	–0.28
0.40	–0.25	–0.23	–0.20	–0.18	–0.15	–0.13	–0.10	–0.08	–0.05	–0.03
0.50	0.00	0.03	0.05	0.08	0.10	0.18	0.15	0.18	0.20	0.23
0.60	0.25	0.28	0.31	0.33	0.36	0.39	0.41	0.44	0.47	0.50
0.70	0.52	0.55	0.58	0.61	0.64	0.67	0.71	0.74	0.77	0.81
0.80	0.84	0.88	0.92	0.95	0.99	1.04	1.08	1.13	1.18	1.23
0.90	1.28	1.34	1.41	1.48	1.55	1.64	1.75	1.88	2.05	2.33

Special Values

P	**0.001**	**0.005**	**0.010**	**0.025**	**0.050**	**0.100**
z_P	–3.090	–2.576	–2.326	–1.960	–1.645	–1.282
P	**.999**	**.995**	**.990**	**.975**	**.950**	**.900**
z_P	3.090	2.576	2.326	1.960	1.645	1.282

7. **Degrees of Freedom (DOF)**
   A degree of freedom denotes the precision with which a "source of variation" can be determined. Necessarily, then, a degree of freedom is associated with each independent comparison element (or metric) developed from the data. The degree of freedom or DOF will be symbolized by the symbol ∂.
8. **Measures of Dispersion**
   There are two measures of dispersion:
   1. Range
      - The range is typically defined as the difference between the largest and smallest value in any grouping of data points.
      - The range often increases with increasing sample size, so care must be exercised when comparing sample sizes. Typically range data should have a sample size of at least 10.
      - The range is widely used in control chart applications.
   2. Standard Deviation
      - The standard deviation is a measure of the data spread, scatter or distribution. A change in the standard deviation widens or narrows the normal curve without changing the curve center location.
      - The variance is the square of the standard deviation. Hence

$$\text{Sample variance} = S^2 = \frac{\left[\sum_{i=1}^{n}(x_i - \bar{x})^2\right]}{n-1} = \frac{\left(n\sum_{i=1}^{n} x_i^2 - \left(\sum_{i=1}^{n} x_i\right)^2\right)}{n(n-1)} \quad (4.2)$$

$$\text{Population variance} = \sigma^2 = \frac{(x_i - \mu)^2}{n} \quad (4.3)$$

Where

x_i = individual data values
n = number of data values
μ = population mean
$\bar{x}$ = sample mean

9. **Central Limit Theorem or Distribution of the Mean**
   If sampling from a population of unknown distribution, either finite or infinite, the sampling distribution of the mean $(\bar{x})$ for a large sample will be approximately normal approaching the mean μ and variance σ^2/n. This approximation is applicable only where sample size is large. For our purposes we will take n "large" to be $n \geq 30$.
10. **Z-score**
   Given a distribution that is known to be normal, then the distance from the population mean (μ) in units of the population standard deviation (σ) is given by

$$Z = \frac{x - \mu}{\sigma} \quad (4.4)$$

4.4 KNOWN NORMAL DISTRIBUTION

We use the Z score (or Z "value") to establish the probability that a value chosen randomly from a known normal distribution will have a value of x or less. Consequently, the probability of occurrences of more than x can be determined by subtracting the value obtained from 1 (which, as noted earlier, is the total area under the normal curve). This approach requires that we know the population mean (μ) and population standard deviation (σ).

EXAMPLE

The number of threaded fastener defects in an automobile assembly is absolutely known to be normally distributed over a monthly period with a mean of 12 and a standard deviation of 4 defects, what is the probability that 20 defects occurring "Solely to Chance" will be observed in a given month?

Given: $\mu = 12$, $\sigma = 4$, $x = 20$

Find: Z

Note: The term "solely to chance" is interpreted to mean that 20 defects will originate from the same population which generated the $\mu = 12$ and $\sigma = 4$ defects.

Approach:

$$Z = \frac{x - \mu}{\sigma}$$

$$= \frac{20 - 12}{4} = 2 \Rightarrow 2\sigma \text{ units}$$

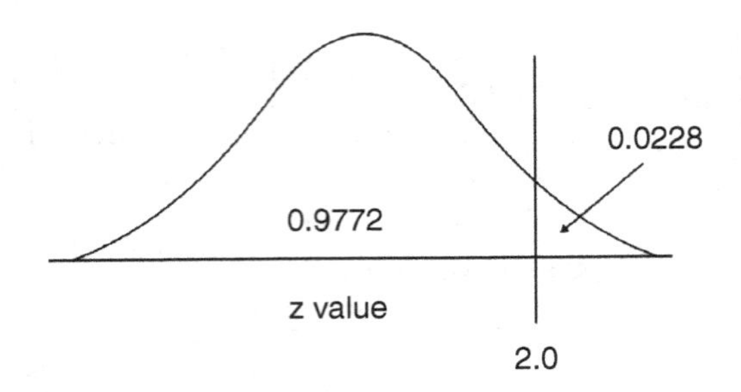

FIGURE 4.5 Example of normal curve distribution displaying values where $Z = 2\sigma$ units.

Summary: Table 4.1 is now used to find the desired values of the normal distribution for 2σ units. When $Z = 2.0$ the probability of Z or less is 0.9772, therefore, there is 1–0.9772 or only a 0.0228 (2.28%) chance of 2 or more σ units, or 20 or more defects occurring only to chance.
Figure 4.5 illustrates the findings.

4.5 DISTRIBUTION NOT KNOWN TO BE NORMAL

If a given distribution is not known to be normal, it can be treated as an assumed normal distribution based on the central limit theorem, which is based on a "large" number of samples; where as noted earlier n large is taken to be: $n \geq 30$. Then z is:

$$z = \frac{\bar{x} - \mu}{\sigma/\sqrt{n}} \tag{4.5}$$

Where

$\bar{x}$ = Sample Average
μ = Population Mean
σ = Population Standard Deviation
n = Sample Size (for this approach n must be $n \geq 30$)

Therefore, this approach requires:

1. Knowledge of the population mean (μ)
2. Knowledge of the population standard deviation (σ)
3. Sample size must be $n \geq 30$

Note on the term "large": non-normal distributions are sometimes characterized by the goodness of their "fit" or match to a normal curve. The better the fit, then the fewer number of samples are required for analysis. However, for the purposes of our discussion we shall assume that we have a "bad fit," and as noted previously, we shall use $n \geq 30$ as a definition of large n. Typically, if the mean is unknown, then σ will also often be unknown. However, for large n ($n \geq 30$), the sample standard deviation "S" will be close to the value of σ. Hence in the case where n is large and where σ is unknown, we can replace σ by the value S and still obtain reasonable accuracy. Then z may be expressed as:

$$z = \frac{\bar{x} - \mu}{S/\sqrt{n}} \quad \text{for} \quad n \geq 30 \tag{4.6}$$

EXAMPLE

An automotive assembly plant uses powered fastener tools that have a length of life that is approximately normally distributed with mean equal to 8,000 h and a standard deviation of 1,600 h.

Find the probability that a random sample of 36 tools will have an average life of less than 7,500 h.

SOLUTION:

1. Distribution: approximately normal
2. μ (Pop. Mean) = 8,000 h
3. σ (Pop. Std. Dev.) = 1,600 h
4. $n = 36$
5. $\bar{x}$ (Sample Avg.) = 7,500 h

Note: we have an approximately normal distribution, n is large, and σ is known. Therefore, using Eq. (4.6) we obtain z as:

$$z = \frac{\bar{x} - \mu}{\sigma/\sqrt{n}} = \frac{7500 - 8000}{1600/\sqrt{36}} = -1.875$$

Or equivalently in terms of probabilities,

$$P(\bar{x} < 7500) = P(z < -1.875)$$

Referring to Table 4.1, for the z-value, we obtain:

$$P(\bar{x} < 7500) = 0.0304 \quad \text{or} \quad 3.04\%$$

Graphically in terms of hours (the raw data) we have

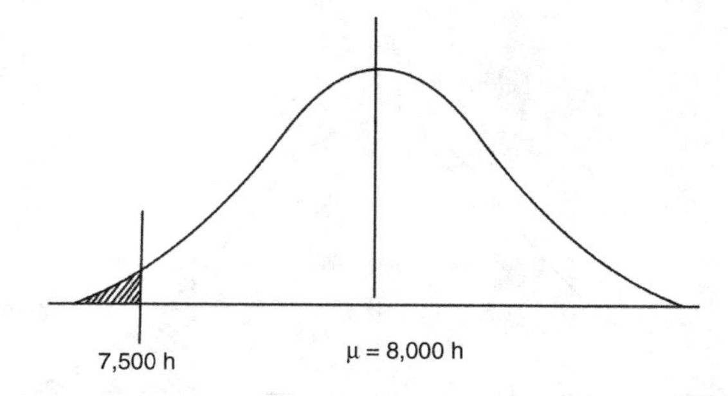

Correspondingly, the z-score may be illustrated as

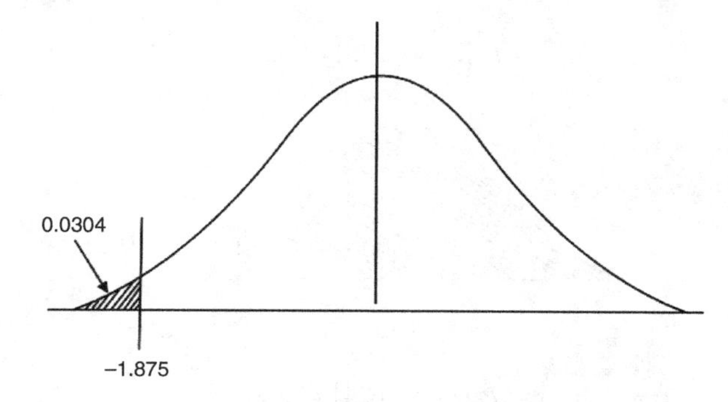

Summary: The probability is 3.04% that a random sample of 36 fastener tools will have an average life of less than 7,500 h.

4.6 t-STATISTIC OR THE STUDENT'S STATISTIC

This procedure, also known as the "Student t-test" was developed by W. S. Gosset in 1908 [4] who, while working in the Guinness Brewery, published his work under the title "Student."

In the "approximately normal distribution" situation discussed earlier, knowledge of σ and a large n are required. In the "real world" this data is often unavailable or difficult to obtain. These shortcomings are addressed by the τ distribution given by:

$$\tau = \frac{\bar{x} - \mu}{s/\sqrt{n}} \tag{4.7}$$

where

[Note: Knowledge of σ (population standard deviation is not required)]
$\bar{x}$= Sample Mean
n = Sample Size: n (sample size) need **not** be large
μ = Population Mean
s = Sample Standard Deviation

The τ statistic is a symmetric random variable that has a different distribution for each value of n. When $n = \infty$ the τ statistic is normally distributed. Therefore, at $n = \infty$ the entries in the table of the τ statistic and those for the z statistic are identical. Significantly, as noted above, for the case of the τ statistic, knowledge of the population standard deviation (σ) is not required.

The τ statistic or τ test has a degree of freedom "∂" equal to (n–1). See Table 4.3.

EXAMPLE

Using n = 15 for the purposes of this example and that data from Table 4.3, we find the following:

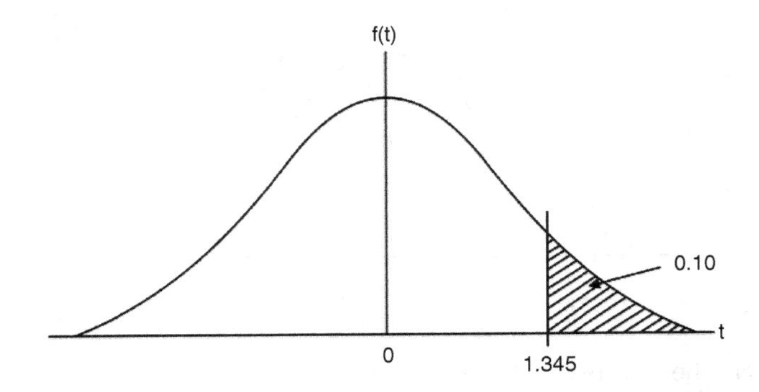

$$P_R\left[t_{15} > 1.345\right] = 0.10$$

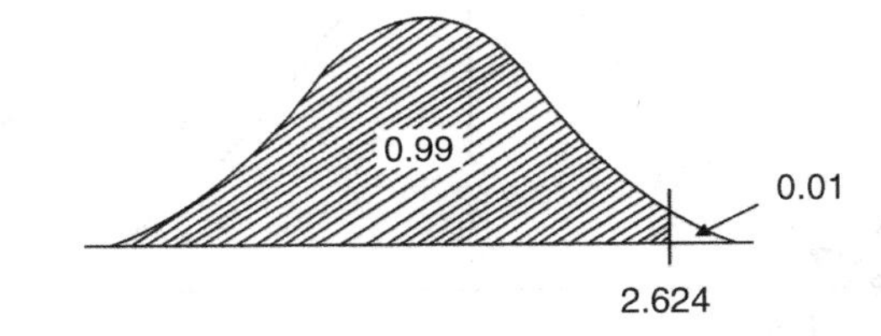

$$P_R\left[t_{15} < 2.624\right] = 0.99$$

For two-tailed situation

TABLE 4.3
Percentile Values for Student's *t* Distribution [5]

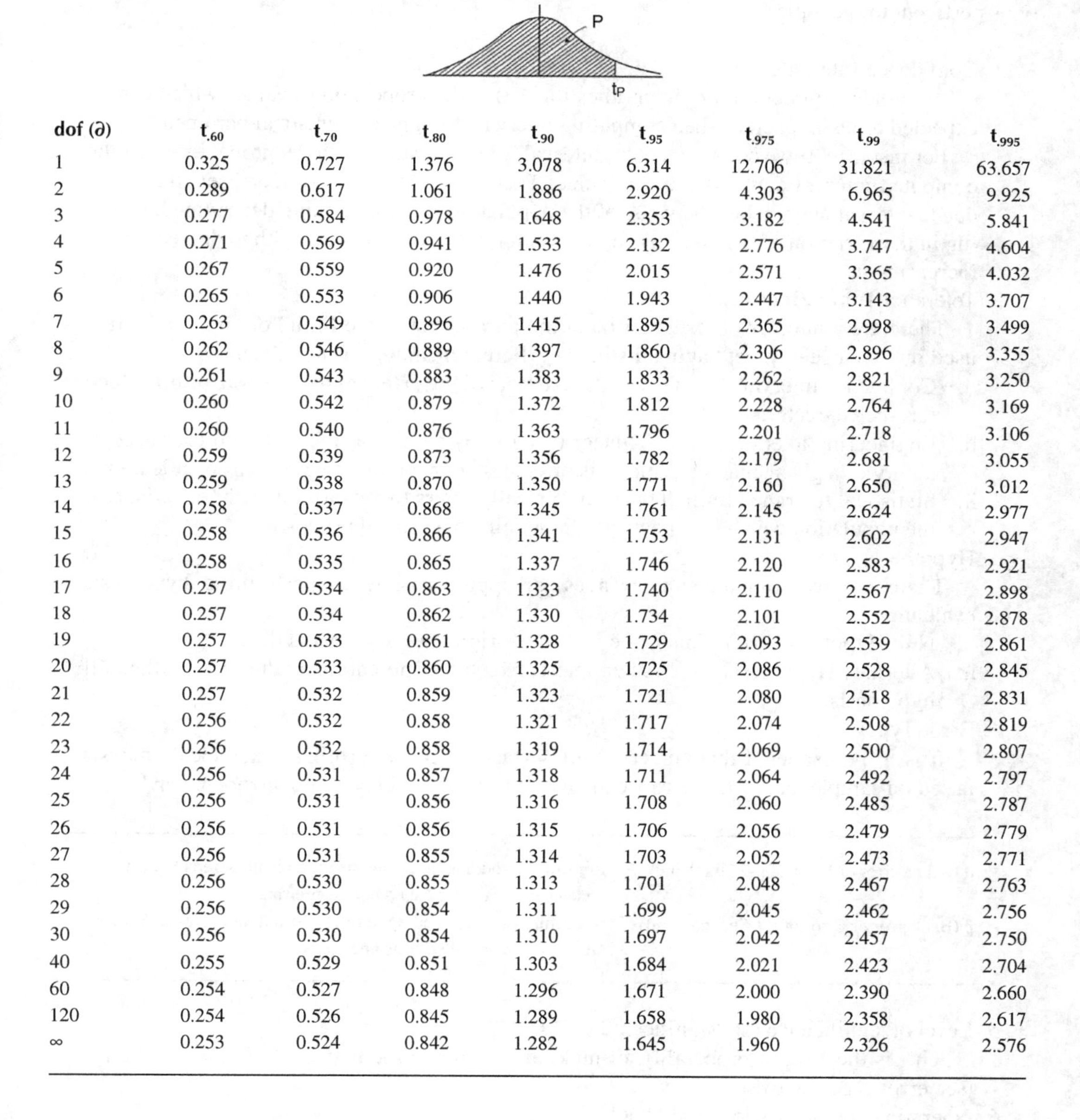

dof (∂)	$t_{.60}$	$t_{.70}$	$t_{.80}$	$t_{.90}$	$t_{.95}$	$t_{.975}$	$t_{.99}$	$t_{.995}$
1	0.325	0.727	1.376	3.078	6.314	12.706	31.821	63.657
2	0.289	0.617	1.061	1.886	2.920	4.303	6.965	9.925
3	0.277	0.584	0.978	1.648	2.353	3.182	4.541	5.841
4	0.271	0.569	0.941	1.533	2.132	2.776	3.747	4.604
5	0.267	0.559	0.920	1.476	2.015	2.571	3.365	4.032
6	0.265	0.553	0.906	1.440	1.943	2.447	3.143	3.707
7	0.263	0.549	0.896	1.415	1.895	2.365	2.998	3.499
8	0.262	0.546	0.889	1.397	1.860	2.306	2.896	3.355
9	0.261	0.543	0.883	1.383	1.833	2.262	2.821	3.250
10	0.260	0.542	0.879	1.372	1.812	2.228	2.764	3.169
11	0.260	0.540	0.876	1.363	1.796	2.201	2.718	3.106
12	0.259	0.539	0.873	1.356	1.782	2.179	2.681	3.055
13	0.259	0.538	0.870	1.350	1.771	2.160	2.650	3.012
14	0.258	0.537	0.868	1.345	1.761	2.145	2.624	2.977
15	0.258	0.536	0.866	1.341	1.753	2.131	2.602	2.947
16	0.258	0.535	0.865	1.337	1.746	2.120	2.583	2.921
17	0.257	0.534	0.863	1.333	1.740	2.110	2.567	2.898
18	0.257	0.534	0.862	1.330	1.734	2.101	2.552	2.878
19	0.257	0.533	0.861	1.328	1.729	2.093	2.539	2.861
20	0.257	0.533	0.860	1.325	1.725	2.086	2.528	2.845
21	0.257	0.532	0.859	1.323	1.721	2.080	2.518	2.831
22	0.256	0.532	0.858	1.321	1.717	2.074	2.508	2.819
23	0.256	0.532	0.858	1.319	1.714	2.069	2.500	2.807
24	0.256	0.531	0.857	1.318	1.711	2.064	2.492	2.797
25	0.256	0.531	0.856	1.316	1.708	2.060	2.485	2.787
26	0.256	0.531	0.856	1.315	1.706	2.056	2.479	2.779
27	0.256	0.531	0.855	1.314	1.703	2.052	2.473	2.771
28	0.256	0.530	0.855	1.313	1.701	2.048	2.467	2.763
29	0.256	0.530	0.854	1.311	1.699	2.045	2.462	2.756
30	0.256	0.530	0.854	1.310	1.697	2.042	2.457	2.750
40	0.255	0.529	0.851	1.303	1.684	2.021	2.423	2.704
60	0.254	0.527	0.848	1.296	1.671	2.000	2.390	2.660
120	0.254	0.526	0.845	1.289	1.658	1.980	2.358	2.617
∞	0.253	0.524	0.842	1.282	1.645	1.960	2.326	2.576

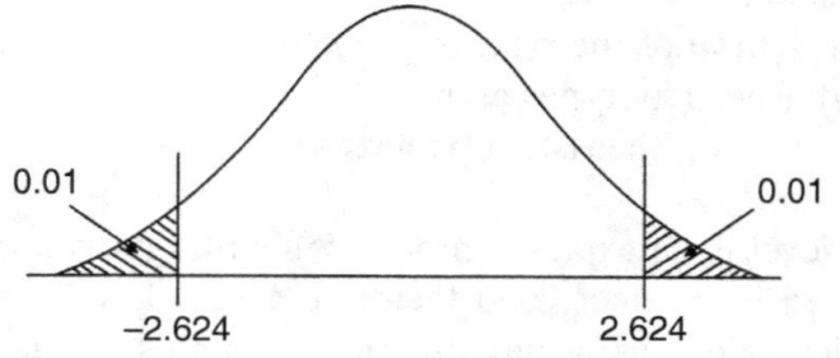

$P_R[-2.624 \leq t_{15} \leq 2.624]$

4.7 HYPOTHESIS TESTING

As an introduction to the topic of hypothesis testing, we will first review a number of basic definitions pertinent to the topic:

1. Confidence Intervals

   A confidence interval (or "interval estimate") is the proportion of values which can be expected to be bracketed when computing intervals for a given statistical parameter.

   For instance, if we compute a 90% interval we will assume with 0.9 probability that the computed interval contains the true value. Then as the sample size increases, the confidence interval approaches the value of the population parameter being determined, resulting in the variation of the confidence interval for both position and width to decrease with increasing n.
2. Tolerance Limits/Intervals

   There are a number of terms utilizing the phrase "tolerance limit" or "interval" often used interchangeably, yet having distinctly different meanings as noted here:
   i. Confidence interval: This term is the estimate of the extent of the interval within which lies the characteristic of interest.
   ii. Engineering tolerance limit: This term is usually defined as the outer limits of acceptability of a given characteristic. It is often based on engineering design specifications.
   iii. Statistical tolerance limit: This term typically refers to a measurable characteristic of the population which is expected to lie within some stated proportion.
3. Hypothesis

   There are two fundamental avenues, or approaches, when performing hypothesis evaluations.
   i. Null Hypothesis (H_0): This is the baseline, original or anticipated data.
   ii. Alternate Hypothesis (H_A): A hypothesis which is different than (alternate to) the null hypothesis.
4. Error Types

   It is to be expected that "errors" will surface when performing a statistical analysis based on sample data, these errors can be categorized as either α or β errors, given by:

α (Alpha error or Error of the First Kind)	Rejecting Good: i.e., rejecting the null hypothesized (H_0) when it is indeed correct. Also called producer error.
β (Beta error or Error of the Second Kind)	Accepting Bad: i.e., Failing to reject the null hypothesis (H_0) when it is false. Also called consumer error.

5. Level of significance (or "significant level")

   This is the risk, or probability, of making an error of the first kind; that is, an α-error (see error types, above).
6. Operating characteristic curve ("OC")

   This is a summary curve of the risk, or probability of making an error of the second kind: that is, a β-error (see error types above).

   As the risk of α decreases, the risk of β increases.
7. Significance level options

   The significance level is an expression of our unwillingness to reject the null hypothesis, i.e., we do not wish to reject good therefore a very low (or soft) significance level (say $\alpha = 0.001$) indicates our great unwillingness to unjustly reject the null hypothesis. However, since α is very low, then β increases, hence the probability of not rejecting the null hypothesis (H_0) when it is false will be large.

Typically values of $\alpha = 0.050$ and $\alpha = 0.01$ have been frequently used in research and development analysis. When $\alpha = 0.05$, the confidence interval is 95%. When $\alpha = 0.01$ there is a wider, 99% confidence interval. The wider the confidence interval, the higher the probability that the unknown parameter is contained within the given interval.

8. Tests of hypothesis

   A test for a hypothesis can be performed as follows:

   i. Formulate the hypothesis
   ii. Perform an experiment concerning the hypothesis
   iii. Classify the experimental results
   iv. Accept or reject the hypothesis

9. A more detailed protocol for hypothesis testing is presented as follows [6]:

   i. State the hypothesis—usually by stating H_0
   ii. State the alternate hypothesis H_A
   iii. Specify the value of α
   iv. Determine the statistic to be used usually based on knowledge of s and size of n

      Note that for n large ($n > 30$):

      - More accuracy, also more cost
      - Often it is impractical to obtain a large n
      - Choose n to be as large as possible within practical constraints

   v. Determine the acceptance and rejection regions of H_0
   vi. Display the resulting statistical data determined, on an appropriate sketch, following the format shown in Table 4.4.
   vii. Analyze data—Draw appropriate conclusions based upon previously determined acceptance/rejection results.

   Throughout the application of our statistical analysis of "test of hypothesis," we shall attempt to follow the approach given above for hypothesis testing. We shall initiate our studies of hypothesis testing by first reviewing and discussing the testing of means from one population.

4.8 DESIGN STUDIES

In this section we present a number of mechanical joint design-studies, and their analyses, to illustrate the use of the aforecited statistical procedures.

4.8.1 Design Study 4.1

The torque output of a fastener tool is required to be at least 75 Nm. It has been determined previously that the standard deviation of torque for this line of tools is $\sigma = 4$ Nm. A random sample of thirty-four tools is tested; the sample yields an average of $\bar{x} = 75.87$ Nm.

a. State the appropriate hypothesis for this experiment.
b. Test the hypothesis using $\sigma = 0.01$. State the conclusions.
c. Construct a 99% confidence interval on the torque output.

4.8.1.1 Solution

a.

1. H_0: $\mu = 75$ Nm
2. H_α: $\mu > 75$ Nm (Note the terminology: the term "at least" in the problem statement indicates "greater than")
3. $\alpha = 0.01$ (Level of significance)

TABLE 4.4
Parameters for the Testing of Means from One Population [7]

Constraints	Statistical Test	Hypothesis	Sketch of Hypothesis
For σ known and $n \geq 30$	Use $z = \frac{\bar{x} - \mu}{\sigma / \sqrt{n}}$	$H_0 : \mu = \mu_0$ $H_a : \mu < \mu_0$	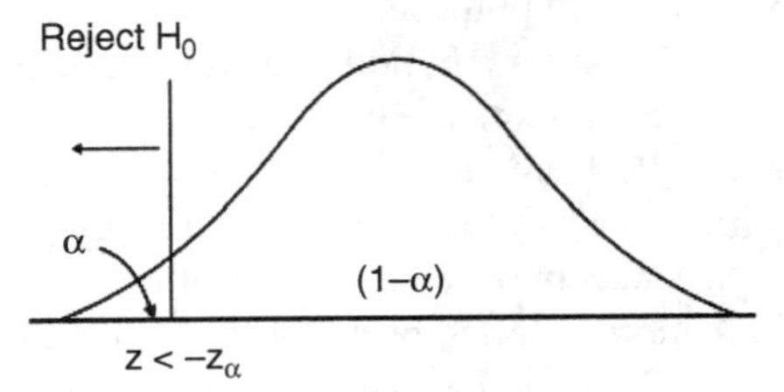
		$H_0 : \mu = \mu_0$ $H_a : \mu > \mu_0$	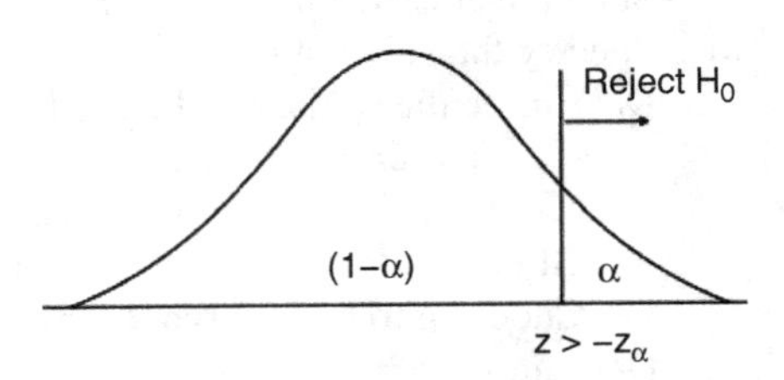
		$H_0 : \mu = \mu_0$ $H_a : \mu \neq \mu_0$ (2 tailed)	
For σ unknown and $n \leq 30$	Use $\tau = \frac{\bar{x} - \mu}{s/\sqrt{n}}$ $\vartheta = (n-1)$	$H_0 : \mu = \mu_0$ $H_a : \mu < \mu_a$	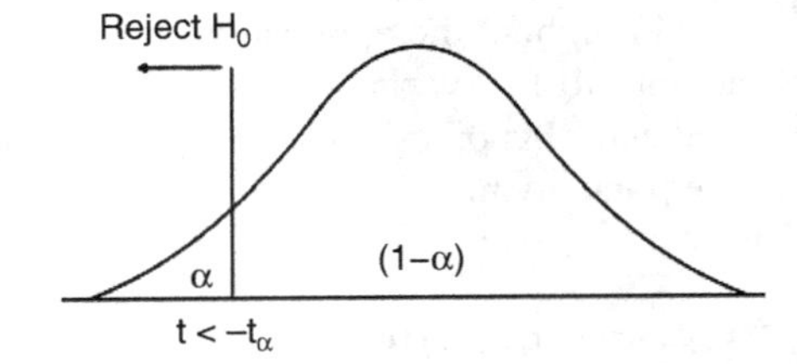
		$H_0 : \mu = \mu_0$ $H_a : \mu > \mu_0$	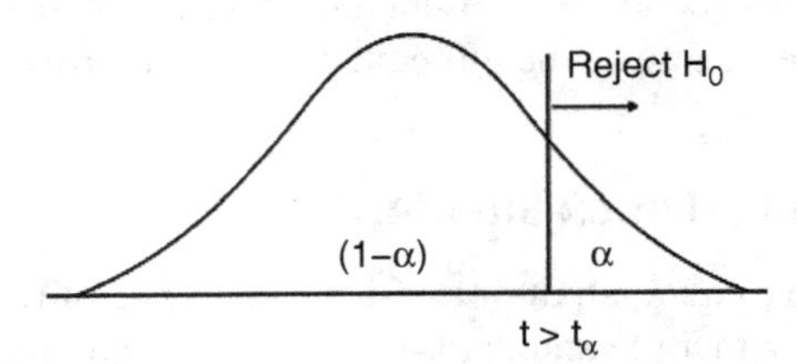
		$H_0 : \mu = \mu_0$ $H_a : \mu \neq \mu_0$ (2 tailed)	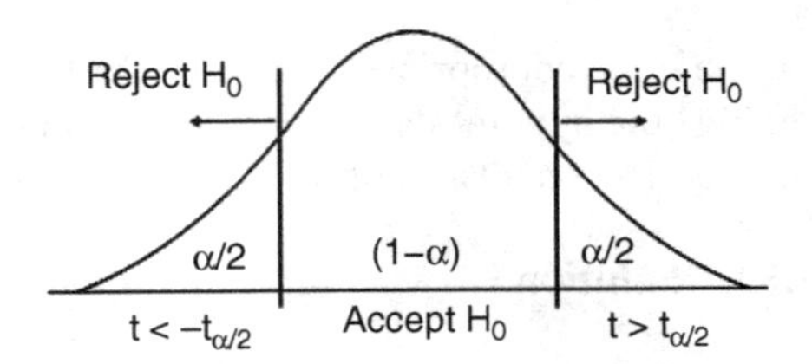

4. n is large ($n = 34$)
   σ is known, $\sigma = 4$ A Hence we use the "z score"

$$\bar{x} = 75.87$$

5. $n = 34$
6. From Table 4.2, for $\alpha = 0.01$, we obtain $Z_{0.01} = 2.326$.

$$\text{Also, we find } Z_0 = \frac{\bar{x} - \mu}{\sigma/\sqrt{n}} = \frac{75.87 - 75}{4/\sqrt{34}} = 1.268$$

We note that Z_0 is equal to 1.268, is less than 0.0, and equals 2.326.
Displaying this data on a sketch yields:

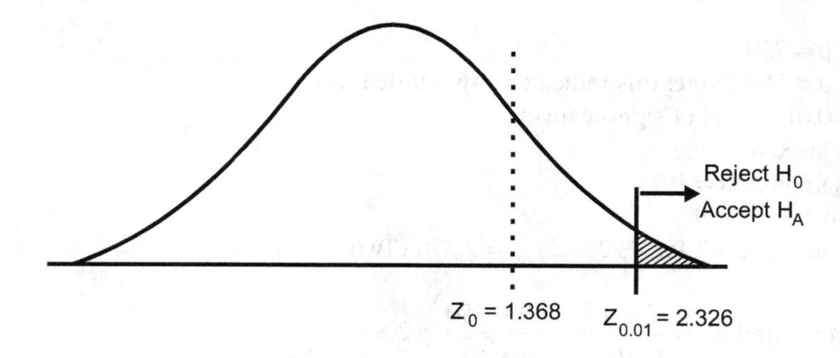

b. Hence, we do not have sufficient data to reject H_0.
   Therefore, we can say that at a level of significance of $\alpha = 0.01$ the torque output does not exceed 75 Nm.
c. The confidence interval is given by

$$\bar{x} - Z_{\alpha/2}\frac{\sigma}{\sqrt{\eta}} \leq \mu \leq \bar{x} + Z_{\alpha/2}\frac{\sigma}{\sqrt{\eta}}$$

Making use of Table 4.2 where we find for 99% $\alpha = 1.28$ and $\alpha/2 = 1.645$ yields the following:

$$75.87 - 1.645\frac{4}{\sqrt{34}} \leq \mu \leq 75.87 + 1.645\frac{4}{\sqrt{34}}$$

$$74.74 \leq \mu \leq 77$$

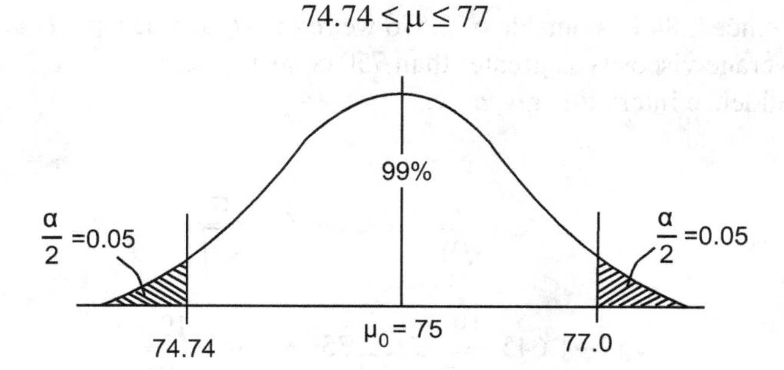

Therefore, at a 99% confidence level we can anticipate that the torque output of this fastener tool to range from 74.74 to 77.0. Stated otherwise, we are 99% confident that the unknown population-mean torque output lies within the interval of 74.74 Nm to 77.0 Nm.

4.8.2 Design Study 4.2

It is desired that the viscosity of a bolt lubricant average 750 centistokes at plant-floor temperatures. Thirty-six random samples of lubricant are collected and the average viscosity is found to be 759 centistokes. The standard deviation of the lubricant is known and given as $\sigma = 19$ centistokes. Determine if the bolt lubricant statistical average is 750 centistokes.

a. State the appropriate hypothesis for this experiment.
b. Test the hypothesis using $\alpha = 0.01$. State the conclusions.
c. Find a 99% interval on the mean.

4.8.2.1 Solution

a.
1. H_0: $\mu = 750$
2. H_α: $\mu \neq 750$ [Note: this indicates a two-tailed test]
3. $\alpha = 0.01$ (Level of significance)
4. n is large ($n = 36$)
   σ is known ($\sigma = 19$)
   $\bar{x}$ given $\bar{x} = 759$

b. Find from Table 4.2 for $Z_\alpha/2 = Z_{0.005} = 2.576$ (Two-tailed test)

$$\text{Then we find } Z_0 = \frac{\bar{x} - \mu}{\sigma/\sqrt{n}} = \frac{759 - 750}{19/\sqrt{36}} = 2.84$$

Display this data on a sketch

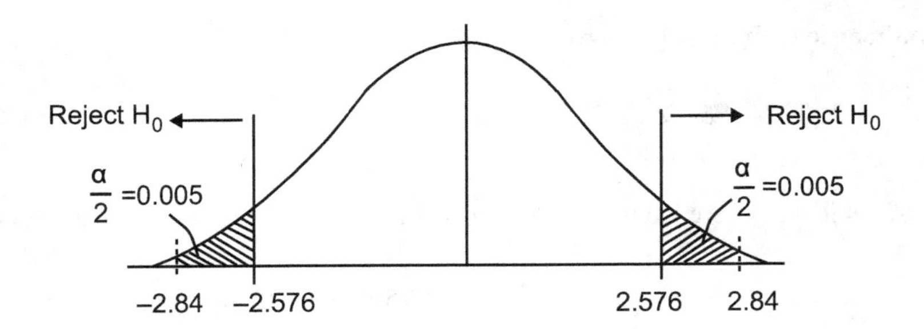

Therefore since 2.84 lies outside of 2.576 we reject H_0 and accept H_A and conclude that the sample average viscosity is greater than 750 centistokes.

c. The 99% confidence interval is given by:

$$\bar{x} - Z_{\alpha/2}\frac{\sigma}{\sqrt{\eta}} \leq \mu \leq \bar{x} + Z_{\alpha/2}\frac{\sigma}{\sqrt{\eta}}$$

$$759 - 1.645\frac{19}{\sqrt{36}} \leq \mu \leq 759 + 1.645\frac{19}{\sqrt{36}}$$

or:

$$753.79 \leq \mu \leq 764.21$$

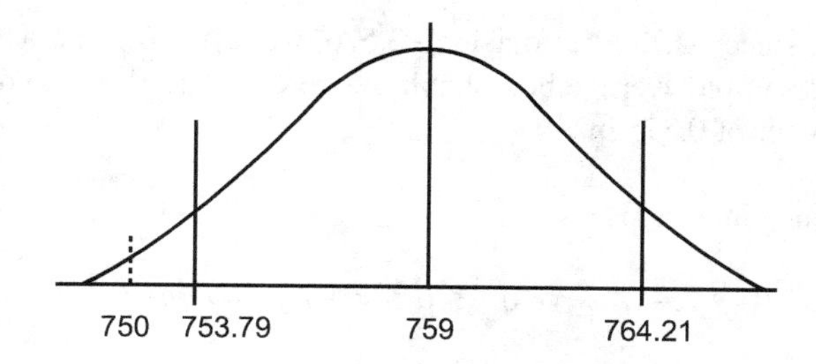

Therefore, at a confidence level of 0.99 (or we can say we are 99% confident that) the viscosity mean of the bolt lubricant samples will range from 753.79 to 764.21 centistokes.

We note that the bolt lubricant viscosity of the samples ranges from 753.79 to 764.21 centistokes, which is outside the mean of the desired population viscosity of 750 centistokes.

4.8.3 Design Study 4.3

The diameter of bolt shanks from a manufacturer should have a mean diameter of 0.375 in. The shank diameter is known to have a standard deviation of $\sigma = 0.005$ in. Thirty bolts are randomly sampled and found to have an average diameter of 0.3725 in.

a. State the appropriate hypothesis for this experiment.
b. Test the hypothesis using $\alpha = 0.01$. State the conclusions.
c. Construct a 99% interval on the mean bolt shank diameter.

Solution:

a.

1. H_0: $\mu = 0.375$
2. H_a: $\mu \neq 0.375$ [Note: this indicates a two-tailed test.]
3. $\alpha = 0.01$ (Level of significance)
4. n is large $n = 30$
   σ is *known* $\sigma = 0.005$ (use Z statistic)
   $\bar{x}$ is given $\bar{x} = 0.3725$

b.

5. Find from Table 4.2 for $Z_{\alpha/2} = Z_{0.005} = 2.576$ (Two-tailed test)

$$Z_0 = \frac{\bar{x} - \mu_0}{\sigma/\sqrt{n}} = \frac{0.3725 - 0.375}{0.005/\sqrt{30}} = -2.739$$

Displaying this data on a sketch reveals

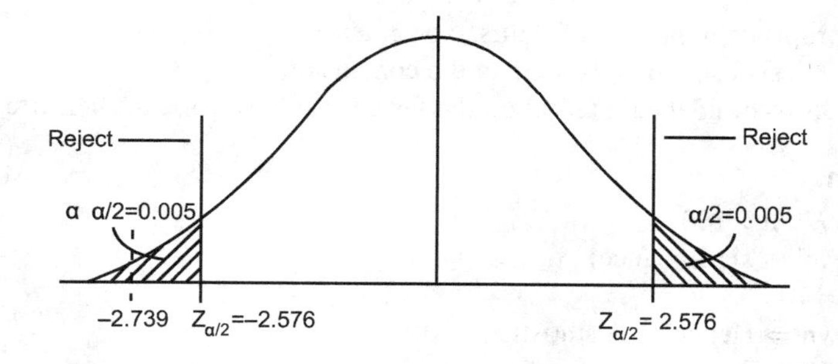

Therefore, since –2.739 lies outside –2.576, we reject H_0 (and accept H_a). We conclude that the sample average bolt shank diameter is less than the desired population average diameter of 0.375 in.

c. The 99% confidence interval is

$$\bar{x} - Z_{\alpha/2}\ \sigma/\sqrt{n} \le \mu \le \bar{x} + Z_{\alpha/2}\ \sigma/\sqrt{n}$$

$$0.3725 - 1.645\frac{0.005}{\sqrt{30}} \le \mu \le 0.3725 + 1.645\frac{0.005}{\sqrt{30}}$$

$$0.371 \le \mu \le 0.374$$

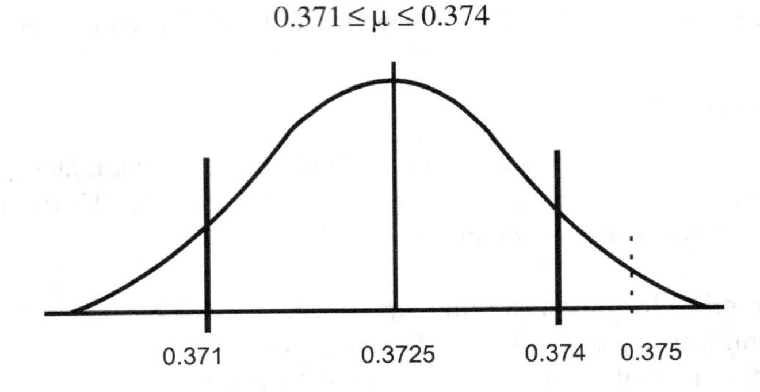

Therefore, at a confidence level of 0.99 (or we can say we are 99% confident that) we can anticipate that the mean diameter of the bolt shank sample, will lie between 0.371 and 0.374 in. We note that the population-mean bolt shank diameter of 0.375 is outside the sample mean range of diameter mean.

4.8.4 Design Study 4.4

The shelf life of an anaerobic fastener adhesive needs determination. Ten random samples of the adhesive are chosen and tested with the following results (data given in days of shelf life).

Days	168	184	186	166	171
	198	223	219	194	209

It is desired to demonstrate that the mean shelf life exceeds 6 months (taken here to be 180 days) following the protocol below.

a. State the appropriate hypothesis for this experiment.
b. Test the hypothesis using $\alpha = 0.05$. State the conclusions.
c. Construct a 95% confidence interval on the fastener adhesive mean shelf life.

4.8.4.1 Solution

1. H_0: $\mu = 180$; H_a: $\mu > 180$
2. $\alpha = 0.05$ (Level of significance)
3. n is small $n = 10$

   σ is not known ⇒ Hence use t statistic

   $\partial = n - 1 = 9$ (DOF)

4. From Table 4.3 we find for $\tau_{\alpha,\partial} = \tau_{\alpha,(n-1)} = \tau_{0.05,9} = 1.833$
5. From the raw data given we find the average

$$\bar{x} = 191.8$$

To determine the sample standard deviation, we shall first compute the sample variance (S^2):

$$S^2 = [(168-101.8)^2 + (184-191.8)^2 + (186-191.8)^2 + (166-191.8)^2 + (171-191.8)^2$$

$$+(198-191.8)^2 + (223-191.8)^2 + (219-191.8)^2 + (194-191.8)^2 + (209-191.8)^2]/(10-1)$$

$$S^2 = 3811.64/9 = 423.51$$

$$S = \sqrt{423.51} = 20.58$$

$$\tau_0 = \frac{\bar{x}-\mu}{S/\sqrt{n}} = \frac{191.8-180}{20.58/\sqrt{10}} = 1.81$$

Showing this data on a sketch reveals the following:

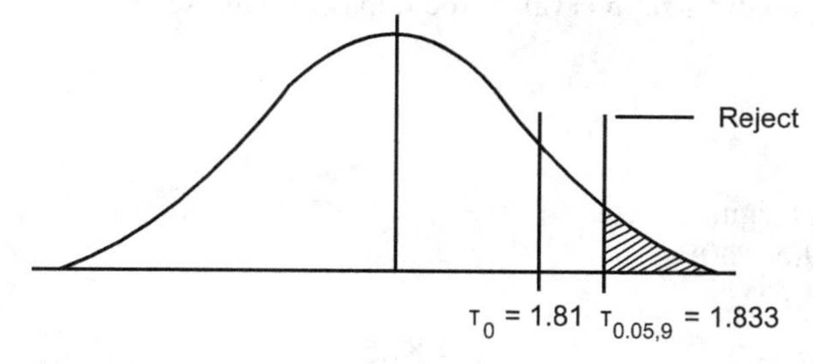

Therefore, since 1.81 lies within the bounds of 1.833, we do not reject H_0, indicating that the adhesive shelf life is less than 180 days (is not >180 days).

c. The 95% confidence interval on the adhesive shelf life is given by

$$\bar{x} = \tau_{\alpha/2,(n-1)}\ S/\sqrt{n} \le \mu \le \bar{x} + \tau_{\alpha/2,(n-1)}\ S/\sqrt{n}$$

Where we find in Table 4.3 $t_{\alpha/2,(n-1)}$ where $\alpha = .05$ and $t_{\alpha/2,(n-1)} = t_{0.025,9}$ and $t_{0.025,9} = 2.262$.

$$191.8 - (2.262)\left(20.58/\sqrt{10}\right) \le \mu \le 191.8 + (2.262)(20.58)/\sqrt{10}$$

$$177.08 \le \mu \le 206.52$$

Showing the data on a sketch reveals the following

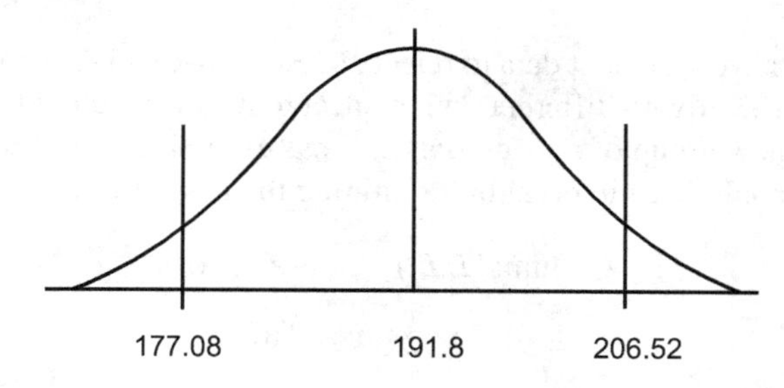

Therefore, at a confidence level of 0.95, we can anticipate that shelf life sample mean will fall between 177.08 and 206.52 days. Stated otherwise, we are 95% confident that the shelf life sample mean will range from 177.08 to 206.52 days.

4.8.5 Design Study 4.5

Historical data on impact-type fastener-tools from a manufacturer has indicated that the noise level of the tools is 89.0 dB with a standard deviation of 4.7 dB. This has been causing noise level issues in the manufacturing plant floor. The tool manufacturer's sale representative, Art Snakeoil, swears that his company has improved upon the noise problem of their impact tools and he submits a group of purportedly improved tools for evaluation.

Being dubious of Art Snakeoil's unsupported assertions, the Purchasing Department sent a sample of 49 tools to their Engineering Department for test and evaluation. This testing yielded a sample mean noise level of $\bar{x} = 87.5$ dB.

Using a significance level of $\alpha = 0.01$, the Engineering Department was tasked to demonstrate that these (supposedly) newly designed tools are indeed less noisy than their presently used tools, by following the format below:

a. State the appropriate hypothesis for this experiment
b. Test the hypothesis using $\alpha = 0.01$. State the conclusions.
c. Construct a 99% confidence interval on the impact tool noise level.

4.8.5.1 Solution

1. H_0: $\mu = 89$
2. H_a: $\mu < 89$
3. $\alpha = 0.01$ (Level of significance)
4. n is Large $n = 49(n > 30)$
   σ is known $\sigma = 4.7$

$$\bar{x} = 87.5$$

5. $n = 49$
6. Find for $\alpha = 0.01 \rightarrow Z0.01 = 2.326$

$$z = \frac{\bar{x} - \mu_0}{\sigma/\sqrt{10}} = \frac{87.5 - 89}{4.7/\sqrt{49}} = -2.23$$

Displaying the data in a sketch reveals:

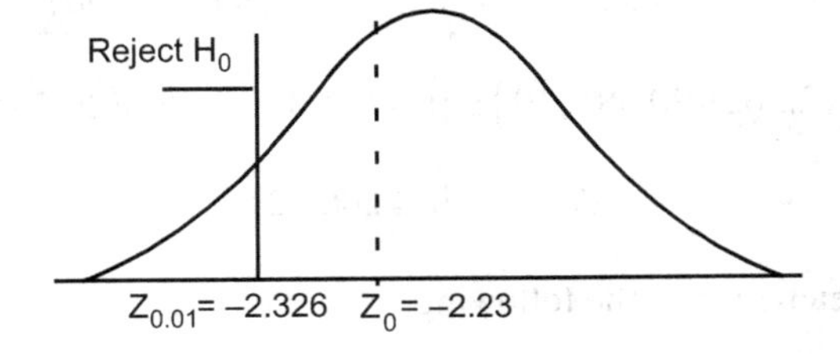

Therefore, we do not have sufficient data to reject H_0. Hence we can say that the tool sample submitted for testing is statistically no different with respect to its noise output than the original group of impact tools and the new group of impact tools still has excessive noise level issues.

Constructing a 99% confidence interval by examining the lower limit (L.L.) yields:

$$\text{Lower limit}\,(L.L.) = \mu_0 - Z_{0.01}\ \sigma/n$$

$$= 89 - 2.326\ 4.7/\sqrt{49}$$

or

$$L.L. = 87.44$$

Showing this data on a sketch yields

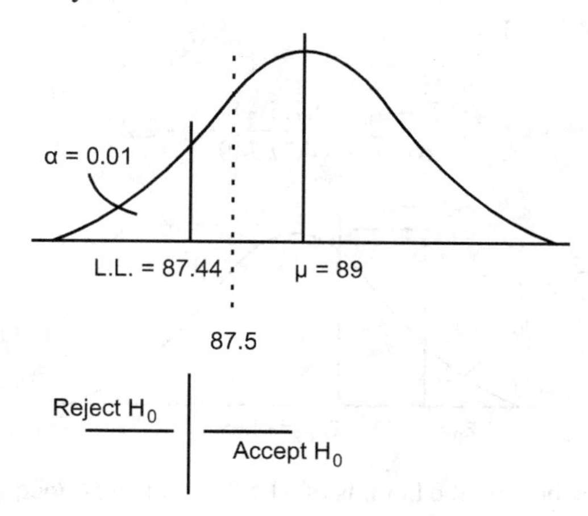

Hence, the L.L. noise level of the original impact tools (or the original data set) is 87.44. This means that at a significance level of 0.01, it is statistically anticipated that the noise level of the original tools could be as low as L.L. = 87.44 dB. The submitted purported new tool group has an average noise level of $\bar{x} = 87.5\,\text{dB}$, which is within the statistical range.

Thus, we must accept H_0: $\mu = 89$
reject H_0: $\mu < 89$

Therefore, based on a significance level of 0.01 it can be asserted that the group of tools submitted for test is statistically no better nor different in noise output than from the original group of impact tools being used.

4.8.6 Design Study 4.6

Art Snakeoil, the tool manufacturer's sale representative, after consulting his engineering group, now suggests that his new group of tools would really "pass" if only purchasing would "broaden" the significance level to $\alpha = 0.05$.

Both the Purchasing and Engineering Departments are adamant and insist on performing the analysis on the historical plant significance level of $\alpha = 0.01$. They point out to Snakeoil, the manufacturer's representative, that in a sense it is unethical or illegal in a technical sense to "shop around" for an α value so that an otherwise rejected sample can now be made to pass. If all plant or shop analyses, previously performed, traditionally used $\alpha = 0.01$, then one must ethically maintain that significance level in all analyses performed. Hence they refuse to reevaluate the tools using a $\alpha = 0.05$ and they state emphatically that the tools fail the test and should not be purchased.

As a matter of interest, however, let us determine if the tool manufacturer's sale representative is indeed correct when he alleges that his tool will "pass" at an $\alpha = 0.05$.

Reviewing the data yields:

1. H_0: $\mu = 89$
2. H_a: $\mu < 89$
3. $\alpha = 0.05$ (the "new" suggested significance level)

4. n is large ($n = 49$)
   σ is known $\sigma = 4.7$

$$\bar{x} = 87.5$$

5. $n = 49$
6. Find for $\alpha = 0.05 \rightarrow Z_{0.05} = 1.645$

$$Z = \frac{\bar{x} - \mu_0}{\sigma/\sqrt{n}} = \frac{87.5 - 89}{4.7/\sqrt{49}} = -2.23$$

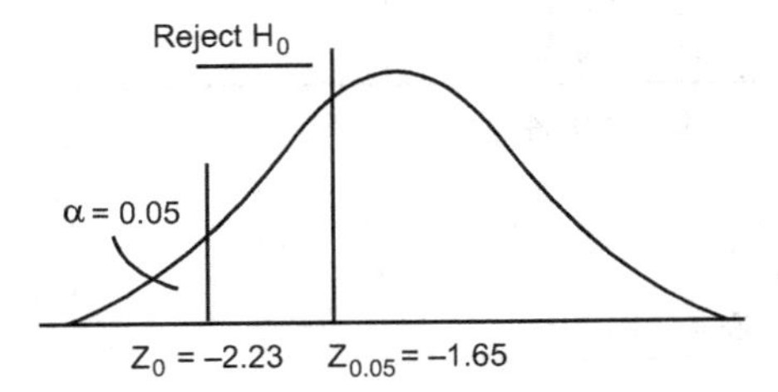

Hence, since −2.23 is beyond the bounds of −1.65, we reject H_0 (accepting H_a) and find that at $\alpha = 0.05$ the sample set of tools has a lower noise output.

Approaching the problem by determining the L.L., we find:

$$L.L. = \mu - Z_{0.05}\ \sigma/\sqrt{n}$$

$$= 89 - 1.645\ 4.7/\sqrt{49}$$

$$\text{L.L.} = 87.9$$

Displaying this data on a sketch reveals:

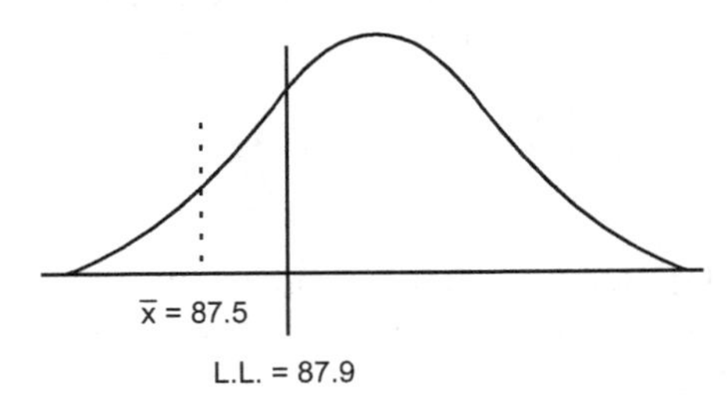

7. Analysis

   Here, we note that at a significance level of $\alpha = 0.05$ the L.L. noise level of the original tools (or original data set) is L.L. = 87.9. The submitted new tools have a noise level of $\bar{x} = 87.5$ dB (A), which is less than and outside the range of the original tool group.

   Hence, we must reject H_0: $\mu = 89$

$$\text{Accept } H_a\text{: } \mu < 89$$

Therefore, by following the tool manufacturer's sales representatives ethically challenged suggestion, and changing the level of acceptance from 0.01 to 0.05 one finds the support to assert that the "new" group of tools are indeed less noisy than the original group of impact tools.

It must be noted that further testing and analysis is usually dictated when it is found that the numbers are as "close" as represented in the examples (i.e., 87.5 v. 87.9).

4.8.7 Design Study 4.7

Art Snakeoil, the tool manufacturer's sales representative, now wishes to introduce a new line of pulse tools to the Purchasing Department. He indicates that historically his company has determined that the mean output torque of his line of pistol grip pulse tools is $\bar{x} = 42.2$ Newton-Meters (Nm) with a standard deviation $\sigma = 5.0$ Nm. However, Art Snakeoil can provide no data on the population mean.

The Purchasing Department requests that Engineering provide information on the statistical range of torque output for these pulse tools, using, in this case, a level of significance of $\alpha = 0.01$ and a sample size of $n = 32$.

Approach:

1. H_0: $\mu \neq \mu_0$ (Note that this indicates a two-tailed test.)
2. H_a: $\mu = \mu_0$
3. $\alpha = 0.01$ (Level of significance)
4. n is large ($n = 32$)
   σ is known $\sigma = 5.0$

$$\bar{x} = 42.2$$

   μ is unknown
5. $n = 32$
6. Find for $\alpha = 0.01 \rightarrow \alpha/2 = 0.005 = Z_{0.005} = 2.576$
   From Table 4.2 we find:
   $Z_{0.005} = 2.576$
   The range of tool output will be determined by the relationship:

$$\bar{x} - z_{\alpha/2} \times \sigma/\sqrt{n} < \mu < \bar{x} + z_{\alpha/2} \times \sigma/\sqrt{n}$$

The general form of the sketch for use above is given by:

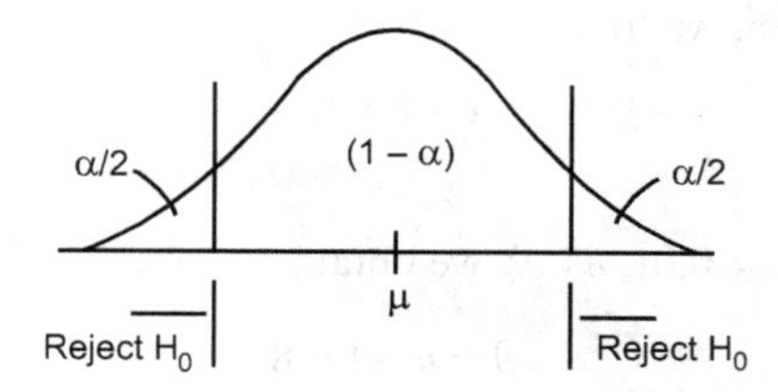

And we find

$$42.2 - (2.576)\frac{5.0}{\sqrt{32}} < \mu < 42.2 + (2.576)\frac{5.0}{\sqrt{32}}$$

or

$$39.92 < \mu < 44.48$$

Sketching this data yields:

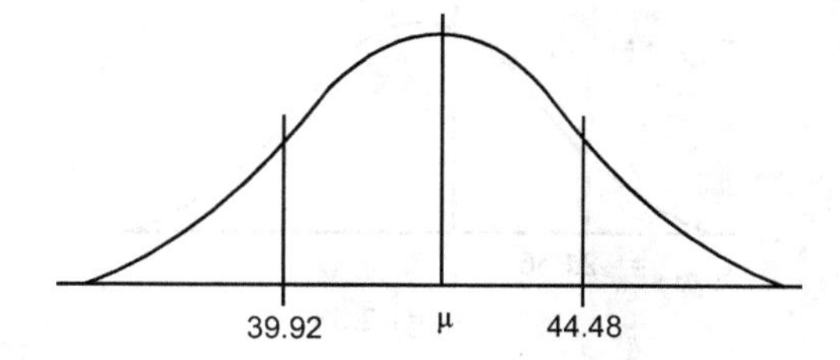

That is, based on the data provided the mean will range between 39.92 and 44.48 Nm, as shown in the sketch.

Therefore, at a confidence level of 0.01 we can anticipate the torque output of this group of pistol grip pulse tools to range from 39.92 to 44.48 Nm. Stated differently, we can say we are 99% confident that the unknown population mean output torque lies within the interval of 39.92–44.48 Nm.

4.8.8 Design Study 4.8

In an effort to overcome the noise problem presented by his company's impact tools (recall that $\mu = 89$ dB) the manufacturer's sale representative Art Snakeoil now suggests replacing the old line of noisy impact tools with his new line of quieter pistol grip impulse tools. Unfortunately, he does not have many samples to provide for testing. Nine tools are tested with the following data provided as follows (in units of dB): 85.1, 90.1, 86.3, 85.4, 89.3, 88.4, 86.3, 89.4, and 86.2.

The Purchasing Department now asks the Engineering Department to determine if the sound level of the pulse tool samples is significantly reduced from that of the impact tools presently being used.

The Engineering Department, using a significance level of $\alpha = 0.01$ analyzes the noise level of the sample of nine tools by following the format:

a. State the appropriate hypothesis
b. Test the hypothesis, using $\alpha = 0.01$
c. Construct a 99% confidence interval on the tool noise level.

1. H_0: $\mu = 89$
2. H_a: $\mu < 89$
3. $\alpha = 0.01$ (level of significance)
4. n is small ($n < 30$, $n = 9$)
5. σ is unknown $\Rightarrow$ use the t test
6. $n = 9$
7. From the sample data, we find

$$\bar{x} = 87.39$$

$S = 1.91$

From Table 4.3 for $\alpha = 0.01$, $n = 9$, we obtain

$$\partial = n - 1 = 8$$

and

$$t_{0.01,8} = 2.896$$

Also we find

$$t_0 = \frac{\bar{x} - \mu_0}{S/\sqrt{n}} = \frac{87.39 - 89}{1.91/\sqrt{9}} = -2.53$$

Displaying this data on a sketch reveals

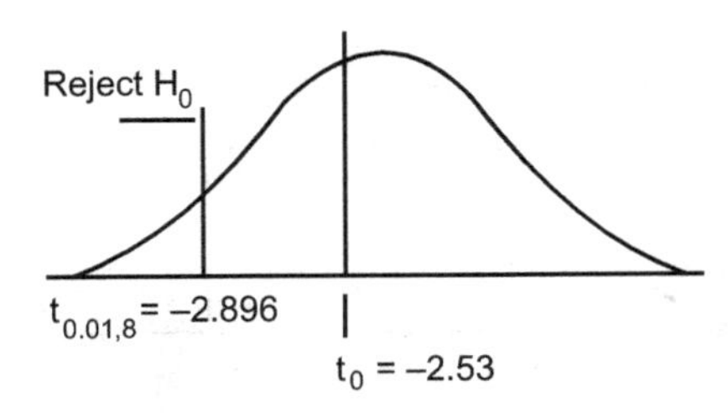

Hence, since the value of –2.53 lies within the bounds of –2.896, we accept H_0, and therefore reject H_A which indicates that the new (albeit small) sample of tools, at a significance level of 0.01, is statistically similar to the original noisy tools with respect to noise. Evaluating the same problem by examining its L.L. yields:

$$LL = \mu_0 - t_{0.01,8}\,\frac{S}{\sqrt{n}} = 89 - 2.896\,\frac{(1.91)}{\sqrt{9}}$$

$$LL = 87.16$$

Displaying the data on a sketch yields

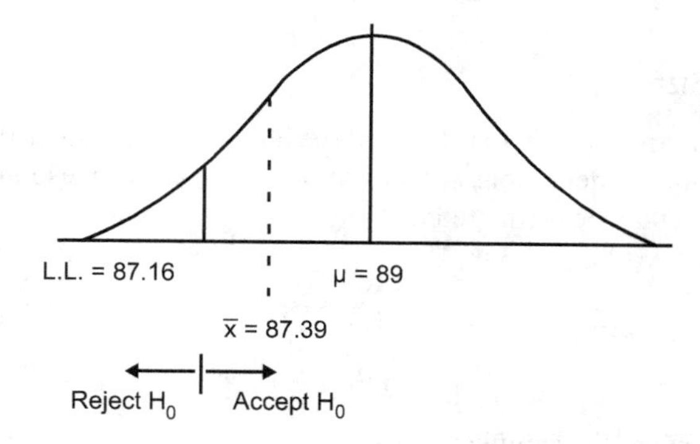

8. Since $\bar{x} = 87.39 > LL = 87.16$
   Accept H_0: $\mu = 89$
   Reject H_0: $\mu < 89$

We conclude, therefore, that the means of the population from which the nine samples were randomly taken (i.e., the pulse tool samples) will have no improvement in the noise attenuation and is therefore no different or better statistically than the noise level of the original population of noisy impact tools.

4.9 HYPOTHESIS TESTING BASED ON A TWO-POPULATION DISTRIBUTION

In the foregoing sections of this chapter, we considered various parameters for testing a single population. Often, however, we will find it of interest to compare two, or more, processes, groups, or populations. In this section, we consider a methodology for hypothesis testing of two different populations or distributions.

Often, when comparing two processes, especially in the instance of before and after testing, it is useful to define a new distribution by combining parameters of the two processes.

Specifically, when "adding" two large distributions, the mean μ and standard deviation σ of the combined population are:

$$\mu = \mu_1 + \mu_2 \tag{4.8}$$

and

$$\sigma^2 = \sigma_1^2 + \sigma_2^2 \tag{4.9}$$

where the subscripts 1 and 2 refer to the two respective distributions.

Similarly, when subtracting two large distributions, the mean and standard distribution of the resulting distribution are:

$$\mu = \mu_1 - \mu_2 \tag{4.10}$$

and

$$\sigma^2 = \left|\sigma_1^2 - \sigma_2^2\right|$$

$$\sigma_{x_1 - x_2} = \sqrt{\frac{\sigma_1^2}{n_1}} - \sqrt{\frac{\sigma_2^2}{n_2}} \tag{4.11}$$

4.9.1 Large Sample Sizes

For large sample sizes (n_2 and $n_2 > 30$) the standard deviations of the two distributions σ_1 and σ_2 are well estimated by the standard deviations s_1 and s_2 of the samples. Correspondingly the Z-score for the combined distribution may be estimated as:

$$Z = \frac{\bar{x}_1 - \bar{x}_2}{\sqrt{\frac{s_1^2}{n_1} + \frac{s_2^2}{n_2}}} \tag{4.12}$$

where, as before, $\bar{x}_1$ and $\bar{x}_2$ are the sample means.

Also, we have:

$$\text{Lower Limit} = \text{LL} = Z_{1-\alpha}\sqrt{\frac{s_1^2}{n_1} + \frac{s_2^2}{n_2}} \quad (\text{always negative}) \tag{4.13}$$

and

$$\text{Upper Limit} = \text{UL} = Z_{\alpha}\sqrt{\frac{s_1^2}{n_1} + \frac{s_2^2}{n_2}} \quad (\text{always positive}) \tag{4.14}$$

4.9.2 Small Sample Sizes

For small samples n_1 and $n_2 < 30$

$$t = \frac{\bar{x}_1 - \bar{x}_2}{\sqrt{(n_1 + 1)s_1^2 + (n_2 - 1)s_2^2}}\sqrt{\frac{n_1 n_2 (n_1 n_2 - 2)}{n_1 + n_2}}$$

Where the sample size is defined as $n_1 n_2 - 1$

$$\text{Lower Limit L.L.} = t_{n_1 + n_2 - 1, \alpha}\sqrt{\frac{(n_1 + n_2)\left[(n_1 - 1)s_1^2 + (n_2 - 1)s_2^2\right]}{n_1 n_2 (n_1 + n_2 - 2)}}$$

$$\text{Upper Limit U.L.} = t_{n_1 + n_2 - 1, \alpha}\sqrt{\frac{(n_1 + n_2)\left[(n_2 - 1)s_1^2 + (n_2 - 1)s_2^2\right]}{n_1 n_2 (n_1 + n_2 - 2)}}$$

The criteria for the testing of means from two populations are shown in Table 4.5. Applications of these criteria will be demonstrated in the following design sets.

TABLE 4.5
Summary Table of Tests Concerning Means

H_0	Value of Test Statistic	H_1	Critical Region
$\mu = \mu_0$	σ known or $n \geq 30$ $z = \dfrac{\bar{x} - \mu_0}{\sigma/\sqrt{n}}$	$\mu < \mu_0$	$z < -z_\alpha$
		$\mu > \mu_0$	$z > z_\alpha$
		$\mu \neq \mu_0$	$z < -z_{\alpha/2}$ and $z > z_{\alpha/2}$
$\mu = \mu_0$	σ unknown and $n < 30$ $t = \dfrac{\bar{x} - \mu_0}{s/\sqrt{n}};\ \nu = n - 1$	$\mu < \mu_0$	$t < -t_\alpha$
		$\mu > \mu_0$	$t > t_\alpha$
		$\mu \neq \mu_0$	$t < -t_{\alpha/2}$ and $t > t_{\alpha/2}$
$\mu_1 - \mu_2 = d_0$	σ_1 and σ_2 known $z = \dfrac{(\bar{x}_1 - \bar{x}_2) - d_0}{\sqrt{(\sigma_1^2/n_1) + (\sigma_2^2/n_2)}}$	$\mu_1 - \mu_2 < d_0$	$z < -z_\alpha$
		$\mu_1 - \mu_2 > d_0$	$z > z_\alpha$
		$\mu_1 - \mu_2 \neq d_0$	$z < -z_{\alpha/2}$ and $z > z_{\alpha/2}$
$\mu_1 - \mu_2 = d_0$	$\sigma_1 = \sigma_2$ but unknown $t = \dfrac{(\bar{x}_1 - \bar{x}_2) - d_0}{s_p\sqrt{(1/n_1) + (1/n_2)}}$ $\nu = n_1 + n_2 - 2$ $s_p^2 = \dfrac{(n_1 - 1)s_1^2 + (n_2 - 1)s_2^2}{n_1 + n_2 - 2}$	$\mu_1 - \mu_2 < d_0$	$t < -t_\alpha$
		$\mu_1 - \mu_2 > d_0$	$t > t_\alpha$
		$\mu_1 - \mu_2 \neq d_0$	$t < -t_{\alpha/2}$ and $t > t_{\alpha/2}$
$\mu_1 - \mu_2 = d_0$	$\sigma_1 \neq \sigma_2$ and unknown $t' = \dfrac{(\bar{x}_1 - \bar{x}_2) - d_0}{\sqrt{(s_1^2/n_1) + (s_2^2/n_2)}}$ $\nu = \dfrac{(s_1^2/n_1 + s_2^2/n_2)^2}{\dfrac{(s_1^2/n_1)^2}{n_1 - 1} + \dfrac{(s_2^2/n_2)^2}{n_2 - 1}}$	$\mu_1 - \mu_2 < d_0$	$t' < -t_\alpha$
		$\mu_1 - \mu_2 > d_0$	$t' > t_\alpha$
		$\mu_1 - \mu_2 \neq d_0$	$t' < -t_{\alpha/2}$ and $t' > t_{\alpha/2}$
$\mu_D = d_0$	Paired observations $t = \dfrac{\bar{d} - d_0}{s_d\sqrt{n}};\ \nu = n - 1$ $\bar{d}$ = mean of the normally distributed differences s_d = standard deviation of the Normally distributed differences	$\mu_D < d_0$	$t < -t_\alpha$
		$\mu_D > d_0$	$t > t_\alpha$
		$\mu_D \neq d_0$	$t < -t_{\alpha/2}$ and $t > t_{\alpha/2}$

4.10 SUMMARY TABLE OF TESTS CONCERNING MEANS [8]

Table 4.5 presents a summary of the test statistic approach, given various criteria of the two populations being analyzed. Also included are the criteria of testing for one population mean previously discussed, presented again here for comparison.

4.11 DESIGN STUDIES OF TWO POPULATION MEANS

In this section, as in Section 4.8, we present a small number of design studies focusing on the analysis of two population means.

4.11.1 Design Study 4.9

Art Snakeoil, the tool manufacturer's sale representative, insists that the tools he sells are more accurate than those of his competitor and he challenges the Purchasing Department to compare the fastener tools from two plants. Plant "A" where the competitor's tools are used and Plant "B" where the Art Snakeoil's tools are used. Fastening defect data has been collected in both plants on a monthly basis for the past 3 years. However, Plant "B" started collecting data 4 months later than Plant "A." The monthly number of defects was recorded as:

Plant A Fastening Defects (Competitor's Tools)	Plant B Fastening Defects (Snakeoil's Tools)
$n_1 = 36$	$n_2 = 32$
$\bar{x}_1 = 16.1$	$\bar{x}_2 = 14.3$
$S_1 = 4.3$	$S_2 = 5.1$

The Purchasing Department notes that an initial review of this data as shown above provides insufficient information to make a buying decision. Therefore, Purchasing requested that the Engineering Department provide a statistical analysis to properly evaluate this comparison by following the format below:

a. State the appropriate hypothesis
b. Test the hypothesis using $\alpha = 0.05$
c. Construct a 95% confidence interval on the defect comparison

Approach:

1. H_0: $\mu_1 = \mu_2$
2. H_a: $\mu_1 > \mu_2$
3. $\alpha = 0.05$ Art Snakeoil, the manufacturer's rep, requests analysis at $\alpha = 0.05$ ($Z_{0.05} = 1.645$)
4. n_1 and n_2 are Large (n_1 and $n_2 > 30$)
5. $n_1 = 36$ and $n_2 = 32$ where n is large, σ can be estimated by S
6. To disprove $H_0 : (\bar{x}_1 - \bar{x}_2) > 0$
   Use the Upper Limit (U.L.) to describe the comparison difference confidence interval.

$$U.L. = Z_{\alpha}\sqrt{\frac{\sigma_1^2}{n_1} + \frac{\sigma_2^2}{n_2}} = 1.645\sqrt{\frac{4.3^2}{36} + \frac{5.1^2}{32}} = 1.645\sqrt{\frac{18.49}{36} + \frac{26.01}{32}} = 1.895$$

7. Note that $(\bar{x}_1 - \bar{x}_2) = 16.1 - 14.3 = 1.8$

   Displaying the data above yields:

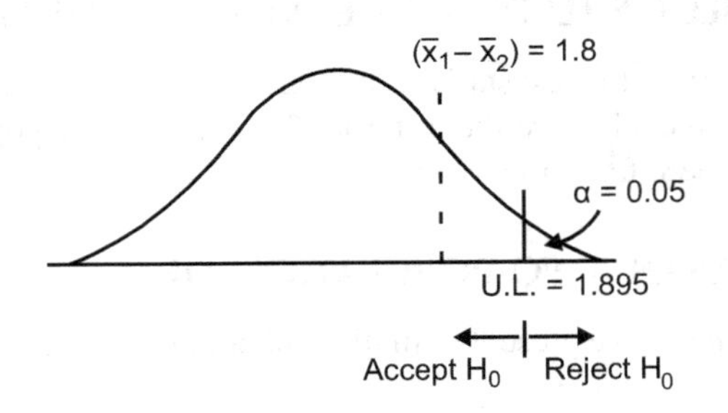

We must, therefore, conclude that the test statistic $(\bar{x}_1 - \bar{x}_2) = 1.8$ is less than the U.L. = 1.895 and therefore is in the acceptance region and we therefore accept H_0. Hence, at the significance level of $\alpha = 0.05$, there is no statistical difference between the creation of fastening defects of the two representative tool data sets. However, when the numbers are as "close" as shown, further testing and analysis is typically warranted.

4.11.2 Design Study 4.10

Two separate assembly lines in the same manufacturing plant are utilized to apply 18 Nm torque to threaded fasteners. The torque process can be assumed to be normal, with standard deviations of $\sigma_1 = 0.019$ and $\sigma_2 = 0.021$. Upper management and the legal office insist that both assembly lines must torque to the same value, whether or not this torque value is 18 Nm. Due to a number of serious fastener failures occurring in the plant's product, an outside expert is retained to evaluate the torque process reliability. He takes random samples from each line and evaluates the torque on each line to find the following:

Assembly Line 1: $\bar{x}_1 = 18.259$; $\sigma_1 = 0.019$; $n_1 = 32$
Assembly Line 2: $\bar{x}_2 = 18.250$; $\sigma_2 = 0.021$; $n_2 = 32$

He follows the format below when performing the statistical experiment:

a. State the appropriate hypothesis for this experiment.
b. Test the hypothesis using $\alpha = 0.01$. State the conclusions.
c. Determine the 99% confidence interval on the difference in mean torque level for the two assembly lines.

Approach:

1. H_0: $\mu_1 = \mu_2$ H_a: $\mu_1 \neq \mu_2$ (Indicates a Two-Tailed test)
2. $\alpha = 0.01$
3. $\bar{x}_1 = 18.259$ $\bar{x}_2 = 18.250$ From the given data
   $\sigma_1 = 0.019$ $\sigma_2 = 0.021$ σ_1, σ_2 known
   $n_1 = 32$, $n_2 = 32$ n large A use Z statistic
   From Table 4.2 find for $Z_{a/2} = Z_{0.05} = 1.645$

$$Z_0 = \frac{\bar{x}_1 - \bar{x}_2}{\sqrt{\frac{\sigma_1^2}{n_1} + \frac{\sigma_1^2}{n_2}}} = \frac{18.259 - 18.250}{\sqrt{\frac{0.019^2}{32} + \frac{0.021^2}{32}}} = 1.798$$

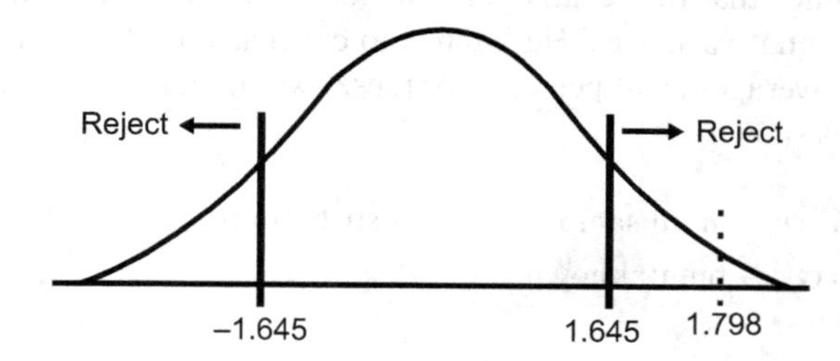

Hence, since z_0 = 1.798 is outside the bound of 1.645, we reject H_0. Hence, we note that there is a statistical difference between the two assembly lines, contrary to the insistence of upper management and the corporate attorneys. The 99% confidence interval on the difference on the torque level for the lines is given by

$$(\bar{x}_1 - \bar{x}_2) - Z_{\alpha/2}\sqrt{\frac{\sigma_1^2}{n_1} + \frac{\sigma_2^2}{n_2}} \le (\mu_1 - \mu_2) \le (\bar{x}_1 - \bar{x}_2) + Z_{\alpha/2}\sqrt{\frac{\sigma_1^2}{n_1} + \frac{\sigma_2^2}{n_2}}$$

Substituting for values then yields

$$(18.259 - 18.250) - 1.645\sqrt{\frac{0.019^2}{32} + \frac{0.021^2}{32}} \le (\mu_1 - \mu_2) \le (18.259 - 18.250) + 1.645\sqrt{\frac{0.019^2}{32} + \frac{0.021^2}{32}}$$

$$0.0008 \le (\mu_1 - \mu_2) \le 0.0174$$

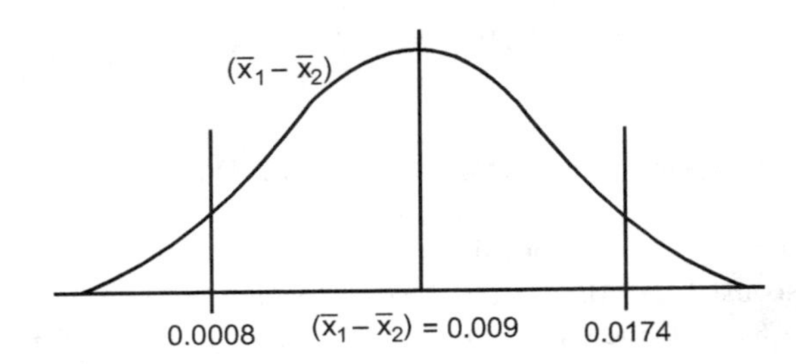

Hence, at a confidence level of 0.01, we can anticipate that the sample combined mean will fall between 0.0008 and 0.0174. Stated otherwise, we are 99% confident that the combined mean difference will range from 0.0008 to 0.0174.

4.11.3 Design Study 4.11

Dorf Motors, a vehicle manufacturer, is interested in performing a fastener integrity reliability study of their vehicles. Dorf Motors retains a consultant to compare their fastener recall data against Fota Motors fastener recall data as presented in the NHTSA website. Eight Dorf and five Fota vehicle platforms were chosen from 20 years of NHTSA recall data for analysis. A review of the data indicates that the eight Dorf vehicle types averaged 7 recalls per year with a standard deviation of 1.5 recalls per year and the five Fota vehicle platforms averaged 4 recalls per year with a standard deviation of 0.75 recalls per year.

The consultant assumes that the recalls per year for each vehicle type are approximately normally distributed with equal variances. He intends to construct a 90% confidence interval for the difference between the average recall per year for these two manufacturers' vehicle types. He then sets forth his analysis as shown:

1. $n_1 = 8 = n_D$ $n_2 = 5 = n_F$ (small sample—use t distribution)
2. $\bar{x}_1 = 7$ $\bar{x}_2 = 4$ $\sigma_1 = \sigma_2 \rightarrow$ but unknown
3. $S_1 = 1.5$ $S_2 = 0.75$
4. $\alpha = 0.1$
5. $\partial = n_1 + n_2 - 2 \quad 8 + 5 - 2 = 11$

6. $S_p^2 = \frac{(n_1-1)S_1^2+(n_2-1)S_2^2}{n_1+n_2-2} = \frac{(8-1)(1.5)^2+(5-1)(0.75)^2}{8+5-2} = \frac{15.75+2.25}{11} = 1.636$

   $S_p = 1.279$

7. $t_{\alpha/2,\partial} = t_{0.05,11} = 1.796$ (From Table 4.3)

8. $$(\bar{x}_1-\bar{x}_2)-t_{\alpha/2,\partial}S_p\left[\frac{1}{n_1}+\frac{1}{n_2}\right]^{\frac{1}{2}} < (\mu_1-\mu_2) < (\bar{x}_1-\bar{x}_2)-t_{\alpha/2,\partial}S_P\left[\frac{1}{n_1}+\frac{1}{n_2}\right]^{\frac{1}{2}}$$

   $$(7-4)-1.796(1.279)\left[\frac{1}{8}+\frac{1}{5}\right]^{\frac{1}{2}} < (\mu_1-\mu_2) < (7-4)+1.796(1.279)\left[\frac{1}{8}+\frac{1}{5}\right]^{\frac{1}{2}}$$

9. $1.691 < (\mu_1-\mu_2) < 4.309$

Therefore, the average vehicle recall difference for the vehicle samples from the two manufacturers at a 90% confidence level will statistically range from 1.691 to 4.309.

4.11.4 Design Study 4.12

A defense contractor has submitted to their customer, samples of a vehicle model for evaluation. The defense contractor was soon informed by their customer that most of the vehicles submitted were found to have many of the critical fasteners loosen as a result of some preliminary off-road testing. A consultant is engaged and he suggests that submitting the prototype vehicles to whole body shaker tests will cause the initially defectively tightened critical fasteners to loosen an average of 4.5 N-m. This, the consultant claims, will minimize the "infant mortality" problem by identifying the incorrectly tightened fasteners prior to delivery to the customer. The consultant is authorized "to prove his allegation" and he subjects seven vehicles to whole body testing and he records the critical fastener torque levels before and after the whole body vibration test. He intends to perform a statistical analysis based on a "difference" approach.

	Vehicles Tested						
	1	2	3	4	5	6	7
Critical-Fastener torque before shaker tests (Nm)	58.5	60.3	61.7	69.0	64.0	62.6	56.7
Critical-Fastener torque after shaker tests (Nm)	60.0	54.9	58.1	62.1	58.5	59.9	54.4

The consultant intends to support his claim by comparing a 95% confidence interval for the mean "difference" of fastener torque.

Assume the difference of fastener torques to be approximately normally distributed.

1. Approach: Vehicles 1 2 3 4 5 6 7
   Differences (D) −1.5 5.4 3.6 6.9 5.5 2.7 2.3
2. Data Yields: $D_{AVG} = \bar{D} = 3.557$

   $S_D = 2.776$
3. $\partial = n-1 = 6$ $\alpha = 0.05; \rightarrow \alpha/2 = 0.025$
4. $t_{\alpha/2,\partial} = t_{0.025,6} = 2.447$ (From Table 4.3)

5. $\bar{D} - t_{\alpha/2,D}\dfrac{S_D}{\sqrt{n}} < \mu_D < \bar{D} + t_{\alpha/2,D}\dfrac{S_D}{\sqrt{n}}$

$$3.557 - 2.447\frac{2.776}{\sqrt{7}} < \mu_D < 3.557 + 2.447\frac{2.776}{\sqrt{7}}$$

6. $0.990 < \mu < 6.124$

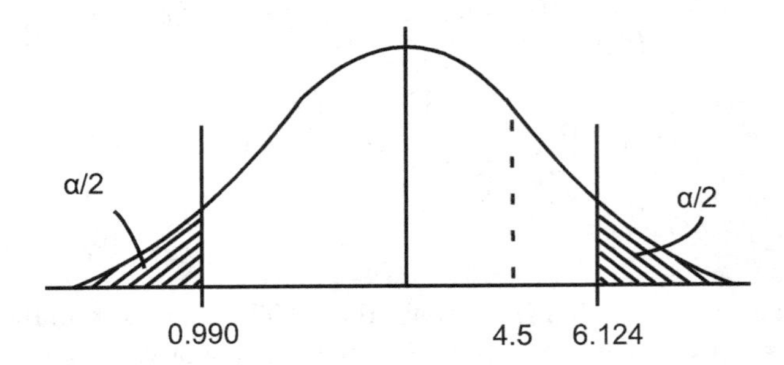

Since the estimated 4.5 Nm torque reduction (clamp load loss) falls within the calculated range, it can be asserted that the consultant's claim is valid. That is, there will be a measureable loss of clamp load due to whole body shaker tests, which will permit identifying the improperly tightened critical fasteners.

4.12 TESTING VARIANCES FROM ONE POPULATION: USE OF THE CHI-SQUARE (χ^2) DISTRIBUTION

Instead of examining the differences in the mean, we are often interested in examining the variability in the data. This will permit us to examine tests of hypothesis and confidence intervals for variances of normal distributions. It should be noted that the tests on variances are sensitive to the assumptions of normality, unlike the Z and t test, which are somewhat insensitive to normality deviations.

Given a random sample of size n, taken from a *normal* population variance σ^2, then the chi-square distribution is given by

$$\chi^2 = \left[(n-1)S^2\right]/\sigma^2 \tag{4.16}$$

Where the DOF $\partial = n - 1$

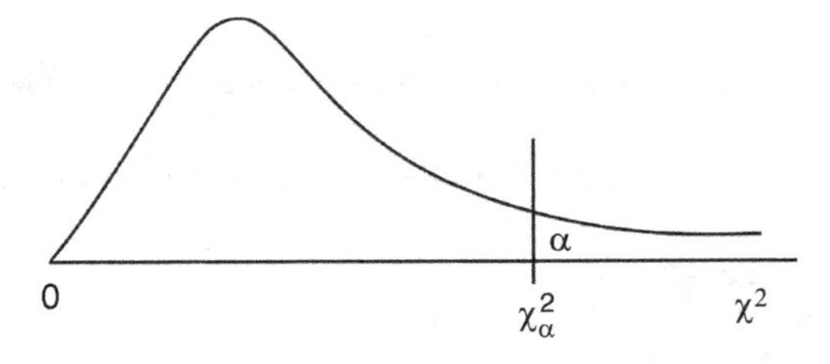

The chi-square (χ^2) distribution is useful to test the variance from a *single* population. (It is important to observe that unlike the Z and t distributions, the chi-square distribution as seen above is not symmetric.)

Table 4.6 provides a list of numerical values for the chi-square distribution.

TABLE 4.6
Critical Values of the Chi-Square Distribution [9] Values of χ^2_P Corresponding to *P*

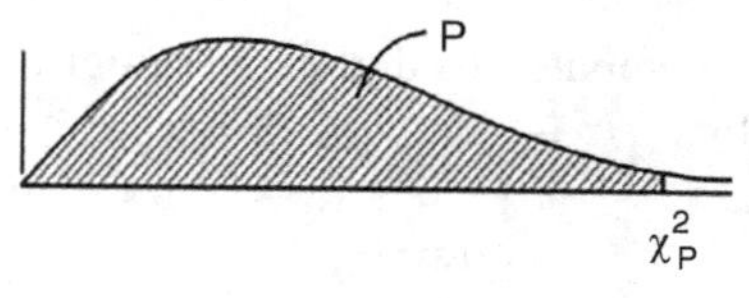

Df	$\chi^2_{.005}$	$\chi^2_{.01}$	$\chi^2_{.025}$	$\chi^2_{.05}$	$\chi^2_{.10}$	$\chi^2_{.90}$	$\chi^2_{.95}$	$\chi^2_{.975}$	$\chi^2_{.99}$	$\chi^2_{.995}$
1	.000039	.00016	.00098	.0039	.0158	2.71	3.84	5.02	6.63	7.88
2	.0100	.0201	.0506	.1026	.2107	4.61	5.99	7.38	9.21	10.60
3	.0717	.115	.216	.352	.584	6.25	7.81	9.35	11.34	12.84
4	.207	.297	.484	.711	1.064	7.78	9.49	11.14	13.28	14.86
5	.412	.554	.831	1.15	1.61	9.24	11.07	12.83	15.09	16.75
6	.676	.872	1.24	1.64	2.20	10.64	12.59	14.45	16.81	18.55
7	.989	1.24	1.69	2.17	2.83	12.02	14.07	16.01	18.48	20.28
8	1.34	1.65	2.18	2.73	3.49	13.36	15.51	17.53	20.09	21.96
9	1.73	2.09	2.70	3.33	4.17	14.68	16.92	19.02	21.67	23.59
10	2.16	2.56	3.25	3.94	4.87	15.99	18.31	20.48	23.21	25.19
11	2.60	3.05	3.82	4.57	5.58	17.28	19.68	21.92	24.73	26.76
12	3.07	3.57	4.40	5.23	6.30	18.55	21.03	23.34	26.22	28.30
13	3.57	4.11	5.01	5.89	7.04	19.81	22.36	24.74	27.69	29.82
14	4.07	4.66	5.63	6.57	7.79	21.06	23.68	26.12	29.14	31.32
15	4.60	5.23	6.26	7.26	8.55	22.31	25.00	27.49	30.58	32.80
16	5.14	5.81	6.91	7.96	9.31	23.54	26.30	28.85	32.00	34.27
18	6.26	7.01	8.23	9.39	10.86	25.99	28.87	31.53	34.81	37.16
20	7.43	8.26	9.59	10.85	12.44	28.41	31.41	34.17	37.57	40.00
24	9.89	10.86	12.40	13.85	15.66	33.20	36.42	39.36	42.98	45.56
30	13.79	14.95	16.79	18.49	20.60	40.26	43.77	46.98	50.89	53.67
40	20.71	22.16	24.43	25.51	29.05	51.81	55.76	59.34	63.69	66.77
60	35.53	37.48	40.48	43.19	46.46	74.40	79.08	83.30	88.38	91.95
120	83.85	86.92	91.58	95.70	100.62	40.23	146.57	152.21	158.95	163.64

4.13 TESTING VARIANCES FROM TWO POPULATIONS: USE OF THE *F*-DISTRIBUTION

Since the chi-square distribution is useful for testing the variance of samples taken from a single population, the question arises as to how to test the variances of samples taken from *two* populations. The *F*-distribution is utilized to test the variances from two populations. Here, the null hypothesis will be tested against one of the usual alternatives. That is:

$$H_a: \sigma_1^2 = \sigma_2^2$$

or

$$H_a: \sigma_1^2 < \sigma_2^2$$

or

$$H_a: \sigma_1^2 > \sigma_2^2$$

or

$$H_a: \sigma_1^2 \neq \sigma_2^2$$

The F-value for testing $\sigma_1^2 = \sigma_2^2$ is:

$$F = S_1^2 / S_2^2 \tag{4.17}$$

where S_1^2 and S_2^2 are the variances computed from the two samples.

The DOF ∂_1 and ∂_2 are given by:

$$\partial_1 = n_1 - 1$$

$$\partial_2 = n_2 - 1$$

n_1 and n_2 are independent random samples, respectively, from two populations.

If the two populations are approximately normally distributed and the null hypothesis is true, then the ratio $F = \dfrac{S_1^2}{S_2^2}$ is a value of the F-distribution.

The upper and lower points of the distribution are related by the expression

$$F_{(1-\alpha),\partial_1,\partial_2} = \frac{1}{F_{\alpha,\partial_2,\partial_1}} \tag{4.18}$$

Percentiles of the F-distribution are provided in Table 4.7 [10].

4.14 DESIGN STUDIES OF VARIANCES

In this section, we present a number of design studies illustrating the statistical analyses of the chi-square and F-statistic.

4.14.1 Design Study 4.13

You are asked to evaluate the statements of a manufacturer of adhesives who claims that the shelf life of his bolt thread adhesive is approximately normally distributed with a standard deviation equal to 50 weeks. A random sample of 10 of these adhesive samples was found to have a standard deviation of 60 weeks. You wish to test the hypothesis that $\sigma > 50$ weeks. You decide to use a 0.01 level of significance for this analysis. You proceed as follows using a χ^2 statistic.

Approach:

1. $H_a : \sigma^2 = 2500 \quad H_a : \sigma^2 > 2500$
2. $\alpha = 0.01 \quad S = 60 \quad n = 10$
3. From the Table 4.6, we find for

   $(n-1) = 9$ and $\alpha = 0.01 \Rightarrow \chi^2 = 21.67$

   Displaying the above data on a sketch reveals that the null hypothesis is rejected, when $\chi^2 > 21.67$.

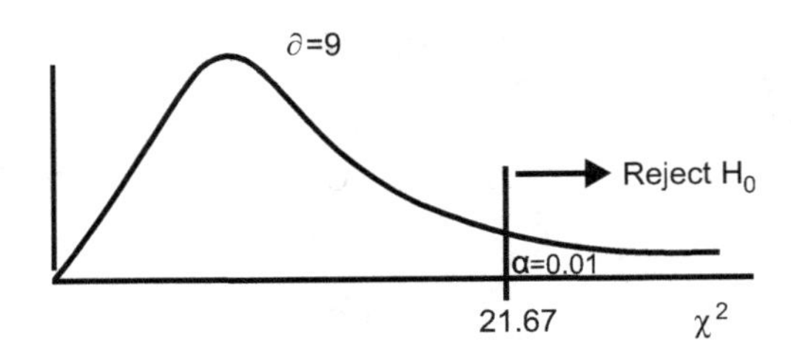

TABLE 4.7

Percentiles of the *F*-Distribution $F_{.95}(\partial_1, \partial_2)$

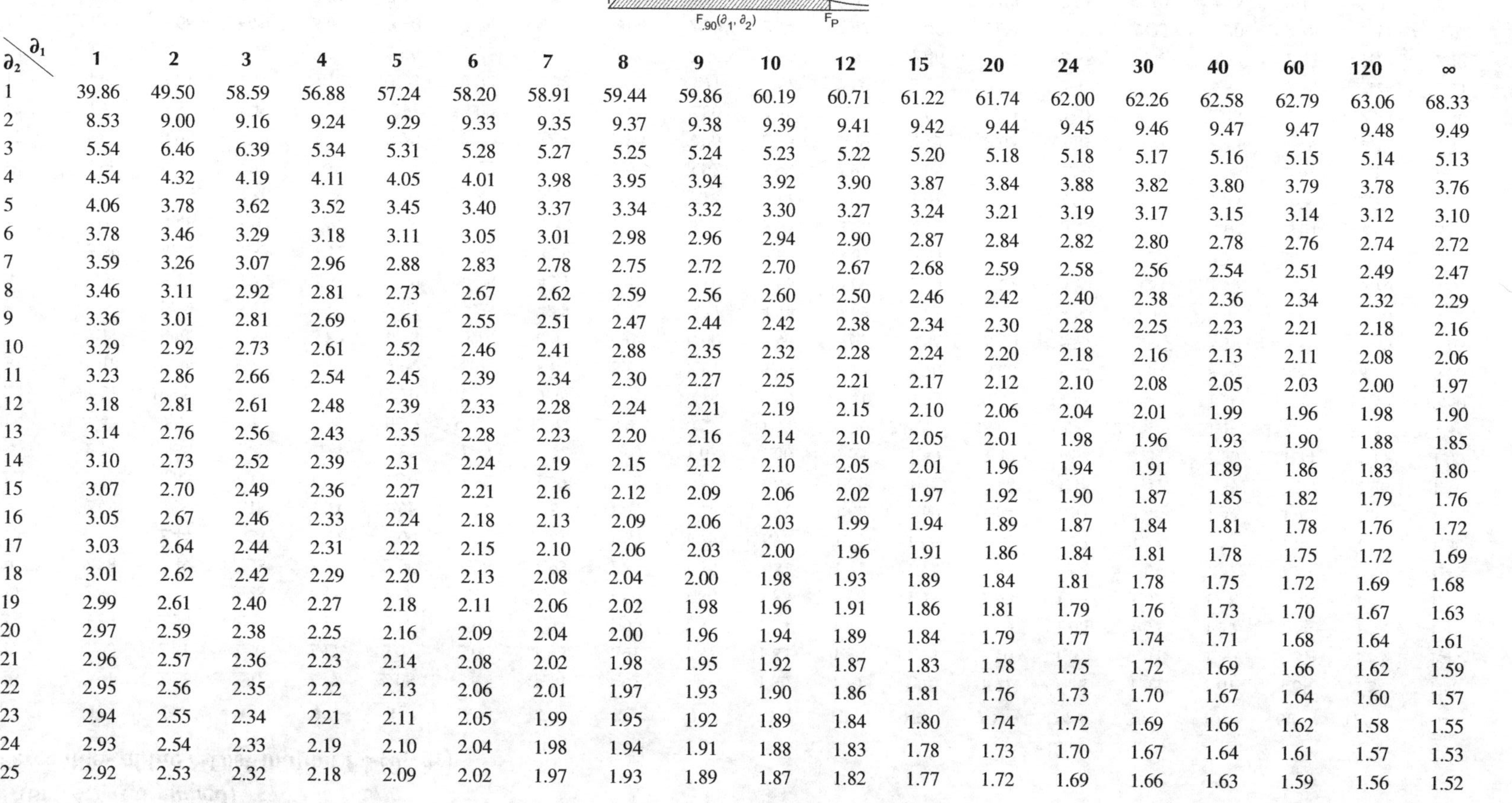

∂_2 \ ∂_1	1	2	3	4	5	6	7	8	9	10	12	15	20	24	30	40	60	120	∞
1	39.86	49.50	58.59	56.88	57.24	58.20	58.91	59.44	59.86	60.19	60.71	61.22	61.74	62.00	62.26	62.58	62.79	63.06	68.33
2	8.53	9.00	9.16	9.24	9.29	9.33	9.35	9.37	9.38	9.39	9.41	9.42	9.44	9.45	9.46	9.47	9.47	9.48	9.49
3	5.54	6.46	6.39	5.34	5.31	5.28	5.27	5.25	5.24	5.23	5.22	5.20	5.18	5.18	5.17	5.16	5.15	5.14	5.13
4	4.54	4.32	4.19	4.11	4.05	4.01	3.98	3.95	3.94	3.92	3.90	3.87	3.84	3.88	3.82	3.80	3.79	3.78	3.76
5	4.06	3.78	3.62	3.52	3.45	3.40	3.37	3.34	3.32	3.30	3.27	3.24	3.21	3.19	3.17	3.15	3.14	3.12	3.10
6	3.78	3.46	3.29	3.18	3.11	3.05	3.01	2.98	2.96	2.94	2.90	2.87	2.84	2.82	2.80	2.78	2.76	2.74	2.72
7	3.59	3.26	3.07	2.96	2.88	2.83	2.78	2.75	2.72	2.70	2.67	2.68	2.59	2.58	2.56	2.54	2.51	2.49	2.47
8	3.46	3.11	2.92	2.81	2.73	2.67	2.62	2.59	2.56	2.60	2.50	2.46	2.42	2.40	2.38	2.36	2.34	2.32	2.29
9	3.36	3.01	2.81	2.69	2.61	2.55	2.51	2.47	2.44	2.42	2.38	2.34	2.30	2.28	2.25	2.23	2.21	2.18	2.16
10	3.29	2.92	2.73	2.61	2.52	2.46	2.41	2.88	2.35	2.32	2.28	2.24	2.20	2.18	2.16	2.13	2.11	2.08	2.06
11	3.23	2.86	2.66	2.54	2.45	2.39	2.34	2.30	2.27	2.25	2.21	2.17	2.12	2.10	2.08	2.05	2.03	2.00	1.97
12	3.18	2.81	2.61	2.48	2.39	2.33	2.28	2.24	2.21	2.19	2.15	2.10	2.06	2.04	2.01	1.99	1.96	1.98	1.90
13	3.14	2.76	2.56	2.43	2.35	2.28	2.23	2.20	2.16	2.14	2.10	2.05	2.01	1.98	1.96	1.93	1.90	1.88	1.85
14	3.10	2.73	2.52	2.39	2.31	2.24	2.19	2.15	2.12	2.10	2.05	2.01	1.96	1.94	1.91	1.89	1.86	1.83	1.80
15	3.07	2.70	2.49	2.36	2.27	2.21	2.16	2.12	2.09	2.06	2.02	1.97	1.92	1.90	1.87	1.85	1.82	1.79	1.76
16	3.05	2.67	2.46	2.33	2.24	2.18	2.13	2.09	2.06	2.03	1.99	1.94	1.89	1.87	1.84	1.81	1.78	1.76	1.72
17	3.03	2.64	2.44	2.31	2.22	2.15	2.10	2.06	2.03	2.00	1.96	1.91	1.86	1.84	1.81	1.78	1.75	1.72	1.69
18	3.01	2.62	2.42	2.29	2.20	2.13	2.08	2.04	2.00	1.98	1.93	1.89	1.84	1.81	1.78	1.75	1.72	1.69	1.68
19	2.99	2.61	2.40	2.27	2.18	2.11	2.06	2.02	1.98	1.96	1.91	1.86	1.81	1.79	1.76	1.73	1.70	1.67	1.63
20	2.97	2.59	2.38	2.25	2.16	2.09	2.04	2.00	1.96	1.94	1.89	1.84	1.79	1.77	1.74	1.71	1.68	1.64	1.61
21	2.96	2.57	2.36	2.23	2.14	2.08	2.02	1.98	1.95	1.92	1.87	1.83	1.78	1.75	1.72	1.69	1.66	1.62	1.59
22	2.95	2.56	2.35	2.22	2.13	2.06	2.01	1.97	1.93	1.90	1.86	1.81	1.76	1.73	1.70	1.67	1.64	1.60	1.57
23	2.94	2.55	2.34	2.21	2.11	2.05	1.99	1.95	1.92	1.89	1.84	1.80	1.74	1.72	1.69	1.66	1.62	1.58	1.55
24	2.93	2.54	2.33	2.19	2.10	2.04	1.98	1.94	1.91	1.88	1.83	1.78	1.73	1.70	1.67	1.64	1.61	1.57	1.53
25	2.92	2.53	2.32	2.18	2.09	2.02	1.97	1.93	1.89	1.87	1.82	1.77	1.72	1.69	1.66	1.63	1.59	1.56	1.52

(Continued)

TABLE 4.7 (*Continued*)
Percentiles of the *F*-Distribution $F_{.95}(\partial_1, \partial_2)$

∂_2 \ ∂_1	1	2	3	4	5	6	7	8	9	10	12	15	20	24	30	40	60	120	∞
26	2.91	2.52	2.31	2.17	2.08	2.01	1.96	1.92	1.88	1.86	1.81	1.76	1.71	1.68	1.65	1.61	1.58	1.54	1.50
27	2.90	2.51	2.30	2.17	2.07	2.00	1.95	1.91	1.87	1.85	1.80	1.75	1.70	1.67	1.64	1.60	1.57	1.53	1.49
28	2.89	2.50	2.29	2.16	2.06	2.00	1.94	1.90	1.87	1.84	1.79	1.74	1.69	1.66	1.63	1.59	1.56	1.52	1.48
29	2.89	2.50	2.28	2.15	2.06	1.99	1.93	1.89	1.86	1.83	1.78	1.73	1.68	1.65	1.62	1.58	1.55	1.51	1.47
30	2.88	2.49	2.28	2.14	2.06	1.98	1.93	1.88	1.85	1.82	1.77	1.72	1.67	1.64	1.61	1.57	1.54	1.50	1.46
40	2.84	2.44	2.23	2.09	2.00	1.93	1.87	1.83	1.79	1.76	1.71	1.66	1.61	1.57	1.54	1.51	1.47	1.42	1.38
60	2.79	2.39	2.18	2.04	1.95	1.87	1.82	1.77	1.74	1.71	1.66	1.60	1.54	1.51	1.48	1.44	1.40	1.35	1.29
120	2.75	2.35	2.13	1.99	1.90	1.82	1.77	1.72	1.68	1.65	1.60	1.55	1.48	1.45	1.41	1.37	1.32	1.26	1.19
∞	2.71	2.30	2.08	1.94	1.85	1.77	1.72	1.67	1.63	1.60	1.55	1.49	1.42	1.38	1.34	1.30	1.24	1.17	1.00
1	161.4	199.5	215.7	224.6	280.2	234.0	236.8	288.9	240.5	241.9	243.9	245.9	248.0	249.1	250.1	251.1	252.2	253.3	254.3
2	18.51	19.00	19.16	19.25	19.30	19.33	19.35	19.37	19.38	19.40	19.41	19.43	19.45	19.45	19.46	1..47	19.48	19.49	19.50
3	10.18	9.55	9.28	9.12	9.01	8.94	8.89	8.85	8.81	8.79	8.74	8.70	8.66	8.64	8.62	8.59	8.57	8.55	8.53
4	7.71	6.94	6.59	6.89	6.28	6.16	6.09	6.04	6.00	5.96	5.91	5.86	5.80	5.77	5.75	5.72	5.89	5.66	5.63
5	6.61	5.79	5.41	6.19	5.05	4.95	4.88	4.82	4.77	4.74	4.68	4.62	4.58	4.53	4.50	4.46	4.43	4.40	4.36
6	5.99	5.14	4.76	4.53	4.39	4.28	4.21	4.15	4.10	4.06	4.00	3.94	3.87	3.84	3.81	3.77	3.74	3.70	3.67
7	5.59	4.74	4.35	4.12	3.97	3.87	3.79	3.73	3.68	3.64	3.67	3.51	3.44	3.41	3.38	3.34	3.30	3.27	3.23
8	5.32	4.46	4.07	3.84	3.69	3.58	3.50	3.44	3.39	3.35	3.28	3.22	3.15	3.12	3.08	3.04	3.01	2.97	2.93
9	5.12	4.28	3.86	3.68	3.48	3.37	3.29	3.23	3.18	3.14	3.07	3.01	2.94	2.90	2.86	2.83	2.79	2.75	2.71
10	4.96	4.10	3.71	3.48	3.33	3.22	3.14	3.07	3.02	2.98	2.91	2.85	2.77	2.74	2.70	2.66	2.62	2.58	2.54
11	4.84	3.98	3.59	3.36	3.20	3.09	3.01	2.95	2.90	2.85	2.79	2.72	2.65	2.61	2.57	2.53	2.49	2.45	2.40
12	4.75	3.89	3.49	3.26	3.11	3.00	2.91	2.85	2.80	2.75	2.69	2.62	2.54	2.51	2.47	2.43	2.38	2.34	2.30
13	4.67	3.81	3.41	3.18	3.08	2.92	2.88	2.77	2.71	2.67	2.60	2.53	2.46	2.42	2.38	2.34	2.30	2.25	2.21
14	4.60	3.74	3.34	3.11	2.96	2.85	2.76	2.70	2.65	2.60	2.53	2.46	2.39	2.35	2.31	2.27	2.22	2.18	2.13
15	4.54	3.68	3.29	3.06	2.90	2.79	2.71	2.64	2.59	2.54	2.48	2.40	2.33	2.29	2.25	2.20	2.16	2.11	2.07
16	4.49	3.63	3.24	3.01	2.85	2.74	2.66	2.59	2.54	2.49	2.42	2.35	2.28	2.24	2.19	2.15	2.11	2.06	2.01
17	4.45	3.59	3.20	2.96	2.81	2.70	2.61	2.55	2.49	2.46	2.38	2.31	2.23	2.19	2.16	2.10	2.06	2.01	1.96
18	4.41	3.55	3.16	2.93	2.77	2.66	2.58	2.51	2.46	2.41	2.34	2.27	2.19	2.15	2.11	2.06	2.02	1.97	1.92
19	4.38	3.52	3.13	2.90	2.74	2.63	2.54	2.48	2.42	2.38	2.31	2.23	2.16	2.11	2.07	2.03	1.98	1.93	1.88

(Continued)

TABLE 4.7 (*Continued*)

Percentiles of the *F*-Distribution $F_{.95}(\partial_1, \partial_2)$

∂_2 \ ∂_1	1	2	3	4	5	6	7	8	9	10	12	15	20	24	30	40	60	120	∞
20	4.35	3.49	3.10	2.87	2.71	2.60	2.51	2.45	2.39	2.35	2.28	2.20	2.12	2.08	2.04	1.99	1.95	1.90	1.84
21	4.32	3.47	3.07	2.84	2.68	2.57	2.49	2.42	2.37	2.32	2.25	2.18	2.10	2.05	2.01	1.96	1.92	1.87	1.81
22	4.30	3.44	3.05	2.82	2.66	2.55	2.46	2.40	2.34	2.30	2.23	2.15	2.07	2.03	1.98	1.94	1.89	1.84	1.78
23	4.28	3.42	3.03	2.80	2.64	2.53	2.44	2.37	2.32	2.27	2.20	2.13	2.05	2.01	1.96	1.91	1.86	1.81	1.76
24	4.26	3.40	3.01	2.78	2.62	2.51	2.42	2.36	2.30	2.25	2.18	2.11	2.03	1.98	1.94	1.89	1.84	1.79	1.73
25	4.24	3.39	2.99	2.76	2.60	2.49	2.40	2.34	2.28	2.24	2.16	2.09	2.01	1.96	1.92	1.87	1.82	1.77	1.71
26	4.23	3.37	2.98	2.74	2.59	2.47	2.39	2.32	2.27	2.22	2.15	2.07	1.99	1.95	1.90	1.85	1.80	1.75	1.69
27	4.21	3.35	2.98	2.73	2.57	2.46	2.37	2.31	2.25	2.20	2.13	2.06	1.97	1.93	1.88	1.84	1.79	1.73	1.67
28	4.20	3.34	2.95	2.71	2.56	2.45	2.36	2.29	2.24	2.19	2.12	2.04	1.96	1.91	1.87	1.82	1.77	1.71	1.65
29	4.18	3.33	2.93	2.70	2.55	2.43	2.35	2.28	2.22	2.18	2.10	2.03	1.94	1.90	1.85	1.81	1.75	1.70	1.64
30	4.17	3.32	2.92	2.69	2.53	2.42	2.33	2.27	2.21	2.16	2.09	2.01	1.93	1.89	1.84	1.79	1.74	1.68	1.62
40	4.08	3.23	2.84	2.61	2.45	2.34	2.25	2.18	2.12	2.08	2.00	1.92	1.84	1.79	1.74	1.69	1.64	1.58	1.51
60	4.00	3.15	2.76	2.53	2.37	2.26	2.17	2.10	2.04	1.99	1.92	1.84	1.75	1.70	1.65	1.59	1.53	1.47	1.39
120	3.92	3.07	2.68	2.45	2.29	2.17	2.09	2.02	1.96	1.91	1.83	1.75	1.66	1.61	1.55	1.50	1.43	1.35	1.25
∞	3.34	3.00	2.60	2.37	2.21	2.10	2.01	1.94	1.88	1.83	1.75	1.67	1.57	1.52	1.46	1.39	1.32	1.22	1.00
1	647.8	799.5	864.2	899.6	921.8	937.1	948.2	956.7	963.3	968.6	976.7	984.9	993.1	997.2	1001	1006	1010	1014	1018
2	38.51	39.00	39.17	39.25	89.3	39.33	39.36	39.37	39.39	39.40	39.41	39.43	89.45	39.46	39.46	39.47	39.48	39.49	39.50
3	17.44	16.04	15.44	15.10	14.88	14.73	14.62	14.54	14.47	14.42	14.34	14.25	14.17	14.12	14.08	14.04	13.99	13.95	13.90
4	12.22	10.65	9.98	9.60	9.36	9.20	9.07	8.98	8.90	8.84	8.75	8.66	8.56	8.51	8.46	8.41	8.36	8.31	8.26
5	10.01	8.43	7,76	7.39	7.15	6.98	6.85	6.76	6.68	6.62	6.52	6.43	6.38	6.28	6.23	6.18	6.12	6.07	6.02
6	8.81	7.26	6.60	6.23	5.99	5.82	5.70	5.60	5.52	5.46	5.37	5.27	5.17	5.12	5.07	5.01	4.96	4.90	4.85
7	8.07	6.54	5.89	5.52	5.29	5.12	4.99	4.90	4.82	4.76	4.67	4.57	4.47	4.42	4.36	4.31	4.25	4.20	4.14
8	7.57	6.06	5.42	5.05	4.82	4.65	4.53	4.43	4.36	4.30	4.20	4.10	4.00	3.95	3.89	3.84	3.73	3.73	3.67
9	7.21	5.71	5.08	4.72	4.48	4.32	4.20	4.10	4.08	3.96	3.87	3.77	3.67	3.61	3.56	3.51	3.45	3.39	3.33
10	6.94	5.46	4.83	4.47	4.24	4.07	3.95	3.85	3.78	3.72	3.62	3.52	3.42	3.37	3.31	3.26	3.29	3.14	3.08
11	6.72	5.26	4.63	4.28	4.04	3.88	3.76	3.66	3.69	3.58	3.48	3.33	3.23	3.17	3.12	3.06	3.00	2.94	2.88
12	6.55	5.10	4.47	4.12	3.89	3.78	3.61	3.61	3.44	3.37	3.28	3.18	3.07	3.02	2.96	2.91	2.85	2.79	2.72
13	6.41	4.97	4.35	4.00	3.77	3.60	3.48	3.39	3.31	3.25	3.15	3.05	2.95	2.89	2.84	2.78	2.72	2.66	2.60

(Continued)

TABLE 4.7 (*Continued*)

Percentiles of the *F*-Distribution $F_{.95}(\partial_1, \partial_2)$

∂_2 \ ∂_1	1	2	3	4	5	6	7	8	9	10	12	15	20	24	30	40	60	120	∞
14	6.30	4.86	4.24	3.89	3.66	3.50	3.38	3.29	3.21	3.15	3.05	2.95	2.84	2.79	2.73	2.67	2.61	2.55	2.49
15	6.20	4.77	4.15	3.80	3.58	3.41	3.29	3.20	3.12	3.06	2.98	2.86	2.76	2.70	2.64	2.59	2.52	2.46	2.40
16	6.12	4.69	4.08	3.73	3.50	3.34	3.22	3.12	3.05	2.99	2.89	2.79	2.68	2.63	2.57	2.51	2.45	2.38	2.32
17	6.04	4.62	4.01	3.66	3.44	3.28	3.16	3.06	2.98	2.92	2.82	2.72	2.62	2.56	2.50	2.44	2.38	2.32	2.25
18	5.98	4.56	3.95	3.61	3.38	3.22	3.10	3.01	2.93	2.87	2.77	2.67	2.56	2.50	2.44	2.38	2.32	2.26	2.19
19	5.92	4.51	3.90	3.56	3.33	3.17	3.05	2.96	2.88	2.82	2.72	2.62	2.51	2.45	2.39	2.33	2.27	2.20	2.13
20	5.87	4.46	3.86	3.51	3.29	3.13	3.01	2.91	2.84	2.77	2.68	2.57	2.46	2.41	2.35	2.29	2.22	2.16	2.09
21	5.83	4.42	3.82	3.48	3.25	3.09	2.97	2.87	2.80	2.73	2.64	2.53	2.42	2.37	2.31	2.25	2.18	2.11	2.04
22	5.79	4.38	3.78	3.44	3.22	3.05	2.93	2.84	2.76	2.70	2.60	2.50	2.39	2.33	2.27	2.21	2.14	2.08	2.00
23	5.75	4.35	3.75	3.41	3.18	3.02	2.90	2.81	2.73	2.67	2.57	2.47	2.36	2.30	2.24	2.18	2.11	2.04	1.97
24	5.72	4.32	3.72	3.38	3.15	2.99	2.87	2.78	2.70	2.64	2.54	2.44	2.33	2.27	2.21	2.15	2.08	2.01	1.94
25	5.69	4.29	3.59	3.35	3.13	2.97	2.85	2.75	2.68	2.61	2.51	2.41	2.30	2.24	2.18	2.12	2.05	1.98	1.91
26	5.66	4.27	3.67	3.33	3.10	2.94	2.82	2.73	2.65	2.59	2.49	2.39	2.28	2.22	2.16	2.09	2.03	1.95	1.88
27	5.63	4.24	3.65	3.31	3.08	2.92	2.80	2.71	2.63	2.57	2.47	2.36	2.25	2.19	2.13	2.07	2.00	1.93	1.85
28	5.61	4.22	3.63	3.29	3.06	2.90	2.78	2.68	2.61	2.55	2.45	2.34	2.23	2.17	2.11	2.05	1.98	1.91	1.83
29	5.59	4.20	3.61	3.27	3.04	2.88	2.76	2.67	2.59	2.53	2.43	2.32	2.21	2.15	2.09	2.03	1.96	1.89	1.81
30	5.57	4.18	3.59	3.25	3.03	2.87	2.75	2.65	2.57	2.51	2.41	2.31	2.20	2.14	2.07	2.01	1.94	1.87	1.79
40	5.42	4.05	3.46	3.13	2.90	2.74	2.62	2.53	2.45	2.39	2.29	2.18	2.07	2.01	1.94	1.88	1.80	1.72	1.64
60	5.29	3.93	3.34	3.01	2.79	2.68	2.51	2.41	2.33	2.27	2.17	2.06	1.94	1.88	1.82	1.74	1.67	1.58	1.48
120	5.15	3.80	3.23	2.89	2.67	2.52	2.39	2.30	2.22	2.16	2.05	1.94	1.82	1.76	1.69	1.61	1.58	1.43	1.31
∞	5.02	3.69	3.12	2.79	2.57	2.41	2.29	2.19	2.11	2.05	1.94	1.88	1.71	1.64	1.57	1.48	1.39	1.27	1.00
1	4052	4999.6	5403	5625	5764	5859	5928	5982	6022	6056	6106	6157	6209	6235	6261	6287	6313	6339	6366
2	98.50	99.00	99.17	99.25	99.30	99.33	99.36	99.37	99.39	99.40	99.42	99.43	99.45	99.46	99.47	99.47	99.48	99.49	99.50
3	34.12	30.82	29.46	28.71	28.24	27.91	27.67	27.49	27.35	27.23	27.05	26.87	26.69	26.60	26.59	26.41	26.32	26.22	26.13
4	21.20	18.00	16.69	15.98	15.52	15.21	14.98	14.80	14.66	14.55	14.37	14.20	14.02	13.93	13.84	13.75	13.65	13.56	13.46
5	16.26	13.27	12.06	11.39	10.97	10.67	10.46	10.29	10.16	10.05	9.89	9.72	9.55	9.47	9.38	9.29	9.20	9.11	8.02
6	13.75	10.92	9.78	9.15	8.75	8.47	8.26	8.10	7.98	7.87	7.72	7.56	7.40	7.31	7.23	7.14	7.06	6.97	6.88
7	12.25	9.55	8.45	7.85	7.46	7.19	6.99	6.84	6.72	6.62	6.47	6.31	6.16	6.07	5.99	5.91	5.82	5.74	5.65

(*Continued*)

TABLE 4.7 (*Continued*)

Percentiles of the *F*-Distribution $F_{.95}(\partial_1, \partial_2)$

∂_2 \ ∂_1	1	2	3	4	5	6	7	8	9	10	12	15	20	24	30	40	60	120	∞
8	11.26	8.65	7.59	7.01	6.63	6.37	6.18	6.03	5.91	5.81	5.67	5.52	5.36	5.28	5.20	5.12	5.03	4.95	4.86
9	10.56	8.02	6.99	6.42	6.06	5.89	5.61	5.47	5.35	5.26	5.11	4.96	4.81	4.73	4.65	4.57	4.48	4.40	4.31
10	10.04	7.56	6.55	5.99	5.64	5.39	5.20	5.06	4.94	4.85	4.71	4.56	4.41	4.38	4.25	4.17	4.08	4.00	3.91
11	9.65	7.21	6.22	5.67	5.32	5.07	4.89	4.74	4.63	4.54	4.40	4.25	4.10	4.02	3.94	3.86	3.78	3.69	3.60
12	9.33	6.98	5.95	5.41	5.06	4.82	4.64	4.50	4.39	4.30	4.16	4.01	3.86	3.78	3.70	3.62	3.54	3.45	3.36
13	9.07	6.70	5.74	5.21	4.86	4.62	4.44	4.30	4.19	4.10	3.98	3.82	3.66	3.59	3.51	3.43	3.34	3.25	3.17
14	8.86	6.51	5.56	5.04	4.69	4.46	4.28	4.14	4.08	3.94	3.80	3.66	3.51	3.43	3.35	3.27	3.18	3.09	3.00
15	8.68	6.86	5.42	4.89	4.56	4.32	4.14	4.00	3.89	3.80	3.67	3.52	3.37	3.29	3.21	3.13	3.05	2.96	2.87
16	8.53	6.28	5.29	4.77	4.44	4.20	4.03	3.89	3.78	3.69	3.55	3.41	3.26	3.18	3.10	3.02	2.93	2.84	2.75
17	8.40	6.11	5.18	4.67	4.34	4.10	3.93	3.79	3.68	3.59	3.46	3.31	3.16	3.08	3.00	2.92	2.83	2.75	2.65
18	8.29	6.01	5.09	4.58	4.25	4.01	3.84	3.71	3.60	3.51	3.37	3.23	3.08	3.00	2.92	2.84	2.75	2.66	2.57
19	8.18	5.98	5.01	4.50	4.17	3.94	3.77	3.63	3.52	3.48	3.30	3.15	3.00	2.92	2.84	2.76	2.67	2.58	2.49
20	8.1	5.85	4.94	4.43	4.10	3.87	3.70	3.56	3.46	3.37	3.23	3.09	2.94	2.86	2.78	2.69	2.61	2.52	2.42
21	8.02	5.78	4.87	4.37	4.04	3.81	3.64	3.51	3.40	3.31	3.17	3.03	2.88	2.80	2.72	2.64	2.55	2.46	2.36
22	7.95	5.72	4.82	4.31	3.99	3.76	3.59	3.45	3.35	3.26	3.12	2.98	2.83	2.75	2.67	2.58	2.50	2.40	2.31
23	7.88	5.66	4.76	4.26	3.94	3.71	3.54	3.41	3.30	3.21	3.07	2.93	2.78	2.70	2.62	2.54	2.45	2.35	2.26
24	7.82	5.61	4.72	4.22	3.90	3.67	3.50	3.36	3.26	3.17	3.03	2.89	2.74	2.66	2.58	2.49	2.40	2.31	2.21
25	7.77	5.57	4.68	4.18	3.85	3.63	3.46	3.32	3.22	3.13	2.99	2.85	2.70	2.62	2.54	2.45	2.38	2.27	2.17
26	7.72	5.53	4.64	4.14	3.82	3.59	3.42	3.29	3.18	3.09	2.96	2.81	2.66	2.58	2.50	2.42	2.33	2.23	2.13
27	7.68	5.49	4.60	4.11	3.78	3.56	3.39	3.26	3.15	3.06	2.93	2.78	2.63	2.55	2.47	2.38	2.29	2.20	2.10
28	7.64	5.45	4.57	4.07	3.75	3.53	3.36	3.23	3.12	3.03	2.90	2.75	2.60	2.52	2.44	2.35	2.26	2.17	2.06
29	7.60	5.42	4.54	4.04	3.73	3.50	3.33	3.20	3.09	3.00	2.87	2.73	2.57	2.49	2.41	2.33	2.23	2.14	2.03
30	7.56	5.39	4.51	4.02	3.70	3.47	3.30	3.17	3.07	2.98	2.84	2.70	2.55	2.47	2.39	2.30	2.21	2.11	2.01
40	7.31	5.18	4.31	3.83	3.51	3.29	3.12	2.99	2.89	2.80	2.66	2.52	2.37	2.29	2.20	2.11	2.02	1.92	1.80
60	7.08	4.98	4.19	3.65	3.34	3.12	2.95	2.82	2.72	2.63	2.50	2.35	2.20	2.12	2.03	1.94	1.84	1.73	1.60
120	6.85	4.79	3.95	3.48	3.17	2.96	2.79	2.66	2.56	2.47	2.34	2.19	2.03	1.95	1.86	1.76	1.66	1.58	1.38
∞	6.63	4.61	3.78	3.32	3.02	2.80	2.64	2.51	2.41	2.32	2.18	2.04	1.88	1.79	1.70	1.59	1.47	1.32	1.00

∂_1 = degrees of freedom for numerator

∂_2 = degrees of freedom for denominator

4. Given $\chi^2 = \left[(n-1)s^2\right] / (\sigma_0)^2$ where $n = 10$, $S = 60$, $\sigma_0 = 50$

   We find $\chi^2 = \dfrac{(10-1)(60)^2}{(50)^2} = 12.96$

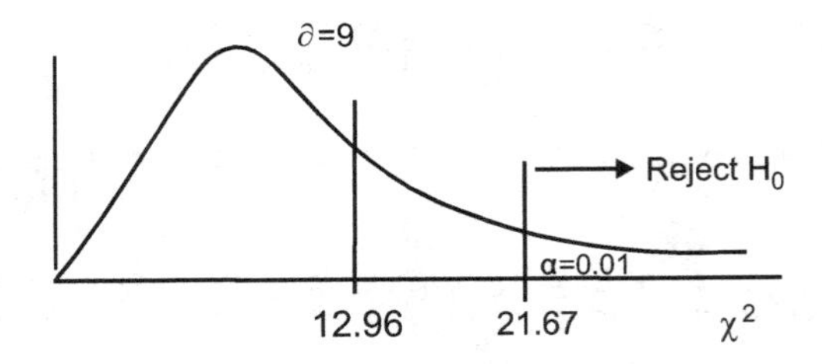

5. Therefore, since the value of $\chi^2 = 12.96$ is within the bounds of $\chi^2 = 21.67$ we do not reject H_0.

You find that there is insufficient data to reject the hypothesis that the standard deviation is 50 weeks. Therefore, we accept H_0 and we can conclude that the standard deviation of the adhesive shelf life is 50 weeks.

4.14.2 Design Study 4.14

A consultant is asked to evaluate some bolt failures found during testing. He is informed that the yield point of a high-strength steel (HSS) bolt is known to be normally distributed with a variance of 1.6. He decides to test the hypothesis that $\sigma^2 = 1.6$ against the alternative that $\sigma^2 \neq 1.6$. He manages to "pull" a random sample of 5 of these fasteners for testing, and he finds that they have a standard deviation $S = 2.1$. He decides to use a 0.01 level of significance for testing.

He utilizes the following approach:

1. $H_0 : \sigma^2 = 1.6$, $H_a : \sigma^2 \neq 1.6$ (two-tailed)
2. $\alpha = 0.01$, $n = 5$
3. $\chi^2 = (n-1)S^2/(\sigma_0)^2$ where $S = 2.1$ and $\sigma_0^2 = 1.6$

   which yields: $\chi^2 = \dfrac{(n-1)5^2}{\sigma_0^2} = (5-1)\dfrac{(2.1)^2}{(1.6)^2} = 6.89$
4. From Table 4.6

   For $\alpha / 2 = \dfrac{0.01}{2} = 0.005 \quad \partial = n - 1 = 5 - 1 = 4$

   $$\left(\chi_{(\alpha/2)}\right)^2 = \chi^2{}_{0.005} = 0.207$$

   $$\left(\chi_{(1-\alpha/2)}\right)^2 = \chi^2{}_{0.995} = 14.86$$
5. Analysis

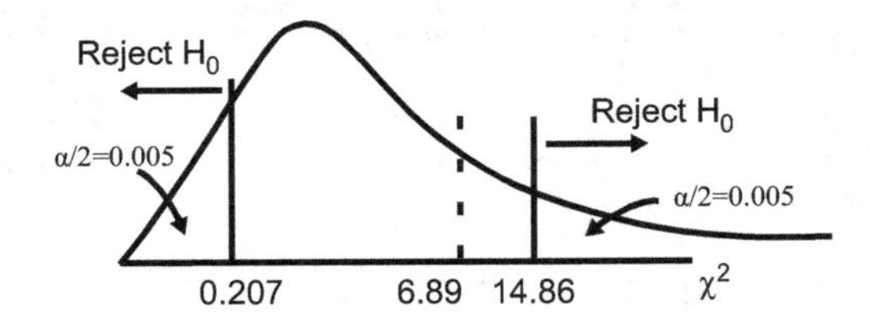

Therefore, since $\chi^2 = 6.89$, which is greater than $\chi^2_{0.005} = 0.207$ and less than $\chi^2_{0.995} = 14.86$, the consultant concludes that at a significance level of 0.01 the null hypothesis is accepted (accept H_0) and he, therefore, concludes that the bolt variance is $\sigma^2 = 1.6$.

4.14.3 Design Study 4.15

A fastener tool lab conducts a study to compare the amount of powder coating of dry thread adhesive applied to prevailing torque fasteners from two different manufacturers. Historical data shows that the distribution of powder coating of the adhesive from both manufacturers is approximately normal but the variances differ. A random sample of adhesive coatings is tested and yields the following:

Manufacturer A	Manufacturer B
$n_1 = 16$	$n_2 = 18$
$S_1 = 7.2$	$S_2 = 6.4$

The tool lab decides to test the hypothesis that $\sigma_1^2 = \sigma_2^2$ against the alternative that $\sigma_1^2 > \sigma_2^2$, using a 0.05 level of significance.

Approach:

1. $H_0: \sigma_1^2 = \sigma_2^2 \; H_A: \sigma_1^2 > \sigma_2^2$
2. $a = 0.05 \; \partial_1 = n_1 - 1 = 16 - 1 = 15 \; \partial_2 = n_2 - 1 = 18 - 1 = 17$
3. From the data and Table 4.7 the lab finds $f_\alpha(\partial_1, \partial_2) = f_{0.05}(15,17) = 2.31$
4. Analysis $F = S_1^2 / S_2^2 = \dfrac{(7.2)^2}{(6.4)^2} = 1.27$

Displaying this data on a sketch reveals the following:

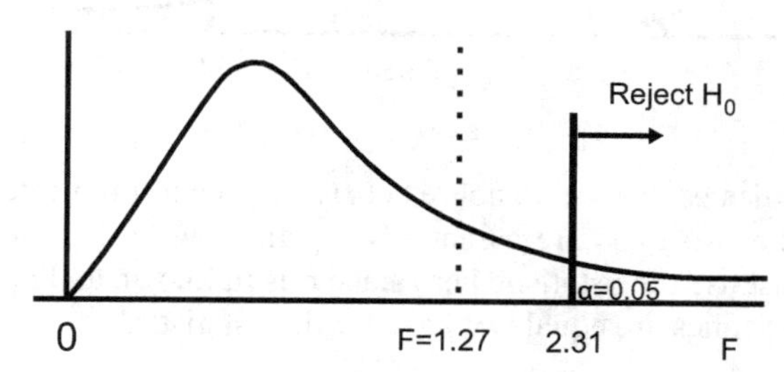

Since we find $F = 1.27$ to be less than $F_{0.05}(15,17) = 2.31$ there is insufficient information to reject H_0 hence $\sigma_1^2 = \sigma_2^2$.

Therefore, the lab concludes that there is no variation in the distribution of dry adhesive thread coating from the two manufacturers.

4.14.4 Design Study 4.16

Two different fastener tool manufacturers have submitted their tools to the Dorf fastening tool lab for testing. The tool engineers are interested in evaluating the time required to tighten the bolt on a given test bracket. Data is collected on run-down times (from snug to fully tightened) given in milliseconds for the two different manufacturers. The Dorf engineers are interested in reviewing the analysis of both the mean and variance of the run-down times. The Dorf engineers perform 16 run-down tests on each of two representative tools and find the following for run-down times:

Tool 1: $n_1 = 16$; $\bar{x}_1 = 67.7$; $S_1 = 8.46$

Tool 2: $n_2 = 16$; $\bar{x}_2 = 67.4$; $S_2 = 8.73$

The Dorf engineers decide to:

A. Test the hypothesis that the variances are equal. Using $\alpha = 0.02$.
B. Using the results of (A) above, they will test the hypothesis that the mean run-down times are equal. Using $\alpha = 0.02$.

Approach:

1. $H_0: \sigma_1^2 = \sigma_2^2 \quad H_A: \sigma_1^2 \neq \sigma_2^2$ (Note: This results in a two-tailed distribution)
2. $a = 0.02$; $\frac{\alpha}{2} = 0.01$ σ Unknown—Use t test
3. From the data: $S_1 = 8.46 \quad S_2 = 8.73 \quad \bar{x}_1 = 67.7 \quad \bar{x}_2 = 67.4$
4. $F_0 = S_1^2/S_2^2 = (8.46)^2/(8.73)^2 = 0.939$
5. $n_1 = 16 \quad n_2 = 16$
6. $\partial_1 = \partial_2 = 16 - 1 = 15$

   From Table 4.7 the Dorf engineers find $F_{0.01,15,15} = 3.52$; $F_{0.99,15,15} = \frac{1}{F_{0.01,15,15}} = \frac{1}{3.52} = 0.284$

   Displaying this data on a sketch reveals the following:

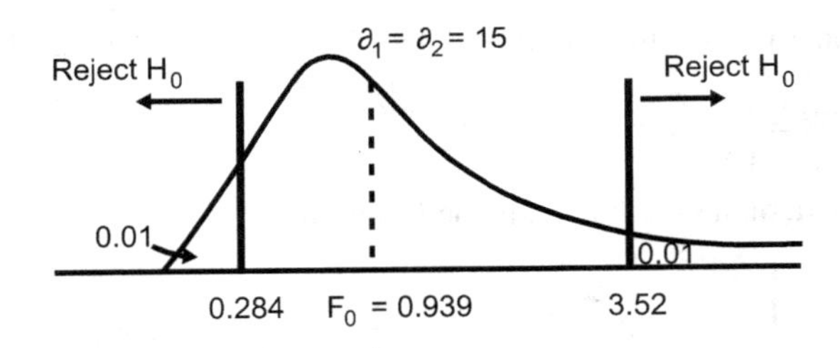

The data shown indicates that we do not reject H_0, and the test indicates that the variances of the output of the two tools are essentially equal. Now to determine the mean "run-down" time we must use the t statistic here since σ is unknown and n_1 and n_2 are small. If the mean run-down times are equal, we have for the t statistic:

7. $t_0 = \dfrac{\bar{x}_1 - \bar{x}_2}{s_p\sqrt{\dfrac{1}{n_1} + \dfrac{1}{n_2}}}$; $\partial = n_1 + n_2 - 1$ where $S_p^2 = \dfrac{(n_1 - 1)S_1^2 + (n_2 - 1)S_2^2}{n_1 + n_2 - 2}$

8. $S_p^2 = \dfrac{(n_1 - 1)S_1^2 + (n_2 - 1)S_2^2}{n_1 + n_2 - 2} = \dfrac{(16-1)(8.46)^2 + (16-1)(8.73)^2}{16 + 16 - 2} = 73.892$; $S_p = 8.596$

9. $t_0 = \dfrac{\bar{x}_1 - \bar{x}_2}{s_p\sqrt{\dfrac{1}{n_1} + \dfrac{1}{n_2}}} = \dfrac{67.7 - 67.4}{8.596\left[\dfrac{1}{16} + \dfrac{1}{16}\right]^{1/2}} = 0.0987$

   From Table 4.3 we find: $t_{\alpha/2,(n_1+n_2-2)} = t_{0.01,30} = 2.457$

   Displaying this data on a sketch reveals the following:

Reject H_0
Reject H_0
0.01
0.01
–2.457
t = 0.0987
$t_{0.01,30}$ = 2.457

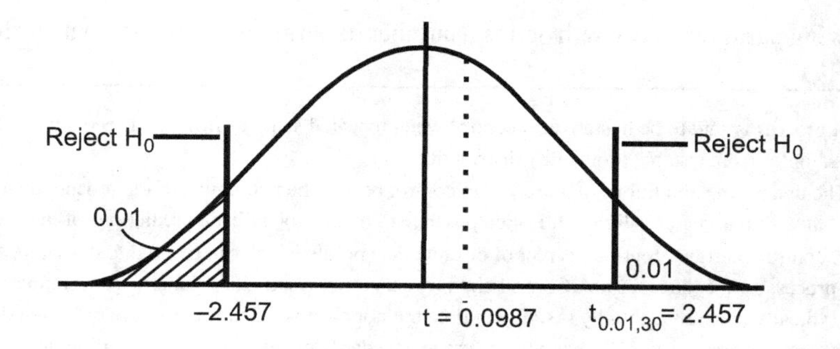

Note that the t test requires normality. However, small departures from normality have little effect on the t test.

Given the data, the Dorf engineers do not reject H_0, and therefore conclude that the run-down times of the tools from the two manufacturers are essentially equal.

4.15 CONTROL CHARTS

The foregoing sections show that a fastening process can have considerable variation. As we have noted previously, variations can occur due to many manufacturing conditions such as: tool differences, machine differences, material differences, operator differences, supplier differences, etc.

The statistical tests presented in the foregoing sections have evaluated process variation at a single instance in the process (essentially a "snapshot" in time). A single instance evaluation, however, neglects the effect of the passage of time in a process. Control charts, or more specifically, "statistical process control charts" (or methods) monitor a process variation from samples taken sequentially (Figures 4.6 and 4.7).

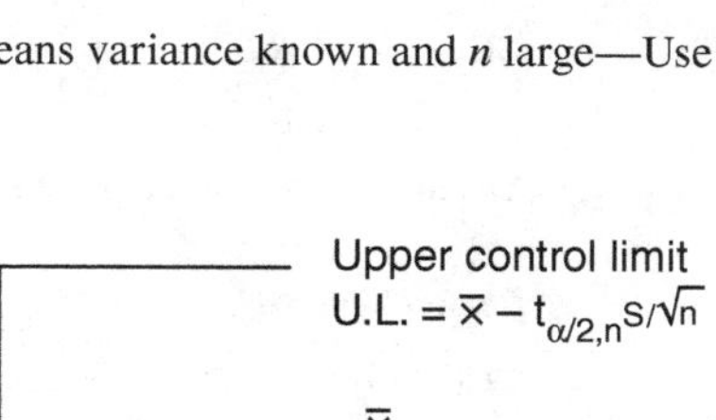

Upper control limit
U.L. = $\bar{x} + Z_{\alpha/2}\sigma$
$\bar{x}$
Lower control limit
L.L. = $\bar{x} - Z_{\alpha/2}\sigma$

FIGURE 4.6 Control chart for means variance known and n large—Use Z statistic.

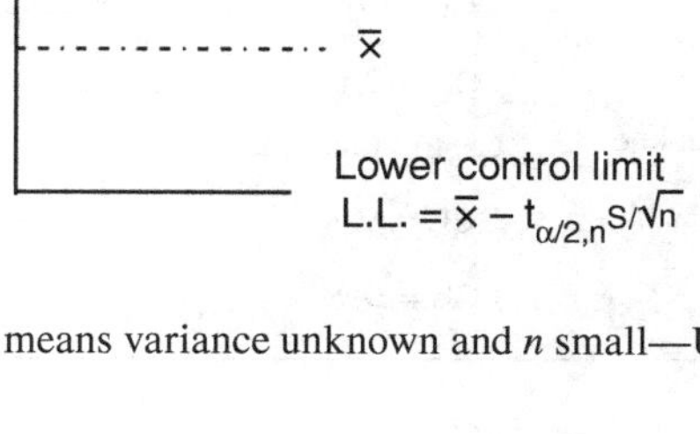

Upper control limit
U.L. = $\bar{x} - t_{\alpha/2,n} s/\sqrt{n}$
$\bar{x}$
Lower control limit
L.L. = $\bar{x} - t_{\alpha/2,n} s/\sqrt{n}$

FIGURE 4.7 Z control chart for means variance unknown and n small—Use t Distribution.

In the following paragraphs, we will define a number of parameters used in control chart methods:

Statistical control	A process is said to be in statistical control when repeated samples from the process behave as random samples from a stable probability distribution.
In control	The underlying conditions of a process in control occur when all points of the process being measured fall within acceptable limits. It is then possible to make probabilistic predictions of the process.
Control chart	A control chart is a continuous plot of discrete data points which can be used to determine whether a process is in control, or not. Control charts are a plot of a summary statistic from samples that are typically taken sequentially (usually in time but not always). Control charts may be based on any of various characteristics, for example, average, standard deviation, range, proportion defective, etc. The construction of a chart consists of graphically applying hypothesis testing on a continuous basis. The advantage is that a visual perception is maintained of the random variable as it develops in the process. It is important to note that actual defect causes cannot be specified from the chart alone. Only their effects can be determined.
Control limits	Control charts include bounds, or "control limits," which assist in determining whether the particular metric (e.g., range, average, proportions, etc.) is within acceptable limits. A point falling above or below the control limits is indicative of an out-of-control situation which often can ultimately be traced to an assignable cause. Control limits are typically computed by using formulas which utilize information from sampling data previously collected.
Not in control	If any points shown on the control chart are not within the control limits, the process is said to be not in control or "out of control." Often the state of "out of control" occurs in one direction only. Other criteria that may indicate an "out of control" situation include the following: 1. Four or more points in a row on one side of the mean line. 2. Identifiable cycles. Here many cycles of data may require collecting in order to identify irregularities. 3. Several points in a row, either monotonically increasing or decreasing away from the mean line.
Control chart for means	The control chart for means, perhaps the most utilized control chart, is accomplished by plotting the means of process samples, this is also known as an $\overline{x}$ chart. Figures 4.4 and 4.5 illustrate control charts for the mean for both known and unknown variances. Other popular chart approaches are the R chart (range) and σ chart (standard deviation).

Additional criteria for developing control charts include the following:

Grand means of sample means

$$\overline{\overline{x}} = \frac{1}{K}\sum_{i=1}^{K} \overline{x}_i$$

Mean sample range

$$\overline{R} = \frac{1}{K}\sum_{i=1}^{K} R_i$$

Control-Chart values for a $\overline{x}$ chart

$$\text{Central Line} = \overline{\overline{x}}$$

$$\text{UCL} = \overline{\overline{x}} + A_2\overline{R}$$

$$\text{LCL} = \overline{\overline{x}} - A_2\overline{R}$$

The constant values shown, e.g., A_2, D_1, D_2, d_2, d_3, etc. are found in standard statistics tables, and are presented in Table 4.8 [11].

Control-Chart values for an R chart (τ known)

$$\text{Central Line} = d_2\tau$$

$$\text{UCL} = D_2\tau \qquad D_2 = d_2 + 3d_2$$

$$\text{LCL} = D_1\tau \qquad D_1 = d_2 - 3d_2$$

Control-Chart values for an R chart (τ unknown)

$$\text{Central Line} = \overline{R}$$

$$\text{UCL} = D_4\overline{R}$$

$$\text{LCL} = D_2\overline{R}$$

TABLE 4.8
Factors for Computing Control Chartlines

Number of Observations in Sample, n	Chart for Averages			Chart for Standard Deviations							Chart for Ranges						
	Factors for Control Limits			Factors for Central Line		Factors for Control Limits				Factors for Central Line			Factors for Control Limits			
	A	A_1	A_2	c_2	$1/c_2$	B_1	B_2	B_3	B_4	d_2	$1/d_2$	d_3	D_1	D_2	D_3	D_4
2	2.121	3.760	1.880	0.5642	1.7725	0	1.843	0	3.267	1.128	0.8865	0.853	0	3.686	0	3.267
3	1.732	2.394	1.023	0.7236	1.3820	0	1.858	0	2.568	1.693	0.5907	0.888	0	4.358	0	2.575
4	1.500	1.880	0.729	0.7979	1.2533	0	1.808	0	2.266	2.059	0.4857	0.880	0	4.698	0	2.282
5	1.342	1.596	0.577	0.8407	1.1894	0	1.756	0	2.089	2.326	0.4299	0.864	0	4.918	0	2.115
6	1.225	1.410	0.483	0.8686	1.1512	0.026	1.711	0.030	1.970	2.534	0.3946	0.848	0	5.078	0	2.004
7	1.134	1.277	0.419	0.8882	1.1259	0.105	1.672	0.118	1.882	2.704	0.3698	0.833	0.205	5.203	0.076	1.924
8	1.061	1.175	0.373	0.9027	1.1078	0.167	1.638	0.185	1.815	2.847	0.3512	0.820	0.387	5.307	0.136	1.864
9	1.000	1.094	0.337	0.9139	1.0942	0.219	1.609	0.239	1.761	2.970	0.3367	0.808	0.546	5.394	0.184	1.816
10	0.949	1.028	0.308	0.9227	1.0837	0.262	1.584	0.284	1.716	3.078	0.3249	0.797	0.687	5.469	0.223	1.777
11	0.905	0.973	0.285	0.9300	1.0753	0.299	1.561	0.321	1.679	3.173	0.3152	0.787	0.812	5.534	0.256	1.744
12	0.866	0.925	0.266	0.9359	1.0684	0.331	1.541	0.354	1.646	3.258	0.3069	0.778	0.924	5.592	0.284	1.716
13	0.832	0.884	0.249	0.9410	1.0627	0.359	1.523	0.382	1.618	3.336	0.2998	0.770	1.026	5.646	0.308	1.692
14	0.802	0.848	0.235	0.9453	1.0579	0.384	1.507	0.406	1.594	3.407	0.2935	0.762	1.121	5.693	0.329	1.671
15	0.775	0.816	0.223	0.9490	1.0537	0.406	1.492	0.428	1.572	3.472	0.2880	0.755	1.207	5.737	0.348	1.652
16	0.750	0.788	0.212	0.9523	1.0501	0.427	1.478	0.448	1.552	3.532	0.2831	0.749	1.285	5.779	0.364	1.636
17	0.728	0.762	0.203	0.9551	1.0470	0.445	1.465	0.466	1.534	3.588	0.2787	0.743	1.359	5.817	0.379	1.621
18	0.707	0.738	0.194	0.9576	1.0442	0.461	1.454	0.482	1.518	3.640	0.2747	0.738	1.426	5.854	0.392	1.608
19	0.688	0.717	0.187	0.9599	1.0418	0.477	1.443	0.497	1.503	3.689	0.2711	0.733	1.490	5.888	0.404	1.596
20	0.671	0.697	0.180	0.9619	1.0396	0.491	1.433	0.510	1.490	3.735	0.2677	0.729	1.548	5.922	0.414	1.586
21	0.655	0.679	0.173	0.9638	1.0376	0.504	1.424	0.523	1.477	3.778	0.2647	0.724	1.606	5.950	0.425	1.575
22	0.640	0.662	0.167	0.9655	1.0358	0.516	1.415	0.534	1.466	3.819	0.2618	0.720	1.659	5.979	0.434	1.566
23	0.626	0.647	0.162	0.9670	1.0342	0.527	1.407	0.545	1.455	3.858	0.2592	0.716	1.710	6.006	0.443	1.557
24	0.612	0.632	0.157	0.9684	1.0327	0.538	1.399	0.555	1.445	3.895	0.2567	0.712	1.759	6.031	0.452	1.548
25	0.600	0.619	0.135	0.9696	1.0313	0.548	1.392	0.565	1.435	3.931	0.2544	0.709	1.804	6.058	0.459	1.541
Over 25	$\frac{3}{\sqrt{2n}}$	$\frac{3}{\sqrt{2n}}$				*	**	*	**							

$^{*}1-\frac{3}{\sqrt{2n}}$; $^{**}1+\frac{3}{\sqrt{2n}}$

4.16 DESIGN STUDY FOR CONTROL CHARTS

4.16.1 Design Study 4.17

Control charts can be utilized to keep track of manufacturing failures. Presented below are data collected over time showing fastening torque defects, evaluating both over and under-torque conditions found per month on a vehicle assembly line. Twelve months of data were obtained for the first year and 10 months of data have been collected to date for the second (or current) year. It is important to note that each data point should be posted as soon as its value is determined.

W	J	F	MA	AP	MY	JN	JL	AU	S	O	N	D
Year 1	92	70	81	87	89	93	97	79	105	86	89	81
Year 2	70	57	80	60	79	65	74	62	61	61		

You are asked to develop a control chart to assess the reliability of the fastening torques.

Approach:

Since there is no previous data upon which to base your analysis, you decide to establish a control chart based on Year 1 data. Year 2 (and subsequent) are to be charted and evaluated based upon Year 1 guidelines. The charting shall be performed on the basis that n is large. Note that for n large, we consider the value of s to approach the value of σ.

From Year 1 data you find: $\bar{x}_{YR1} = 87.42 \quad \sigma_{YR1} = 9.13$

Yielding:

$$UCL\ (95\%) = 87.42 + 1.96(9.13) = 105.13$$

$$LCL\ (95\%) = 87.42 - 1.96(9.13) = 69.52$$

$$UCL\ (99\%) = 87.42 + 2.576(9.13) = 110.94$$

$$UCL\ (99\%) = 87.42 - 2.576(9.13) = 63.90$$

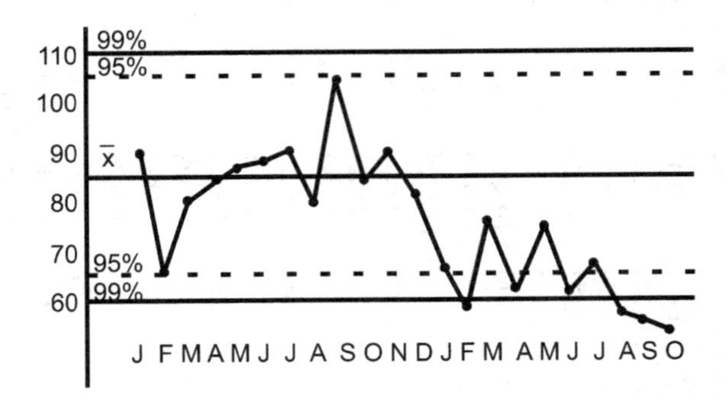

Analysis:

In Year 1, we note from the chart that there are four points in a row monotonically increasing above the mean line. This should be the cause for some concern. Indeed, the data for the remainder of Year 1 indicate significant swings, leading ultimately to "out of control" points in February of Year 2.

The Year 2 data appears to be entirely different than that of Year 1 data. Year 2 data appears to have been displaced "downward" to out of control conditions and/or to form a new base. This data would require either a careful examination of the process to determine the cause of the shift or a reevaluation of the data, or both. Additionally, the Year 2 data displays a different cycle characteristic than that of Year 1 data.

REFERENCES

1. Natrella, M. G. 1963. *Experimental Statistics*, NBS Handbook 91. National Bursar of Standards, United States Department of Commerce. pp. 1–9.
2. Natrella, M. G. 1963. *Experimental Statistics*, NBS Handbook 91. National Bureau of Standards & United States Department of Commerce. p. T-2.
3. Natrella, M. G. 1963. *Experimental Statistics*, NBS Handbook 91. National Bureau of Standards, United States Department of Commerce. p. T-3.
4. Box, G. E. P., W. G. Hunter and J. S. Hunter. 1978. *Statistics for Experimenters.* Wiley-Interscience. p. 49.
5. Natrella, M. G. 1963. *Experimental Statistics*, NBS Handbook 91. National Bureau of Standards, United States Department of Commerce, p. T-5.
6. Walpole, R. E. and R. H. Myers. 1985. *Probability & Statistics for Engineers and Scientists*, 3rd Ed. MacMillan Publishing. p. 271–272.
7. Walpole, R. E. and R. H. Myers. 1985. *Probability and Statistics for Engineers and Scientists*, 3rd Ed. Macmillan Publishing.
8. Walpole, R. E. and R. H. Myers. 1985. *Probability and Statistics for Engineers and Scientists*, 3rd Ed. MacMillan Publishing, p. 270.
9. Natrella, M. G. 1963. *Experimental Statistics*, NBS Handbook 91. National Bureau of Standards, United States Department of Commerce. p. T-4.
10. Natrella, M. G. 1963. *Experimental Statistics*, NBS Handbook 91. National Bureau of Standards, United States Department of Commerce. pp. T6–T9.
11. Natrella, M. G. 1963. *Experimental Statistics*, NBS Handbook 91. National Bureau of Standards, United States Department of Commerce. p. 18–13.

SYMBOLS

D Mean of normally distributed differences
F Testing distribution, see Eq. (4.17)
H Hypothesis
H_0 Null hypothesis
H_A Alternative hypothesis
LCL Lower control limit
LL Lower Limit
m Mean
n Number of data values
OC Operating characteristic
P Probability
s Sample standard deviation
s^2 Sample variance
S_D Standard deviation of normally distributed differences
t Student distribution
UCL Upper control limit
UL Upper Limit
$\overline{x}$ Sample mean
Z Defined by Eq. (4.4)
∂ Number of degrees of freedom
α Level of significance

ζ	Testing number
μ	Population mean
σ	Population standard deviation
σ^2	Population variance
τ	Student distribution
χ^2	Chi-square

5 Riveted and Bolted Joints

5.1 BASIC FORMULAS FOR RIVETS

Several fundamental rules concerning the design of single rivets and riveted joints have already been given in Chapters 1 and 2. In addition, the brief historical account in Chapter 1 shows how the popularity of rivet fastening has declined over the years in favor of high-strength bolts, adhesives and welded connections. Nevertheless, the concept of a riveted joint is still of interest in industrial applications. And, with regard to shear transfer, the mechanical behavior of bolts and rivets is quite similar.

A riveted connection can be made in several ways: impact, spinning, or squeeze forming. An obvious disadvantage of a riveted joint, however, is that the connected components are not easily separated.

While conventional rivet components are usually less expensive than bolts per se, their strength in both tension and shear is usually lower than comparable size bolts. Moreover, bolts can be inserted and tightened without the specialized tooling required for rivets.

Traditionally, riveted joints are either of lap or butt type (see Figures 1.2 and 1.3) and arranged in either a chain or a symmetrical pattern.

Figure 5.1 (a reproduction of Figure 2.1) depicts a rivet subjected to double shearing. From Eqs. (2.1) and (2.2) the average and maximum shear stresses are:

$$\tau_{avg} = 0.64\frac{F}{d^2} \tag{5.1}$$

and

$$\tau_{\max} = 0.85\frac{F}{d^2} \tag{5.2}$$

When, however, the connected plates instead of the rivet, are highly stressed, the plate metal around the rivet hole becomes progressively distorted and elongated. The load F_b where failure can be expected is:

$$F_b = tdS_C \tag{5.3}$$

where F_b is the maximum bearing load, t is the plate thickness, S_C defines the ultimate compressive strength of the plate, and d is the rivet diameter, usually taken as equal to the hole diameter.

In general d is often taken as the nominal diameter of the rivet prior to the driving operation. In the case of an aluminum plate, a rivet hole may begin to distort at a bearing load corresponding to the stress that is lower than the expected bearing yield strength of the aluminum alloys. The bearing strength is that value of stress at which the permanent set in the rivet hole is about 2% of the hole diameter. The analysis of bearing strength is about the same for countersunk and protruding-head rivets. However, countersunk rivets extending completely through the plate thickness should be avoided because of lower efficiency of the bearing area.

The strength of a plate in tension between the rivets can be calculated on the basis of the expression:

$$F_t = S_u(w - nd)t \tag{5.4}$$

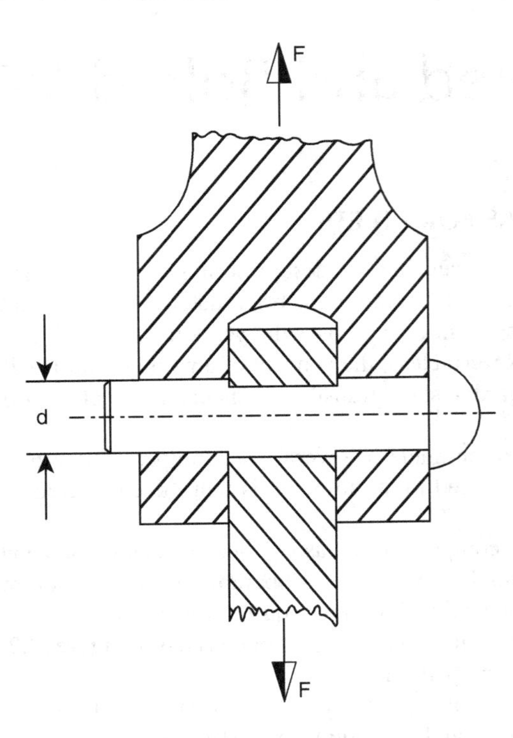

FIGURE 5.1 A rivet in double shear.

where F_t is the tensile load corresponding to the ultimate tensile strength, S_u, of the plate, w is the total width of the plate, n is the number of rivets in a transverse row, and as before, t is the plate thickness and d is the rivet diameter.

Essentially, Eq. (5.4) says that the ultimate strength multiplied by the net area of the plate gives the working tensile load. Hence, if the rivet pitch is small, the diameter big, and the plate thin, we can expect a tear-out model along the rivet row. The calculations should be based on the minimum net area of the cross-section.

5.2 FORMULAS FOR ECCENTRIC SHEAR

The formulas given by Eqs. (5.3) and (5.4) apply to symmetrical designs with all the rivet loads being equal. When, however, an external load is applied to a group of rivets, not through the centroid of the entire joint, the distribution of the forces and geometry can be illustrated as in Figure 5.2.

If the connected plates can be assumed to be rigid and the resisting force on each rivet proportional to the applied stress, then the problem can be solved as follows. Due to the external force F shown in Figure 5.2, the direct force component, one each of the four rivets in this case would be $F/4$. The shear component on each rivet caused by the external moment F_e is proportional to its distance r measured from the centroid, shown in Figure 5.2.

Figure 5.3 shows the shear component, represented by F_s and its components in the horizontal (X) and vertical (Y) directions for the lower right corner of the joint illustrated in Figure 5.2. In this case, the reaction identified in Figure 5.3 corresponds to the torsional effect on one rivet only, and it should be noted that the direction of F_s is perpendicular to the line connecting the rivet center and the centroid (CG).

This procedure can be used to resolve all the direct and torsional loads into components. These can then be added vectorially at each fastener to determine which fastener is likely to be critical for a particular joint design and direction of the external load F.

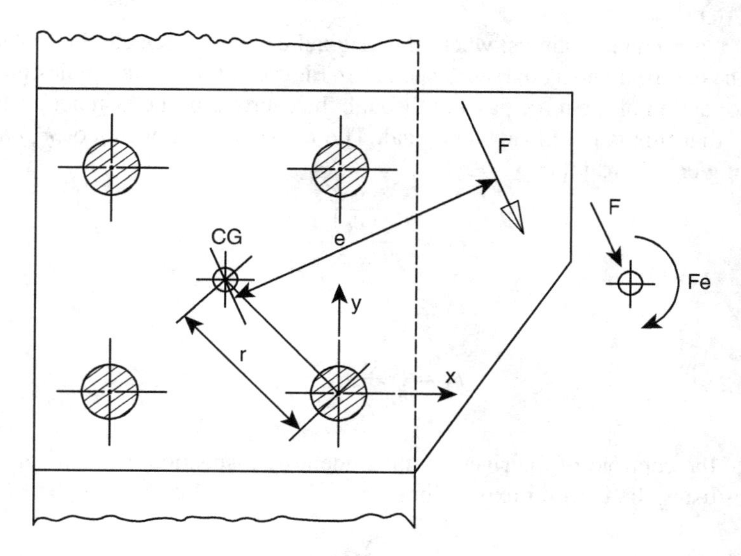

FIGURE 5.2 Example of an eccentric shear load.

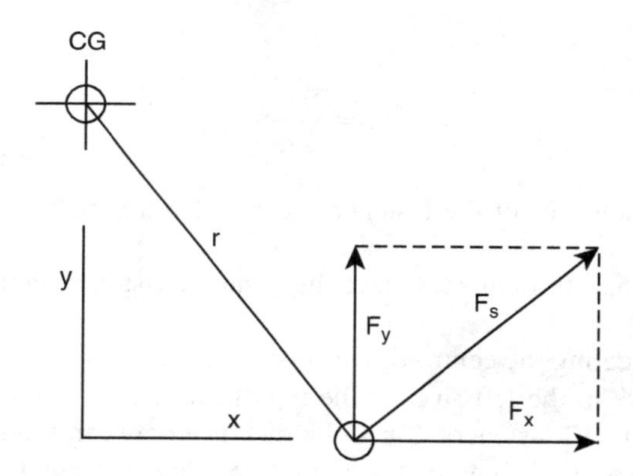

FIGURE 5.3 Shear forces on the lower right rivet by the joint of Figure 5.2.

If n is the number of fasteners in an eccentric shear joint then the following expressions can be used for calculating the shearing forces from the equivalent applied force and couple systems:

$$F_a = \frac{F}{n} = \text{ direct load on the individual rivet} \tag{5.5}$$

and

$$F_S = \frac{(F_e)_r}{\Sigma\left(x^2 + y^2\right)} = \text{ shear load on rivet} \tag{5.6}$$

where r is the distance shown in Figure 5.2.

In the conventional design literature the equivalent, reactive force may be either axial or shear. By locating the appropriate vectors and examining the positions of the fasteners with respect to the direction of loading, it is possible to identify the critical fastener. The method of the joint analysis is then based on a principle of statics; where the original external force can be replaced by a force acting through the centroid and a couple F_e, such as in Figure 5.2. In optimum design, it is desirable to have the line of action of the force passing through the centroid of the fastener group. Experience shows, however, that this is not always practical. The relevant components of F_S can be obtained from the following equations [1]:

$$F_X = \frac{(F_e)_y}{\Sigma(x^2 + y^2)} \tag{5.7}$$

and

$$F_y = \frac{(F_e)_x}{\Sigma(x^2 + y^2)} \tag{5.8}$$

If the location of the centroid of the rivets is not evident by inspection, we can determine the centroid coordinates using the familiar expressions:

$$\overline{y} = \frac{\Sigma Ay}{\Sigma A} \tag{5.9}$$

and

$$\overline{x} = \frac{\Sigma Ax}{\Sigma A} \tag{5.10}$$

where A is the cross-section area of the fastener and x and y are the respective coordinates of the rivets.

The formulas of Eqs. (5.5) through (5.8) have the inherent assumption that all the rivets have the same geometries.

In a typical joint, in locating the centroid, the coordinates used in Eqs. (5.9) and (5.10) may conveniently be taken relative to the left rivet of the bottom row. For example, Figure 5.4 depicts an eccentric joint system with 10 fasteners. Since the distance between the second and third rows is relatively large, the centroid is likely to be located in the space between the second and third rows, as indicated in Figure 5.4. Also, the choice of the reference lines for the determination of the position of the centroid is, of course, arbitrary.

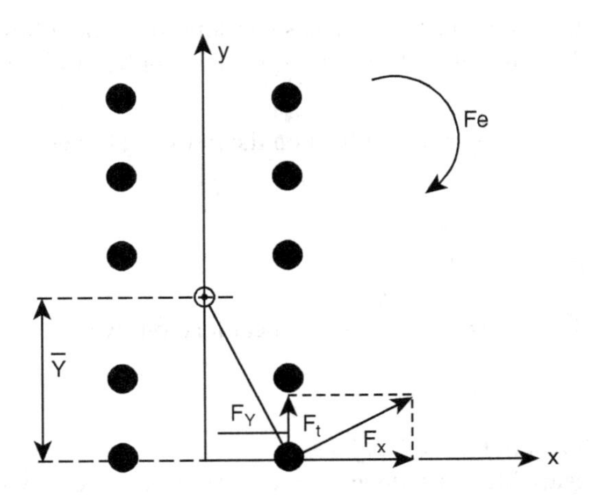

FIGURE 5.4 Centroid notations.

5.3 PERFORMANCE OF RIVETED JOINTS

Riveted joints should be designed so that the rivet shaft is not in tension. That is, the objective of a rivet joint is to maintain the connection through shear forces in the rivet shaft. However, the designers should be cautioned that failure of the joint may occur by shearing of the rivets at a lower average shear stress than the ultimate shear strength of a single rivet. Nevertheless, tests on hot-driven, low-carbon rivets indicate a favorable equalization of the rivet forces resulting from the deformation capacity of the rivet material. This is particularly true in the case of shorter connections using rivets made out of A36 steel.

It is generally recognized that riveted joints can fail under certain cyclic loads, even though rivet holes can act as crack propagation inhibitors. At the same time, however, the holes represent stress concentrations which tend to reduce the endurance limit of the plate by as much as about one-third. Also, the joint failure can be influenced by the physical condition of a rivet hole. For example, a punching operation can set up a system of stresses making the endurance limit of a drilled hole higher than that of a punched hole.

There are also a number of helpful hints which can lead to a better joint design for fatigue. These can be summarized by the following two headings:

1. Improved Fatigues Strength
   - Butt-riveted joint
   - Double-splice
   - Double shear
2. Interior Resistance to Fatigue
   - Lap joint
   - Single-splice
   - Single shear

The fatigue life of a riveted joint can be enhanced by increasing the number of rivet rows. The method of fabrication can also be important in the case of steel rivets. For example, cold-driven rivets appear to have a better fatigue resistance than hot rivets. At the same time, aluminum rivets are not sensitive to the hot or cold driving operations.

It is obvious that a riveted joint cannot be as strong as a solid plate. The appropriate ratio of a riveted joint strength to that of a solid plate determines the efficiency. Usually, the efficiency calculations are made with respect to bearing, shear, or tension. Table 5.1 provides a quick summary of the typical efficiencies [2].

Since the cost of a fabricated structure is governed to a large degree by the joint efficiency, the choice of the material and the method of production are always important. However, with recent developments in adhesives, bolt, and welding technology, the frequency of shop riveting is

TABLE 5.1
Design Efficiencies of Riveted Joints

Type of Joint	Average Efficiency (%)
Single butt joint	65
Double butt joint	79
Triple butt joint	84
Single lap joint	55
Double lap joint	65
Triple lap joint	75

decreasing and modern field riveting is almost nonexistent. Hence, the remainder of this chapter will be devoted to bolted connections. In some respects, the behavioral characteristics of a number of mechanical joints are quite similar for riveted and bolted constructions.

5.4 RIGID BOLTED CONNECTIONS

In a bolted joint there are two principal forces of interest: 1) the initial tensile load W_i on the bolt and 2) the external load W_e on the joint. Assuming that the joint has not separated under the applied load W_e, the decrease in the deformation of the connected parts must be equal to the increase in bolt extension.

For an ideal metal-to-metal contact of a typical bolted joint, there must be simultaneous axial tensions and frictional effects as represented in Figure 6.6.

In general, the actual force carried by the bolt can be expressed in terms of the applied external load and the initial preload as follows:

$$W = W_i + CW_e \tag{5.11}$$

where C represents the overall stiffness coefficient of the entire joint. Specifically, C may be expressed as:

$$C = \frac{k_b}{k_b + k_c} \tag{5.12}$$

where k_b is the stiffness of the bolt and k_c is the stiffness of the connected members in the joint. In the sketch of Figure 5.5, the resultant spring constant of the entire joint thus depends only on the stiffness of the bolted parts, excluding the effect of frictional forces F_f.

Observe in Eq. (5.12) we have the following idealized bracketing conditions on the material stiffnesses:

$k_c >> k_b$ (soft bolt clamps rigid components) $W = W_i$
$k_c = k_b$ (all components have equal stiffness) $W = W_i + 0.5W_e$
$k_c << k_b$ (rigid bolt clamps soft components) $W = W_i + W_e$

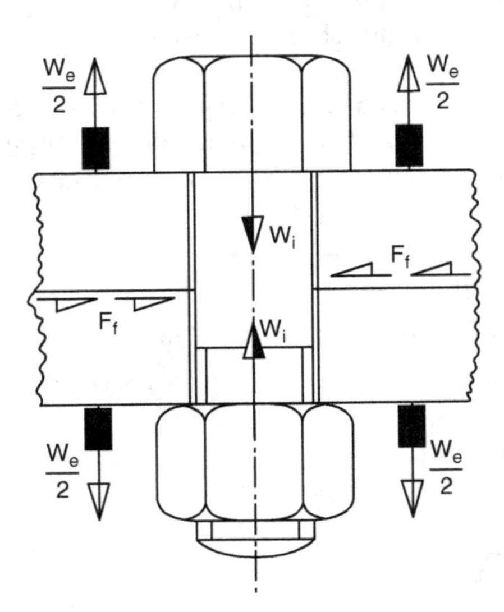

FIGURE 5.5 Sketch of a bolted joint.

5.5 GASKET EFFECTS

In many cases, a bolted joint involves a gasket so that the spring constant of the assembly K_c becomes a function of two basic stiffnesses: k_j and k_g, where k_j denotes the stiffness of the connected plates or flanges and k_g is the gasket stiffness. Figure 5.6 provides a sketch of a bolted joint with a gasket.

The spring constant k_c of the assembly may then be expressed as:

$$k_c = \frac{1}{\dfrac{1}{k_g} + \dfrac{1}{k_j}} \tag{5.13}$$

By substituting Eq. (5.13) into (5.12) we obtain a general formula for estimating the joint stiffness C of a gasketed joint:

$$C = \frac{k_b\left(k_g + k_j\right)}{k_b k_j + k_b k_g + k_g k_j} \tag{5.14}$$

Equation (5.14) can be used when the magnitude of k_j is comparable to that of the gasket stiffness k_g. In the case of the so-called flexible joint or soft joint, where a relatively soft gasket is placed between the rigid plates or flanges, the parameter k_g may govern the design entirely, since the effect of a softer gasket is to increase the portion of the external load in the bolt. Under such conditions $k_g << k_j$ so that Eq. (5.14) reduces to:

$$C = \frac{k_b}{k_b + k_g} \tag{5.15}$$

In Eq. (5.15) the value of C can vary between 0 and 1, depending on the ratio of k_g/k_b. As a general guide recommended for the selection of the stiffness parameter C in terms of bolt and gasket flexibilities [3], the summary in Table 5.2 may be of use.

The foregoing discussion is intended to gather some of the rules and formulas necessary for sizing the bolt properly in relation to the external load and stiffness characteristics of the joint.

Prior to the application of the external load, the bolt is tensioned in proportion to a preload W_i. Then, any additional load in a gasketed joint is expected to increase the bolt deformation in the axial

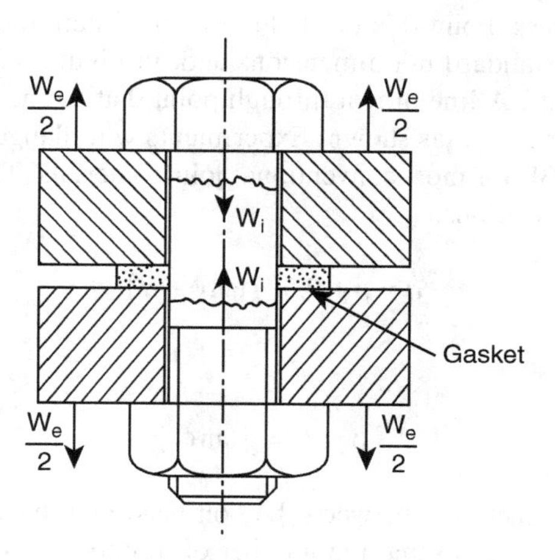

FIGURE 5.6 Sketch of a bolted joint with a gasket.

TABLE 5.2
Guide to Joint Stiffnesses

	Parameters	
Type of Bolted Joint	$\frac{k_g}{k_b + k_g}$	$\frac{k_b}{k_b + k_g}$
Soft gasket joint held by studs	0.00	1.00
Soft gasket joint with through bolts	0.25	0.75
Asbestos gasket joint	0.40	0.60
Soft-copper gasket with long through bolts	0.50	0.50
Hard-copper gasket with long through bolts	0.75	0.25
Metal-to-metal rigid joint with long through bolts	1.00	0.00

direction. At the same time, the application of the external separating load W_e causes the gasket and the connected members to undergo a decrease in deformation. This can be accomplished through a partial relief of the compression of the flanges and the gasket, originally induced by W_i. Furthermore, in a special case when the external load W_e becomes so large as to make the force on the gasket equal to zero, the joint becomes separated and the bolt assumes the entire external load W_e.

In terms of the formulas given in this section, the mathematical reasoning can be summarized as follows. When k_g is small compared with k_j, Eq. (5.13) gives $k_c - k_g$. Hence, Eq. (5.12) reduces to Eq. (5.15). Finally, for $k_g < k_b$, Eq. (5.15) yields $C = 1$ and Eq. (5.11) shows that $W = W_i + W_e$.

Where relatively large washers and gaskets are used, the joints develop greater flange areas under compression, mitigating local deformation of the assembled parts but increasing the portion of the external load to be carried by the bolt.

Experience shows that by changing the stiffness of the individual washers and gaskets, the overall spring constant of the assembled joint and the resultant bolt load can be significantly changed.

5.6 STIFFNESS OF A FLANGE IN COMPRESSION

The problem of determining the stiffness constant of a compressed flange or abutment is not trivial because it requires knowledge of the effective cross-sectional area of the flange undergoing deformation. Although in a complex joint the stiffness constants are more likely to be found experimentally, a semi-graphical procedure for finding the effective area of deformation can also be tried. Figure 5.7 provides the relevant notation.

The procedure is as follows: Point 0 is located at the maximum radius of the bolt contact area. This can be deduced from standard nut dimensions and, in Figure 5.7, this radius corresponds to half the width across the flats. A line drawn through point 0 at the half-cone angle ϕ establishes a point of reference at a distance $h/2$, as shown. Experiments with flange stiffness indicate [4] that ϕ may vary between 25 and 33E for most conventional joint materials. The required annulus width a follows from the geometric relations:

$$a = \rho + 0.5(h \tan\phi - d) \tag{5.16}$$

where ρ is

$$\rho = R = g \tan\phi \tag{5.17}$$

The maximum radius of contact area between the bolt head and the washer is R. This dimension is known for any given bolt size, so that the number of unknowns in the equations above can be reduced to three.

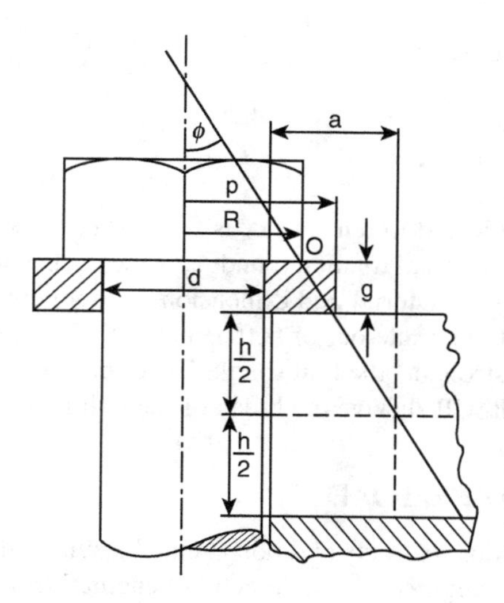

FIGURE 5.7 Notation for the effective area method.

An effective calculation sequence is to start with an assumed value of ρ or g, defining the washer as shown in Figure 5.7. In this manner, the solution of Eqs. (5.16) and (5.17) establishes the necessary washer dimensions and the annulus width. This, in turn, permits the calculation of the approximate compressive area and the relevant spring constants.

5.7 CONVENTIONAL DIAGRAM FOR THE BOLT PRELOAD

The initial elastic response of a complete bolted joint can be seen with the aid of a simplified diagram, such as that shown in Figure 5.8. As the tightening torque on the bolt is increased, the shank elongates according to the linear force-deformation relation of Hooke's law. The flange, representing a typical connected member, undergoes elastic compression. Since the amount of preload acts on the bolt and the flange simultaneously, the two force-deformation characteristics can be assembled into one diagram, as in Figure 5.8.

This type of a complete diagram is given in many books on machine design as a basis for defining the elastic response of a typical bolted connection. The relevant spring constants can be expressed as $k_b = \tan \psi_0$ and $k_c = \tan n_0$ for the bolt and clamped member, respectively. When $k_c = k_g$, the spring constant or the compressive stiffness of the gasket is:

$$k_g = \frac{A_g E_g}{L_g} \tag{5.18}$$

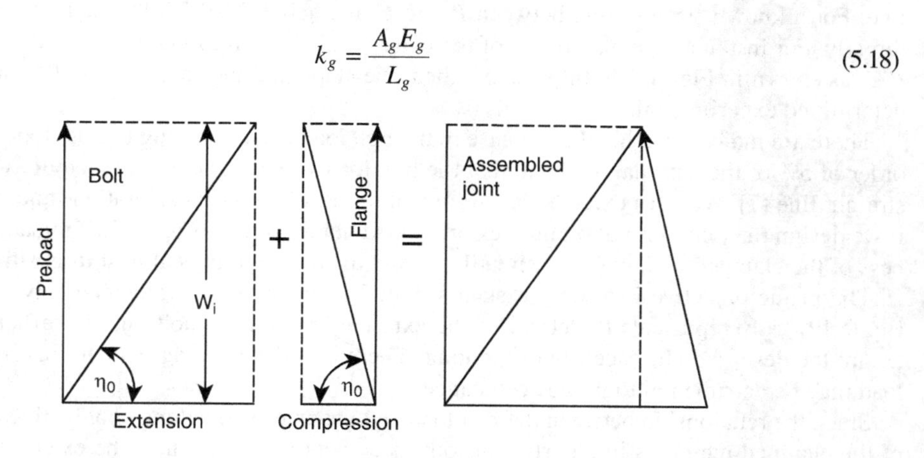

FIGURE 5.8 Diagram for bolt preload.

Similarly for the bolt shank, we have:

$$k_b = \frac{A_b E_b}{L_b} \tag{5.19}$$

where L_b is the bolt length subjected to tension, L_g is the gasket thickness, A_b is the cross-section area of the bolt based upon its nominal diameter, and A_g is the gasket area under compression.

It should be noted that the nut material and dimensions seldom enter the consideration of stiffness, even though the nuts can be made out of softer material. There is generally no requirement for the nut to be as hard or as strong as the bolt shank. Experience shows that a nut with a material strength of 75% that of the bolt will survive the break of the bolt in tension.

5.8 THEORIES OF JOINT PRELOAD

Although in some circles the topic of "nuts and bolts" may be synonymous with boring details and monotonous work, bolted joint engineering has recently benefited from increased attention to classical ideas which were developing in the early 1900s [5]. The emphasis in Swedish and German industries, for example, continues to be on a "system's" approach, precise measurements, high-strength bolt design and a significant joint rigidity to assure minimum bending, misalignment, and prying action. The bolt preload then should be as high as possible followed by retightening of the joint after proof testing or after a brief service life of the assembled joint.

The current view of preload exceeding the amount normally required by design is based on the premise that a bolted joint seldom fails under purely static conditions.

It was previously shown that when very rigid components are clamped together by the conventional bolts, the actual load in the bolt is very nearly equal to the theoretical preload. During his recent studies of metal-to-metal contact flanges in Sweden, Webjörn [6] considered an intriguing interpretation of behavior of the bolted joint in terms of the preload and the external load. For a tensile rod observed in a tensile testing machine, the external load is felt directly by the rod. Knowing the relevant elongation and the modulus of elasticity of the rod material, the load experienced by the rod can be calculated.

If the analogy is pursued between the test rod and a joint where a bolt clamps together, say, two pieces of steel, the bolt preload W_i and the external load W_e can be plotted as shown in Figure 5.8.

As the bolt preload W_i develops from point A to B, the external load W_e is still zero with the joint carrying no applied load. In service, W_e takes place and the actual load felt by the bolt progresses from B to C. Beyond point C, the preload has no additional effect and the curve follows the C-D line. For a known straight line between B and C, the actual load felt by the bolt can be obtained directly. For instance, the magnitude of the external load equal to AG corresponds to the bolt load $B'G$, as shown in Figure 5.9. In principle then, the slope and the shape of the BC curve should be determined experimentally.

There are indications that the increase in the bolt load corresponding to this slope may be on the order of 5% of the external load and that the bolt load versus the external load does not follow the straight line [7]. Webjörn extends this argument to the stiffness considerations and concludes that, if we design the joint for a maximum external load not to exceed 67% of the preload, then the stiffness of the clamped members is fully utilized and the bolted joint is almost insensitive to fatigue.

The prime objective is to design such a bolted joint that the parameter C given originally by Eq. (5.12) truly represents the effect of the external load on the bolt, and is sufficiently small to assure the desired endurance life of the joint. There is widespread agreement that insufficient preload may be detrimental to fatigue resistance.

Since the relationship between the bolt load and the external load may not be linear, the stiffness of the clamped members in a particular joint is evidently a function of the external load when its value approaches the preload in magnitude. The limitation of the external load to 2/3 of the preload,

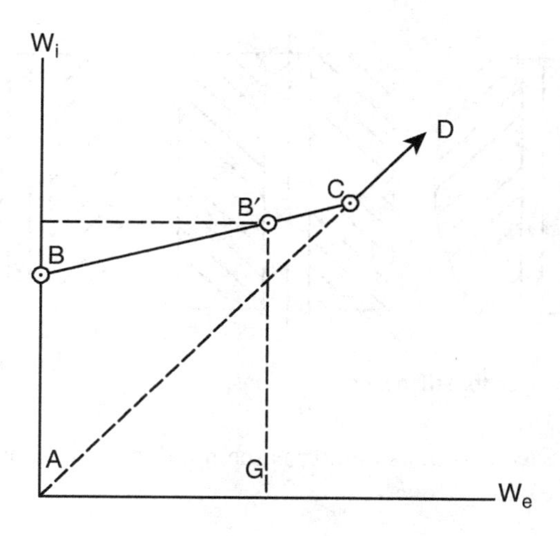

FIGURE 5.9 Preload diagram for a rigid joint.

as suggested by Webjörn [5], gives assurance that for all practical purposes the joint stiffness in this region of bolt response is invariant. Such reasoning, of course, is possible only under such conditions as "no-gasket" design and other circumstances which can be summarized as follows:

- Clamped components are rigid.
- Clamped length is large compared to bolt diameter.
- Bolts are made from high-strength material.
- Solid, flat, and hard washers are used.
- Maximum allowable torque is permitted.

The above statements imply that, under the very special circumstances of rigid joint design, there is virtually no need to be concerned with the elasticity and deformation characteristics of the bolted connection.

5.9 EVALUATION OF SPRING CONSTANTS

Various arguments can be made for and against the calculations of spring stiffness parameters in a bolted joint. One of the more convincing against computational techniques is, perhaps, our inability to accurately model a bolted assembly. For this reason, many investigators elect to use experimental approaches instead of the calculation of the spring constants.

This brief section is intended to be a general guide for those who might wish to ponder over the nonlinear effects between the joint stiffness and the external load.

In calculating joint stiffness, the portion of the joint compressed by the bolt can be defined as a barrel with a central hole, a hollow cylinder, or a pair of truncated cones. Figure 5.10 depicts the relevant cross-sections of these axisymmetric solids.

In using the idea of compression cones, it is often assumed that the cone sides intersect the bearing surfaces under the nut and the bolt head at approximately 45E. If the cone is replaced by a hollow cylinder for the purpose of the calculations, using a technique similar to that shown in Figure 5.7, we can define the effective cross-sectional area as follows:

$$A_C = \pi\left(a^2 + ad - 0.75d^2\right) \tag{5.20}$$

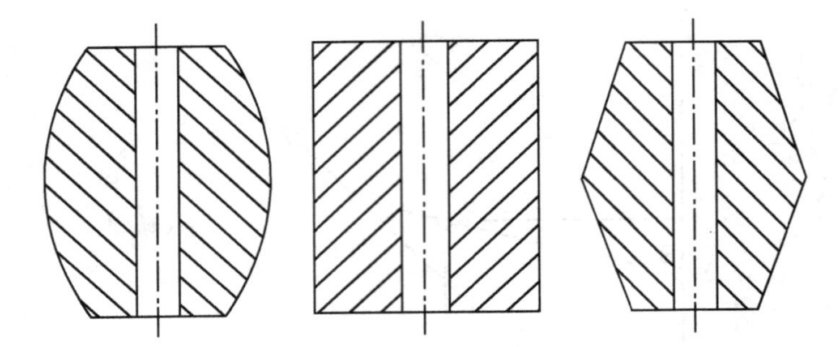

FIGURE 5.10 Equivalent shapes for stiffness calculations.

where a is given by Eq. (5.16) and d, as before, is the nominal bolt diameter. The stiffness of the connected members can be expressed as:

$$K_j = \frac{EA_C}{L_b} \quad (5.21)$$

If the assembled plates or flanges include a gasket, as in Figure 5.6, then the spring constant of the assembly K_C can be determined using Eqs. (5.13), (5.18), and (5.21) as:

$$K_C = \frac{A_C A_g E E_g}{A_C E L_g + A_g E_g L_b} \quad (5.22)$$

Once K_C is evaluated, the load carried by the bolt in a given joint can be found from Eq. (5.11) with the help of Eqs. (5.12) and (5.19). This procedure, however, requires knowledge of the amount of preload assigned to the bolt. (The amount of preload is denoted in this chapter by W_j.)

5.10 SYMMETRICAL SPLICE CONNECTION

Figure 5.11 provides an illustration of a simple tension-type splice held together by a number of fasteners. Even though the tensile force is transmitted across the splice, the bolts or rivets holding the joint together may, or may not, respond in shear.

If the compression of the connected plates is sufficiently high, then a *friction*-type joint develops which has a reduced probability of slip during the life of the structure provided the total external load is not exceeded. This type of design assures that the load is primarily transferred by frictional forces acting on the contact area of the plates, angles, or other structural shapes that may constitute a portion of the joint.

Observe in Figure 5.11 that the joint has geometrical symmetry. The ultimate capacity of this connection is then reached when the total frictional resistance is overcome with the attendant overall slip and bearing against the bolts. A slip-resistant design is recommended in the following situations:

- Slippage cannot be tolerated.
- Spliced joint is subjected to stress fluctuations.
- Spliced joint is designed for fatigue.

The externally applied load F_{s1} where slip can occur may be expressed as:

$$F_{s1} = mnW_iR_s \quad (5.25)$$

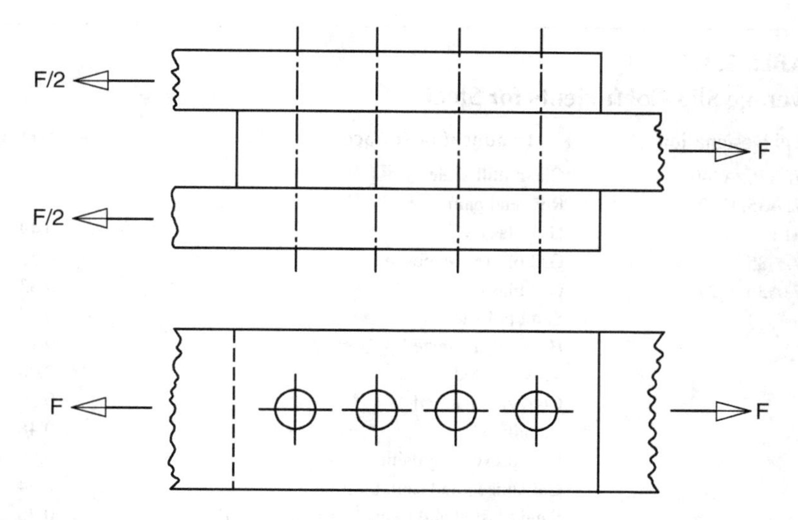

FIGURE 5.11 Example of a simple splice joint.

where m is the number of slip planes and n is the number of fasteners (typically bolts). As before, W_1 is the bolt (or rivet) preload and R_S is the slip resistance coefficients.

For a given geometry of the splice, the slip load F_{s1} is proportional to the product of slip resistance factor and bolt tensile load. Again, we are faced with a relatively simple mathematical model which, however, contains a number of limitations and variables that are not easy to quantify. This seems to be the general characteristic applicable to many mechanical joints where deceptive simplicity of the design formulas often prevails against the usual complexity of actual practice.

In the case of Eq. (5.23), it is necessary to recognize that the slip resistance coefficient and bolt tension can vary from joint to joint. In particular, the parameter R_S is highly dependent on the surface condition and treatment, which are very difficult to define and control. Standard experimental techniques have been developed [8] for a slip test using symmetric bolt joints with two and four bolt configurations typically utilizing A325 and A490 bolts. However, since there is still considerable uncertainty with the problem of bolt preload, it is necessary to calibrate each bolt prior to the test. The alternative is to utilize some of the special bolt systems available now on the market which include automatic indicators of the bolt force [9].

5.11 SLIP COEFFICIENTS

Precise mechanics of "slip" is difficult to describe and quantify: The difficulty arises since sudden slip is essentially the cumulative effect of multiple microslips. Since sites of microslip can occur seemingly at random between the contacting surfaces it is virtually impossible to have a theoretical model predicting the global slip (or friction) coefficient. Therefore, we need to rely upon empirical (or testing) results. Table 5.3 provides a list of such results for steel.

The effects of the number and pattern of bolts as well as the number of clamped surfaces have been studied and have shown to be less important than the influence of the type of material and surface condition on the parameter R_S. The summary of R_S values, given in Table 5.3, is based on the compilation provided by Fisher and Struik [9].

Although a wide range of slip coefficients is theoretically available, substantial scatter in all of them should be expected due to the variations of fabrication procedures. Hot dip galvanizing treatment and organic zinc-rich paints are not recommended for slip-resistant joints.

TABLE 5.3
Average Slip Coefficients for Steel

Steel Designation	Treatment of Surface	R_S, Average
A7, A36, A440	Clean mill scale	0.32
A7, A35, F_e37	Red lead paint	0.07
A514	Grit blasted	0.49
A7, A35	Grit blasted, exposed	0.53
A7, A36, F_e37	Grit blasted	0.33
	Semi polished	0.28
	Hot dip galvanized	0.18
	Vinyl treated	0.28
	Cold zinc painted	0.30
	Metallized	0.48
	Rust preventing paint	0.60
	Galvanized and sand blasted	0.34
	Sand blasted and treated with linseed oil	0.26
	Sand blasted	0.47

5.12 LOAD TRANSFER BETWEEN BEARING AND SHEAR

When the joint illustrated in Figure 5.11 loses its frictional resistance and a major slip occurs, the hole clearance is taken up to the point where the joint load F is transferred by means of shearing and compression of bolt shanks. This type of a connection is called a "bearing type" joint where the shear strength of the fastener is the critical parameter. This is quite different from the slip-resistant design where the bolt preload plays the decision role. Nevertheless, it is the practice to appropriately tighten all bolts in the "bearing type" joint, to increase the overall joint rigidity, assure more even load transfer, and to safeguard against nut loosening. The effect of the axial stress in the bolt on the shear strength of the bolt is minimal.

As the load on the bolted splice increases, the slip zone moves inward from the ends toward the center of the joint. Finally, major slip occurs, and gradually causes the succeeding bolts to experience horizontal (shear) loadings. Figure 5.12 presents a schematic illustration of the deformation of the "bearing type" splice.

In the representation of Figure 5.12 the curved lines correspond, in an exaggerated way, to the center lines of the bolts (five assumed here).

The bolt deformations are greater at the ends of the joint suggesting that the end bolts are carrying higher loads. Only in short connections with a small number of fasteners, load equalization is likely to take place prior to bolt failure in shear.

In longer joints, the end bolts may reach their critical shear deformations and break prior to the remaining bolts achieving their full strength. The break of the end bolt is eventually followed by a sequential failure of other bolts known as the "unbuttoning." This particular feature of a longer

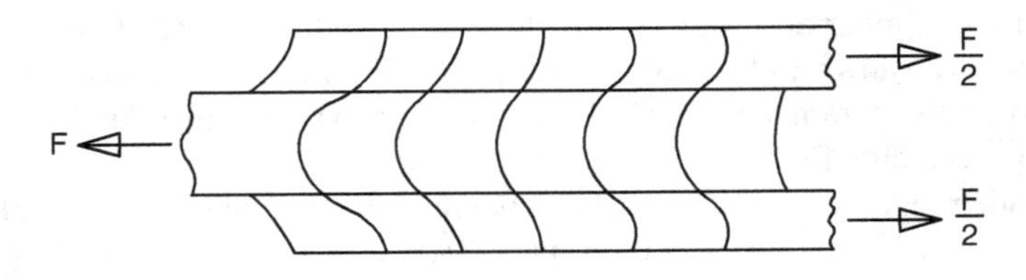

FIGURE 5.12 Approximate deformation lines for a "Bearing type" joint experiencing major slip.

splice is, of course, the result of uneven load distribution which should have been accounted for in the original design.

Experiments show that large shearing deformations of the bolts cause the clamping force to be relaxed. The simultaneous reduction in the frictional resistance can develop in the vicinity of the bolts carrying higher loads. The problem of predicting load transfer in a long, bolted splice is therefore not trivial. The solution of the equilibrium and compatibility equations for long joints with many fasteners is laborious, and is often impractical for conventional design purposes. However, such studies have been documented for the yield strength of plate materials ranging from about 30 to 100 ksi [9] and can be used to get some idea as to the distribution of the frictional forces and loading on bolts. The studies have been verified for the case of large steel joints.

Figure 5.13 provides a representation of an approximate distribution of the friction forces along the joint, here curve A represents a virtually no-slip condition and only the ends of the splice are involved. As the slip develops under increasingly higher loads, the maximum frictional resistance is reached as shown by curve B. Theoretically, with further deformation a uniform maximum resistance is overcome over the entire length of the joint. This condition must then be consistent with full slip.

Figure 5.14 depicts the load distribution on bolts corresponding to the ultimate strength of the joint. The figure provides a representation of the onset of yield and also a representation of the ultimate shear stress on the fastener bolts:

This illustration, although symbolic, suggests that under no condition is each fastener carrying an equal share of the load. As the load is increased, the fastener forces change, until an end fastener fails because of the overstrain. Hence, the end fasteners reach a critical shear deformation and stress before the full strength of each fastener in the joint can be utilized. This phenomenon has been observed in tests of riveted, as well as bolted connections.

The foregoing brief discussion of spliced connections shows that material and geometrical parameters must govern the joint behavior. As far as load distribution is concerned, the joint length is an important factor and it influences the ultimate load carrying capacity of the joint. For example, Fisher and Struik [9] point to the case of A36 material for plates and A325 bolts, as the ideal combination of properties consistent with the degradation of shear strength in longer joints.

Figure 5.15 provides an illustration of the effect of joint length on the shearing within the joint. It is interesting to note that joints up to about 10 in. in length had minimal effect on the shear strength. As the length increased, the decrease in the shear strength continued, although at a diminishing rate.

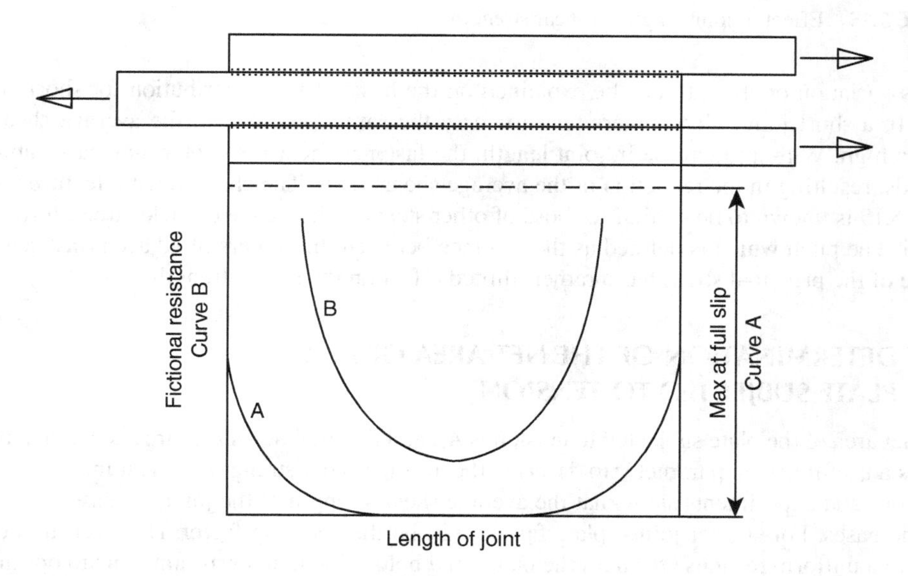

FIGURE 5.13 Approximate distribution of frictional resistance.

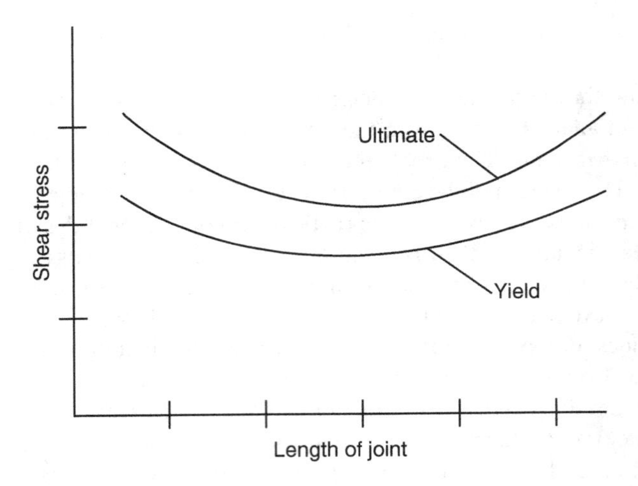

FIGURE 5.14 Load distribution along the joint.

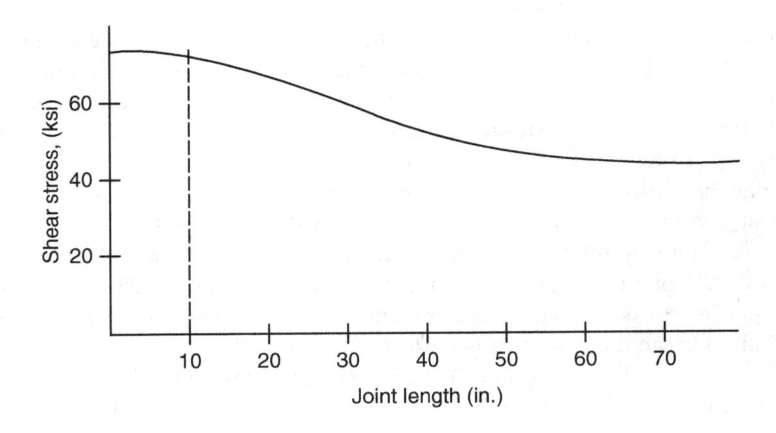

FIGURE 5.15 Effect of joint length on shear strength.

This variation of strength can be explained on the basis of load distribution for short and long joints. In a short joint, all fasteners tend to carry the same load; hence, the average shear value must be high. With an increase in joint length, the fasteners near the center may carry about half the loads, resulting in the reduction of the average shear stress. The characteristic feature shown in Figure 5.15 is known to be similar to those of other steels with a range of yield strength from 36 to 100 ksi. The pitch which is defined as the distance between the centers of adjacent fasteners along the line of the principal stress has a rather limited effect on the shear strength.

5.13 DETERMINATION OF THE NET AREA OF A PLATE SUBJECTED TO TENSION

If the net area of the plate subjected to tension is A_n, and the total bolt shear area is A_S, then the ratio A_n/A_S is a useful design parameter to describe the average shear strength of the joint.

Theory and experiments show that the average shear strength of the joint increases as the A_n/A_S ratio increases. For shorter joints, plate failure will be the deciding factor. However, it is difficult to assure a uniform response for both the plates and bolts which, in a way, amounts to optimizing a system involving several variables. It is perhaps sufficient to keep in mind the bracketing conditions

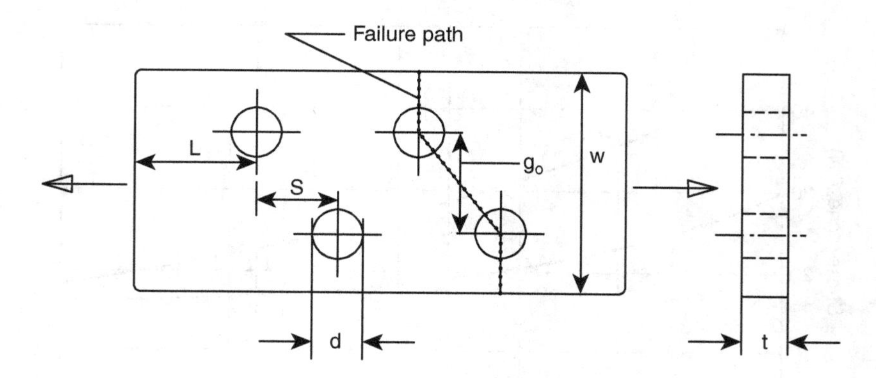

FIGURE 5.16 Notation for Eq. (5.24)

indicating the effect of the A_n/A_S ratio on the load distribution. For example, the theoretically equal participation of all the bolts is expected when the parameter A_n/A_S tends to infinity. The opposite is true of course when A_n/A_S approaches a small fraction.

Studies of spliced connections have also shown an interesting phenomenon concerning the ultimate strength of joints. It is known, for example, that the ultimate strength of perforated plates at the net section was higher than that of the plate calibration. This is explained by the fact that free lateral contraction, normally observed in a standard tensile specimen, could not develop because of a bi-axial stress pattern.

In ideal joint design of perforated plates, it would be desirable to yield the gross section before failure is initiated at the net section. The net section of the plate can be obtained by subtracting the term *ndt* from the nominal plate cross-section, as shown previously in Eq. (5.4) where, as before, *n* is the number of bolts, *d* is the bolt diameter, and *t* is the plate thickness.

Since the hole pattern in a plate influences the net area required in the design of a tension connection, a general formula developed for practical use [9] can be stated as follows:

$$A_n = t\left(w - nd + \sum \frac{s^2}{4g_0}\right) \tag{5.24}$$

where Figure 5.16 illustrates the dimensions for the formula.

In Figure 5.16, a staggered hole pattern is shown because it is more difficult to define the net section when the direction of loading is not perpendicular to the anticipated failure path.

Equation (5.24) is still widely used in design, although it was originally based on test results which date back to the early 1900s.

5.14 PLATE EDGE EFFECTS

Other variables affecting the performance of a spliced connection are the bolt grip length and the distance to the end of the plate. Since a bolt bends due to the slippage of the plates, as indicated in Figure 5.12, the number of plies in the joint is an important factor. The bending appears to increase with the increase of A_n/A_S ratios, enhancing the ultimate strength of the joint itself based on the assumption that failure takes place in the fasteners.

To prevent the fastener from splitting out of the plate material, the radio L/d, shown in Figure 5.17, should exceed the value calculated from the following expression:

$$\frac{L}{d} \geq 0.5 + 0.72\frac{S_b}{S_u} \tag{5.25}$$

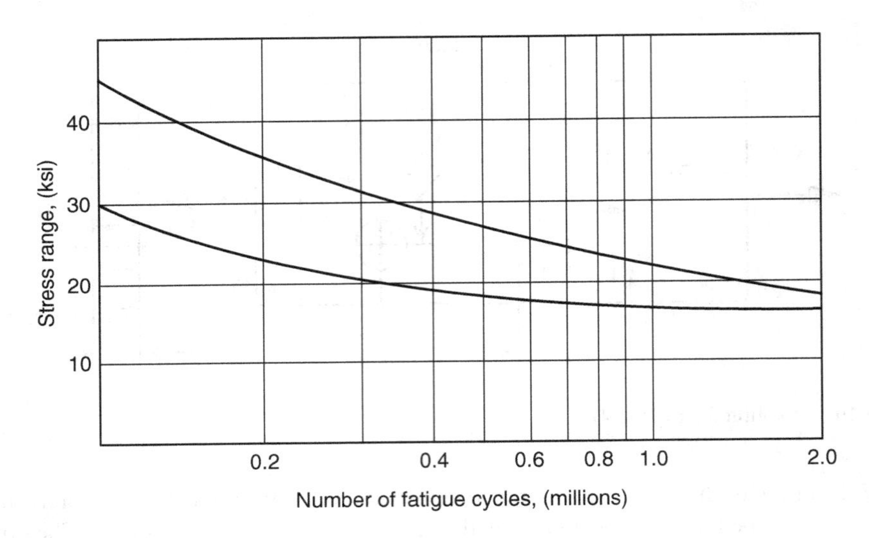

FIGURE 5.17 Design envelope for joint fatigue.

where S_b is the bearing stress in the plate material adjacent to the hole, as well as on the bolt surface. This stress follows from Eq. (5.1) as we make $S_c = S_b$, on the assumption that F_b is the bearing load transmitted by the fastener.

In the discussion of spliced connections, the term "fastener" can describe a rivet and a bolt alike. The agreement between the results based on Eq. (5.25) and test results is reported to be good for lower ranges of *L*/*d*. The results diverge, however, for *L*/*d* values higher than about 3. This is not surprising, because the failure mode away from the plate edge should change from the failure mode closer to the edge.

5.15 ALLOWABLE SHEAR IN SPLICED CONNECTIONS

Experience shows that the ultimate capacity of both "slip-resistant" and "bearing-type" joints is limited by the failure of one or more components of the connection. In general, the design can be based on stress or load criteria.

In the case of a slip-resistant joint, either the shear strength of the fasteners or the slip will govern the selection of geometry and materials. Current design practice tends to treat the mechanically fastened joints on the basis of allowable stresses in the plates and fasteners. The concept of joint design based on shear stress in a fastener has been successfully used for over a century. The load criterion practice is still based on the idea that the structure is designed to a maximum strength compatible with the safety factored load. This load is normally determined by multiplying the working load by a factor greater than 1.0. Essentially, the joint is designed for strength and then checked for performance under operating conditions.

A popular choice for the factor of safety is based on the ratio of the yield point to the tensile strength of the material, and in the area of spliced connections may vary between 2 and 2.5. However, as indicated earlier, the actual factor of safety against failure of the joint is also a function of joint length.

In the past, the largest and the more critical joints have probably had the lowest factors, and still the joints operated satisfactorily. It appears, therefore, that a minimum factor of 2.0 for bolts in shear is also in line with current practice for all steels, including the quenched and tempered alloy steels. Current experience covers the material strength values between 36 and 115 ksi and the joint lengths up to about 100 in. [9]. The allowable shear strength for the high-strength bolts can be expressed as:

$$\tau_a = \beta\tau_b = \text{ allowable shear strength} \tag{5.26}$$

where

τ_b = bolt basic shear strength
β = correction factor

For the case of a no slip-resistant joint design, $\beta = 1.0$. However, when the joint length exceeds about 50 in., $\beta = 0.8$. Hence, the allowable shear load for a spliced connection, symmetrically loaded, such as that shown in Figure 5.11, is:

$$F = mn\tau_a A_b \tag{5.27}$$

where

M = number of slip planes
n number of fasteners
A_b = bolt nominal cross-section
τ_a = allowable shear

The formula given by Eq. (5.27) assumes that the shear plane passes through the shank. When the bolts are sheared across the threaded portion, assume the joint load to be 75% of the value computed form Eq. (5.27). Fisher and Struik [9] suggest the following shear allowable for slip-resistant joint design:

$\tau_b = 30$ ksi A325 bolts
$\tau_b = 40$ ksi A490 bolts

5.16 FATIGUE AND SHEAR LAG EFFECTS

The effect of cycling loading on joint integrity can be looked upon from two aspects: If the load is transmitted by a frictional resistance, failure can be expected in the gross section. If, however, slip is present, the load is supported by bearing and shear reactions with the likely failure through the net section. The allowable stresses in fatigue, applied to the plate material with 95% survival rate, are given in Figure 5.17. This diagram is based on conservative assumptions [9]. The area between the two bounding curves contains stress range relationships currently supported by experimental data.

The basic criterion for determining a safe distance from the edge of plate is given by Eq. (5.26). For a properly installed bolt or rivet in a joint, a minimum distance from the center of the fastener to the edge of plate should be governed by $L/d = 1.5$.

Experience shows that when the L/d ratio tends to be 3.0, the mode of failure changes from a purely "shearing type" to that which also involves the deformation of the fastener hole and upsetting the material in front of the fastener. There is also a potential for a local instability of the plate which may limit the ultimate capacity of the end zone. This condition is possible when the lap plates are critical in bearing.

Other recommendations for the allowable stress criteria includes S_b/S_u ratio to be less than 1.5. With the provision of a suitable factor of safety, Eq. (5.25) becomes:

$$\frac{L}{d} \geq 0.5 + 1.44\frac{S_b}{S_u} \tag{5.28}$$

The foregoing considerations are applicable to design of flat plate joints where the shear planes are parallel to one another.

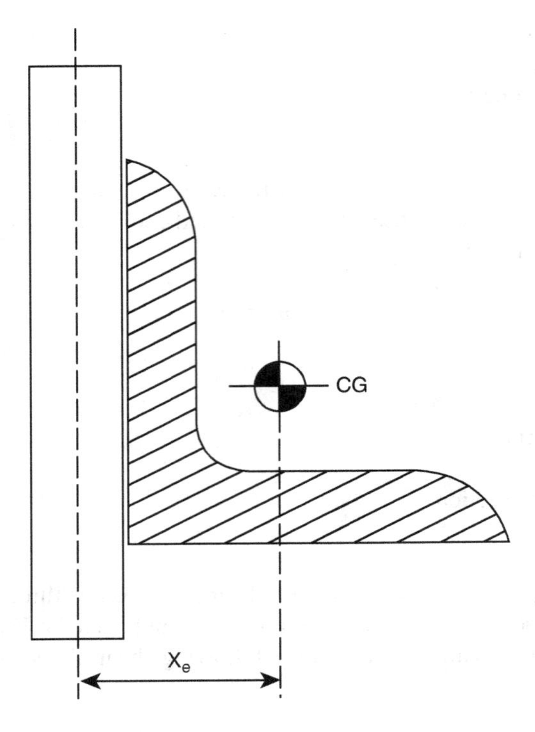

FIGURE 5.18 Definition of eccentricity for a built-up joint.

The behavior of a truss-type connection involving eccentricity of forces, built-up sections, angles and gusset plates creates a significant variance in correlating the theoretical and experimental efficiencies. The loss of efficiency is often attributed to nonuniform distribution of material and stresses in built-up sections. The so-called shear lag theory is used to explain the dominant effect of the key parameters on the predictions of the effective net area of the joint. The revised net area then becomes [9]:

$$A_{nr} = A_n\left(1 - \frac{x_e}{L_j}\right) \tag{5.29}$$

where x_e denotes the eccentricity measured from the gusset plate to the centroid of the connected member as shown in Figure 5.18, and where L_j is the length of the joint which influences the load distribution among the fasteners in a similar manner so that observed in a symmetric splice.

Current standard specifications cover single angle and T configurations and provide rules for the calculation of the net section A_{nr} [10].

The recommendations made for fatigue assessment in symmetric joints also apply to built-up sections, provided there is sufficient restraint. Ideally, slip-resistant joints assure higher joint fatigue strength.

5.17 ECCENTRIC JOINTS

In our brief comments on the behavior of riveted joints, it was instructive to indicate the main principles of mechanics used in the evaluation of the forces acting on a multi-rivet joint when the external force does not pass through the centroid of the entire group of fasteners. Under such conditions, the fastener group is carrying a shear force and also a twisting moment as shown in Figure 5.2 and as shown again here in Figure 5.19.

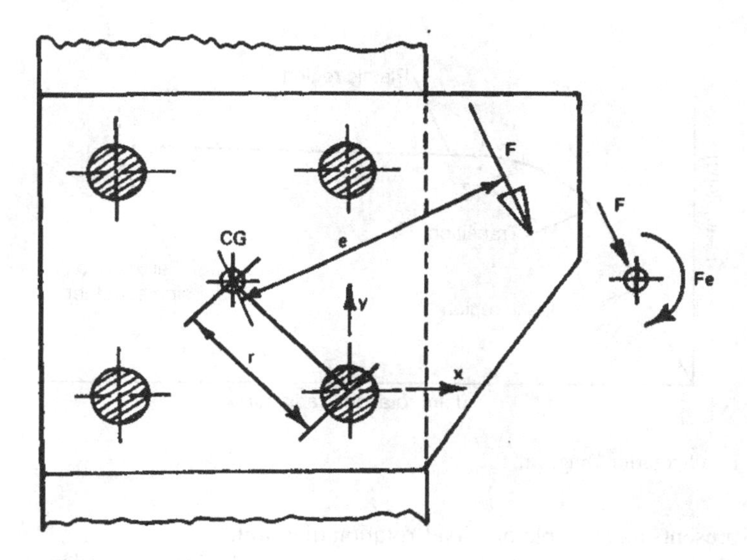

FIGURE 5.19 Example of an eccentric shear load.

The effects of these forces and the twisting moment have been studied extensively. In many cases the joint of interest, often a bracket connection joint, has only one or two lines of fasteners. Figure 5.20 depicts a typical such joint.

In a typical analysis for the joint design it is convenient to assess individual fasteners to determine the performance of an entire joint. There are, however, several variables which may affect the load-deformation behavior of a given joint.

A useful practical procedure is to develop a load-rotation diagram increasing the theoretic load up to the point of failure of the joint. The failure point should be consistent with the failure load for a single fastener.

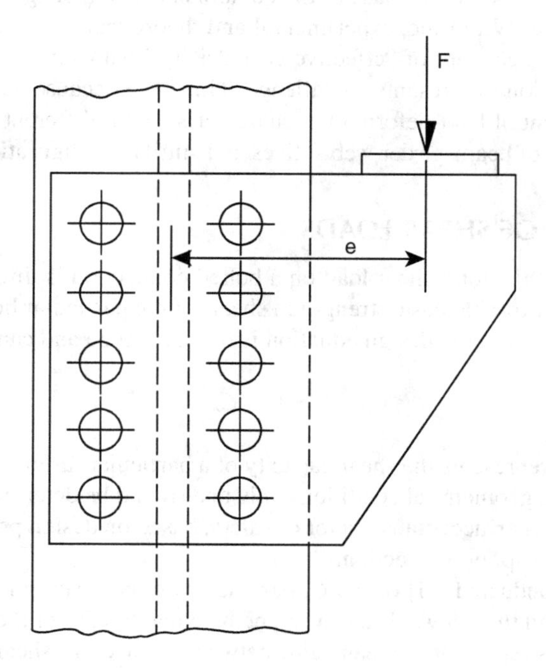

FIGURE 5.20 Example of an eccentrically loaded bracket joint.

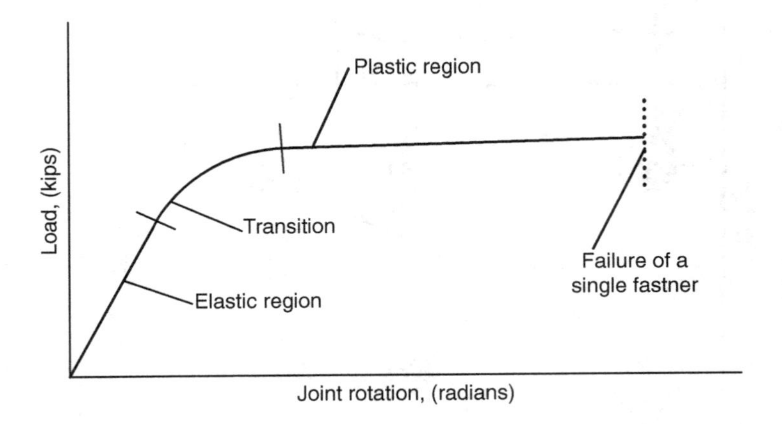

FIGURE 5.21 Load-rotation diagram.

Figure 5.21 presents an example of a load-rotation diagram.

It is usual to find the first segment of the curve to be a straight line followed by a transition region into the plastic region. These types of diagrams have been developed for riveted as well as bolted connections. Specific examples of load-rotation diagrams can be found in the literature [9].

It is normally assumed that the hole clearance is typically sufficiently small to prevent slip. Since, in practice, bolts are placed with less than 1/16 in. clearance on the diameter. Often bolt holes are match-drilled and the slip is, therefore, too small to have any appreciable effect on the joint performance.

For many years, eccentrically loaded joints were designed for elastic condition with the fastener stresses assumed to be varying linearly as a function of the distance from the particular fastener to the center of rotation. It was further assumed that the center of rotation was the same as the centroid of the entire group of fasteners. The elastic method also assumed that the connected plates were quite rigid and that any frictional response of the components could be ignored. These assumptions were indeed conservative. With time, experimental and theoretical results have led to a number of modifications such as the concept of "effective eccentricity," but even today the prediction of the ultimate strength of the joint represents a challenge. This is so because additional research is still needed for the assessment of load-deformation characteristics for different groups of fasteners and geometry, such as cases of beam-girder web splices and similar configurations.

5.18 PREDICTION OF SHEAR LOADS

In Eq. (5.27) we see that the total shear load on a bolted connection is directly proportional to the bolt shear area and the allowable basic strength in shear for the particular bolt. In the case of eccentric joints, the general form of the design equation is the same type and can be expressed as:

$$F = \tau_a A_b C_e \tag{5.29}$$

While the product $\tau_a A_b$ represents the shear capacity of a particular fastener, the factor for eccentric joints, C_e depends on the geometrical conditions only and should be determined for a given bolt pattern. The term τ_a contains an acceptable factor of safety, based on design practice with simple shear splices as discussed in the previous section.

Studies have been conducted [11] of the C_e parameter on the assumption of hypothetical joints involving A490 bolts, and the allowable shear, τ_a, being equal to 65% of the ultimate shear strength of the bolt material. This represents a mean value between the average shear strength obtained from the tension and compression-type shear tests.

TABLE 5.4
Selected C_e Values for One-Line Fasteners

E	Number of Fasteners										
In.	2	3	4	5	6	7	8	9	10	11	12
4	0.74	1.46	2.28	3.20	4.21	5.29	6.45	7.68	8.97	10.32	11.73
6	0.49	1.05	1.74	2.56	3.48	4.51	5.64	6.86	8.18	9.58	11.07
8	0.36	0.80	1.37	2.06	2.85	3.75	4.75	5.84	7.04	8.32	9.70
10	0.28	0.64	1.12	1.69	2.37	3.14	4.01	4.97	6.02	7.15	8.38
12	0.23	0.54	0.94	1.43	2.02	2.69	3.45	4.29	5.21	6.21	7.29
14	0.19	0.46	0.81	1.24	1.75	2.35	3.05	3.76	4.58	5.47	6.44
16	0.17	0.40	0.71	1.10	1.55	2.08	2.68	3.35	4.09	4.89	5.76
18	0.15	0.36	0.64	0.98	1.40	1.88	2.42	3.03	3.70	4.43	5.22
20	0.14	0.32	0.58	0.89	1.27	1.71	2.21	2.77	3.38	4.05	4.78

Source: Based on Reference 5.8.

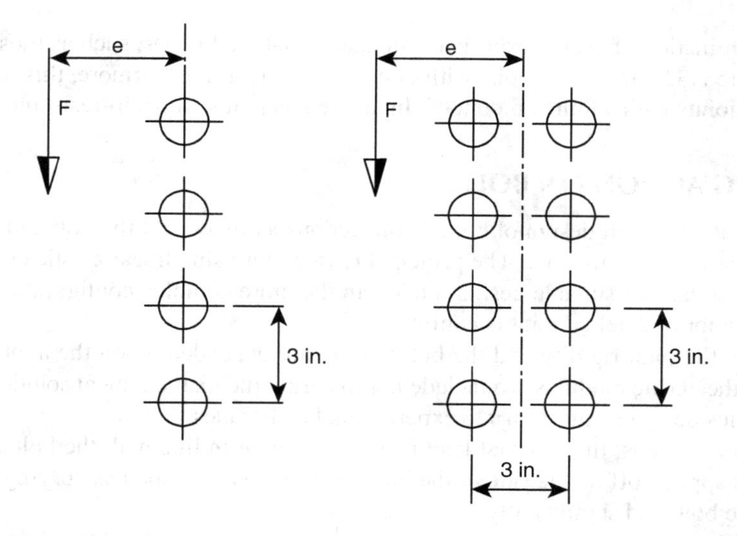

FIGURE 5.22 Typical configurations for eccentric joints.

C_e values have been determined for the two eccentric configurations such as those shown in Figure 5.22. For one line of fasteners shown, Fisher and Struik [9] recommend a number of C_e values. Table 5.4 provides a list of some of these values.

It appears that, for a given number of fasteners in a particular one-line joint, the values of C_e can be interpolated with good accuracy for any eccentricity between 4 and 20 in. The safe joint capacity F can then be calculated by using Eq. (5.29) for A490 bolts as well as other grades of fasteners. The method is conservative, since the C_e numbers have been derived for the A490 bolts which exhibit the least ductility.

Table 5.5 lists a number of C_e values for the two-line group of fasteners—as illustrated in the right side of Figure 5.22.

When joints have more than two vertical lines of fasteners, it is acceptable to use the values from Table 5.5, although a theoretical verification of such an approach does not appear to be presently available. Various specifications propose different factors of safety which, are reasonably close considering the complexity of the problem. A range of factors between, say, 2.5 and 3.4 may be recommended for A325 bolts.

TABLE 5.5
Selected C_e Values for Two-Line Fasteners

E	Number of Fasteners										
In.	2	3	4	5	6	7	8	9	10	11	12
4	1.88	3.15	4.67	6.43	8.38	10.53	12.84	15.31	17.93	20.69	23.59
6	1.34	2.37	3.68	5.24	7.04	9.06	11.29	13.72	16.35	19.16	22.16
8	1.02	1.86	2.95	4.28	5.83	7.60	9.58	11.76	14.14	16.71	19.47
10	0.82	1.51	2.43	3.56	4.89	6.43	8.15	10.06	12.16	14.44	16.89
12	0.68	1.27	2.06	3.03	4.19	5.53	7.04	8.72	10.58	12.59	14.77
14	0.58	1.10	1.78	2.64	3.65	4.85	6.19	7.68	9.33	11.14	13.08
16	0.51	0.97	1.58	2.34	3.26	4.32	5.52	6.87	8.36	9.98	11.74
18	0.46	0.87	1.42	2.11	2.94	3.90	5.00	6.22	7.58	9.06	10.67
20	0.41	0.79	1.29	1.92	2.68	3.57	4.57	5.70	6.94	8.31	9.79

Source: Based on Reference 5.8.

To date, examination of design experience indicates that load tables, such as those given above, can be applied to A325 and A490 bolts with only a small error. Furthermore, this information can be used for the joints with a finite amount of slip as well as joints with slip-resistant characteristics.

5.19 PRYING ACTION ON BOLTS

The conventional and usual design of bolted connections assumes that the bolt and joint members behave in a linearly elastic manner. The principal reasons for using linear elastic equations is their simplicity and the lack of suitable design models in the more complex configurations. In practice, the overall behavior of a bolted joint is nonlinear.

For example, the joint rigidity and the bolt tension are dependent upon the applied loads, joint geometry, and the elastic modulus. To include nonlinearity, the mathematical solutions are difficult to obtain and thus designers must turn to experimental techniques.

In elementary analysis, the external load is assumed to be in line with the bolt axis. When the external load is applied off to one side of the bolt, the effect is described as "prying action," which can increase the bolt load significantly.

Figure 5.23 illustrates the principle of prying action via a lever analogy: After the bolt is tightened, the bolt must resist the external load, plus the full value of the reaction force developed at the edge.

In Figure 5.23, F is the external load, Q is the reaction of the lever arm and F_p defines the bolt force. Hence, the bolt will have to develop an additional force determined by some mechanism such as the lever ratio $(a_0+b_0)/a_0$ from the moments about the point of application of Q.

If the additional load on the bolt could be calculated from the simple moment equation, then the problem of finding the prying load would not present any difficulties. Unfortunately, this problem depends upon the flange or plate rigidity as well as on rather complex distributions of stresses and strains.

Figure 5.24 presents an exaggerated view of a bolted connection subjected to prying forces.

Despite a number of ongoing attempts, there is still no general mathematical equation available for the design of such systems. The apparent simplicity of Figures 5.23 and 5.24, however, suggests an analytical approach to the prying action in more general terms. This generalization states that ultimately a bolt resisting the prying action should carry the load equal to the sum of the external load F and the prying force Q. Using the above symbols, this statement gives [12]:

$$F_p = F + Q \tag{5.31}$$

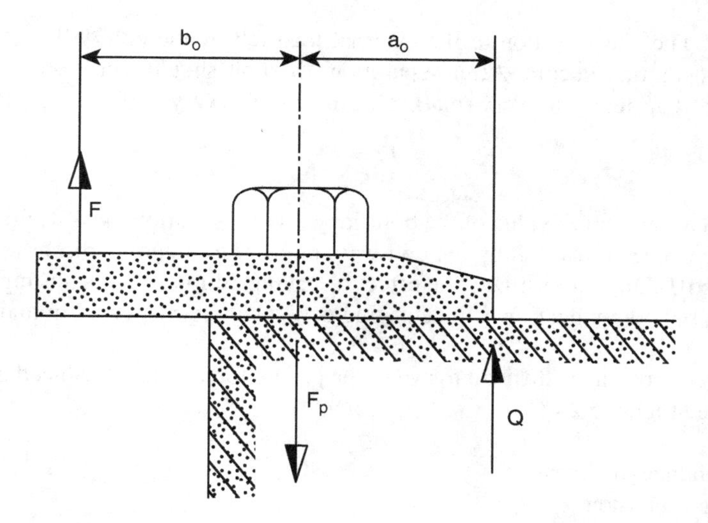

FIGURE 5.23 Illustration of prying action.

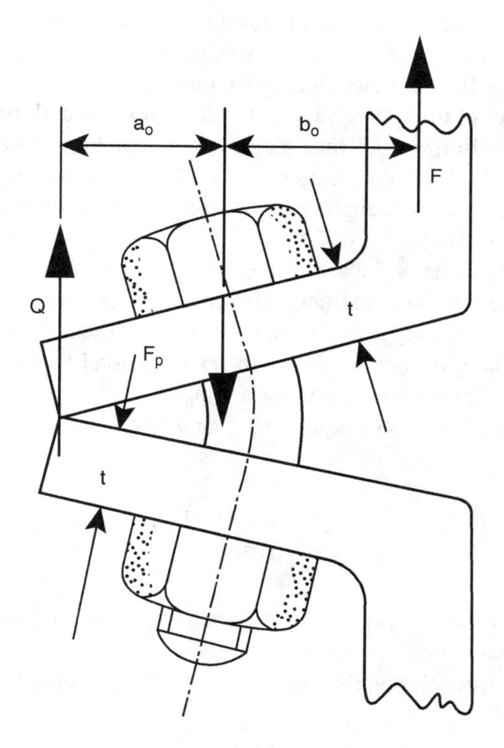

FIGURE 5.24 A bolted joint subjected to prying forces.

To date, experiments show that in tension-type T-connections, the ratio Q/F seldom exceeds 20 to 25%. Furthermore, if a T-stub is rigid, very little prying force will be developed.

5.20 CONTROL OF JOINT SEPARATION

Observe that Eq. (5.31) does not include any dimensional terms indicating the extent of preload. It is clear that for moderate values of the external load, the bolt will not sense the full impact of the component F. The exact portion of the external load felt by the bolt will depend on the ratio k_b/k_j until joint separation occurs. After separation, the bolt should sense all the applied loads. Furthermore, if the k_b/k_j tends to a very small value indicating a very stiff joint, Eq. (5.13) reduces to:

$$F_p = F \tag{5.32}$$

We saw previously that a small value of the bolt/flange stiffness ratio k_b/k_j is desirable because the percentage of the external load felt by the bolt was small. The foregoing discussion also suggests that a relatively stiff flange minimizes the effect of prying action. This is an important observation in analyzing bolted connections, since prying action can be suspected in many conventional joints.

When the effect of prying is difficult to avoid, the prying forces can be reduced using the following techniques (see Figure 5.24):

- Increasing flange thickness t
- Decreasing dimension b_0
- Increasing dimension a_0

Addition of outboard bolts does not help significantly because the first row of bolts, nearer to the line of action of F, carries the bulk of the load. Alternate interpretations of the prying effects for flexible and rigid flanges are further discussed by Bickford [12].

While there are a number of mitigating factors for designing around the prying phenomenon, the prying action is never truly eliminated. Thus, the design calculations based on linear behavior may prove to be unacceptable. The German practice is that almost all connections are considered to be eccentrically loaded and the German engineering society, VDI, has developed detailed guidelines for the design of nonlinear joints [13].

An important feature of design is the assessment of the external critical load required to reduce the contact pressure between the joint members to zero under the point of application of the maximum external load. Bickford [12] suggests a design formula for the bolt preload necessary to prevent a joint separation at the point of the application of the external load F as shown in Figure 5.25. When the radius of gyration of the contact area corresponding to one bolt is calculated on the basis of a rectangular geometry having longer side equal to $2u$, Bickford's theory of the minimum preload can be approximated as:

$$F_p = \frac{F(j-i)}{0.34j+i} \tag{5.33}$$

The minimum preload F_p given by Eq. (5.33) does not imply that any specific part of the joint is yielding, since nonlinear behavior does not necessarily mean that a part must deform plastically.

When the dimension j approaches u, the ratio F_p/F can be represented by the following simplified relation:

$$\frac{F_p}{F} = \frac{1-\gamma}{0.34+\gamma} \tag{5.34}$$

where γ is the ratio i/j, defining the position of the bolt center relative to the line of action of the external load F. Figure 5.26 provides a theoretical interpretation of the effect of γ on the load ratio.

The relations provided by Eqs. (5.33) and (5.34) together with the illustration of Figure 5.25 and the graph of Figure 5.26 are intended here primarily as an application illustration. For a more comprehensive analysis of this problem, the reader is referred to the original study of Bickford [12].

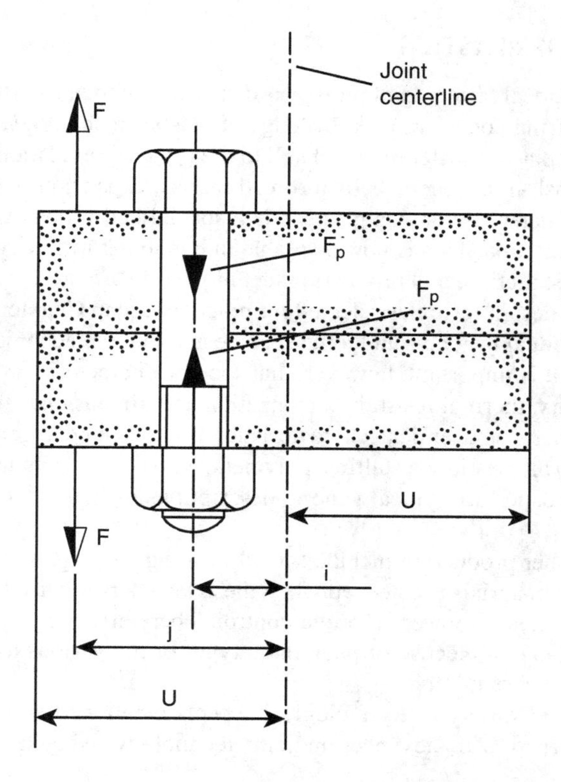

FIGURE 5.25 Joint equilibrium for minimum preload of eccentrically loaded joint.

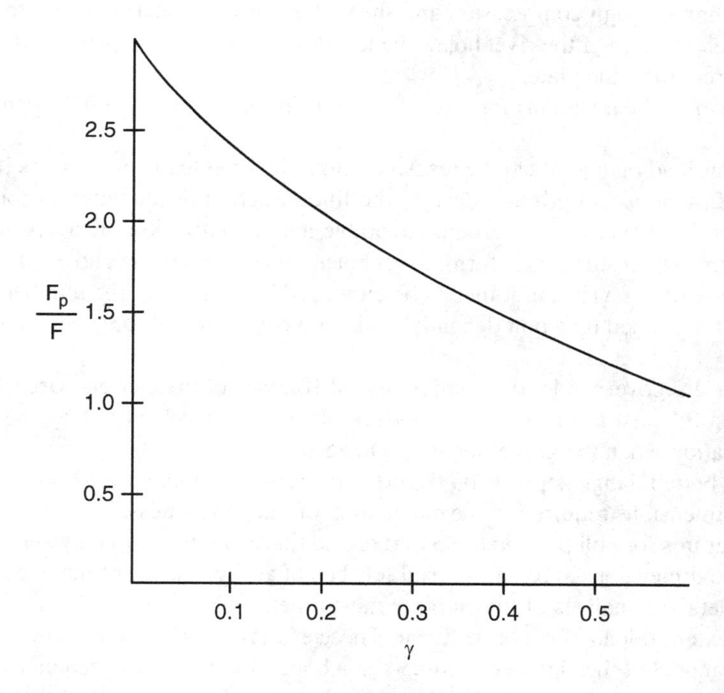

FIGURE 5.26 Bolt preload criterion for mitigating joint separation.

5.21 FASTENING OF PLASTICS

The entire field of mechanical fastening of plastics is concerned with plastic-to-plastic and metal-to-plastic assemblies, requiring considerable knowledge of plastics technology. Much of the available information and data on plastics fastening is related to the typical applications of special screws for automotive assembly, push-in fasteners, helical thread inserts, expansion inserts, and similar components. The choice of materials for this purpose is a specialized problem of collecting the right information in one central area. This is now available in handbook form covering the full range of fasteners, equipment, installation, and production techniques [14].

A number of specific design formulas selected for practical use in Chapters 3 and 5, dealing with more conventional individual fasteners and joints, can be applied to mechanical connections involving metals and plastics. It is important, however, that specific characteristics of plastics are known; such as strength, elasticity, thermal resistance, elongation and dimensional stability; under stress.

There is a great variety of chemical characteristics grouped under generic names and trade names. Polyolefins, thermoplastic crystalline polymers, vinyls, amorphous polymers, structural foams, and thermoset plastics are typical generic descriptions of the materials suitable for plastic fastening needs.

The methods of fastener production include thread forming, cutting, or rolling. Precise fastener specifications for plastic materials are needed when these fasteners are used in automatic assembly and when they are subjected to powered torque control. There are many choices of power screwdriving equipment available of electric or pneumatic type for the various assembly speeds, torque control levels, and accuracy required.

Timely coverage of the variety of technological aspects mentioned above belongs on the desk of a practitioner concerned with advances in joint technology using high-performance plastic materials [14].

5.22 SUMMARY

Rivets resist loading through compression and shear. The ultimate bearing strength is governed by the permanent deformation of the rivet hole. The tensile strength of the lap and butt-type joints may depend on the strength of the plate.

The mechanism of shear support for rivets is similar for bolts, although bolts can tolerate higher clamping forces.

Eccentric shear load on a joint can be resolved into individual load components in order to identify the critical fasteners. In optimum design, the line of action of the external force should pass through the centroid of the fastener group. Favorable load equalization on rivets and bolts can be experienced when contact surfaces deform. Rivet holes act as stress risers and crack inhibitors at the same time. Fatigue life of a riveted joint can be enhanced by increasing the number of rivet rows.

The actual force carried by a bolt depends on the preload, external force, and spring constant of the entire joint.

Flexible joints are governed by the compressive stiffnesses of the gaskets. Usually, the presence of a gasket or "soft" washer increases the portion of the external load felt by the bolt. The joint undergoes separation when the gasket load goes to zero.

Stiffness of a bolted flange depends on the effective cross-sectional area. Formulas are given for estimating the dimensions required in the calculation of flange stiffness.

Classical diagrams for bolt preload are based on the linear relationship between bolt preload and deformation. The dimensions and the material for the nut are typically not included in the stiffness analysis. More detailed analysis of the entire metal-to-metal joint suggests that the function of bolt load versus the external load may not be linear. These effects, however, are relatively small.

The stiffness model of the clamped members in a bolted joint can be based on a barrel, cylinder, or a pair of frustum cones, each containing a central hole equal to the nominal diameter of the bolt.

A spliced joint held by a row of fasteners can be of a friction or a bearing type. In a friction type joint, the ultimate capacity is governed by the total frictional resistance. The relevant coefficient of slip resistance is best determined experimentally.

In a bearing-type splice, as the slip occurs, the succeeding bolts come into contact and develop compressive stresses. The deformations are greater at the ends so that the end bolts must carry higher loads. This effect increases with the increase in the length of the joint. Conversely, in a short joint, all fasteners tend to carry similar loads.

As the average shear strength of the entire joint increases, the connected plate can fail in tension along the net area, particularly in the case of shorter joints. Formulas are available for estimating the net areas for different bolt hole patterns. It is noted from the experiments, that the ultimate strength of a perforated plate in tension can be as high, or higher, than that for the solid plate.

The plate-end distance may be critical in a spliced connection. The desired ratio of this distance to the fastener diameter depends on the bearing and tensile strength of the plate material.

In slip-resistant joints, either slip or the shear strength of a fastener will determine the design. For steels, a factor of safety of two on bolts in shear is typically acceptable.

Design for joint resistance to fatigue can be based on two modes of potential failure. For the case of a fully developed bearing and shear resistance, the failure can take place through the net section. When the entire load is carried by frictional resistance, the gross section may be subject to overload. When the plate-end distance to fastener diameter ratio approaches 3, the mode of failure can change from shear to hole deformation. In eccentric joints, the net area of the cross-section should then be modified.

In the case of a bracket connection subjected to an eccentric load, the force versus bracket rotation curve is best obtained experimentally. The elastic method of analysis of this connection usually results in being conservative.

The shear load capacity of an eccentric joint is normally based on the allowable shear for individual fasteners and on the experimental parameter given in terms of the eccentricity and the total number of fasteners involved.

The behavior of a bolted joint is often nonlinear. One of the main causes of nonlinearity is prying action. While simple static equilibrium illustrates the nature of bolt response to prying, the problem is statically indeterminate and no general closed-form solution is available. The prying action, however, can often be controlled by proper design.

The general approach to nonlinear design of a joint is also directed toward the control of joint separation. Design formulas are available for defining the preload necessary to prevent such a separation. Nonlinearity of joint behavior does not necessarily mean that a given part of the joint deforms plastically.

A brief reference is made to the field of mechanical fastening of plastics and a convenient source of information for practical purposes is presented.

5.23 DESIGN ANALYSES

EXAMPLE ANALYSIS 1

Consider a typical bolted joint as represented in Figure a by a line drawing. It consists of two plates A and B being held together by a threaded bolt and tightened nut as shown.

A common issue arising with these configurations is: What is the stiffness of the connecting bolt?

SOLUTION

Unfortunately, such problems, although common, are statically indeterminate. The solution requires knowledge of the statics, the deformation, and the stress–strain relation (Hooke's law) for the various components. Also, although the configuration may appear to be relatively simple, it

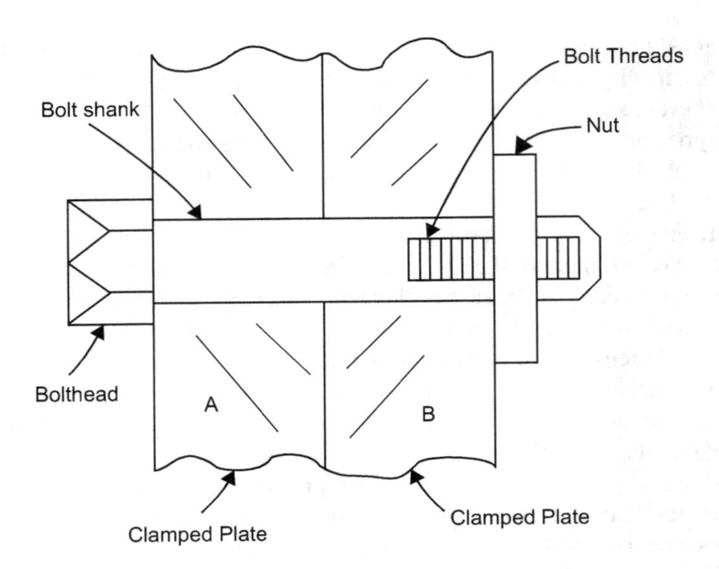

FIGURE A A typical bolted joint.

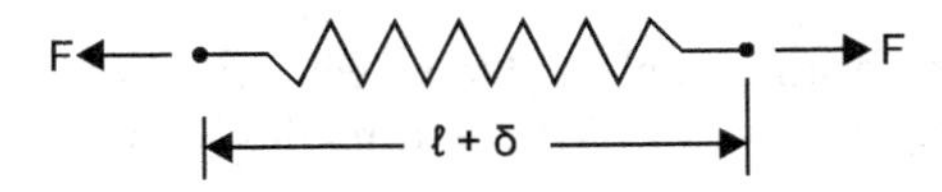

FIGURE B A stretched spring.

nevertheless has six components: 1) the bolt head; 2) the bolt shaft, or "shank"; 3) the bolt threads; 4) the nut; 5) clamped plate A; and 6) clamped plate B.

The analysis of such systems requires physical and geometric approximations and assumptions. Also, we need to know physical properties (specifically, the elastic modulii) of the various components.

First, we can think of the clamped plates A and B as being in compression as the nut is tightened and then consequently the bolt shaft will be in tension.

It is convenient to regard the components as being linear in force-deformation (that is, the deformation is proportional to the applied force). With this assumption we can model the components as linear springs, as in Figure b where a spring with natural length P is stretched by a tensile force F producing a spring stretch: δ.

If the spring has modulus, or stiffness, k, we have the simple relation:

$$F = k\delta \tag{a}$$

Next, in modeling the bolt shaft of Figure a, we can regard the shaft as having two parts: the shank and the threaded portion. Modeled as springs these two parts then become springs in series as represented in Figure c, where the subscripts s and t on the parameters designate the shank and threaded parts, with k_s and k_t being the shank and threaded part stiffnesses.

k_s k_t

F ← → F

$\ell_s + \delta_s$ $\ell_t + \delta_t$

FIGURE C Spring model of a bolt shaft.

By inspection of Figure c a static analysis shows that each spring has a tension force *F*. That is:

$$F_s = F_t = F \tag{b}$$

From the force-deformation relation of Eq. (a) we then have:

$$F_s = k_s\delta_s \quad \text{and} \quad F_t = k_t\delta_t \tag{c}$$

Suppose now we seek to obtain an equivalent stiffness k_{eq} of the series spring model relating the applied force *F* with the overall displacement δ as:

$$F = k_{eq}\delta \tag{d}$$

From Figure b, δ is simply:

$$\delta = \delta_s + \delta_t \tag{e}$$

Then by substituting from Eqs. (b), (c), and (e) into (d) we have:

$$F = k_{eq}\delta = k_{eq}\left(\delta_s + \delta_t\right) = k_{eq}\left[\left(F_s/k_s\right) + \left(F_t/k_t\right)\right]$$

or

$$F = k_{eq}F\left[\left(1/k_s\right) + \left(1/k_t\right)\right] \tag{f}$$

After simplification, Eq. (f) yields:

$$1/k_{eq} = \left(1/k_s\right) + \left(1/k_t\right) \tag{g}$$

or

$$k_{eq} = \frac{k_s k_t}{\left(k_s + k_t\right)} \tag{h}$$

Finally, recall from elementary strength of materials analyses that if a bar with length ℓ, uniform cross-section *A*, and elastic modulus *E*, is subjected to a tensile force *F*, as represented in Figure d, the elongation δ is:

$$\delta = F\ell / AE \tag{i}$$

If we rewrite Eq. (i) as:

$$F = \left(AE/\ell\right)\delta = k\delta \tag{j}$$

We immediately see that the bar stiffness *k* is:

$$k = AE/\ell \tag{k}$$

We can use Eq. (k) in our analysis of the bolted joint of Figure a—and specifically, to estimate the stiffness of the bolt.

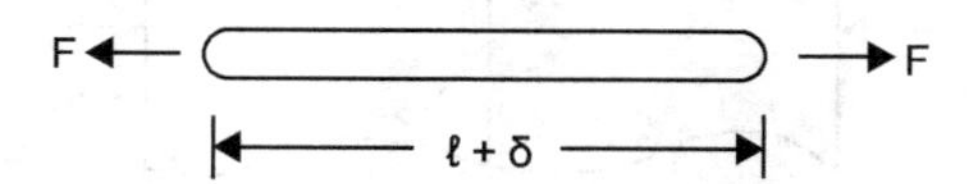

FIGURE D A bar elongated by tension.

To this end, and to numerically illustrate the procedure, suppose we have a bolt whose shaft is 3 in. long with a thread length of 1¼ in. Let the shank (unthreaded portion of the shaft) have a diameter of 3/8 in., and let the threaded portion have an equivalent diameter of 5/16 in. Let the bolt head be ¼ in. thick and let the nut thickness be 5/16 in. Figure 3 shows these dimensions.

Next in view of Figure a let the grip length ℓ_g be 2¼ in. as in Figure f. The grip length is the combined widths of the undeformed plates being clamped.

Observe in Figure f, that as the nut is tightened, with the bolt head and nut made of the same material, say steel, the deformable head effectively lengthens the bolt shank and the deformable nut effectively lengthens the threaded portion of the bolt. A question arising then is: How much are the lengthenings?

Here again, we need to make some assumptions: Perhaps the easiest is to assume that the additional lengths are simply half the widths of the bolt head and nut.

With this assumption, the effective length of the shank ℓ_{se} is:

$$\ell_{se} = \ell_s + (\ell_h/2) = \left(3 - 1\tfrac{1}{4}\right) + \left(\tfrac{1}{4}\right)/2 = 1.875 \text{ in} \tag{l}$$

where ℓ_s is the shank length obtained by inspection of Figure e.

Similarly, the effective length of the threaded portion ℓ_{te} is:

$$\ell_{te} = \ell_g - \ell_s + (t_n/2) = 2\tfrac{1}{4} - 1\tfrac{3}{4} + \left(\tfrac{5}{16}\right)/2 = 0.656 \text{ in} \tag{m}$$

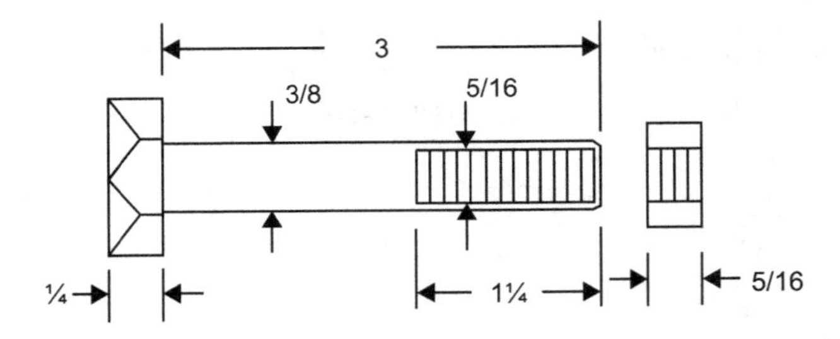

FIGURE E Bolt and nut dimensions.

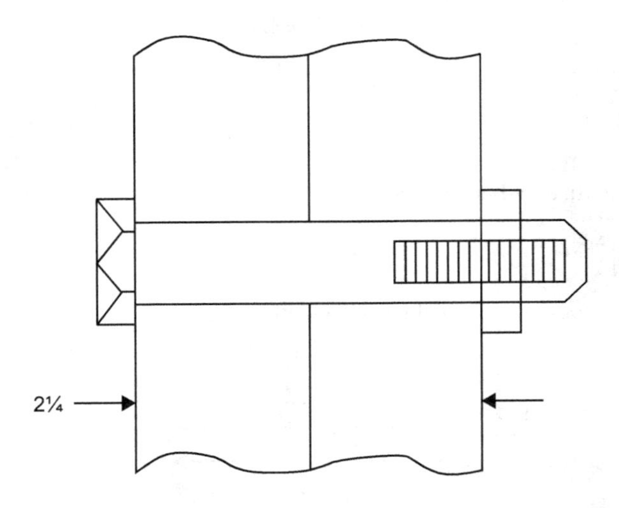

FIGURE F Grip length of the bolted joint.

Referring now to Eq. (k) the stiffnesses of the effective shank k_s, and the effective threaded portion k_e are:

$$k_s = A_s E/\ell_{se} \quad \text{and} \quad k_t = A_t E/\ell_{te} \tag{n}$$

where from the given data the cross-section areas of the shank A_s and the threaded portion A_t of the bolt are:

$$A_s = (\pi/4)D_s^2 = (\pi/4)(3/8)^2 = 0.1104 \text{ in}^2 \tag{o}$$

and

$$A_t = (\pi/4)D_t^2 = (\pi/4)(5/16)^2 = 0.0767 \text{ in}^2 \tag{p}$$

Using the results of Eqs. (P), (m), (o), and (p) and by letting E be 30×10^6 lb/in^2 (steel), and by substituting the values into Eq. (n), k_s and k_t become:

$$k_s = (0.1104)(30)(10^6)/(1.875) = 1.766\times10^6 \text{ lb/in} \tag{q}$$

and

$$k_t = (0.0767)(30)(10^6)/(0.656) = 3.5\times10^6 \text{ lb/in} \tag{r}$$

Finally, by substituting the results of Eqs. (q) and (r) into Eq. (h) the equivalent bolt stiffness is:

$$k_{eq} = \frac{(1.766)(3.5)(10^{12})}{(1.766+3.5)(10^6)} = 1.1738\times10^6 \text{ lb/in} \tag{s}$$

Comment: In view of Eq. (k), suppose we simply estimate the equivalent bolt stiffness as:

$$k_{eq} = A_s E/\ell_{shaft} \tag{t}$$

Then k_{eq} becomes:

$$k_{eq} = (0.1104)(30)(10^6)/(3) = 1.104\times10^6 \text{ lb/in} \tag{u}$$

Although the results of Eqs. (s) and (u) differ by only about 6% the value of Eq. (s) is closer to the actual bolt stiffness due to the inclusions of the clamped plate thicknesses and the thicknesses of the bolt head and nut. Nevertheless, Eq. (t) provides a quicker and more convenient estimate.

EXAMPLE ANALYSIS 2

Consider a gasketed joint consisting of a pair of plates, separated by a gasket and held together by a bolt as represented in Figure a.

Knowing the stiffnesses of the plates, the bolt, and the gasket determine the overall stiffness of the joint.

Solution

We can model the components of the joint as linear springs—and more than that—as springs in series.

As a review, consider again two linear springs in series as represented in Figure b, where k_1 and k_2 are the respective spring module, ℓ_1 and ℓ_2 are the natural lengths of the springs, and δ_1 and δ_2 would represent the shortening of the springs.

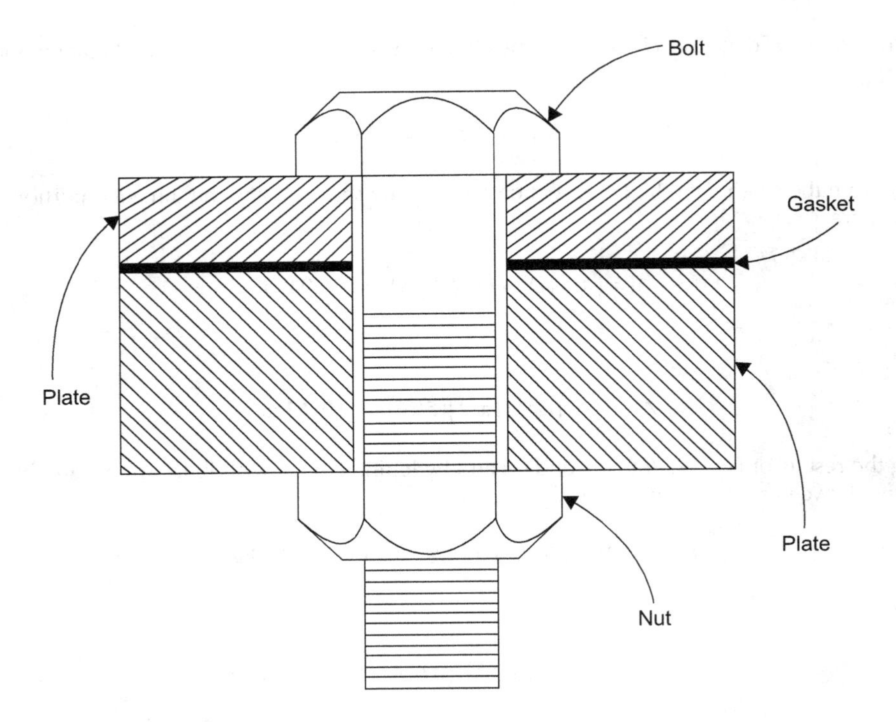

FIGURE A A bolted joint with a gasket.

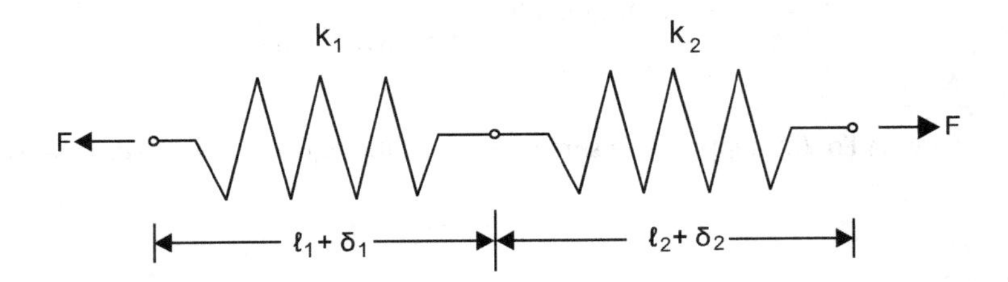

FIGURE B Springs in series.

Suppose now that we want to obtain an equivalent spring constant k for the two springs, such that F may be expressed as:

$$F = k\delta \tag{a}$$

where δ is the overall elongation. That is,

$$\delta = \delta_1 + \delta_2 \tag{b}$$

With the springs in series, each spring will experience the same axial force F. That is

$$F_1 = F_2 = F \tag{c}$$

where F_1 and F_2 are the forces in the individual springs. Since the springs are linear F_1 and F_2 are related to the individual spring elongation δ_1 and δ_2 by the expressions:

$$F_1 = k_1\delta_1 \text{ and } F_2 = k_2\delta_2 \tag{d}$$

or

$$\delta_1 = F_1/k_1 \text{ and } \delta_2 = F_2/k_2 \quad \text{(e)}$$

By substituting from Eqs. (e) into (b) and by using Eqs. (a) and (c) we have:

$$\delta = F/k = \delta_1 + \delta_2 = (F_1/k_1) + (F_2/k_2) = (F/k_1) + (F/k_2) \quad \text{(f)}$$

Then, by inspection we see that the reciprocal of the overall spring constant is

$$(1/k) = (1/k_1) + (1/k_2) \quad \text{(g)}$$

or

$$k = k_1 k_2 / (k_1 + k_2) \quad \text{(h)}$$

Returning now to the gasket joint of Figure A, if the nut, bolt, bolt head, and clamped plates together have stiffness k_b, and if the gasket has stiffness k_g, then using Eq. (g), the overall joint stiffness k is:

$$k = k_b k_g / (k_b + k_g) \quad \text{(i)}$$

Consider two special cases: 1) $k_b = k_g$; and 2) $k_g << k_b$:

Case 1) $k_b = k_g$

This is a somewhat unusual case in that most gasket material is considerably softer than the bolt and clamped plates. Nevertheless, by substituting into Eq. (b), we have:

$$k = k_b/2 = k_g/2 \quad \text{(j)}$$

Thus, even with a relatively stiff gasket the overall stiffness of the joint is reduced by a factor of two.

Case 2) $k_g << k_b$

This is a more realistic case than Case 1). By dividing the numerator and denominator by k_b in the right side of Eq. (h), we have:

$$k = k_g / \left[1 + (k_g/k_b)\right] = k_g \left[1 + (k_g/k_b)\right]^{-1} \quad \text{(k)}$$

By using the binomial theorem and expanding $[1+k_g/k_b]^{-1}$ and by neglecting products of small terms, we have

$$k = k_g \left[1 - (k_g/k_b)\right] \quad \text{(P)}$$

EXAMPLE ANALYSIS 3

Consider again a gasketed joint as in the foregoing example, and as shown again in Figure a.

Suppose it is known that the joint stiffness k_b without a gasket is: 125 kN/mm. Suppose further that it is known that the gasket is an O-ring which after being compressed deforms so that it is stiffer if it is reused after the joint is disassembled. Let the initial gasket stiffness k_{g1} be: 75 kN/mm and then the deformed stiffness k_{g2} be: 100 kN/mm.

Determine the overall joint stiffnesses for the two cases: 1) Undeformed O-ring and 2) Deformed O-ring.

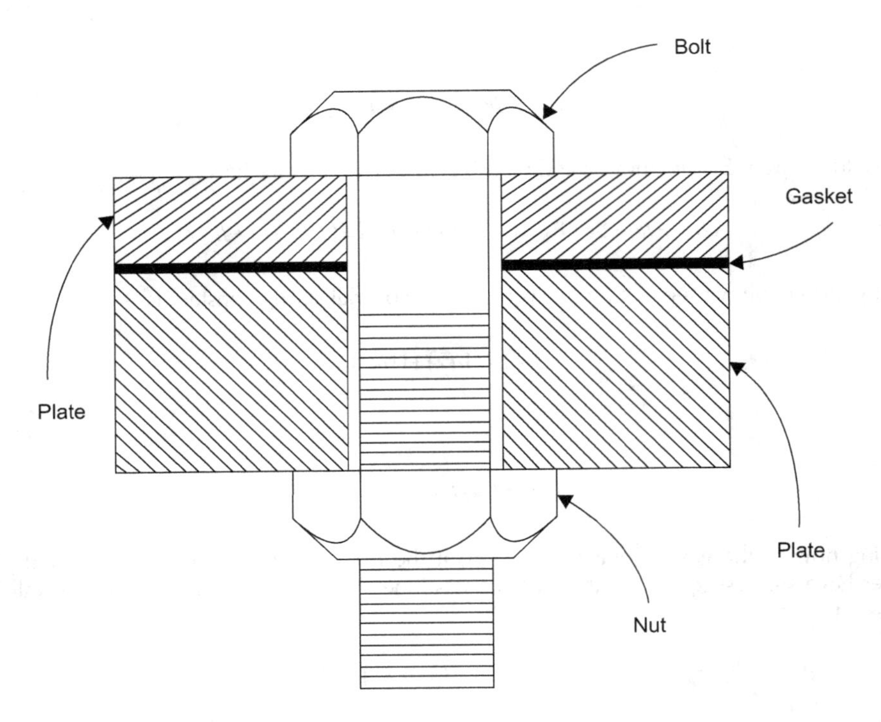

FIGURE A A bolted joint with a gasket.

Solution

The given data are:

Case 1)

Ungasketed joint stiffness: $k_b = 125$ kN/mm (714 kip/in)

Gasket (O-ring) stiffness: $k_{g1} = 75$ kN/mm (428 kip/in)

From Eq. (h) of Example 3, the overall joint stiffness k_1 is then:

$$k_1 = k_b k_{g1} / \left(k_b + k_{g1}\right) = \frac{(125)(75)}{(125+75)} = 46.875 \text{ kN/mm} = 268 \text{ kip/in}$$

Case 2)

Ungasketed joint stiffness: $k_b = 125$ kN/mm (714 kip/in)

Gasket (O-ring) stiffness: $k_{g2} = 100$ kN/mm (576 kip/in)

From Eq. (b) of Example 3, the overall joint stiffness k_2 is then:

$$k_2 = k_2 k_{g2} / \left(k_b + k_{g2}\right) = (125)(100)/(125+100) = 55.56 \text{ kN/mm } (317.2) \text{ kip/in}$$

Comment: Observe that in both cases the presence of a gasket greatly reduces the overall stiffness of the joint. The deformed (or "stiffened") gasket, however, can ease this reduction to some extent.

EXAMPLE ANALYSIS 4

Consider yet again a gasketed joint as in the foregoing example, and as shown again in Figure a.

Suppose in this case the gasket is relatively soft. For instance, let the ungasketed joint stiffness k_b be 125 kN/mm, as before, but let the gasket stiffness k_g be only 12.5 kN/mm.

Use Eq. (h), of Example 3, to determine the overall joint stiffness, and then use Equation (k) to estimate the overall joint stiffness.

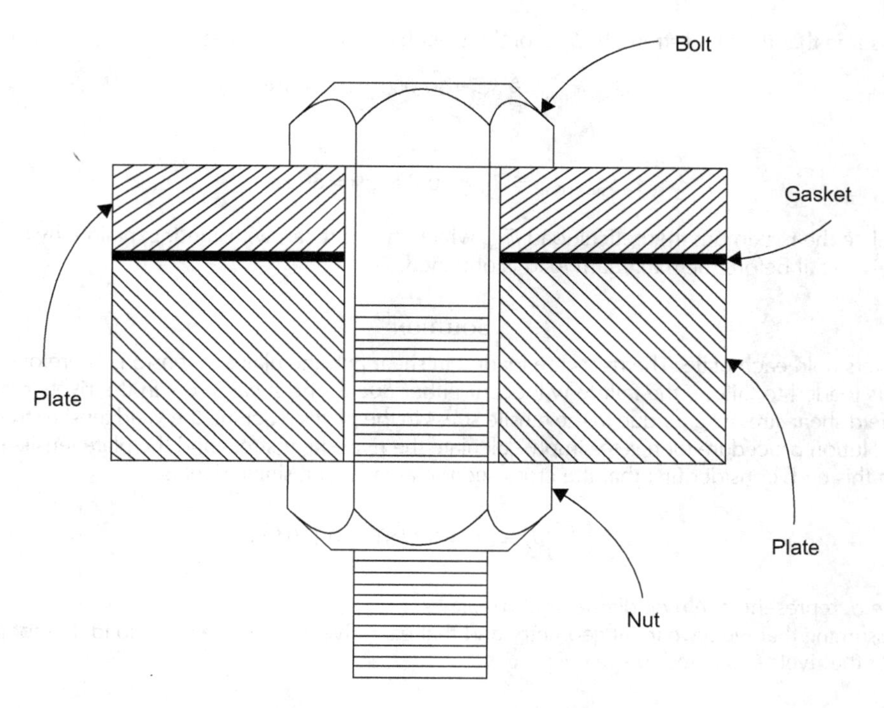

FIGURE A A bolted joint with a gasket.

Solution

From Eq. (h) of Example 3, the overall joint stiffness k is:

$$k = k_b k_g / \left(k_b + k_j\right) \quad \text{(a)}$$

By substituting the given data, we have:

$$k = (125)(12.5)/(125 + 12.5) = 11.36 \text{ kip/mm (64.9 kip/in)}$$

Next for soft gaskets, Eq. (k) of Example 3 estimates the overall joint stiffness k to be:

$$k = k_g \left[1 - k_k / k_b\right] \quad \text{(b)}$$

By substituting the given data we have:

$$k = 12.5\left[1 - (12.5/125)\right] = 11.25 \text{ kip/mm (64.2 kip/in)}$$

The difference between the exact and estimated values is less than 1%.

Comment: Although in this example, the approximate expression is not especially easier to use than the exact expression, the closeness of the results shows that when approximate expressions are available and simpler, there can be considerably less computational effort and the results can be within the accuracy of the material properties of the components.

EXAMPLE ANALYSIS 5

Figure a provides a line drawing of 12 rivets holding two plates together. Both the plates and the rivets are made of steel. The rivets all have the same geometry, with each having 3/8 in. diameter. The plates, A and B, are each 4 in. wide and ¼ in. thick.

Assume that the yield strengths S_{yield} of the steel used in the joint are:

$$S_{yield-tension} \triangleq S_{yt} = 50{,}000 \text{ psi} = 50 \text{ ksi} \tag{a}$$

and

$$S_{yield-shear} \triangleq S_{ys} = 30{,}000 \text{ psi} = 30 \text{ ksi} \tag{b}$$

Calculate the maximum theoretical load F_{max} which the joint can withstand. Explain why the joint is likely to fail before the maximum load is obtained.

Solution

Six rivets hold each plate. The rivets are loaded in shear and the plates in tension. Therefore, if the joint is loaded to failure, the failure will occur either due to the shear stress in the rivets reaching the yield shear stress: S_{ys}, or due to the tensile stress in the plates reaching the yield tensile stress: S_{yt}. The solution procedure then is to simply calculate the rivet shear stress and the plate tensile stress.

To this end, consider first that the cross-section area A_r of a single rivet is:

$$A_r = \left(\frac{\pi}{4}\right)d_r^2 = (\pi/4)(3/8)^2 = 0.1104 \text{ in}^2 \tag{a}$$

where d_r represents the rivet diameter (3/8 in.).

Assuming that we have ideal geometry and that each rivet carries the same load, the total area A_{rtot} of the rivets subjected to shear is:

$$A_{rtot} = (6)(A_r) = (6)(0.1104) = 0.6627 \text{ in}^2 \tag{b}$$

Thus, the theoretical maximum load F_{max} that the six rivets can sustain is:

$$F_{max} = (S_{ys})(A_{rtot}) = (30)(0.6627) = 19.88 \text{ kip} \tag{c}$$

Next, again assuming ideal geometry, the maximum load F_{max} that the plate can sustain is:

$$F_{max} = (S_{yt})(A_{plate}) \tag{d}$$

where A_{plate} refers to the cross-section of the plate being exposed to the external load. From (a) we see that A_{plate} is:

$$A_{plate} = (\text{plate width} \times \text{plate thickness}) \text{ minus } (\text{rivet projected area}) \tag{e}$$

The rivet projected area is simply the rivet diameter d_r multiplied by the plate thickness t. By substituting the given data into Eq. (e) we have:

$$A_{plate} = (4)(1/4) - (3)(3/8)(1/4) = 0.7188 \text{ in}^2 \tag{f}$$

Thus, the failure load F_{max} for the plate is

$$F_{max} = S_{yt}A_{plate} = (50)(0.7188) = 35.93 \text{ kip} \tag{g}$$

The results of Eqs. (d) and (g) show that the yielding will occur first in the rivets, and then from Eq. (c) the maximum theoretical load F_{max} for the joint is:

$$F_{max} = 19.88 \text{ kip} \tag{h}$$

Comment

In an actual physical joint, the geometry is likely to be less than ideal and as a consequence each rivet will not experience the same load. Thus, the multiplier of 6 (six rivets) in Eq. (b) is not realistic.

Instead, failure will occur at the rivet experiencing the greatest load, and once this failure occurs the loading on the remaining five rivets will suddenly increase, leading to a cascade of failures.

To be on the safe side in the design it is probably wise to delete the multiplier of 6 in Eq. (b). This gives the maximum load F_{max} as:

$$F_{\max} = S_{ys} A_r = (30)(0.1104) = 3.313 \text{ kip} \tag{i}$$

A factor of safety, say 2, then reduces F_{max} even more yielding:

$$F_{\max} = 3.313/2 = 1.6565 \text{ kip} = 1656 \text{ lb} \tag{j}$$

Observe how much less this load is than the theoretical maximum load of 19.88 kip [Eq. (c)].

EXAMPLE ANALYSIS 6

See Example Analysis 5. Let the joint configuration be the same as in that configuration but let the plate thickness be reduced to 1/8 in. Again let the objective be to calculate the theoretical maximum load F_{max} that the joint can sustain.

Figure a shows the rivet joint, this time with the plate thickness being 1/8 in

SOLUTION

Again, we assume that the rivets and plate are made of steel, with tension and shear yield stresses (S_{yt} and S_{ys}) being:

$$S_{yt} = 50 \text{ ksi} \quad \text{and} \quad S_{ys} = 30 \text{ ksi} \tag{a}$$

As before we recognize that the rivets are loaded in shear and the plates in tension.

Regarding the rivets, with the rivet diameter d_r being 3/8 in., the area A_r subjected to shear is:

$$A_r = (\pi/4)d_r^2 = (\pi/4)(3/8)^2 = 0.1104 \text{ in}^2 \tag{b}$$

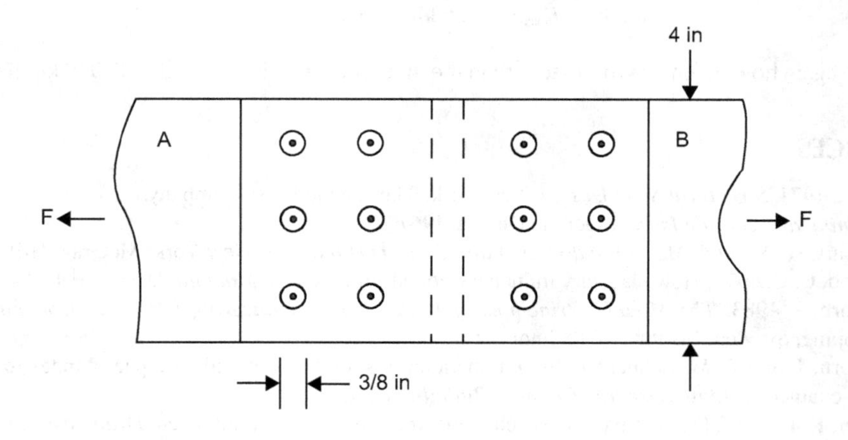

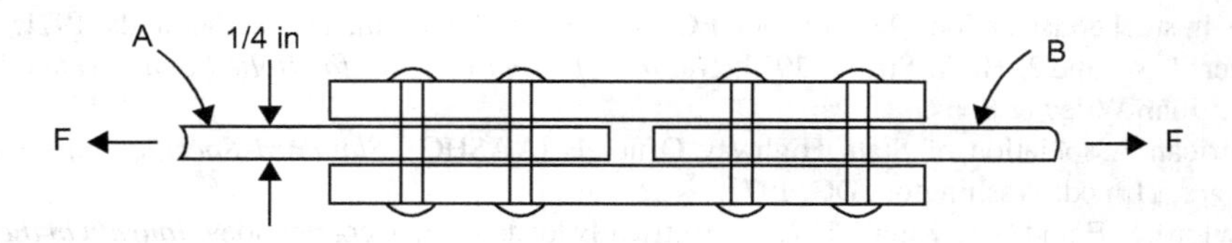

FIGURE A A riveted joint connecting two plates A and B.

Assuming ideal geometry, with each of the rivets carrying the same load, the total rivet area A_{rtot} subjected to shear, in either plate (A or B) with six rivets, is:

$$A_{rtot} = 6A_r = (6)(0.1104) = 0.6627 \text{ in}^2 \tag{c}$$

Then, as before, the maximum load F_{max} that the six rivets can sustain is:

$$F_{max} = (S_{ys})(A_{rtot}) = (30)(0.6627) = 19.88 \text{ kip} \tag{d}$$

By now directing our attention to the plates and by again assuming ideal geometry, the maximum load F_{max} that either plate A or B can sustain is:

$$F_{max} = (S_{yt})(A_{plate}) \tag{e}$$

where A_{plate} is the cross-section area of a plate being subjected to tension.

By inspection of Figure a, we see that A_{plate} is:

$$A_{plate} = (4)(1/8) - (3)(3/8)(1/8) = 0.3593 \text{ in}^2 \tag{f}$$

Then by substituting into Eq. (d), F_{max} becomes:

$$F_{max} = (50)(0.3593) = 17.969 \text{ kip} \tag{g}$$

Comment

The maximum rivet shear load of 19.88 kip [Eq. (d)] is now slightly larger than the maximum plate load of 17.969 kip [Eq. (g)]. This means that theoretically, the joint will fail due to the plate yielding (stretching). But as we observed in Example Analysis 6, in an actual joint, the geometry will undoubtedly be less than ideal. Therefore, it is likely that the failure will again occur at a rivet.

Therefore, in actual joint design, we need to again employ a factor of safety, say 2. Then, by referring again to Example Analysis 6 we see that a practical maximum load is:

$$F_{max} = 1.656 \text{ kip} = 1656 \text{ lb} \tag{h}$$

Observe again how much less this load is than the theoretical maximum load of 17.969 kip [Eq. (g)].

REFERENCES

1. Tall, L. 1974. *Structural Steel Design*. New York: The Ronald Press Company.
2. *Fastening and Joining Issue*, Machine Design, 1967.
3. Rothbart, H. A. 1964. *Mechanical Design and Systems Handbook*. New York: McGraw-Hill.
4. Osgood, C. C. 1972. How elasticity influences bolted-joint-design. *Machine Design*, Feb. 24, 92–95.
5. Webjörn, J. 1983. *The Modern Principles of Bolted Joint Engineering (Personal Communication)*. Linköping, Sweden: Institute of Technology.
6. Webjörn, J. and R. W. Schneider. 1980. Functional test of a vessel with compact flanges in metal-to-metal contact. *Welding Research Council, Bulletin* 262, October.
7. Illgner, K. H. 1967. Das verspannungs-schaubild von schraubenverbindungen. *Draht-Welt* 63.
8. European Convention for Constructional Steelwork, European guidelines on the use of high strength bolts in steel constructions. Document CECM-X-70–80, Roterdam, The Netherlands, 1971.
9. Fisher, J. W. and J. H. A. Struik. 1974. *Guide to Design Criteria for Bolted and Riveted Joints*. New York: John Wiley & Sons.
10. American Association of State Highway Officials (AASHO). *Standard Specifications for Highway Bridges*, 11th ed. Washington, DC, 1973.
11. Crawford, S. F. and G. L. Kulak. 1971. Eccentrically loaded bolted connections. *Journal of the Structural Division* ASCE, 97(ST3), 765–783.

12. Bickford, J. H. 1981. *An Introduction to the Design and Behavior of Bolted Joints*. New York and Basel: Marcel Dekker, Inc.
13. Junker, G. 1974. Principles of the calculation of high duty bolted connections-interpretation of the guildline VDI 2230, VDI Berichte No. 220. An Unbrako technical Thesis, published by SPS, Jenkintown, PA.
14. Lincoln, B., K. J. Gomes and J. F. Braden. 1983. *Mechanical Fastening of Plastics*. New York: Marcel Dekker, Inc.

SYMBOLS

a	Effective width of annulus
a_0	Key joint dimension
A	Cross-section of individual fastener
A_b	Nominal cross-section of bolt
A_c	Effective area
A_g	Area of gasket in compression
A_n	Net area
A_{nr}	Revised net area
b_0	Key joint dimension
C	Coefficient of overall stiffness
C_e	Factor for eccentric joints
d	Fastener diameter
e	General symbol for load eccentricity
E	Modulus of elasticity for flanges
E_b	Modulus of bolt materials
E_g	Modulus of gasket material
F	External force on joint
F_a	Direct load on fastener
F_b	Bearing load
F_f	Frictional load
F_p	Preload in eccentric joint
F_s	Shear load on fastener
F_{s1}	Load at slip
F_t	Tensile load
F_x	Component of shear load
F_y	Component of shear load
g	Thickness of washer
g_0	Distance between rows of bolts
h	Depth of flange
i	Distance to bolt center
j	Distance to line of external force
k_b	Stiffness of bolt
k_c	Stiffness of joint without bolt effects
k_g	Stiffness of gasket
k_j	Stiffness of flanges
L	Distance from end of plate
L_b	Length of bolt in tension
L_g	Thickness of gasket
L_j	Total length of joint
m	Number of slip planes
n	Number of fasteners
Q	Prying load

r	Distance to centroid
R	Average interface radius
R_s	Coefficient of slip resistance
s	Amount of stagger
S_b	Bearing stress
S_c	Ultimate compressive stress in plate
S_u	Ultimate tensile strength of plate
t	Thickness of plate
u	Half width of flange
w	Width of plate
W	Actual load on bolt
W_e	External load on joint
W_i	Initial bolt load
x	Coordinate
x_e	Eccentricity at gusset plate
$\bar{X}$	Centroid coordinate
y	Coordinate
$\bar{Y}$	Centroid coordinate
β	Correction factor
γ	Length ratio
η_0	Auxiliary angle
ρ	Radius of washer
τ_a	Shear allowable
τ_b	Basic shear strength of bolt
ϕ	Half-cone angle
ψ_0	Auxiliary angle

6 Flanges and Stiffeners

6.1 INTRODUCTION

One of the more challenging areas of design technology is the design of pipe flanges and stiffener configurations which are difficult, if at all possible, to describe by a single formula or theory. The experimental and theoretical work since the second World War has led to a number of specifications and design approaches, which can be useful in initiating relevant technical approaches. This knowledge, when combined with a high degree of engineering judgment, should help to determine the minimum acceptable strength of flanges and stiffeners, as well as the characteristics of the overall system.

For example, a design of a pipe flange can influence the response of a bolted joint and the resultant leak rates. The problems of flange design became, over the years, a subject of extensive discussions and publications introducing a rational stress analysis into the commercial field of flanged connections. About the same time, the importance of the longitudinal bending stress in the pipe hub was recognized and a number of different approaches were reviewed with the purpose of improving the design formulas. But the effect of shearing forces, for instance, at the junction of hub and ring was not included in the original deliberations.

There are three principal elements of a bolted pipe joint which have individual characteristics and call for different design approaches:

- Interface and gasket material
- Fasteners
- Flange ring and hub geometry

The material presented in this chapter is intended to be applicable to the majority of circular flanges and bolted joints found in heat exchangers, condensers, piping, and a variety of pressure vessels. We assume that the flange materials are homogeneous and have stable mechanical properties in the elastic regime. Obviously, however, the stress distribution under plastic deformation will be different from that obtained under the conditions of pure elasticity.

The problem of flange design is especially important in the case of large-diameter, high-pressure, and high-temperature equipment where structural integrity is essential.

A design engineer will be faced with an extensive list of analysis and design problems to be compiled when confronted with the need to design the geometry and response of a flanged connection. The matters are made much worse when it becomes necessary to design a circular flange with reinforcing gussets. Such configurations require a three-dimensional analysis, which, even with the help of finite element methods (FEMs), represents a tedious and a costly task. In addition, there may be circumferential stiffeners, radial ribs, and gussets in various structural joints which deserve our attention.

6.2 GASKET EFFECTS

Several details, related to individual fasteners and the mechanical compliance of bolted joints, have been reviewed in Chapters 2 and 5. The role of a gasket in a flanged connection is especially unique because of the relatively low stiffness of the gasket material and its importance in preventing any potential leak paths.

This section summarizes some general gasket technological concerns [1]; a brief overview of the original experience with the gasket concept; and a current view of the subject [2]. This material is designed to serve engineers, designers, and trades people who are chartered to solve day-to-day problems in gasket technology, but who have little time to rummage through the files of articles on joint technology.

The steps in gasket design and analysis include materials selection, configuration design, and the determination of the contact pressure for a given task. There is generally a great variety of gasket materials which are difficult to classify. In flanges designed for low-pressure service, for example, it is customary to use full-face gaskets extending to the outside flange diameter. Such gaskets are relatively soft, requiring the contact pressure of not more than the internal pressure of the system for which the flange is designed.

The design of a flat ring-type gasket is determined by the distance between the inside diameter of the joint and a point within the centerline of the bolt. The degree of the joint-tightness for this particular case depends on the gasket material and the contact pressure. Typical choice of a rubber-type material or metal will determine the minimum bolt preload to assure the correct contact pressure.

For higher pressures and temperatures tongue-and-groove facings have specific advantages with the practical gasket widths on the order of 0.25 to 1.00 inches. The narrow gasket, of course, implies higher unit pressures and higher loads particularly in the case of metals requiring a certain amount of plastic deformation to assure joint tightness. There must also be some concern about the gasket thickness; thinner gaskets appear to have less chance of being blown out.

An important design parameter in gasket selection is the ratio of the gasket contact pressure to the inner working pressure of the vessel. Field experience with rotating machinery, oil refinery, and similar equipment indicates that the desired pressure ratios can be summarized as in Table 6.1.

In some circumstances, it is customary to use a narrow, ring-type joint, which depends entirely on a metal-to-metal line contact and represents a high contact pressure over a small area. This type of a joint has been used successfully with high-pressure and high-temperature vessels and piping. In such situations, it becomes difficult to evaluate the actual contact pressure and the size of the contact area. At best, only an assessment can be made of the force required to cause the plastic flow of the metallic gasket. A word of caution is warranted here because the best prediction of a contact force needed should only be used as an initial guide.

The subject of the actual compressive gasket forces in the various applications remains elusive. Lack of some definite information on the allowable unit pressures of the industrial gaskets and the effect of the gasket forces on bolt design, still create special problems for the designer since the matters of unit pressures and bolt stresses are interrelated.

As stated previously, the role of a gasket in a completely assembled bolted joint is both complex and important. The bolts selected for a joint must exert sufficient contact pressure on the gasket to keep the joint leak-proof without causing overloading of the bolts or the flange. The bolts are preloaded prior to the introduction of the internal fluid pressure, at which time the total bolt effort can be utilized in compressing the gasket.

As the internal pressure is increased, the contact pressure on the gasket is partially relaxed. This decrease in gasket strain will depend on the relative rigidity of the bolt and the gasket as shown by

TABLE 6.1
Approximate Ratios of Gasket Contact Pressure/Internal Pipe Pressure

Wide raised-face joints	2–4
Wide tongue-and-groove joints	3–6
Narrow tongue-and-groove joints	3–8

Eq. (5.15) and Table 5.2. Hence, our immediate task is to obtain the elastic properties of the gasket material and to estimate the relevant spring constants. It was shown in Chapter 5 that when the gasket material is substantially more flexible (soft) than the other joint components, the gasket will dominate the behavior of the entire bolted connection.

Bolted connections with gaskets are often considered to be weak when subjected to cyclic loading. Normally, when fatigue causing loads are applied to a bolted assembly with initial high preload, only a small portion of the stress change is expected to be induced in the bolt. When the bolt is preloaded according to a prescribed initial torque and the gasket is compressed, however, this equilibrium does not stay constant, since temperature, service loads and time can cause gasket creep.

Osman, et al. [3] document an attempt to develop a design procedure for selecting an optimum bolt diameter for a gasketed joint subjected to fatigue. Their study showed that thicker gaskets required larger bolt diameters. It also followed that softer gaskets demanded bigger fasteners. This is consistent with the general principle that the bolt diameter required for a gasketed joint should be greater than that for a metal-to-metal contact. It was also shown that the ratio of the maximum to minimum loads, applied to a bolted connection under a cyclic load, could be correlated with the required bolt diameter for specific fatigue applications. Other variables included the material and dimensional characteristics of the flanges and the bolts.

It should be noted, that under some conditions of a vibration cycle, the decreased bolt diameter and increased nut diameter in a bolted joint can precipitate a radial slippage of the engaged thread. When a relatively soft gasket material causes larger bolt-load variation, special care should be taken to minimize nut loosening due to the thread slippage. To accomplish this, the initial amount of bolt elongation due to the preload should be maximized. This is in line with the current trend to promote higher-strength fasteners.

6.3 TIGHTNESS CRITERION

The tightness of a pipe joint cannot be generalized because the rate of leakage depends upon the type of fluid as well as the fluid pressure contained. For example, tightness criteria will be different for containing water, steam, compressed air, or a vacuum.

In many cases, the joint can be described as "tight" if it passes a conventional bubble test with compressed air in the system. This is evidence of no significant cracks or openings and that the only escaping gas flow possible is via diffusion.

To prevent diffusion, an opening must be decreased below molecular size. It is known that diffusion of gases is possible through essentially all materials, including steel.

Based upon the foregoing observations, we can state that a practical definition of a tight joint is the criterion that there is no fluid flow out of the joint, aside from diffusion.

In practice, we find that a joint passing a hydrostatic test will also pass the soap-bubble test. Such a test, however, is not a proof for high-vacuum applications where the diffusive flow of in-leaking air or other gas can take place.

When a gasket is compressed between two rigid flanges, the rough flange surfaces can cause high local stresses in the gasket material so that voids near the surfaces of harder materials are filled-in and sealed. Local gasket stresses, or course, must exceed the yield strength of the gasket material and in-void deformations will then become permanent. Although we can assume that a gasket stress is reasonably uniform, local surface roughness, manufacturing tolerances, and the deformations of a bolted joint overload may cause uneven pressure patterns and create channels large enough to support a mass fluid flow.

Furthermore, as the gasket loading created by the bolt is decreased through the increase of the internal pressure to the pipe, the areas of lower gasket stresses may separate from the flange surface and enlarge any existing leakage paths. It follows that a higher initial load on the gasket will make the distribution of stresses more uniform and provides a tighter joint.

Another factor influencing joint tightness is the phenomenon of creep of the gasket material. This is represented by a slow process of reduction in bolt and gasket stress, which may fall below a value required for sealing the joint. The immediate remedy for this is retightening of the bolts as long as the flanges are accessible. In some systems, such joints may be covered by insulation—so that the original bolt preload should be higher to avoid the problems with the torque control as the gasketed joint relaxes.

To make such a provision in design, it is necessary to have a series of creep curves for the particular gasket material. According to Roberts [2], the gasket stress S_g can be calculated from the expression:

$$S_{gi} - S_g = \frac{E_b A_b L_g}{A_g L_b}\left(\varepsilon_p - \frac{S_{gi} - S_g}{E_g}\right) \tag{6.1}$$

where

$$\varepsilon_p = \frac{\delta_p}{L_g} \tag{6.2}$$

and where δ_p is the change in gasket thickness due to creep and ε_p is the plastic strain. The initial gasket stress is denoted by S_{gi} and the gasket stress at any time after the bolt assembly is S_g. Other symbols are the same as those used in Chapter 5 for the definition of the bolt gasket spring constants K_b and K_g.

If we denote the change in gasket stress due to creep by ΔS, then rearranging Eq. (6.1) yields:

$$\Delta S = \frac{\varepsilon_p E_g K_b}{K_b + K_g} \tag{6.3}$$

Using the concept of joint stiffness C from Chapter 5, Eq. (5.15) gives:

$$\Delta S = \varepsilon_p E_g C \tag{6.4}$$

In case a constant value of E_g cannot be assured over the entire range of creep data, the gasket modulus may have to be expressed as function of the gasket stress S_g and time.

Equation (6.1) applies to a bolted joint which is assembled, torqued to the required level and then allowed to creep. If the internal pressure is then applied to the joint, the general model appears to be the same: although the initial gasket stress S_{gi} should be reduced by the amount consistent with the effect of the internal pressure load.

6.4 SPACING OF BOLTS

The efficiency of load transfer in a bolted joint depends on the bolt spacing, flange thickness, and the properties of the gasket material. If the internal pressure and the gasket response are known, then the total cross-section of the fastening is known. Hence, the problem at hand should be concerned with the assignment of the cross-sectional area to a particular set of bolts. The choice is essentially between a few larger bolts or a greater number of bolts of smaller size.

The analysis of this problem, based on the theory of beams on elastic foundation utilized by several investigators in the past, is thought to be useful, but it may be of a limited application if one considers the local effects of bolt-heads and the stiffness contribution of the hub.

According to a practical formula of Taylor Forge Company [2], the required bolt spacing to assure the tightness of a bolted joint, is:

$$D_b = 2d + \frac{6H}{m_0 + 0.5} \tag{6.5}$$

Where d is the bolt diameter and H is the thickness of the flange.

While there is evidence that Eq. (6.5) has been used by engineers for many years with good results, the gasket factor m_0, indicating the efficiency of a bolted joint, has been under discussion for many years.

The original rules of the ASME Boiler Code, which was under development for many years, offered a rather simple approach to the choice of the gasket factor. Successful usage of flanges designed according to the ASME Code has, no doubt, been governed by the fact that the bolt stress used in the calculations never exceeded one quarter of the yield strength.

The original gasket factors ranged from 1.0 for soft rubber to 7.0 for stainless steel. No allowance was made for the fact that many m_0 values could approach unity or less when the initial gasket stress was high. This was perhaps of interest not only on account of stress but also due to service conditions.

There is evidence that the actual m_0 values vary over a much larger range than 1 to 7 when joint sealing is required against a nitrogen or water head. German experience in the past suggested that variations in bolt spacing had little effect on the leakage pressure. In general, however, the practice of engineering design to a low m_0 value such as, perhaps, $m_0 = 1$, using wider gaskets could be justified on a statistical basis. This appeared to be possible simply because wider gaskets represented a lower probability for a complete leakage channel to develop.

It may be of interest to note that an H/d ratio equal to 3.0 corresponds roughly to a compact flange design, assuring very little flange deformation. For low-pressure service, this ratio can be on the order of 1.0. Using Eq. (6.5) and the range of gasket factor from 1.0 to 10.0, Figure 6.1 shows the variation of the D_b/d ratio with the gasket factor.

It appears that, for a relatively high value of parameter m_0, both curves in Figure 6.1 tend to the same ratio of 2 which may not be practical.

In addition to the choice of the gasket factor, it is necessary to determine the load per bolt on the basis of the torsional resistance of the flanges. Boardman, for instance, suggested a simple but a rather ingenious formula [4], which can be stated as:

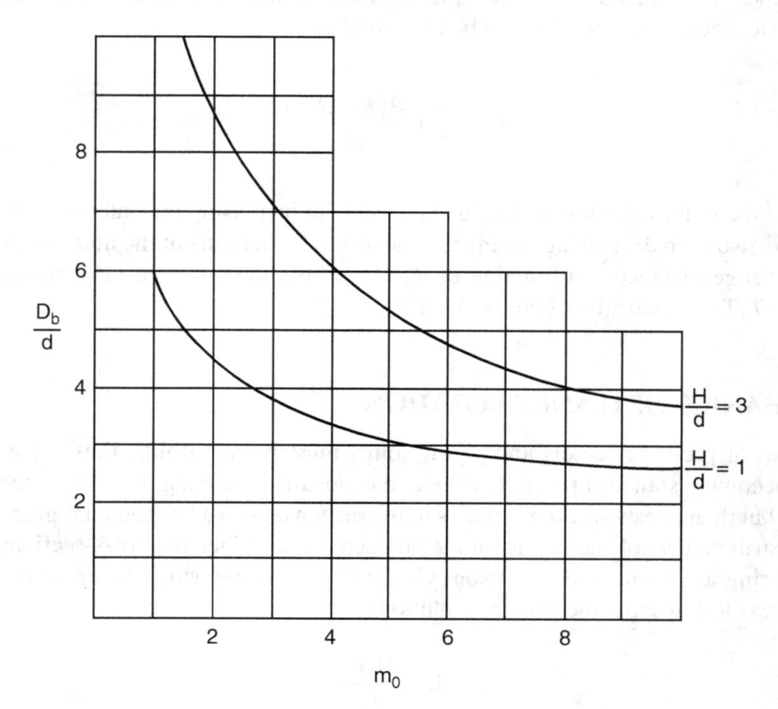

FIGURE 6.1 Bolt spacing characteristics.

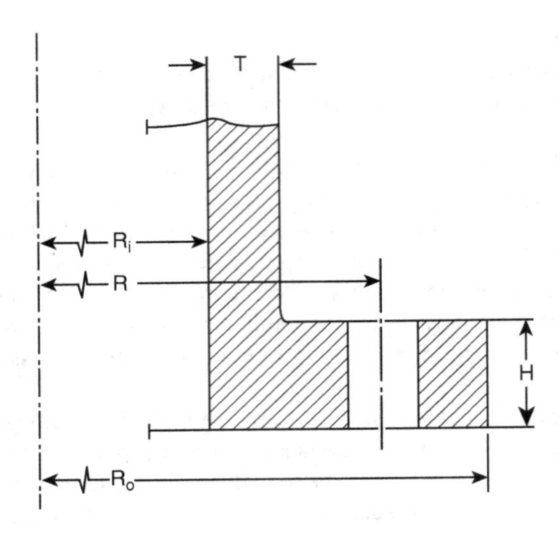

FIGURE 6.2 Simplified flange notation.

$$F = \frac{\pi P R_i^2 (R_0 - R_i)}{N(R_0 - R)} \tag{6.6}$$

where Figure 6.2 illustrates the notation, with N being the number of bolts, F being the load per bolt, and P is the internal pressure applied to the joint.

Due to the end force a bending moment is created about the bolt circle acting against the torsional resistance of the flanges. Using similar reasoning, Boardman [4] also suggested a practical design criterion for the flange thickness H, shown in Figure 6.2. In terms of the notation adopted in this chapter, the relevant design equation can be presented as:

$$H = \frac{7R_i}{4}\left(\frac{P(R - R_i)}{RS_r}\right)^{1/2} \tag{6.7}$$

It appears that the required bolt size should decrease with increasing the outer-edge distance on the bolt circle and also with decreasing the inner-edge distance. In terms of the notation given in Figure 6.2, the outer-edge distance is a function of $R_0 - R$ and the inner-edge dimension depends on the term: $R_0 - R_i - T$. The radial stress here is given by S_r.

6.5 MECHANICS OF FLANGE ROTATION

Many closures of pressure vessels and piping joints must be removable, thus the use of circular flanges has become a standard practice. Hence, the closures and flanges are constructed as bolted connections, but then questions often arise as to the response such a flanged configuration has to the uniformly distributed couple acting about the bolt centerline. When the cross-sectional dimensions of the flange ring are small in comparison with the radius of the ring, the angle of twist and the maximum stress follow from the classical solution [5]:

$$\theta = \frac{M_t R^2}{EI} \tag{6.8}$$

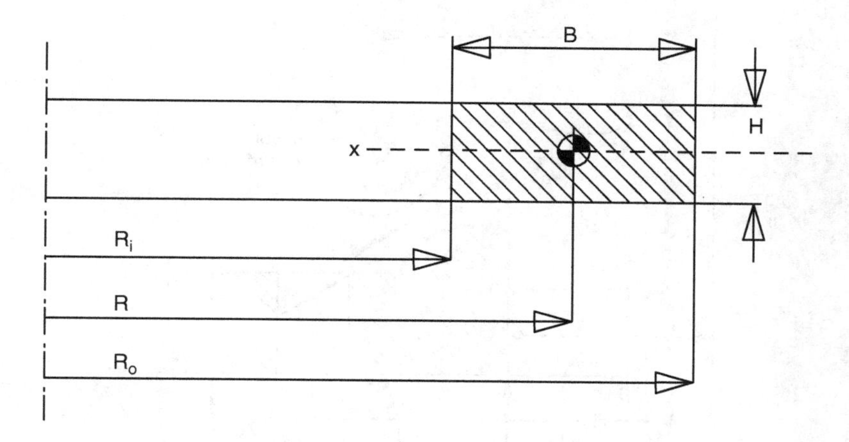

FIGURE 6.3 Rectangular flange ring section.

and

$$S = \frac{M_t R H}{2I} \tag{6.9}$$

or in terms of the angle of twist, the stress becomes:

$$S = \frac{\theta H E}{2R} \tag{6.10}$$

where Figure 6.3 illustrates the symbols used in Eqs. (6.8), (6.9), and (6.10). Also, it should be noted that the second moment of area (I) found in Eqs. (6.8) and (6.9) is taken about the x-x axis.

The average radius R corresponds to the bolt circle radius shown in Figure 6.2, which may or may not be going through the central point of the flange ring in an actual bolted joint. When the cross-sectional dimensions of the flange ring are not small compared to the various radii such as R_i, R, and R_0, then the simplifying assumptions used in Eqs. (6.8) through (6.10) cannot be made [6]. Hence, the more precise formulas become:

$$\theta = \frac{12 M_t R}{E H^3 \log_e (R_0 / R_i)} \tag{6.11}$$

and

$$S = \frac{6 M_t R}{H^2 R_i \log_e (R_0 / R_i)} \tag{6.12}$$

By analogy with Eq. (6.10), the relevant formula based on Eqs. (6.11) and (6.12) becomes:

$$S = \frac{\theta H E}{2 R_i} \tag{6.13}$$

Observe that Eqs. (6.10) and (6.13) are still quite similar and, other dimensions being equal, the maximum stress for the less compact section is smaller.

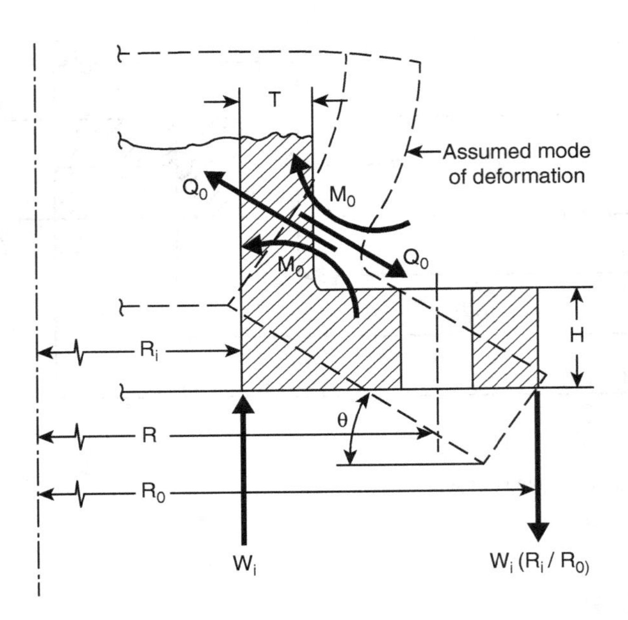

FIGURE 6.4 Ring twist applied to flange.

In analyzing the equations for the bending moment and the shearing force per unit length of the inner circumference of the pipe, where the flange ring and the pipe are joined, radial deflection can be assumed to be zero while the angle of rotation of the edge of the pipe is made equal to the angle of rotation of the flange cross-section. In Figure 6.4, this angle is denoted by θ. Here, the use is made of the theory of edge bending of shells and the toroidal deformation of a circular ring.

Enforcing the continuity relations and using notation given in Figure 6.4 yields:

$$M_0 = \frac{W_i(R_0 - R_i)}{1 + \frac{\beta_s H}{2} + \frac{(1-\nu^2)}{2\beta_s R_i}\left(\frac{H}{T}\right)^3 \log_e\left(\frac{R_0}{R_i}\right)} \tag{6.14}$$

and

$$Q_0 = \beta_s M_0 \tag{6.15}$$

where β_s is

$$\beta_s = \frac{1.285}{(R_i T)^{0.5}} \tag{6.16}$$

The maximum stress S follows from the conventional analysis of plates and shells where we have:

$$S = \frac{6M_0}{T^2} \tag{6.17}$$

Combining Eqs. (6.14), (6.16), and (6.17), the maximum stress can be calculated. It should be noted that the parameter β_S is extremely useful in the analysis of beams on an elastic foundation and it also applies with thin shells. For example, this parameter can stimulate the extent of a stress-affected zone in the vicinity of the edge load.

According to Timoshenko [5], the mechanics of flange rotation involves W_i which is the force per unit length of the inner circumference of the flange corresponding to radius R_i. The external

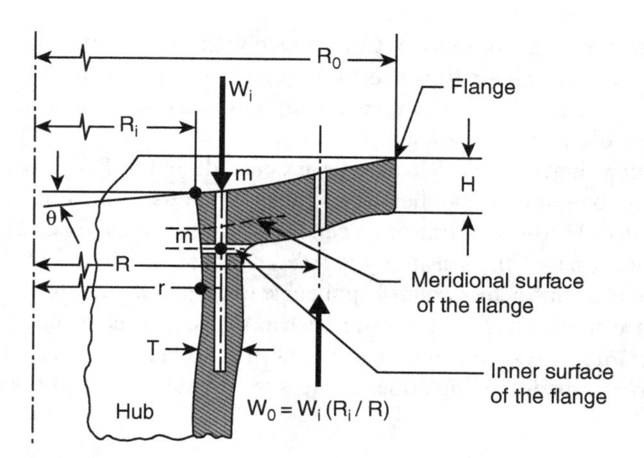

FIGURE 6.5 Flange modeled as a circular plate.

bending moment applied to the flange is based on the moment equal to $R_0 - R_i$ which is likely to overestimate the overall bending effect. In an actual joint, the circumferential loading may be significantly displaced from the inner and outer edges of the flange.

Further, refinement of the foregoing analysis can be achieved by applying the theory of plates where radial and circumferential stresses are taken into account. Radial stresses may be of importance in flanges, integral with thick pipes resisting the angle of rotation of the flange. Figure 6.5 presents a mechanical model of a flange treated as a plate. It is assumed that the bending moment existing at the root of the hub, acts at the meridian plane of the flange instead of at its inner plane. When the flange portion is deformed, the plane surfaces become curved. The cylindrical surface containing such points as $m-m$ does not alter its original curvature and the expansion of the hub due to the internal pressure can safely be ignored.

6.6 DESIGN OF STRAIGHT FLANGES

Figure 6.2 illustrates a common flanged connection. Essentially a pipe or vessel of a cylindrical shape and wall thickness T is considered to be rigidly attached to the flange having uniform thickness H. The bolt circle radius is R. The force acting on the flange joint is assumed to be caused by the hydrostatic pressure acting on the area bounded by the internal diameter of the flange equal to $2R_i$.

Until the early 1900s, most designs were based on the tangential stress at the inner diameter of the flange, ignoring any possibility of the hub stresses existing at the junction of the hub flange identified as shown in Figure 6.5. The discussion of this topic raged over many years and a number of interesting formulas have been developed with the aim of the simplifying the design calculations.

Some of the main contributions to this topic resulted in formulas and regulatory guides in this country and elsewhere [5,7–9]. The majority of formulas for flange design are simple in appearance. However, there are many variables and fairly cumbersome terms which can lead to substantial calculational difficulties, if we wish to design a series of flanges with consistent characteristics.

Essentially, only three basic formulas are required for flange design. These should represent radial stress in the flange ring, the corresponding tangential stress component and the maximum axial stress at the outer surface of the pipe or shell adjoining the flange. Out of these three quantities, the so-called hub stress, measured in the axial direction of the pipe, always appears to be numerically the highest.

There are a number of opinions in industry indicating that, although numerically high, the axial stress in the hub may not be too critical in a well-designed bolted joint. Many practical cases can be found where the calculated hub stresses are in excess of the actual yield point of the material and yet the flanges perform well in service. Unfortunately, this apparent contradiction can only be resolved by extensive tests which are time consuming and costly. One can reason that the hub stress is limited to the outer fiber and, as long as we are dealing with a ductile material, even a slight local deformation should result in redistribution of this stress.

A resulting question, however, is: What can we conclude if the flange material is inherently brittle? Brittleness can be totally unpredictable at times, if we use a material such as some of the mild steels, wherein ductile-to-brittle transition can develop close to the service temperature. Recall that some of the implications of the sensitivity to fracture have been reviewed in Chapter 1.

Based on the principles of fracture control applicable to certain materials, it may be necessary to resort to a conservative approach of flange design such as that suggested by one of the recommended formulas for a critical hub stress. For this reason and in all other situations where a conservative design is not a detriment, the following dimensionless expression for hub stress calculations can be used [10].

$$\frac{ST^2}{W} = \frac{0.318}{2m_i + 1} + \frac{0.614(k+1)(k^2-1)(m_i)^{1/2}}{(0.77+1.43K^2)\left[2.57(m_i)^{1/2}+1.65n\right]+n^3(k^2-1)} \quad (6.18)$$

Where S is the maximum anticipated stress in the hub if W is the external load on the joint caused by the internal pressure when the flange is a part of the pressure vessel or a pipe. This load may or may not be equal to the actual bolt load as indicated in Chapter 5 concerning riveted and bolted joints.

As far as the hub stress is concerned, the term W describes the actual hydrostatic load or other force in the system trying to pull the joint apart. The conventional factor of safety can be applied either to the allowable stress or to the working load, depending on design preference.

When the hub thickness is denoted by T, the relevant nondimensional parameters in Eq. (6.18) are defined as:

$$k = \frac{R_0}{R_i}$$

$$m_i = \frac{R_i}{T}$$

$$n = \frac{H}{T}$$

Figure 5.2 illustrates the remaining symbols. Observe that Eq. (6.18) contains two major parts. The first part involving m_i only defines the effect of uniform tension on the hub wall due to the hydrostatic load. The remaining part, featuring the k, m_i and n ratios, determines the effect of flange rotation and represents local bending of the hub by a couple, distributed per unit length over the inner circumference of the flange. The equilibrium condition at the juncture of the flange and vessel also requires that a shearing force per unit length of the circumference be included in the derivation of formulas for the hub stress [11].

It may be of interest to note that determination of the maximum stresses in the pipe wall adjacent to the flange ring can be accomplished in a rather simple manner, provided we can predict the orientation of the actual broken surface [7]. This approach recognizes the two basic modes of failure depending on the relative thickness of the flange ring and the pipe. This procedure is applicable to tapered as well as straight hubs.

When the flange ring is thicker than the pipe, failure can be anticipated near the hub, with the fractured surface inclined at approximately 20°–30° as shown in Figure 6.6. The reverse would, of course, be expected in unusually heavy pipe walls, but in practice the case illustrated in the sketch

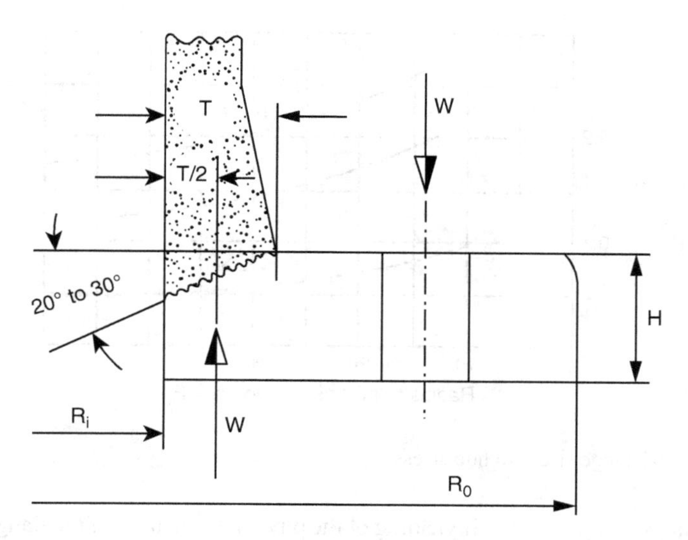

FIGURE 6.6 Assumed hub-failure described in German standard.

reflects the more likely configuration that dominates actual practice. Utilizing the notation adopted for this chapter and assuming the average angle at failure to be 25°, the maximum bending stress in the pipe adjacent to the flange can be expressed as:

$$S = \frac{W(k-1)\left[0.23 + 0.04n\sin^2(\pi/n)\right]}{nT^2} \tag{6.19}$$

Equation (6.19) should be applicable to all cases for which $n \neq 0$ and $k < 1.8$. It is also assumed that the flange ring under strain conforms to a spherical shape.

Once the maximum hub stress is known, a relatively simple calculation can be made to approximate the corresponding circumferential stress in the flange ring [12]. In this approach, we assume that the radial expansion of the flange ring can be neglected, due to the shear force discontinuity and the bulging of the pipe due to internal pressure. This is justified by the fact that during the flange rotation, one can expect to find the maximum tensile stress on the outer surface of the pipe; while pipe bulging due to internal pressure should produce a compressive stress in the same area. Hence, our assumptions should be conservative.

If the maximum bending stress S_b develops in the vicinity of the flange ring, then the corresponding flange stress in circumferential direction can be estimated from a relatively simple formula:

$$\frac{S_F}{S_b} = \frac{n(2m_i + 1)}{3.64m_i n + 4m_i(2m_i + 1)^{1/2}} \tag{6.20}$$

Eq. (6.20) can be simplified when $2m_i \gg 1$, yielding the following:

$$\frac{S_F}{S_b} = \frac{n}{1.82n + 2.83(m_i)^{1/2}} \tag{6.21}$$

Figure 6.7 provides a graphical representation of Eq. (6.21).

The circumferential stress S_F is produced by the dishing effect under the axial bolt load W. The theoretical limits for the stress ratio (S_F/S_b) range between 0.55 and 0 corresponding to zero and infinite m_i values, respectively. The intermediate range of m_i values indicate that the theoretical maximum S_F in the flange ring is always significantly lower than the corresponding value of the hub stress S_b.

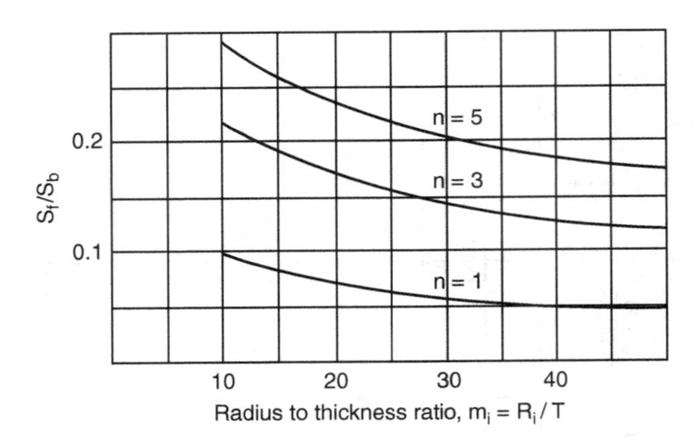

FIGURE 6.7 Ratio of flange stress to hub stress.

Conversely, the analysis suggests that, if yielding of the pipe in the vicinity of the flange is to be avoided, the flange ring would have to be extremely thick and, therefore, unacceptable for all practical purposes.

The major conclusion drawn from this analysis is that, for a truly economic design in ductile materials, plastic deformation of the pipe in the vicinity of the flange ring can be permitted. The reserve of strength beyond the onset of yield can be quite significant. For example, the theoretical collapse load of a beam of a rectangular cross-section subjected to bending is 1.5 times the load causing yield in the outer fibers, if we assume a rigid-plastic stress–strain curve for the material.

6.7 DESIGN OF RIBBED FLANGES

Figure 6.8 depicts a variety of example flange connections and some illustrations of failure sites. The sketches indicate features of progressive complexity and the regions of more significant stress gradients.

A first step in design of these type configurations is to provide an estimate of the local stress gradients based on the theory of elasticity, FEMs, or experimental results. This is probably the most difficult phase of the design since closed-form solutions applicable to ribbed flanges are difficult to find.

One of the most challenging problems in this area is concerned with a heavy duty joint involving straight and tapered ribs together with a back-up ring as illustrated in Figure 6.9. Approximate size and location of the flange reinforcement requires the determination of hub, gusset and flange ring stress levels acceptable in practice.

The design of a heavy duty joint, such as that shown in Figure 6.9, raises a question of flange-ring resistance to out-of-plane deformation, often termed "dishing." In the case of a compound flanging where two concentric rings on a pipe are connected by parallel ribs of constant cross-section, the overall rigidity against the toroidal deformation must be high. To use the hub stress formula given by Eq. (6.18), it is first necessary to calculate an equivalent value of H_e or effective flange depth for the compound flange having the two rings (main and back-up) as illustrated in Figure 6.9.

Obviously, the sum of the two ring thicknesses and the length of the ribs joining the two rings would produce a number too high for determining the ratio n (recall that $n = H/T$). However, using the elementary rules of mechanics, it is possible to calculate the dimensions of a solid rectangular cross-section that would yield the same angle of rotation as the original channel section of depth H.

Figure 6.10 provides the relevant section details needed for equivalent depth calculation.

Maintaining the overall width of the section $(T + B_r)$ constant and calculating the second moment of area with respect to the $x - x$ axis, the necessary expressions can be derived. For example, when the back-up ring is not used and the flange ring is $2T_0$ thick, the formula for calculating the equivalent depth of the ribbed flange assembly can be approximated as:

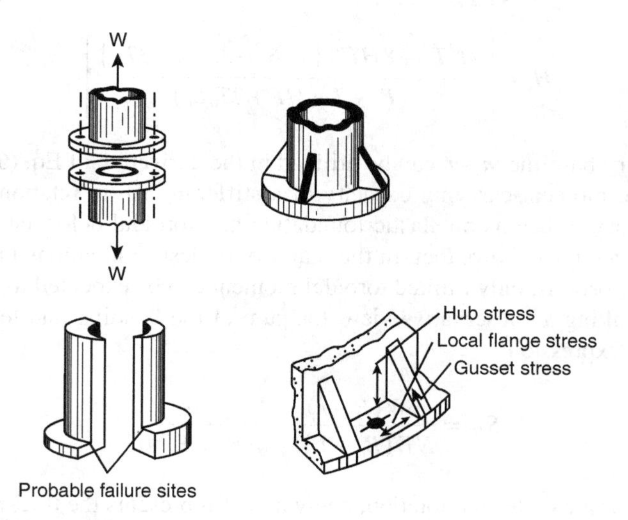

FIGURE 6.8 Examples of flanged connections.

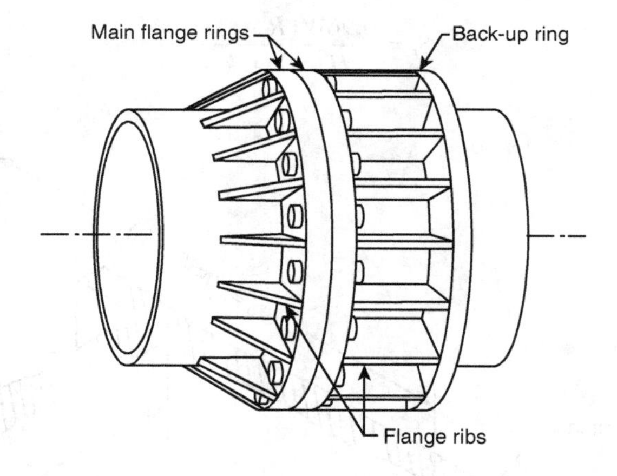

FIGURE 6.9 A complex pipe joint.

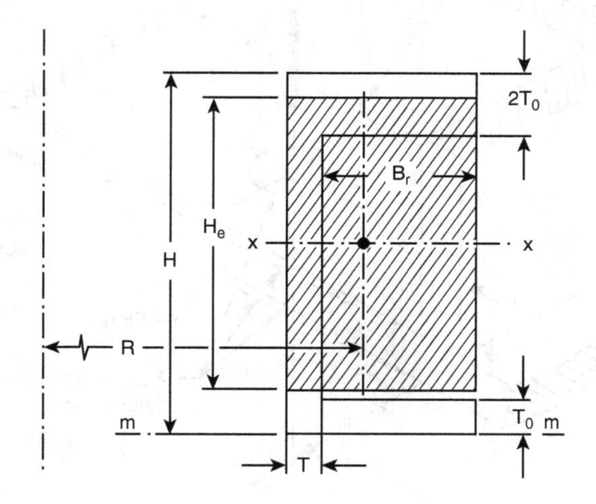

FIGURE 6.10 Section details for equivalent depth calculation.

$$H_e = \left[\frac{H^2T^4 + 8HTT_0\left(T^2B_r - 3HT_0 + 3T_0^2\right)}{(B_r + T)(HT + 2T_0B_r)} \right]^{1/3} \tag{6.22}$$

Note that an arbitrary baseline $m-m$ can be utilized in the derivation of Eq. (6.22) [10].

In evaluating the moment-carrying capacity of a stiffening rib in relation to that of the main flange ring, the theory of beams on elastic foundation and toroidal deformation of a circular ring may be of use. The analysis shows that, in the majority of design situations involving rib stiffened flanges of usual proportions, only limited toroidal moment can be expected to be transferred by the main flange ring. Taking a conservative view, the sum of the bending and tensile stresses can be estimated using the expression:

$$S_{TR} = \frac{6W(R-r)}{NT_r(B_r+T)^2} + \frac{W}{NB_rT_r + 2\pi rT} \tag{6.23}$$

where Figure 6.11 exhibits relevant notation, and where T represents the thickness of the pipe wall.

The joint is assumed to contain N equally distributed bolts and stiffening ribs. The theoretical stress in the flange ring due to the twist may be approximated in a conservative manner as:

$$S_t = \frac{0.96W(R-r)R}{H^2r^2 \log_e k} \tag{6.24}$$

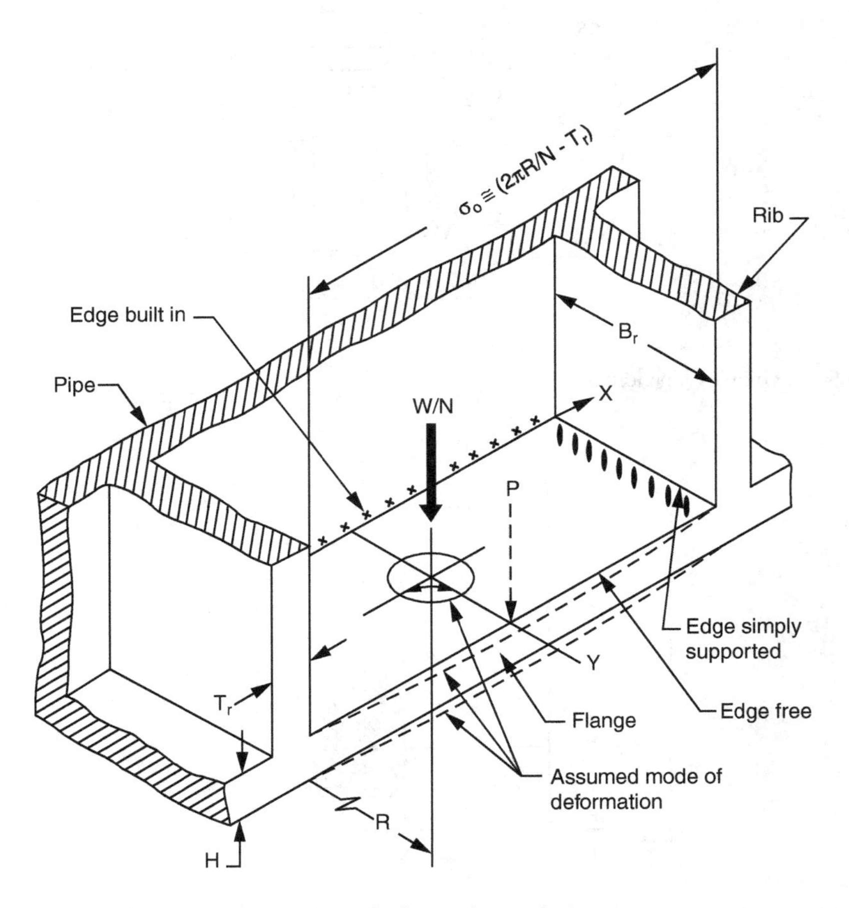

FIGURE 6.11 Approximate plate model for main flange ring analysis.

In Eqs. (6.23) and (6.24), the distance r is:

$$r = \frac{2R_i + T}{2} \tag{6.25}$$

Observe that H is used here for the simple, straight flange and heavy duty flange as shown by Figures 6.10 and 6.11. In the case of estimating the equivalent depth of the flange ring without a backup ring, the term H is made equal to $2T_0$.

When the rib system is relatively rigid and the resulting toroidal deformation rather small, the question of the stresses in the flange under individual bolts can still be unresolved. Due to symmetry, only one portion of the flange, supported by the two consecutive ribs and the pipe wall, needs to be examined. Figure 6.11 shows a portion of the flange where a bolt load W/N causes the deformation of the rectangular plate having the length $(2PR/N)–T_r$ and width B_r.

Assuming that the outer edge is unsupported, the maximum bending stress can be defined as:

$$S_b = \frac{8VWt_0}{NB_rH^2} \tag{6.26}$$

When the outer radius of the flange R_0 is relatively large compared with the mean radius of the pipe, the bending stress S_b in the flange ring can be estimated as:

$$S_b = \frac{0.95Wt_0}{NB_rH^2} \tag{6.27}$$

where the dimension t_0 may be obtained from the relation:

$$t_0 = R - R_i - T \tag{6.28}$$

Figure 6.12 provides a graphical representation for the design parameter V of Eq. (6.26).

The stresses calculated using Eqs. (6.26) and (6.27) decrease with an increase in the number of gussets, dimensions of the gussets, and the thickness of the flange ring.

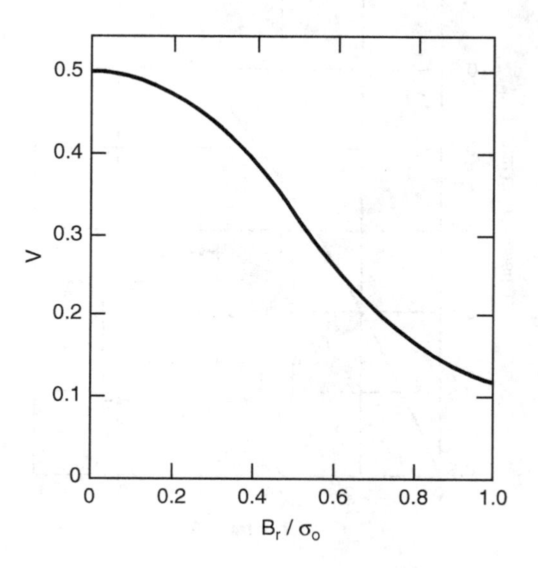

FIGURE 6.12 Design parameter V for Eq. (6.26).

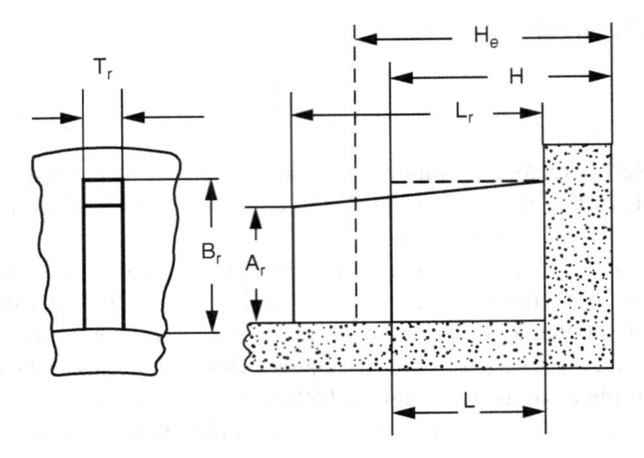

FIGURE 6.13 Tapered gusset.

Figure 6.13 shows a linearly tapered gusset with depth decreasing from B_r to A_r, as shown. In this case, the equations derived for rectangular gussets can also be applied to the tapered rib geometry by determining an equivalent length L for a rib or gusset of the actual length L_r.

Figure 6.14 presents a graphical means for estimating the equivalent gusset length ratio in terms of the actual gusset depth ratio. Once the equivalent length L is known, the overall depth H will be the sum of L and the thickness of the flange ring shown in Fig. 6.13. With this dimension in hand, the equivalent depth H_e and the resulting stresses can be found as for uniform rectangular gussets.

It should be emphasized that the design of ribbed flanges for pipe joints described in this section is based on bending rather than shear as the potential mechanism of structural failure. The approach to the normal tensile component as a criterion also demands that all the calculated stresses should be scrutinized with regard to fracture criteria described in Chapter 3.

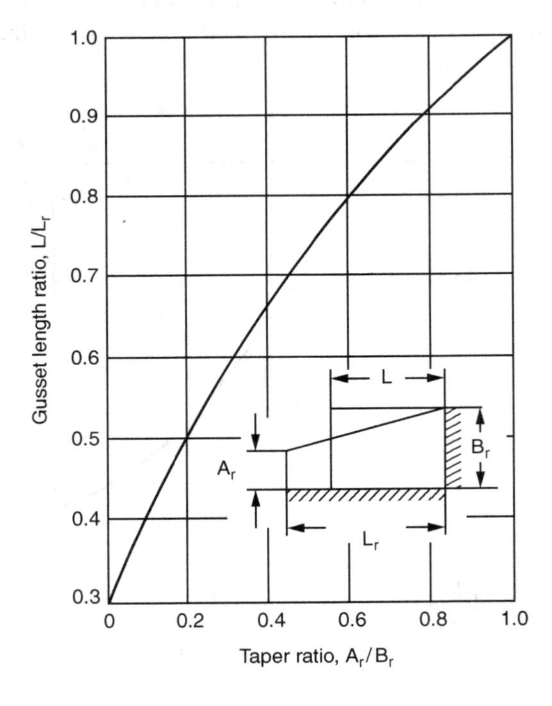

FIGURE 6.14 Design chart for equivalent length of tapered gussets.

6.8 STIFFENERS

This brief section is concerned with a special problem area involving the role of straight and circumferential stiffeners in interfaces. The performance of stiffeners depends on the integrity of a mechanical joint using welds, adhesion bonds, or fasteners. The alternative is to machine or cast the stiffening members together with the main components which typically is not a practical solution when dealing with plates or vessels.

Figure 6.15 depicts a classical example of a cover plate with radial stiffeners. Such reinforcement is often based on a system of orthogonal and equidistant ribs. For a symmetrical system, the plate rigidity can be defined as:

$$D = \frac{Et^3}{12(1-\nu^2)} + \frac{EI}{\lambda} \tag{6.29}$$

where λ is the distance between the consecutive rib centers and I is the second moment of area of a single rib with respect to the middle axis of the plate.

When a rib is placed on one side of the wall only, we obtain a T shape cross-section so that the flexural rigidity D can be reduced to:

$$D = \frac{EI}{\lambda} \tag{6.30}$$

where I is the average second moment of area of the T section with width λ about its centroid.

Other types of plates with reinforcement involve grill-type design or concentric ring stiffening adding local toroidal stiffness. The concentric ring effect can be optimized if we weld this ring at about 0.6 distance of the radius measured from the plate center. To satisfy the deflection, rather than the stress criteria, a relatively thin plate with a significant reinforcement should be designed. The analytical and experimental problems are substantial because of a three-dimensional nature of response and often unmanageable boundary conditions.

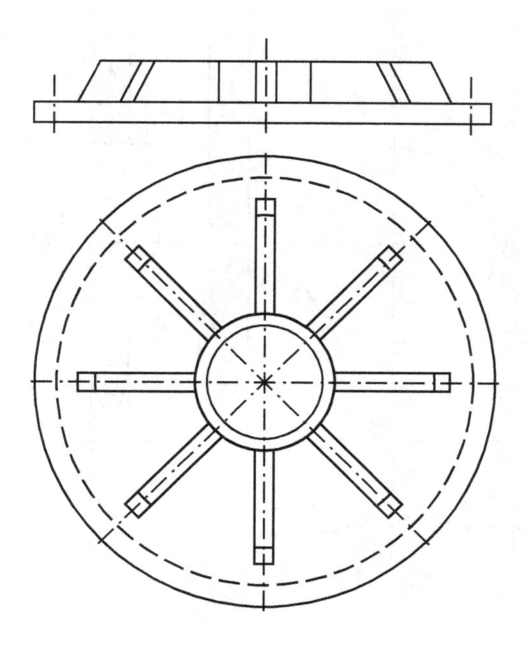

FIGURE 6.15 Example of a reinforced plate.

In the case of a closure plate, such as that shown in Figure 6.15, an individual rib can be regarded as a simple beam subjected to a bending moment at the junction with the concentric stiffener, a reaction of the other end of the rib and a distributed load along the rib length according to some linear function. The ribs behave as *T* beams of variable stiffness with the portions of the adjacent plate acting as flanges.

Some experiments were conducted on cover plates with eight radial ribs under uniform pressure, to determine the plate rigidity and optimum distribution of the material [13]. The empirical formula for the central deflection of an eight-rib configuration was found to be:

$$Y = \frac{21.6qa^{10}}{EV_p^3} \tag{6.31}$$

Equation (6.31) assumes Poisson's ratio to be equal to 0.3, which is good for the majority of conventional metals used in the cover plate construction. The results of experimental work show that the most effective use of material is likely to be achieved with deep and slender welded ribs, in such a way as to make the cumulative mass of the rib system equal to about 40% of the total mass of the plate. For the more heavily reinforced plate, the 40% criterion may not be possible to achieve.

Analysis of the experimental findings for the eight-rib cover plate also suggests that a rough guide can be established for the deflection criteria extended to other numbers of ribs, provided the 40% rule can be maintained.

In a riveted or a bolted connection, a plate member can be loaded in the manner shown in Figure 6.16. This type of loading may be experienced in a number of situations where a pin or a shank-type part is compressed against a curved surface. The stress distribution resulting from this form of loading may require a redesign of the connection, where the $2r/B$ ratio is an important parameter.

One of the solutions may be a circumferential stiffener aimed at lowering the contact pressure. If we denote the ratio of the maximum tensile stress S, at the edge of the hole to the average bearing

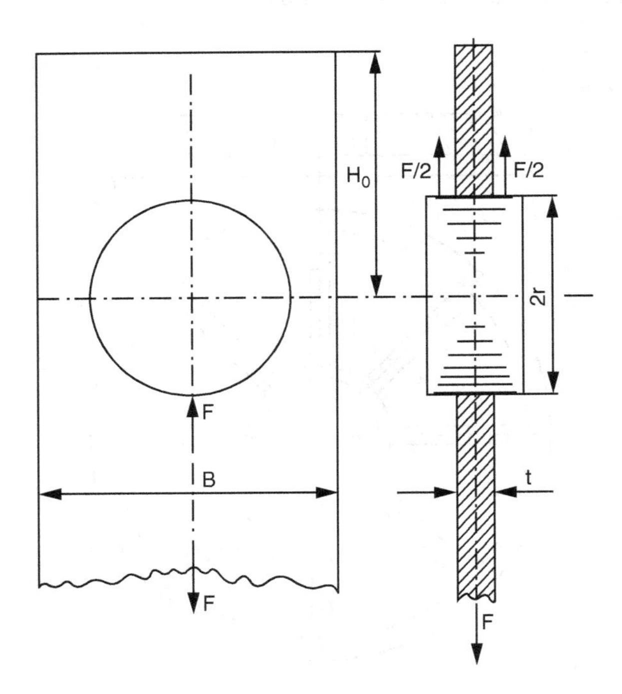

FIGURE 6.16 Pin loaded plate.

stress on the bolt shank or pin and acting in the manner shown in Fig. 6.16, then the following experimentally based design parameter can be established [14].

$$D_p = \frac{2rtS}{F} \tag{6.32}$$

where Figure 6.16 shows the symbols.

Figure 6.17 presents a convenient design graph: If, for example, $2r/B$ is 0.3 then the curve in Figure 6.17 yields D_p as 1.5. Substituting this value into Eq. (6.32) and solving for S, results in $S = 0.75(F/rt)$, which is the maximum tensile stress at the edge of the hole.

The parameter D_p increases with the increase in the pin clearance and depends on the head distance ratio H_0/B. Also, D_p is expected to decrease with an increase in the number of pins and the quality of lubrication.

Reinforcement of cylindrical vessels and large piping often calls for design of circumferential stiffeners fastened firmly to the shell. In the great majority of these applications, the stiffeners are sized to prevent the overall collapse of the vessels subjected to external pressure.

Figure 6.18 shows notation for a circumferential stiffener.

The corresponding design formula for stiffener sizing is:

$$\frac{PL}{ET} = \frac{0.24\phi^2(\varepsilon+1)^4 + 0.39\phi(m)^{1/2}(\varepsilon+1)(4\varepsilon^2+2\varepsilon+1)}{m^3\left[\phi(\varepsilon+1)+1.57(m)^{1/2}\right]} \tag{6.33}$$

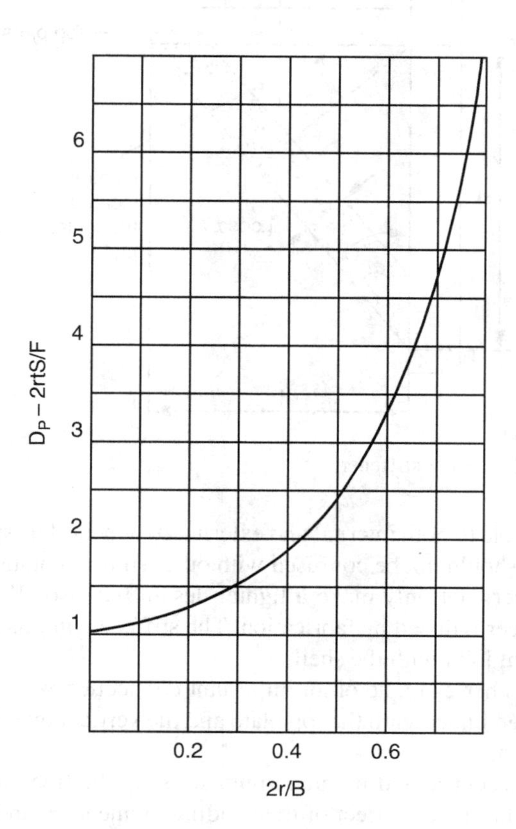

FIGURE 6.17 Experimental stress factor for a pin-loaded plate.

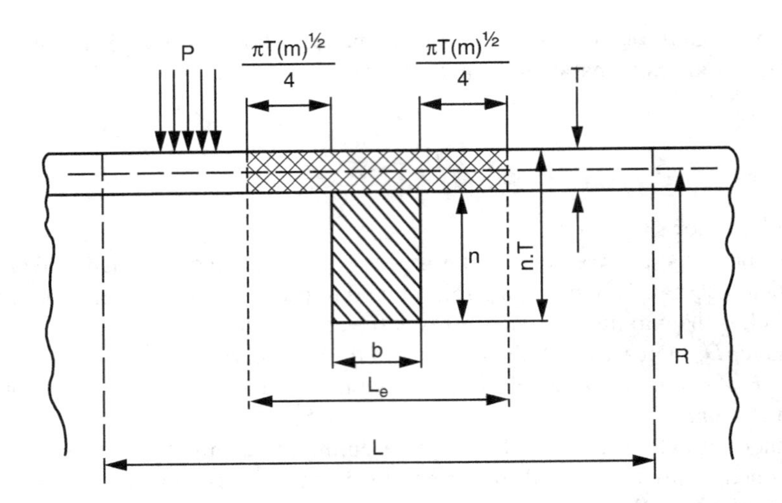

FIGURE 6.18 Notation for a ring stiffener.

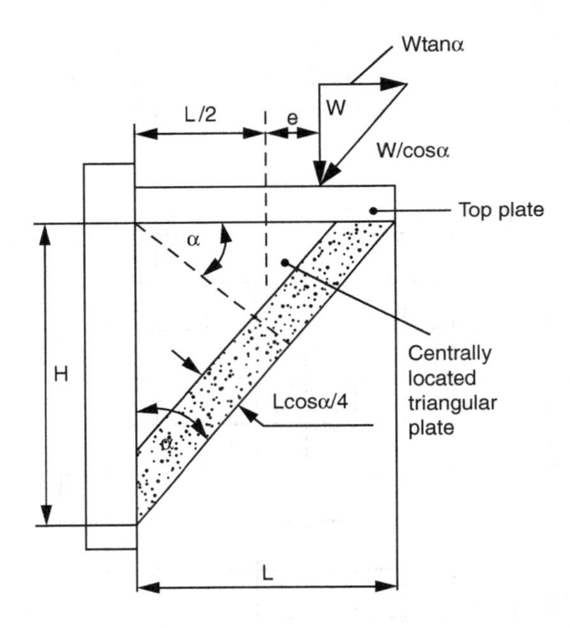

FIGURE 6.19 Example of a corner stiffener.

Equation (6.33) is applicable to both internal and external stiffeners. The symbols given in Eq. (6.33) apply to Figure 6.18 and should not be confused with other similar notation utilized in this chapter.

In general, the stiffeners not only make a lighter design but also allow some relaxation in the out-of-roundness tolerances induced by fabrication. The stiffeners in this case are always welded to form a perfect mechanical joint with the shell.

Figure 6.19 shows another example of an important connection with a stiffener. In this case, a triangular gusset is welded underneath the top plate and the vertical member to form very rigid and efficient structural support.

The external load is eccentric and the maximum stress at the free edge of the triangular plate may be obtained by combining the effect of the bending moment W_e and the compressive load $W/\cos\alpha$ to give the expression:

$$S = \frac{W(L + 6e)}{tL^2 \cos^2 \alpha} \tag{6.34}$$

The elastic stability criterion can be defined as the allowable design ratio L/t according to the expression:

$$\frac{L}{t} \leq \frac{48 + 24(L / H)}{(S_y)^{1/2}} \tag{6.35}$$

The maximum permissible load W_{pl} under fully plastic conditions can be expressed as:

$$W_{pl} = tS_y \cos^2 \alpha \left[\left(L^2 + 4e^2 \right)^{1/2} - 2e \right] \tag{6.36}$$

At the same time, the cross-sectional area of the top plate shown in Figure 6.19 should be sized on the basis of a horizontal component of the external load:

$$A = W_{pl} \frac{\tan \alpha}{S_y} \tag{6.37}$$

Here again is an example of a relatively simple mechanical connection where a single formula cannot describe the structural integrity of the entire system involving strength of the welds, beam strength, or stability of individual components.

6.9 COMPACT FLANGES

So far, in this chapter, we have a few glimpses of flange theory and design including a number of topics and examples related to the general problem of stiffeners. We saw how the design complexity has increased by progressing from a conventional straight flange geometry, through a tapered version, up to the complex configuration involving backup rings and gussets.

Many improvements have been made in the analysis of flanges with tapered hubs. Since the design equations for this last case have proven to be tedious and multi-variable, serious students of this particular topic must be referred to various available regulatory and industrial sources. We believe, however, that by accepting only a moderate error, the design formulas derived for the straight flanges can be applied to tapered hub configurations by assuming average thickness of the hub.

For a more accurate approach, the design engineer is encouraged to use FEMs and his or her own practical knowledge in arriving at a suitable technical decision.

The material presented in this brief, but important, section is based on a long-standing research experience of Webjörn [15] and other concerned with the state of flange design and some of the ideas which, in the past, tended to limit the capacity of bolted joints. There are several indications that it is not particularly difficult to make the joint stronger than the pipe to which the flange is attached and to assure a no-leak condition.

The work done on the flat-face, metal-to-metal flanges has been well documented in the open literature, although only a relatively few papers have addressed experimental studies devoted to a compact flange concept. The potential advantages of the rigid and compact flanges were at one time fully discussed and related to the practice of operating low-pressure steamlines in the nuclear industry [16].

Recent tests by Webjörn were designed to prove the functional characteristics of compact flanges held together by highly preloaded, slender bolts. The purpose of the experiment [15] was to demonstrate that even a heavy, compact flange connection performs as predicted and that heavy bolts indeed can be preloaded above the required amount by using hydraulic tensioners.

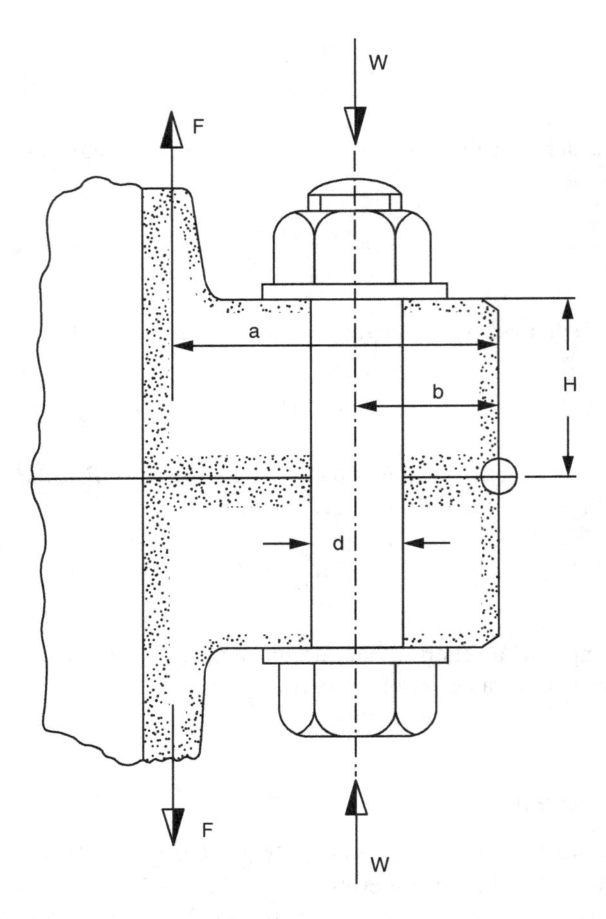

FIGURE 6.20 Force diagram for a compact flange.

The conventional assumptions for a compact flange design might possibly be based on the idea that the joint can deform as a rigid body. The relevant criterion for the flange separation, for the particular case illustrated in Figure 6.20, could then be expressed by a simple relation [17] of the type $wb > Fa$.

Webjörn's research [15], however, indicated clearly that the compact flange has such high rotational stiffness that the conventional lever arm theory may not apply.

In the limiting case when the flange thickness is many times the bolt diameter, it is quite sufficient to calculate the axial, hydrostatic end force; provided one can select the appropriate number of sufficiently strong bolts capable of exceeding the end force. The same research also shows that the flange resistance to warping is a function of the third power of the flange height H and that the minimum criterion for a compact flange design would be $H = 3d$.

Figure 6.21 illustrates principal differences between conventional and compact flange designs, showing the pattern of interaction of the forces and the resulting differences in flange sizes. The orientation of forces and moments appears to be such that the highest stresses are more likely to be in the pipe rather than the flanges in a compact configuration. The only problem here is to select a sufficiently large number of strong bolts, preloaded to at least 80% of the guaranteed minimum yield strength of the bolt material.

A number of tests conducted in Sweden [15] have shown that compact flange configurations maintained a no-separation condition up to the proof test pressure, resisted pulsating loads well and were generally stronger than the pipe itself.

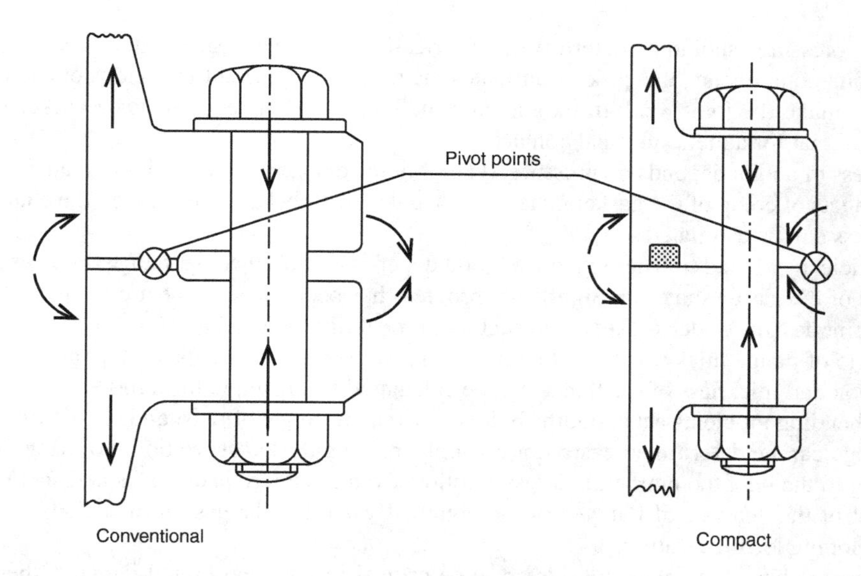

FIGURE 6.21 Comparison of conventional and compact flanges.

It should be emphasized that, in the conventional flange assembly, it is difficult to control the bolt preload. Insufficient preload may result in leakage if the gasket does not set properly. Too much preload, on the other hand, can also cause leaks because of the flange warpage. In addition, one is likely to expect that in a conventional flange design, the combined tensile stress occurs in the fillet between the flange and the hub.

The concept of hub stress related to various flange configurations has already been discussed in the various parts of this chapter. The advantage of a compact flange design comes from realization that the amount of hub bending due to the flange moment in a compact joint must be minimal.

The progress in welding technology, however significant, does not always apply to such important operations as in the construction of pipe-lines on the ocean floor, under deep water, or in large feed lines in the nuclear power plants where large valves may need replacing during a malfunction. It appears that the use of a compact flange design for these applications would be highly desirable.

Nevertheless, the building of such systems with compact flange technology may necessitate special knowledge and refined techniques. It may also be necessary to prefabricate various pipe-line sections at the plant and assemble them in the field according to well-defined tolerances.

The proposed high values of H/d ratios for the compact flange design may strictly be a matter of economics. However, the research quoted in this section, as well as some evidence of older engineering practice with compact flanges, indicate that $(H/d) = 3$, should remain as a recommended minimum criterion. It is anticipated that for the above design ratios lower than about 3, any reduction in cost and weight would indeed be marginal.

Research has been extended to the study of the effect of thermal shock in pipe connections with compact flanges [18]. The conclusions were that the compact flange assembly was less sensitive to thermal shock than the corresponding conventional joint. The study involved temperature measurements and the two-dimensional computer calculations of the heat transfer.

6.10 SUMMARY

Efficient gasket application depends on material selection, configuration, and contact pressure. Typical choice lies between rubber fiber composite or metal gaskets. Gasket factor is defined as the ratio between the contact pressure and the contained vessel pressure. High pressure and temperature applications require metal-to-metal line contact.

Contact pressures should be determined empirically. When the gasket material is more flexible than the joint components, the gasket dominates the response of the entire bolted connection.

Gaskets make the joint weak in fatigue, and the bolt diameter required for a gasketed joint is greater than that for a metal-to-meal contact.

Tightness of a joint depends on pressure, type of fluid contained, uniformity of load distribution, and the effects of creep of the gasket material. Once the creep behavior is known, the change in the gasket stress can be estimated.

The efficiency of load transfer in a bolted joint depends on bolt spacing, flange thickness, and the properties of the gasket material. Significant progress has been made in the area of gasket technology and joint design. Wider gaskets represent less probability of leakage.

The ratio of flange thickness to bolt diameter in low-pressure service is on the order of 1.0. The load per bolt and thickness of the flange can be calculated from simple formulas.

When bending moments act about the bolt centerline, the angle of twist and the maximum stress in the flange can be determined easily for a simple, rectangular cross-section. For a rigidly joined flange ring to the hub, the maximum stresses follow from the analysis of plates and shells. Further refinement of the analysis of flanges can be obtained when the flanges are modeled as a circular plate developing local curvature.

Hub stresses have long been considered to be critical in a typical bolted flange. Other stresses of interest include radial and tangential stresses in the flange ring. The importance of hub stress, however, should be viewed with respect to the local yield effects and redistribution of the original stress gradients, provided the material is ductile over the entire range of the working temperatures.

According to recent design practice, the maximum stress in the pipe wall adjacent to the flange can be expressed as a function of the angle of the fractured surface varying between 20° and 30°. The circumferential (tangential) stress in the flange can be expressed in terms of the hub stress and flange geometry.

Complexity of flange design increases considerably when the gussets and ribs are added for the purpose of controlling the toroidal deformation of the flange ring. The analysis requires some knowledge of load shearing between the conventional elements of a straight flange and the ribs. Additional considerations involve the stresses in ribs and the flange ring. The analysis presented is approximate, at best, because the effects of local shear stresses have been neglected for the purpose of simplifying the calculations procedures. A rigorous analysis of this problem, however, even with the help of FEMs, appears to be prohibitive.

According to approximate equations, local bending stresses in the flange decrease with an increase in the number of gussets, gusset size, and the thickness of the flange ring.

Design of cover plates with radial stiffeners represents a special case of joint technology, if the stiffeners are welded to the base plate. The design of a plate with the combination of radial and circumferential stiffening requires experimental analysis. The optimum use of the material is indicated when the size of the rib system is equal to 40% of the total mass of the plate.

In the case of a plate loaded through a pin in shear, the stress distribution at the edge of the hole depends on the plate and pin dimensions. Other important parameters include clearance between the pin and the hole, as well as the distance between the edge of the hole and the plate.

The design of circumferential stiffeners, for externally loaded cylindrical vessels, can be accomplished with the aid of one formula containing seven parameters. The effect of the circumferential stiffeners is to make a lighter design and to reduce the sensitivity of the vessel to out-of-roundness tolerances.

Another example of a mechanical connection is the bracket where a triangular gusset supports a plate member. Here again, not only the weld strength but also the stresses and stability of the gusset should be evaluated for a successful joint design.

An interesting case of a bolted joint design with a metal-to-metal contact is one with heavy, rigid flanges and relatively slender, high-strength bolts. Joints of this nature can be designed for zero separation and zero leakage. Recent tests, as well as long-standing experience, indicate that the recommended minimum ratio of the flange thickness to bolt diameter would be equal to 3. This compact flange design cannot be achieved with the conventional gaskets and spring washers. The bolts for the compact flange should be preloaded to at least 80% of the minimum yield strength of the bolt material.

6.11 DESIGN ANALYSES

EXAMPLE ANALYSIS 1

Two identical transmission shafts are to be connected end-to-end by bolted circular flanges as represented in Figures a and b. The shafts are securely welded by the plates, and the plates are connected by six evenly and symmetrically placed bolts as in Figure b.

Let the radii of the shafts be r_s and let the shafts shear strength be τ_{max}. Similarly, let the bolts have radius r_b and also with shear strength τ_{max}. Let the bolt centers be mounted on a circle with radius R and whose center is on the shaft axis as in Figure b.

Suppose the shaft radii r_s, the bolt radii r_b and the mounting radius R, are to have values so that if the shaft has a peak transmission torque T_{max}, the bolts and the shafts fail simultaneously in shear.

Determine the relation between r_s, r_b, and R.

Solution

Recall from elementary strength of materials that the maximum shear stress τ_s on a twisted shaft occurs on the shaft surface with value:

$$\tau_s = \frac{Tr_s}{J} \tag{a}$$

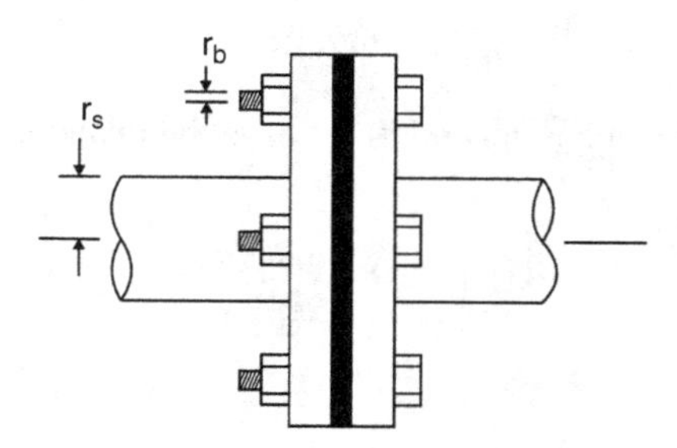

FIGURE A Side view of coupling of a transmission shaft.

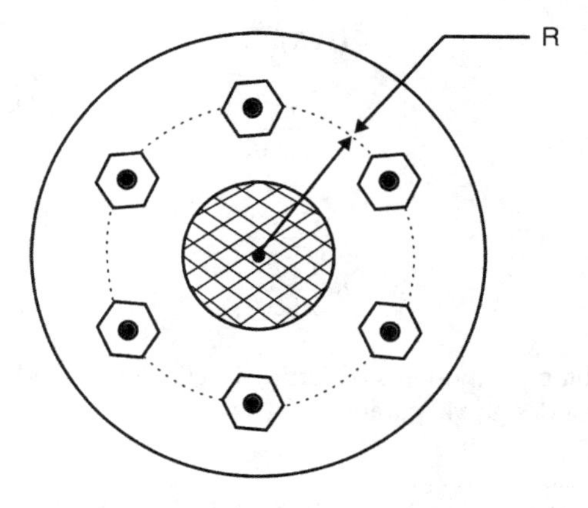

FIGURE B End view of coupling.

where T is the twisting torque, r_s (as before) is the shaft radius, and J is the second axial moment of area (aka polar moment of inertia), given by:

$$J = \left(\frac{\pi}{2}\right) r_s^4 \tag{b}$$

With the maximum shear stress being τ_{max}, and the maximum twisting moment being T_{max}, Eqs. (a) and (b) produce the relation:

$$T_{max} = \tau_{max} \frac{J}{r_s} = \left(\frac{\pi}{2}\right) r_s^3 \tau_{max} \tag{c}$$

Next, for the bolts, for simplicity, we assume that each of the six bolts carries an equal share of the twisting load*. Then by inspection of Figure b, the shear force F_{sh} on each bolt is approximately:

$$F_{sh} = \frac{T}{6R} \tag{d}$$

where, as before, T is the twisting moment. This shear force, in turn, will create an average shear stress τ_b on the bolts given by:

$$\tau_b = \frac{F_{sh}}{A_b} = \frac{F_{sh}}{\pi r_b^2} = \frac{T}{6\pi R r_b^L} \tag{e}$$

where A_b is the cross-section area of the bolt.

From Eq. (a) the maximum shaft twist, T_{max}, that the bolts can support is:

$$T_{max} = 6\pi R r_b^2 \tau_{max} \tag{f}$$

Finally, by equating the results of Eqs. (c) and (f) we have the relation:

$$\left(\frac{\pi}{2}\right) r_s^3 \tau_{max} = 6\pi R r_b^2 \tau_{max}$$

or, after simplification,

$$r_s^3 = 12 R r_b^2 \tag{g}$$

From Eq. (g) we immediately obtain the results:

$$r_s = (12R)^{1/3} r_b^{2/3} \tag{h}$$

$$r_b = \frac{r_s^{3/2}}{2\sqrt{3R}} \tag{i}$$

and

$$R = \frac{r_s^3}{12 r_b^2} \tag{j}$$

Observe that with the multiple exponents occurring in Eqs. (h), (i), and (j), we can easily change the appearance of the joint design via parameter variation.

* This is not a valid assumption since it is impossible to construct the joint with perfect geometry. If, however, the geometry is reasonably precise, the assumption may be reasonable as a first approximation.

EXAMPLE ANALYSIS 2

See Example Analysis 1.

Suppose the bolt diameter d_b is ½ inch, the ring radius R is 2½ inches, and that the shear strength of the material τ_{max} is 20,000 psi.

a. Determine the smallest allowable shaft radius $r_{s\,min}$.
b. Determine the maximum torque T_{max} that can be transmitted through the coupling joint.

Solution

a. From Eq. (h) of Example Analysis 1, r_s is:

$$r_s = (12)^{1/3} R^{1/3} r_b^{2/3} \tag{a}$$

Then, by substituting the given data r_s becomes:

$$r_{s\,\min} = (12)^{1/3}(2.5)^{1/3}(0.25)^{2/3} = (2.289)(1.357)(0.397)$$

or

$$r_{s\,\min} = 1.233 \text{ in} \tag{b}$$

b. From Eq. (c) of Example Analysis 1, T_{max} is:

$$T_{\max} = (\pi/2) r_s^3 \tau_{\max} \tag{c}$$

Then, by substituting the given data, T_{max} becomes:

$$T_{\max} = (\pi/2)(1.233)^3(20{,}000) = (1.571)(1.875)(20{,}000)$$

or

$$T_{\max} = 58{,}904 \text{ in lb} = 4909 \text{ ft lb} \tag{d}$$

As a check, in reviewing Example Analysis 1, we see from Eq. (f) that T_{max} is also:

$$T_{\max} = 6\pi R r_b^2 \tau_{\max}$$

Then, by again substituting the given data, T_{max} becomes:

$$T_{\max} = (6\pi)(2.5)(0.25)^2(20{,}000) = (18.850)(2.5)(0.0625)(20{,}000)$$

or

$$T_{\max} = 58{,}905 \text{ in lb} = 4{,}909 \text{ ft lb}$$

The slight differences in the fifth place is simple due to roundoff error in the calculation, and well within the accuracy of the dimensional data and the given shear strength. Indeed in calculations with experimental data, there is little, if anything, to be gained by making calculations to more decimal places than those of the given data.

EXAMPLE ANALYSIS 3

Review Example Analyses 1 and 2. Suppose that due to minor manufacturing inaccuracies, only two of the six bolts transmit the transmission torque. Repeat the analyses for this two-bolt load carrying assumption.

That is, determine:

a. The relation between r_s, r_b, and R.
b. The smallest allowable shaft radius, $r_{s\ min}$, if (as before) the bolt diameter, d_b, is ½ inch, and the ring radius, R, is 2½ inches.
c. The maximum torque T_{max} that can be transmitted through the coupling joint if, as before, d_b is ½ inch, R is 2½ inches, and that the shear strength τ_{max} of all components is 20,000 psi.

Solution

The problem statement did not say *which two* of the six bolts were to carry the load. There are three possible pairs: i) two diametral bolts, such as A and D in Figure 1; ii) two adjacent bolts, such as A and B; and iii) two intermediate bolts, such as A and C.

For a comprehensive solution, we should consider all three cases. Actually, the only difference between the cases is the distance between any given two bolts. Consider each case in turn:

Case i) Two diametral bolts: A and D

By inspection of Figure a we immediately see that the distance d_{AD} between the diametral bolts is simply: $2R$. That is,

$$d_{AD} = 2R \tag{a}$$

Case ii) Two intermediate bolts: A and B

By inspection of Figure a, we see that the center O of the shaft and the two adjacent bolt centers form an equilateral triangle. Therefore, the distance d_{AB} between the adjacent bolts is simply R. That is,

$$d_{AB} = R \tag{b}$$

Case iii) Two intermediate bolts: A and C

Finally, from Figure a we see that the shaft center O and the centers of bolts A and C form an isosceles triangle with the equal side lengths being R and the central vertex angle being 120E, as in Figure b. Then by elementary trigonometry we see that the distance between the intermediate bolts d_{AC} is: $\sqrt{3}R$. That is:

$$d_{AC} = \sqrt{3}R \tag{c}$$

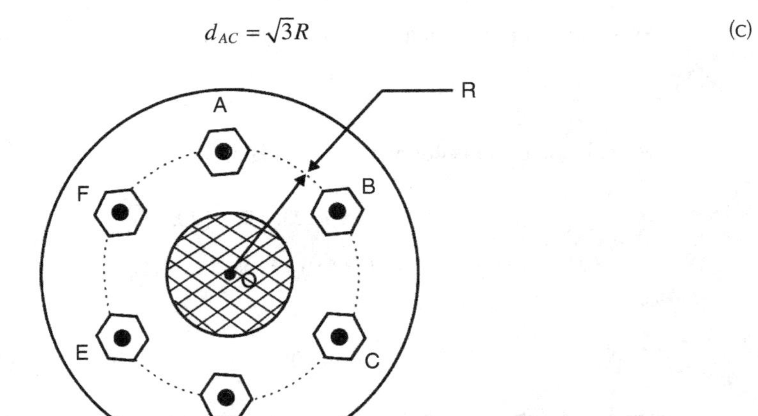

FIGURE A Bolt labeling.

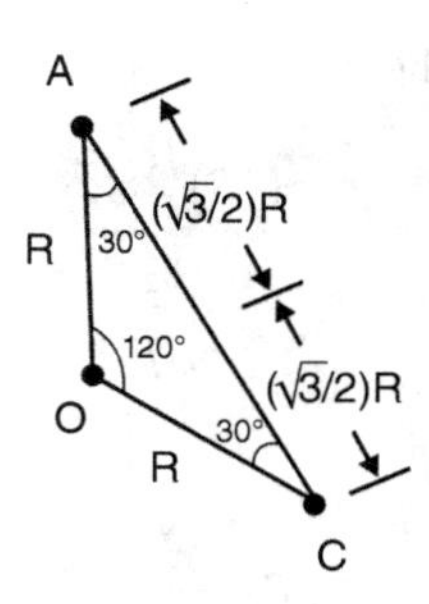

FIGURE B Triangle OAC geometry.

For the analysis we can consider all three cases simultaneously by simply letting the distance between any two bolts be: d. Then in this way we can address all three requests of the problem statement by simply following the solutions of the Example Analyses 1 and 2.

a. The relation between r_s, r_b, and R

In reviewing the Example Analysis 1 we see that with only <u>two</u> supporting bolts, Eq. (d) should be replaced by the simple expression:

$$F_{sh} = \frac{T}{d} \tag{d}$$

where, as before, T is the twisting torque and d is the generic distance between the two supporting bolts.

Following the solution in Example Analysis 2, the average shear stress τ_b on a bolt is then:

$$\tau_b = \frac{F_{sh}}{A_b} = \frac{F_{sh}}{\pi r_b^2} = \frac{T}{d\pi r_b^2} \tag{e}$$

where, as before, A_b is the cross-section area of the bolt.

From Eq. (e) the maximum shaft twist, $T_{\max}$, that a bolt can support is:

$$T_{\max} = d\pi r_b^2 \tau_{\max} \tag{f}$$

From the solution of Example Analysis 1, we see in Eq. (c) that $T_{\max}$ (due to the shaft strength) is also:

$$T_{\max} = \left(\frac{\pi}{2}\right) r_s^3 \tau_{\max} \tag{g}$$

By equating the results of Eqs. (f) and (g) we have:

$$d\pi r_b^2 \tau_{\max} = \left(\frac{\pi}{2}\right) r_s^3 \tau_{\max}$$

or after simplification,

$$r_s^3 = 2dr_b^2 \tag{h}$$

Consider now the three two-bolt cases:

Case i) Two diametral bolts: A and D:

In this case we see from Eq. (a) that d_{AD} is $2R$. Then by substituting $2R$ for d in Eq. (h) we obtain:

$$r_s^3 = 4Rr_b^2 \tag{i}$$

The requested relations are then:

$$r_s = (4R)^{1/3} r_b^{2/3} \tag{j}$$

$$r_b = \frac{r_s^{3/2}}{2\sqrt{R}} \tag{k}$$

and

$$R = \frac{r_s^3}{4r_b^2} \tag{l}$$

Case ii) Two adjacent bolts: A and B

In this case we see from Eq. (b) that d_{AB} is R. Then by substituting R for d in Eq. (h) we obtain:

$$r_s^3 = 2Rr_b^2 \tag{m}$$

The requested relations are then:

$$r_s = (2R)^{1/3} r_b^{2/3} \tag{n}$$

$$r_b = r_s^{3/2}\sqrt{2R} \tag{o}$$

$$R = \frac{r_s^3}{2r_b^2} \tag{p}$$

Case iii) Two intermediate bolts: A and C

In this case we see from Eq. (c) that d_{AC} is $\sqrt{3}R$. Then, by substituting $\sqrt{3}R$ for d in Eq. (h) we obtain:

$$r_s^3 = 2\sqrt{3}Rr_b^2 \tag{q}$$

The requested relations are then:

$$r_s = \left(2\sqrt{3}R\right)^{1/3} r_b^{2/3} \tag{r}$$

$$r_b = \frac{r_s^{3/2}}{\left(2\sqrt{3}R\right)^{1/2}} \tag{s}$$

and

$$R = \frac{r_s^3}{2\sqrt{3}r_b^2} \tag{t}$$

b. The smallest allowable shaft radius, $r_{s\ min}$, with bolt diameter, d_b, being ½ inch, and the ring radius R being 2½ inches.

This request can be met by substituting the given data into Eq. (j), (n), and (r) for the three cases. To this end we have:

Case i) Two diametral bolts: A and D

From Eq. (j):

$$r_{s\min} = (4R)^{1/3} r_b^{2/3} = \left[(4)(2.5)\right]^{1/3} (0.25)^{2/3}$$

or

$$r_{s\min} = 0.855 \text{ in} \tag{u}$$

Case ii) Two adjacent bolts: A and B
From Eq. (n):

$$r_{s\,\min} = (2R)^{1/3} r_b^{2/3} = [(2)(2.5)]^{1/3} (0.25)^{2/3}$$

or

$$r_{s\,\min} = 0.679 \text{ in} \tag{v}$$

Case iii) Two intermediate bolts: A and C
From Eq. (r)

$$r_{s\,\min} = \left(2\sqrt{3}R\right)^{1/3} r_b^{2/3} = [(3.464)(2.5)]^{1/3} (0.25)^{2/3}$$

or

$$r_{s\,\min} = 0.815 \text{ in} \tag{w}$$

c. The maximum torque T_{max} that can be transmitted through the joint with τ_{max} = 200,000 psi.

We can obtain Tmax using either Eqs. (f) or (g), or better, either with the other as a check. Then for the data and results of the three cases we have:

Case i) Two diametral bolts: A and D

For convenience, it may be helpful to first review and list the data for this case:

$$d = d_{AD,} = 2R, \quad R = 2.5 \text{ in}, \quad r_b = 0.25 \text{ in}$$

$$r_{s\,\min} = 0.855 \text{ in}, \quad \tau_{\max} = 20{,}000 \text{ psi}$$

From Eq. (f), T_{max} is then

$$T_{\max} = d_{AD}\pi r_b^2 \tau_{\max} = (2)(2.5)\pi(0.25)^2(20{,}000)$$

or

$$T_{\max} = 19{,}635 \text{ in lb} = 1636.2 \text{ ft lb} \tag{x}$$

From Eq. (g), T_{max} is

$$T_{\max} = \left(\frac{\pi}{2}\right) r_s^3 \tau_{\max} = \left(\frac{\pi}{2}\right)(0.855)^3(20{,}000)$$

or

$$T_{\max} = 19{,}635 \text{ in lb} = 1636.2 \text{ ft lb}$$

Case ii) Two adjacent bolts: A and B

The data for this case are:

$$d = d_{AB} = R, \quad R = 2.5 \text{ in}, \quad r_b = 0.25 \text{ in}$$

$$r_{s\,\min} = 0.679 \text{ in}, \quad \tau_{\max} = 20{,}000 \text{ psi}$$

From Eq. (f), T_{max} is then:

$$T_{max} = d_{AB}\pi r_b^2 \tau_{max} = (2.5)\pi(0.25)^2(20{,}000)$$

or

$$T_{max} = 9817.5 \text{ in lb} = 818.12 \text{ ft lb}$$

From Eq. (g), T_{max} is:

$$T_{max} = \left(\frac{\pi}{2}\right) r_s^3 \tau_{max} = \left(\frac{\pi}{2}\right)(0.679)^3(20{,}000)$$

or

$$T_{max} = 9817.5 \text{ in lb} = 818.12 \text{ ft lb}$$

Case iii) Two intermediate bolts: A and C
The data for this case are:

$$d = d_{AC} = \sqrt{3}R, \quad R = 2.5 \text{ in}, \quad r_b = 0.25 \text{ in}$$

$$r_{s\,min} = 0.815 \text{ in}, \quad \tau_{max} = 20{,}000$$

From Eq. (f), T_{max} is then

$$T_{max} = d_{AC}\pi r_b^2 \tau_{max} = \left(\sqrt{3}\right)(2.5)\pi(0.25)^2(20{,}000)$$

or

$$T_{max} = 17{,}004 \text{ in lb} = 1{,}417 \text{ ft lb}$$

From Eq. (g), T_{max} is:

$$T_{max} = \left(\frac{\pi}{2}\right) r_s^3 \tau_{max} = \left(\frac{\pi}{2}\right)(0.815)^3(20{,}000)$$

or

$$T_{max} = 17{,}004 \text{ in lb} = 1{,}417 \text{ ft lb}$$

EXAMPLE ANALYSIS 4

Consider yet again the coupling joint of Example Analyses 1, 2, and 3, and as shown again in Figure a. This time let the shafts and joint be subjected to a bending moment M as shown.

As before, let the joint consist of circular flanges firmly attached to transmission shafts and connected together by six evenly spaced bolts, named A, B, ..., F, as in Figure b. Also, as before, let the bolts have radius r_b and let them be mounted with their centers on a circle having radius R and center O on the shaft axis.

Objective: Find the relation between the axial (tensile) bolt forces: F_A, F_B, ...,F_F, and the bending moment M, with the bending axis being about the line passing through the centers of bolts C and F as represented in Figures a and b.

For simplicity, assume: i) there is ideal geometry, ii) the flange plates are relatively stiff (and thus able to be thought of as "rigid"); and iii) that the initially uniformly tightened bolt tension F_O is

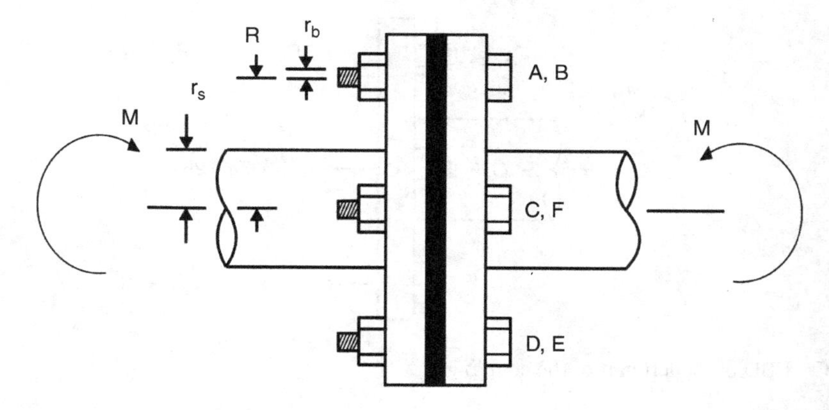

FIGURE A A coupling joint subjected to a bending moment.

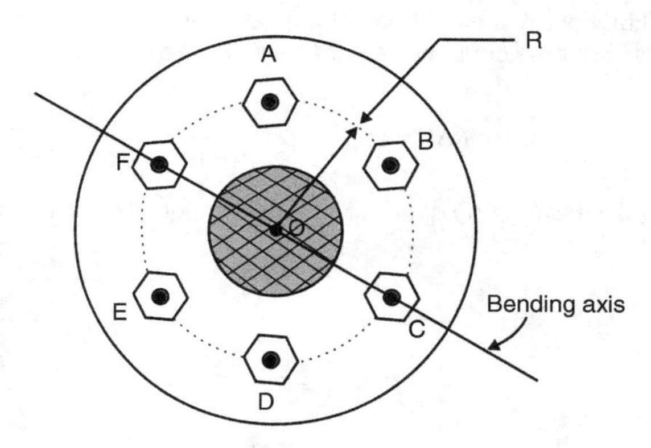

FIGURE B End view of joint bolts and flange.

sufficiently large that during the bending none of the bolts become loose. (That is, the bolts remain in tension throughout the application of the bending moment.)

Solution

We can respond to the request by using the principles of elementary beam theory. Indeed, the system of Figure a is suggestive of a beam, subjected to a bending moment.

Recall from beam theory that cross-sections, planar and perpendicular to the beam axis, before bending remain planar and perpendicular to the axis after bending. Thus with the flanges assumed to be rigid, their flat surfaces can be modeled as planes perpendicular to the shaft axis before and after bending.

With this observation, we see from Figures a and b that during the bending, bolts A and B become shorter, D and E become longer, and C and F are unchanged. By assuming linearity in the force-deformation relations, the amount of shortening of bolts A and B is exactly equal to the amount of lengthening of bolts D and E. Consequently, the decrease in bolt force (ΔF), in bolts A and B, due to their shortening, is exactly equal to the increase in bolt force, in bolts D and E, due to their lengthening.

Therefore, we immediately obtain the following expressions for the bolt forces after bending:

$$
\begin{aligned}
F_A &= F_B = F_O - \Delta F \\
F_C &= F_F = F_O \\
F_O &= F_E = F_O + \Delta F
\end{aligned}
\tag{a}
$$

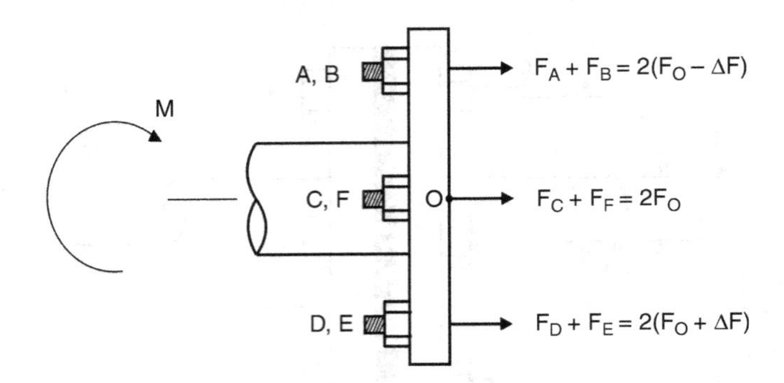

FIGURE C FBD of the left half of the coupling.

Next, observe that the force increments ΔF occurring in bolts A, B, D, and E are in response to the applied bending moment M on the shaft. To determine the relation between ΔF and M, consider a free-body diagram (FBD) of say the right half of the coupling as in Figure c.

By setting moments about the coupling center O equal to zero we obtain (after simplification):

$$4R\Delta F = M \quad \text{or} \quad \Delta F = \frac{M}{4R} \tag{b}$$

Finally, by substituting this result into Eq. (a) we obtain the requested relations for the bolt forces after bending as:

$$\begin{aligned} F_A = F_B &= F_O - \frac{M}{4R} \\ F_C = F_F &= F_O \\ F_D = F_E &= F_O + \frac{M}{4R} \end{aligned} \tag{c}$$

Comment

Observe that neither the bolt radius nor the shaft radius are involved in the analysis. But, these parameters are necessary for stress analysis, as in the following example analysis.

EXAMPLE ANALYSIS 5

Review Example Analysis 4. Suppose the maximum allowable tensile stress $\sigma_{b\max}$ in a bolt is 25,000 psi. If, as before, the bolt diameter d_b is 0.5 in and the bolt center circle radius R is 2.5 in, determine the maximum allowable bending moment $M_{\max}$ that can be applied to the shaft. For specificity, let the initial bolt tension F_O be 2000 lb.

Solution

From Eq. (c) of Example Analysis 4, we see that the maximum bolt force occurs in both D and E (see Figures a and b of Example Analysis 4) and is given by:

$$F_D = F_E = F_O + \frac{M}{4R} \tag{a}$$

By solving Eq. (a) for M we have

$$M = 4R\left(F_{D,E} - F_O\right) \tag{b}$$

Then the maximum moment M_{max} is

$$M_{max} = 4R(F_{max} - F_O) \tag{c}$$

where F_{max} is the bolt force at the yield stress (25,000 psi).
The bolt cross-section area is:

$$A = \left(\frac{\pi}{4}\right)d_b^2 = \left(\frac{\pi}{4}\right)\left(\frac{1}{2}\right)^2 = 0.196 \text{ in}^2 \tag{d}$$

The maximum bolt force F_{max} is then:

$$F_{max} = A\sigma_{b\,max} = (0.196)(25,000) = 4,909 \text{ lb} \tag{e}$$

By substituting the result of Eqs. (e) into (c) we find M_{max} to be:

$$M_{max} = (4)(2.5)(4,909 - 2,000) = 29,090 \text{ in lb} \tag{f}$$

Comment

It is tempting to stop the analysis at this point, but it might be of interest to evaluate the forces in the other bolts (A, B, C, and F): From Eq. (c) of Example Analysis 4, the forces in these bolts are:

$$F_A = F_B = F_O - \frac{M}{4R}$$

or

$$F_C = F_F = F_O \tag{g}$$

The forces in bolts C and F are unchanged by the bending moment. The force $F_{A,B}$ in bolts A and B is:

$$F_{A,B} = F_O - \frac{M_{max}}{4R} = 2,000 - \frac{(29,090)}{A(2.5)}$$

$$= 2,000 - 2,909$$

or

$$F_{A,B} = -909 \text{ lb} \tag{h}$$

The result of Eq. (h) is problematical because we assumed in Example Analysis 4 that the bolt force cannot be negative. Therefore, the result of Eq. (f) is too large. That is, in view of Eq. (g), $M/4R$ must not exceed F_O! This in turn means that the maximum bending moment is simply:

$$M_{max} = 4RF_O = (4)(2.5)(2,000) = 20,000 \text{ in lb}$$

or

$$M_{max} = 1,666.7 \text{ ft lb} \tag{i}$$

EXAMPLE ANALYSIS 6

Review Example Analysis 5. With the joint dimensions being the same and the maximum tensile stress also being 25,000 psi, what should the tightening force F_O be so that the forces in bolts A and B do not become negative—that is so that bolt loosening does not occur? Also, what then is the maximum bending moment M_{max} on the shaft?

Solution

Referring to the solution for Example Analysis 4 in Eq. (c) we see that the forces in the bolts are:

$$F_A = F_B = F_O - \frac{M}{4R} \tag{a}$$

$$F_C = F_F = F_O \tag{b}$$

$$F_D = F_E = F_O + \frac{M}{4R} \tag{c}$$

Notationally, as before, let $F_{A,B}$ refer to the force in either bolts A or B, and similarly for the other bolts with equal forces.

By adding both sides of Eqs. (a) and (c) we then obtain:

$$F_{A,B} + F_{D,E} = 2F_O \tag{d}$$

From the example problem statement, in the limit, as bolts D and E reach their maximum tension we want the forces in bolts A and B to not be negative—thus, in the limit: zero. Therefore, in Eq. (d) $F_{A,B}$ and $F_{D,E}$ are:

$$F_{A,B} = 0 \tag{e}$$

and

$$F_{D,E} = \sigma_{\max} A_b = (25{,}000)\left(\frac{\pi}{4}\right)\left(\frac{1}{2}\right)^2$$

or

$$F_{DE} = 4{,}908.74 \text{ lb} \tag{f}$$

By substituting the results into Eq. (d) we have:

$$0 + 4908.74 = 2F_O$$

or

$$F_O = 2454.36 \text{ lb} \tag{g}$$

Next, regarding the maximum bending moment M_{max} on the shaft, we see from Eqs (a) and (e) that M_{max} is:

$$M_{\max} = (4R)F_O = (4)(2.5)(2454.36) = 24{,}543.7 \text{ in lb}$$

or

$$M_{\max} = 2045.3 \text{ ft lb} \tag{h}$$

Comment

Although not requested in the analysis statement, the stresses in the middle bolts C and F should be checked to be sure the initial tightening is not too much. To this end, from Eq. (b), the stresses in bolts C and F are:

$$\sigma_{C,F} = \frac{F_{C,F}}{A_b} = \frac{F_O}{A_b} = \frac{(2454.36)}{(\pi/4)(1/2)^2}$$

or

$$\sigma_{C,F} = 12{,}500 \text{ psi} \tag{i}$$

These stresses are exactly half the given maximum tensile stress. This result could well have been obtained by inspection, by assuming maximum stresses (25,000 psi) in both D and E, and then zero stresses in both A and B. The stresses in the middle bolts C and F are simply the average of the stresses in the upper and lower bolts, or 12,500 psi.

EXAMPLE ANALYSIS 7

Consider yet again the coupling joint of Example Analyses 1 to 6. Let the dimensions and material strengths be the same as before: that is, let the bolt diameters d_b be ½ in, the bolt circle radius R be 2.5 in, the maximum shear stress τ_{max} be 20,000 psi, and the maximum tensile stress σ_{max} be 25,000 psi.

From Example Analysis 2, we determined that to keep within the maximum shear stress and by assuming ideal geometry, the maximum torque T_{max} that can be transmitted through the joint is 4,909 ft lb. The corresponding minimum shaft radius r_{min} was found to be 1.233 in.

Similarly, we found from Example Analysis 6 that to keep within the maximum tensile stress the maximum bending moment M_{max} in the shaft is 2,045.3 ft lb and that the bolt tightening forces F_O, to prevent bolt loosening during shaft bending, is 2,454.3 lb.

In view of these findings, compute the maximum bending (flexural) stress in the shaft.

SOLUTION

Recall from elementary strength of materials that the maximum flexural stress σ in a beam occurs at the outer beam fibers at a maximum distance c from the neutral axis is given by:

$$\sigma = \frac{Mc}{I} \tag{a}$$

where M is the bending moment and I is the second moment of area of the beam cross-section.

For a beam with a circular cross-section c is simply the beam radius and I is:

$$I = \left(\frac{\pi}{4}\right) r^4 \tag{b}$$

By substituting into Eq. (a), σ becomes:

$$\sigma = \frac{4M}{\pi r^3} \tag{c}$$

Finally, by substituting the values for the maximum bending moment M_{max} and the minimum shaft radius r_{min}, we find the maximum shaft flexural stress to be:

$$\sigma = \frac{(4)(2454.3)(12)}{\pi(1.233)^3}$$

or

$$\sigma = 20{,}000 \text{ psi} \tag{d}$$

Fortunately, this value is well below the allowable 25,000 psi limit.

EXAMPLE ANALYSIS 8

Consider again Example Analysis 4 where a bending moment M is applied to the shaft of the coupling. In that situation, recall that the moment was applied about a line passing through two oppositely positioned bolt centers.

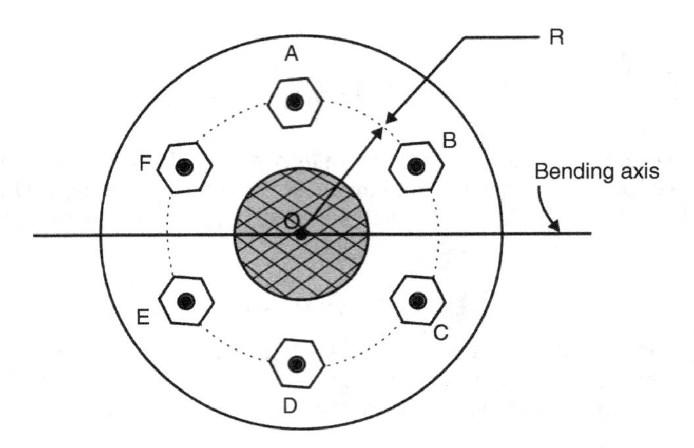

FIGURE A End view of joint bolts and flange.

Consider the same coupling, but instead of a moment about a diametral line through bolt centers, let the moment be applied about a diametral line passing between bolt centers, as represented in Figure a.

As before, let the bolts be mounted on a circle, with radius R and center O, on the shaft axis, as shown.

Objective: As in Example Analysis 4, find the relations between the axial (tensile) bolt forces: F_A, F_B, …, F_F and the bending moment M about the axis shown in Figure a. Again for simplicity, assume: i) there is ideal geometry; ii) the flange plates are relatively stiff (and thus able to be thought of as "rigid"); and iii) that the initial uniformly tightened bolt tension F_O is sufficiently large that during the bending none of the bolts becomes loose. (That is, the bolts remain in tension throughout the application of the bending moment.)

Solution

Consider a side view of the coupling as in Figure b. When the shaft moments are applied and directed as shown, the tension in the lower bolts (C, D, and E) is increased, and the tension in the upper bolts is decreased. With the flanges regarded as being rigid, we can use the principles of elementary beam theory to determine the bolt forces. Specifically, in beam bending plane cross-sections perpendicular to the neutral axis remain plane during the bending. With the faces of the flanges assumed to be rigid, they are thus well-modeled by the beam theory planes. This means that the bolt deformations are proportional to their distance, above or below, the bending axis (the equivalent neutral axis).

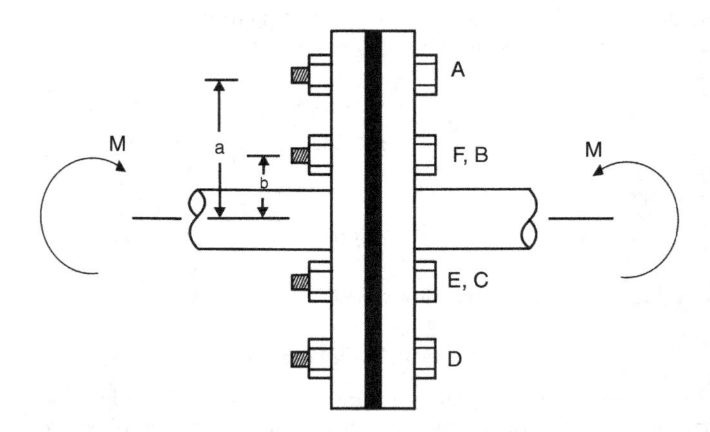

FIGURE B Side view of the coupling

Since the bolt forces are proportional to their deformations, their change in magnitude, different from their initial force F_O, is proportional to their distance above or below the bending axis.

Suppose the change in force of bolt A, during bending, is ΔF, then from Figure b the forces in the bolts are:

$$\begin{aligned} F_A &= F_O - \Delta F \\ F_F &= F_B = F_O - (b/a)\Delta F \\ F_E &= F_C = F_O + (b/a)\Delta F \\ F_D &= F_O + \Delta F \end{aligned} \tag{a}$$

where a and b are dimensions shown in Figure b.

We can find a and b by looking again at an end view of the coupling as in Figure c. Since there are six bolts mounted equally spaced, around the circle, the central angle between bolts is: 60E. Therefore, with the horizontal dividing the space (see Figure c) and with the circle radius being R, the distance of the four bolts: B, C, E, and F, from the bending axis is: $R/2$. This is distance b in Figure b. Also, by inspection of Figure c, we immediately see that the distance of bolts A and D from the bending axis is: R, which is distance a in Figure b. Thus the factor (b/a) in Eq. (a) is simply: ½.

We can now determine the requested relation between the bending moment and the bolt forces via a FBD of say the right half of the coupling, as in Figure d.

By setting moments about O equal to zero, we have:

$$RF_A + \left(\frac{1}{2}\right)R(F_F + F_B) - \left(\frac{1}{2}\right)R(F_E + F_C) - RF_D + M = 0 \tag{b}$$

By substituting from Eq. (a) we obtain:

$$R(F_O - \Delta F) + (1/2)R(2)\left[F_O - \left(\frac{1}{2}\right)\Delta F\right] - \left(\frac{1}{2}\right)R(2)\left[F_O + \left(\frac{1}{2}\right)\Delta F\right] - R(F_O + \Delta F) + M = 0$$

or after simplification,

$$2R\Delta F + R\Delta F = M$$

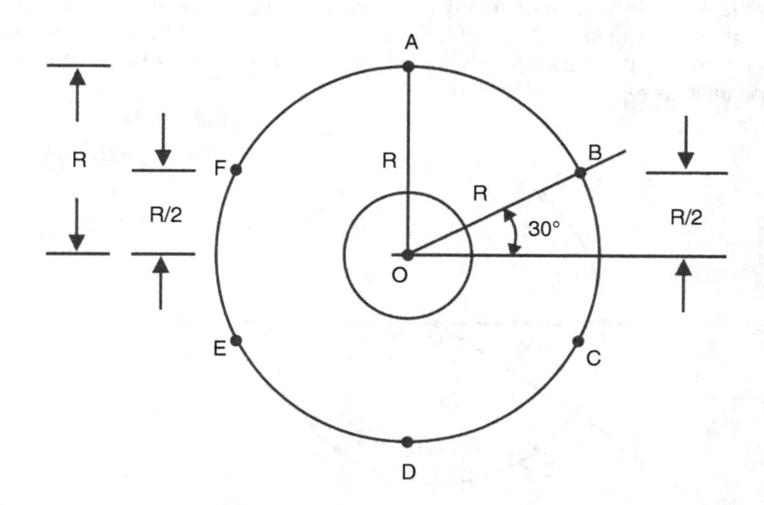

FIGURE C End view of the coupling.

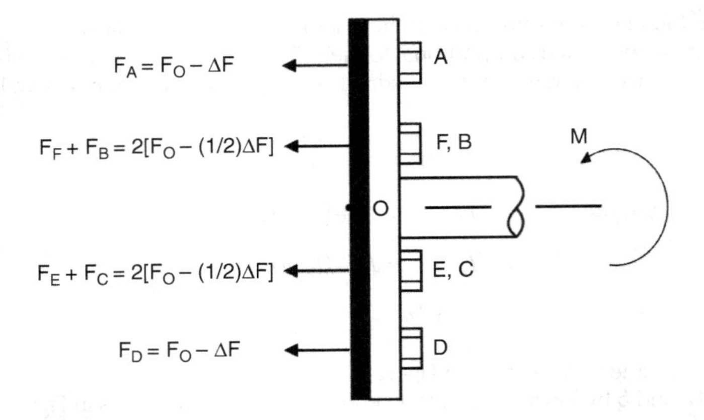

FIGURE D FBD of the right half of the coupling.

or

$$\Delta F = \frac{M}{3R} \tag{c}$$

Finally, by substituting the result of Eqs. (c) into (a) we find the bolt forces to be:

$$\begin{aligned} F_A &= F_O - \frac{M}{3R} \\ F_F + F_B &= F_O - \frac{M}{6R} \\ F_E = F_C &= F_O + \frac{M}{6R} \\ F_D &= F_O + \frac{M}{3R} \end{aligned} \tag{d}$$

EXAMPLE ANALYSIS 9

Review Example Analysis 8 and particularly its solution. Suppose now that the bending axis does not pass through either a diametral pair of bolt centers nor exactly midway between bolt centers. Instead, let the bending axis pass through the center of the coupling, as before, but now inclined at an arbitrary angle as represented in Figure a.

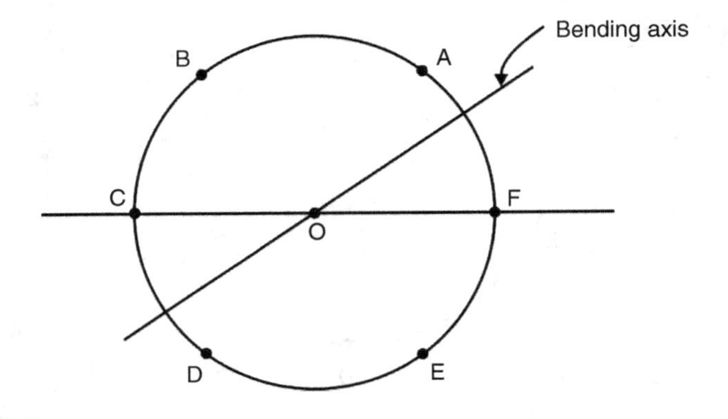

FIGURE A Joint bending about an arbitrarily inclined axis.

Objective: As in Example Analyses 4 and 8, find the relations between the axial (tensile) bolt forces: F_A, F_B, ..., F_F and the bending moment M about the inclined axis shown in Figure a. As before, for simplicity, assume: i) there is ideal geometry; ii) the flange plates are relatively stiff (and thus able to be thought of as "rigid"); and iii) that the initial uniformly tightened bolt tension F_O is sufficiently large that during the bending none of the bolts becomes loose. (That is, the bolts remain in tension throughout the application of the bending moment.)

Also, for simplicity in notation, let the bolt labels (A, B, ..., F) be assigned as in Figure a.

SOLUTION

From the principles of elementary beam theory and from the solutions of Example Analyses 4 and 8, we see that the increments in bolt forces due to the bending moment are proportional to the distances of the bolt centers from the bending axis. For bolts A, B, and C, these distances are a, b, and c respectively, as in Figure b.

Therefore, the forces in bolts A, B, and C may be expressed as:

$$F_A = F_O + \Delta F_A = F_O - ka$$
$$F_B = F_O + \Delta F_B = F_O - kb \qquad \text{(a)}$$
$$F_C = F_O + \Delta F_C = F_O - kc$$

where k is a constant to be determined, and the negative sign is introduced since the direction of the bending moment, as seen in Figure c, reduces the tensile forces in bolts A, B, and C.

By inspection of Figures b and c, we see that bolts E, D, and F are symmetrically positioned about the bending axis relative to bolts B, A, and C, respectively. This means that the distances of bolt centers of D, E, and F from the bending axis are also a, b, and c, respectively, as seen in Figure d.

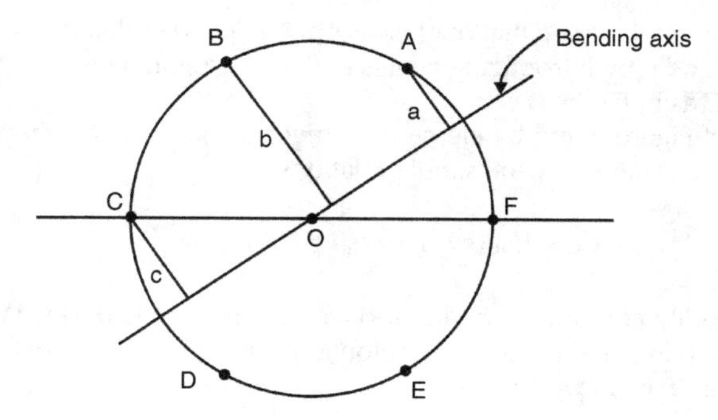

FIGURE B Bolt center distances from the bending axis.

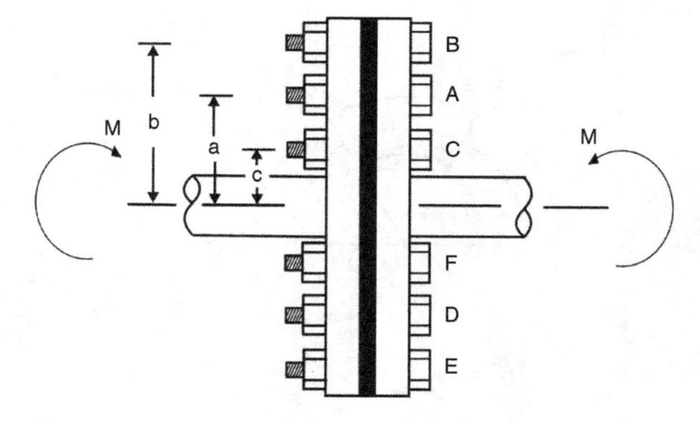

FIGURE C Side view of the coupling showing the bending moment direction and the bolt positions.

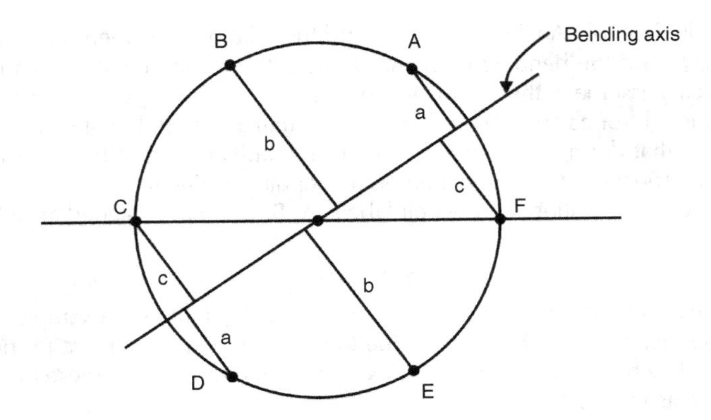

FIGURE D Symmetry of bolt center distances from the bending axis.

Again, from elementary beam theory, with the increments in bolt forces due to the bending moment being proportional to the distances of the bolt centers from the bending axis, we see that the forces in bolts D, E, and F may be expressed as:

$$\begin{aligned} F_D &= F_O + \Delta F_D = F_O + ka \\ F_E &= F_O + \Delta F_E = F_O + kb \\ F_F &= F_O + \Delta F_F = F_O + kc \end{aligned} \qquad \text{(b)}$$

where now the force increments are positive since the direction of the bending moment increases the bolt tensile forces (See Figure c).

In view of Eqs. (a) and (b) we see that our task is now reduced to determining *a, b, c,* and *k*. We can readily find *a, b,* and *c* by introducing angles α, β, and γ defining the angular position of bolt centers of *A, B,* and *C* as in Figure 3.

From inspection of Figure e and by elementary trigonometry, we see that *a, b,* and *c* may be expressed in terms of *α, β,* and γ by the simple relations:

$$a = R\sin\alpha \quad b = R\sin\beta \quad c = R\sin\gamma \qquad \text{(c)}$$

Therefore, we can readily compute *a, b,* and *c* once we know α, β, and γ. With six bolts evenly spaced around the coupling, the angular separation is 60° or π/3 radians. Then again by inspection of Figure 3 we see that α, β, and γ are:

$$\alpha = \left(\frac{\pi}{3}\right) - \theta, \quad \beta = \left(\frac{2\pi}{3}\right) - \theta, \quad \gamma = \pi - \theta \qquad \text{(d)}$$

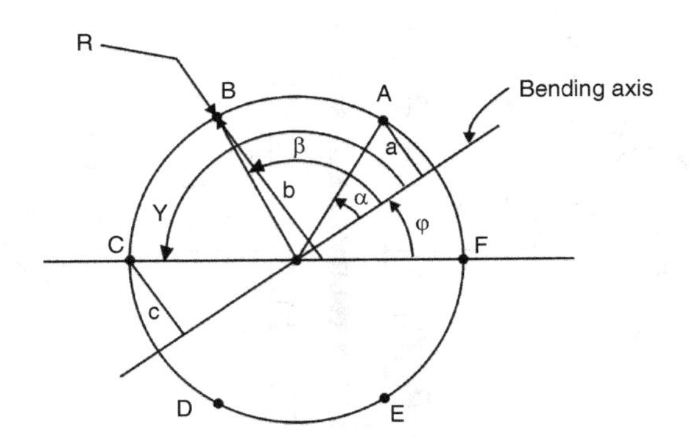

FIGURE E Bolt angular orientation.

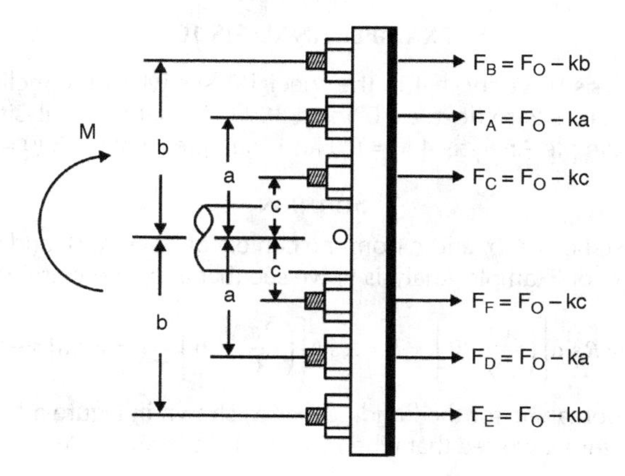

FIGURE F FBD of the left half of the coupling joint.

By substituting from Eqs. (d) into (c) we can express a, b, and c directly in terms of θ as:

$$a = R\sin\left[\left(\frac{\pi}{3}\right) - \theta\right], \quad b = R\sin\left[\left(\frac{2\pi}{3}\right) - \theta\right], \quad c = R\sin[\pi - \theta] \tag{e}$$

Therefore, once θ is given, we can immediately evaluate a, b, and c.

Assuming knowledge of a, b, and c, we can again use a FBD of half of the coupling joint to determine the relation of the bolt forces F_A, F_B, ..., F_F in terms of the bending moment M. To this end, consider a left side isolation of the joint as in Figure f, where as in Figure c the view is normal to the bending axis. Also, in Figure f we used Eqs. (a) and (b) for the force expressions.

By balancing moments about the joint center O we obtain:

$$\begin{aligned} M &= -bF_B - aF_A - cF_C + cF_F + aF_D + bF_E \\ &= -b(F_O - kb) - a(F_O - ka) - c(F_O - kc) + c(F_O + kc) + a(F_O + ka) + b(F_O + kb) \end{aligned}$$

or

$$M = 2k\left(a^2 + b^2 + c^2\right) \tag{f}$$

By solving Eq. (f) for k we have:

$$k = M / 2\left(a^2 + b^2 + c^2\right) \tag{g}$$

Finally, by back substitution into Eqs. (a) and (b) the forces F_A, F_B, ..., F_F are:

$$\begin{aligned} F_A &= F_O - Ma / 2\left(a^2 + b^2 + c^2\right) \\ F_B &= F_O - Mb / 2\left(a^2 + b^2 + c^2\right) \\ F_C &= F_O - Mc / 2\left(a^2 + b^2 + c^2\right) \\ F_D &= F_O + Ma / 2\left(a^2 + b^2 + c^2\right) \\ F_E &= F_O + Mb / 2\left(a^2 + b^2 + c^2\right) \\ F_F &= F_O + Mc / 2\left(a^2 + b^2 + c^2\right) \end{aligned} \tag{h}$$

Comment

Observe the pattern in Eqs. (h).

EXAMPLE ANALYSIS 10

Review Example Analysis 9. Verify that in the special cases when the inclination angle θ of the bending axis has the values zero degrees (0°) and 30 (30°), that the analysis of Example Analysis 9 is consistent with Example Analysis 4 (θ = 0°) and Example Analysis 8 (θ = 30°), respectively.

Solution

Consider again the distances *a*, *b*, and *c* from the centers of bolts A, B, and C to the bending axis as in Figure a. In Eq. (e) of Example Analysis 9, we see that *a*, *b*, and *c* are expressed as:

$$a = R\sin\left[\left(\frac{\pi}{3}\right)-\theta\right],\quad b = R\sin\left[\left(\frac{2\pi}{3}\right)-\theta\right],\quad c = R\sin[\pi-\theta] \tag{a}$$

where θ is the inclination angle for the bending axis as shown in Figure a.

By inspection of Figure a we see that when θ = 0° (as in Example Analysis 4) *a*, *b*, and *c* are:

$$a = \left(\frac{\sqrt{3}}{2}\right)R,\quad b = \left(\frac{\sqrt{3}}{2}\right)R,\quad \text{and}\quad c = 0 \tag{b}$$

Similarly, when θ = 30° (as in Example Analysis 8),*a*, *b*, and *c* are seen to be:

$$a = \left(\frac{1}{2}\right)R,\quad b = R,\quad \text{and}\quad c = \left(\frac{1}{2}\right)R \tag{c}$$

If we assign θ to be 0° in Eq. (a), *a*, *b*, and *c* are:

$$\begin{aligned} a &= R\sin\left(\frac{\pi}{3}\right) = R\sin 60° = \frac{\sqrt{3}}{2R} \\ b &= R\sin\left(\frac{2\pi}{3}\right) = R\sin 120° = \frac{\sqrt{3}}{2R} \\ c &= R\sin\pi = R\sin 180° = 0 \end{aligned} \tag{d}$$

These results are identical to the values in Eq. (b).

If we assign θ to be 30° (or π/6 radians) in Eq. (a), *a*, *b*, and *c* are:

$$\begin{aligned} a &= R\sin\left[\left(\frac{\pi}{3}\right)-\left(\frac{\pi}{6}\right)\right] = R\sin\left(\frac{\pi}{6}\right) = R\sin 30° = \left(\frac{1}{2}\right)R \\ b &= R\sin\left[\left(\frac{2\pi}{3}\right)-\left(\frac{\pi}{6}\right)\right] = R\sin\left(\frac{\pi}{2}\right) = R\sin 90° = R \\ c &= R\sin\left[\pi-\frac{\pi}{6}\right] = R\sin\left(\frac{5\pi}{6}\right) = R\sin 150° = \left(\frac{1}{2}\right)R \end{aligned} \tag{e}$$

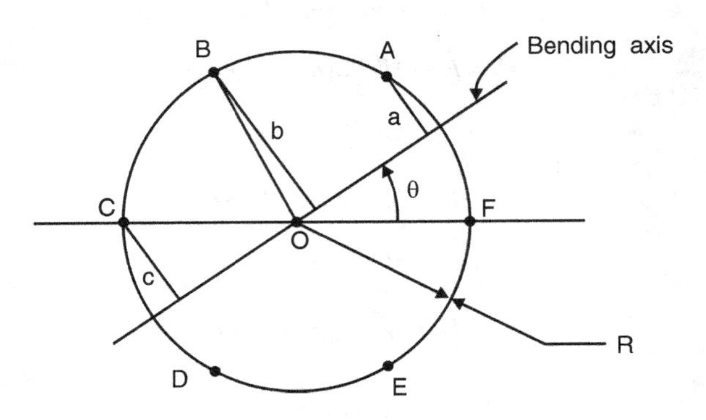

FIGURE A Bolt distances to the bending axis.

These results are identical to the values in Eq. (c).

Finally, by reflective symmetry we see in Figure a that the distances, say d, e, and f, from the bolt centers to the bending axis are equal to a, b, and c, respectively. Therefore, the foregoing analysis is similarly applicable for bolts D, E, and F.

EXAMPLE ANALYSIS 11

Figure a shows a sketch of a bracket attached to a vertical beam structure by two rivets A and B. If a weight W is applied to the bracket as shown, determine the forces and then the axial (tensile) stresses in the rivets. Assume that the bracket and wall are essentially rigid and that the dimensions shown are known.

SOLUTION

Let P be the point at the lower inside corner of the bracket, adjacent to the wall. With the bracket assumed to be rigid, and as the weight is placed on the shelf portion of the bracket, the rivets will be elongated as the bracket pivots about P.

With the bracket portion against the wall being rigid and thus remaining straight during the loading, the deformations of rivets A and B will be directly proportional to their distances from the pivot point P. Therefore, if δ_A and δ_B are the respective rivet elongations, we can express δ_A and δ_B as:

$$\delta_A = k(d+e) \quad \text{and} \quad \delta_B = ke \tag{a}$$

where k is the constant of proportionality.

From the principles of elementary strength of materials we know that the respective rivet forces, F_A and F_B, are proportional to the rivet deformations δ_A and δ_B. Specifically,

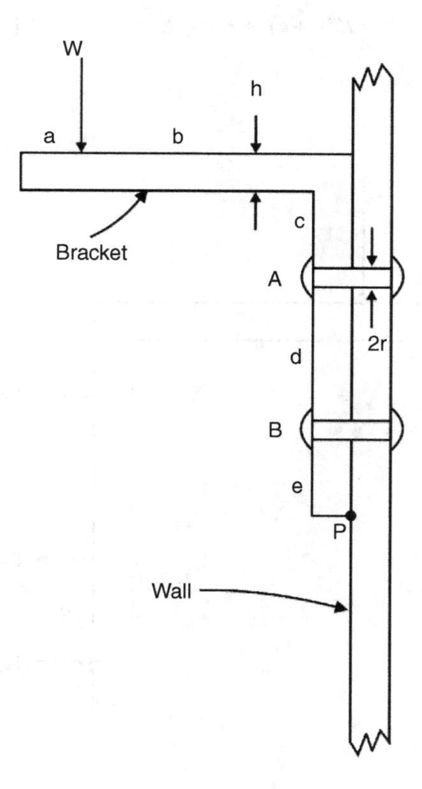

FIGURE A A rivet supported bracket.

$$F_A = \left(\frac{AE}{\ell}\right)\delta_A \quad \text{and} \quad F_B = \left(\frac{AE}{\ell}\right)\delta_B \tag{b}$$

where A is the rivet shaft cross-section area, ℓ is rivet length, and E is the modulus of elasticity of the rivet material.

By substituting from Eqs. (a) into (b) the rivet forces become:

$$F_A = k\left(\frac{AE}{\ell}\right)(d+e) = K(d+e) \tag{c}$$

and

$$F_B = k\left(\frac{AE}{\ell}\right)e = Ke \tag{d}$$

where K is a constant which by inspection is defined as:

$$K \triangleq \frac{AEk}{\ell} \tag{e}$$

Next, envision a FBD of the bracket as in Figure b, where R_X and R_Y are the horizontal and vertical components of the reaction force exerted by the wall on the bracket at P.

By balancing moments about P we obtain

$$Wb = F_A(d+e) + F_B(e) \tag{f}$$

By substituting for F_A and F_B from Eqs. (c) and (d) Eq. (f) becomes:

$$Wb = K(d+e)^2 + Ke^2 = K\left[(d+e)^2 + e^2\right] \tag{g}$$

Then K is seen to be:

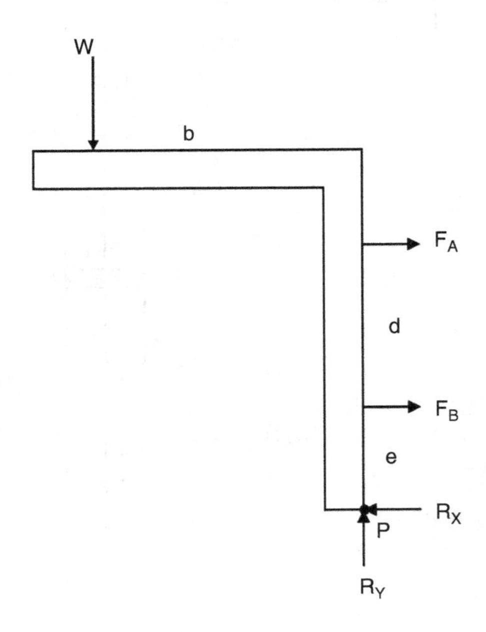

FIGURE B Bracket FBD.

$$K = \frac{Wb}{\left[(d+e)^2 + e^2\right]} \tag{h}$$

Finally, by back substituting from Eq. (h) into Eqs. (c) and (d) the requested rivet forces are:

$$F_A = \frac{Wb(d+e)}{\left[(d+e)^2 + e^2\right]} \tag{i}$$

and

$$F_B = \frac{Wbe}{\left[(d+e)^2 + e^2\right]} \tag{j}$$

The rivet stresses σ_A and σ_B are then

$$\sigma_A = \frac{Wb(d+e)}{(\pi r^2)\left[(d+e)^2 + e^2\right]} \tag{k}$$

and

$$\sigma_B = Wbe / (\pi r^2)\left[(d+e)^2 + e^2\right] \tag{P}$$

Comment

With the use of rivets, as opposed to bolts, we assume no preload tension. If, however, both had been used with a preload of say F_O, then F_O would need to be added to the values for F_A and F_B in Eqs. (i) and (j).

EXAMPLE ANALYSIS 12

Review Example Analysis 114. Consider again the bracket of Figure a and as shown again here in Figure a. Let the dimensions/distances shown have the values:

$$\begin{aligned} &a = 2\text{ in}, \quad b = 12\text{ in}, \quad c = 4\text{ in}, \quad d = 8\text{ in} \\ &e = 4\text{ in}, \quad h = 3/16\text{ in}, \quad 2r = 1/4\text{ in} \end{aligned} \tag{a}$$

Suppose the allowable rivet stress (the tensile stress) is 25,000 psi. Determine the maximum weight (load) W_{max} that the bracket can hold.

Solution

Recall from Eqs. (k) and (P) of the solution for Example Analysis 11 that the rivet tensile stresses are:

$$\sigma_A = \frac{Wb(d+e)}{(\pi r^2)\left[(d+e)^2 + e^2\right]} \tag{b}$$

and

$$\sigma_B = \frac{Wbe}{(\pi r^2)\left[(d+e)^2 + e^2\right]} \tag{c}$$

By solving these expressions for W we have:

$$W = \frac{(\sigma_A)(\pi r^2)\left[(d+e)^2 + e^2\right]}{(b)(d+e)} \tag{d}$$

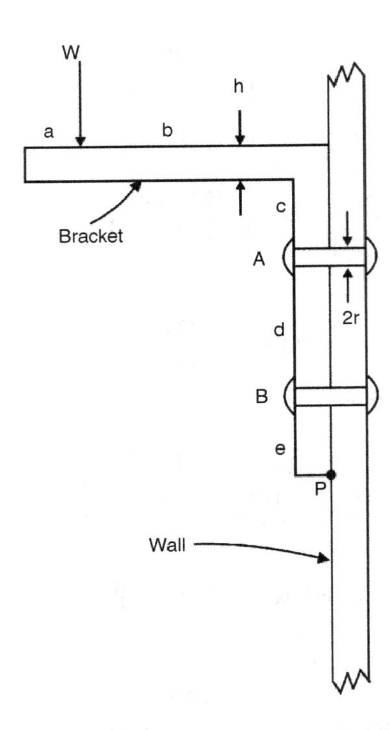

FIGURE A Rivet supported bracket.

and

$$W = \frac{(\sigma_B)(\pi r^2)\left[(d+e)^2 + e^2\right]}{(be)} \tag{e}$$

Observe in these expressions that the value of W in Eq. (d) is smaller than that in Eq. (e). Therefore, Eq. (d) provides the maximum safe load W_{max}. (Also, by inspection of Figure a we see that rivet A will bear the greater portion of the load.)

By substituting the given data W_{max} then is:

$$W_{max} = (25)(10^3)\pi\left(\frac{1}{8}\right)^2\left[(8+4)^2 + (4)^2\right] / (12)(8+4)$$

or

$$W_{max} = 1363 \text{ lb} = 1.363 \text{ kip} \tag{f}$$

Comment

Observe in Eq. (d) that the distances a and c, and the thickness dimension t, do not influence the value of the maximum allowable load.

EXAMPLE ANALYSIS 13

Repeat Example Analysis 12 for the following data:

$$\begin{gathered} a = 0.05 \text{ m}, \quad b = 0.3 \text{ m}, \quad c = 0.1 \text{ m}, \quad d = 0.2 \text{ m} \\ e = 0.1 \text{ m}, \quad h = 4.75 \text{ mm}, \quad 2r = 6 \text{ mm} \end{gathered} \tag{a}$$

Find W_{max} assuming the maximum allowable rivet tensile stress σ_{max} is 170 MPa.

Solution

Recall from Example Analysis 12 that the maximum rivet tensile stress occurs in rivet A, and that the corresponding maximum weight (load) W_{max} is then:

$$W_{max} = (\sigma_{max})(\pi r^2)\left[(d+e)^2 + e^2\right]/(b)(d+e) \tag{b}$$

By substituting the data values from Eq. (a) we have:

$$W_{max} = (170)(10)^6 \pi(3)^2(10)^{-6}\left[(0.3)^2 + (0.1)^2\right]/(0.3)(0.3)$$

or

$$W_{max} = 53.41\ \text{N} = 5.341\ \text{kN} \tag{c}$$

EXAMPLE ANALYSIS 14

Review again Example Analyses 11 to 13.

Let τ_{max} be the maximum allowable shear stress in the rivets. Determine relations between the weight load W and τ_{max}. Let the dimensions and distances be the same as in the foregoing example analyses and as shown again in Figure a.

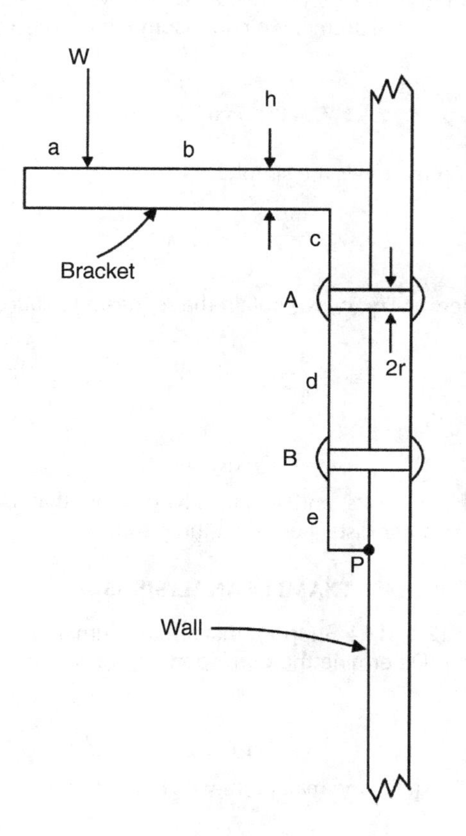

FIGURE A A rivet supported bracket.

Solution

This problem is statically indeterminate: If we envision a FBD of the bracket we immediately see that the downward directed shear forces on the rivets, from the bracket, are resisted by equal upward directed forces on the bracket, by the rivets. If we call these shear forces V_A and V_B, on rivets *A* and *B*, respectively, the vertical force balance yields:

$$W = V_A + V_B \tag{a}$$

Observe from an envisioned FBD that neither a balance of horizontal forces nor a balance of moments will involve either V_A and V_B. Therefore, Eq. (a) is insufficient in itself for determining V_A and V_B.

The usual approach with statically indeterminate systems is to 1) make a force balance; 2) examine the deformation; and 3) relate the forces and deformation via Hooke's law. Steps 2) and 3) then hopefully produce additional equations, enabling the determination of all unknown forces and deformations.

In the current example, however, this procedure requires some additional assumptions before we can determine the unknown forces: If we assume ideal geometry, and if we also assume relatively rigid bracket material, we find by a symmetry argument that V_A and V_B are equal: Then in view of Eq. (a) we have:

$$V_A = V_B = \frac{W}{2} \tag{b}$$

Alternatively, if we assume ideal geometry, but if more realistically, we assume the bracket material is deformable, then with rivet *A* being closer to the applied load *W* than rivet *B*, the deformation of the bracket material in contact with rivet *A* will relieve the force on rivet *B*. Therefore, in the interest of safe design, in the extreme, we can assume that rivet *A* absorbs all the load. Then from Eq. (a) we have:

$$V_A = W \quad \text{and} \quad V_B = 0 \tag{c}$$

Finally, the shear stresses on the rivets are simply:

$$\tau_A = \frac{V_A}{\pi r^2} \quad \text{and} \quad \tau_B = \frac{V_B}{\pi r^2} \tag{d}$$

Then for safe design, in view of Eq. (c), we obtain the requested relations as:

$$\tau_{max} = \frac{W}{\pi r^2} \quad \text{and} \quad W = \pi r^2 \tau_{max} \tag{e}$$

Comment

Observe the simplicity of the results, and specifically, observe that aside from the rivet radius *r*, none of the other dimensions nor distances affect the result.

EXAMPLE ANALYSIS 15

Review Example Analyses 12 and 14. Suppose that a maximum allowable shear stress τ_{max} for the bracket rivets is 20,000 psi. Determine the corresponding maximum allowable safe weight load W_{max}.

Solution

From Example Analysis 14, Eq. (e), the maximum weight load is:

$$W_{max} = \pi r^2 \tau_{max} \tag{a}$$

From Example Analysis 12, the rivet radius r is 1/8 in.

By substituting the given data into Eq. (a) we find the maximum weight load as:

$$W_{max} = \pi\left(\frac{1}{8}\right)^2 (20)10^3 = 981.7 \text{ lb} \tag{b}$$

Comment

Observe that this result is less than the 1363 lb result obtained in Example Analysis 25. Observe also, however, that the only dimension occurring in Eq. (a) is the rivet radius r, whereas in Example Analysis 12, where the tensile stress strength is the quantity of interest, we see from Eq. (d) of Example Analysis 12, that the maximum load W_{max} is governed by the dimensions b, d, and e in addition to r. If, for example, the lateral dimension b is doubled to say 24 in, then the maximum load W_{max} is reduced to 681 lb—less than that of the result in Eq. (b). This shows that both the tensile and shear strengths need to be considered in bracket designs.

EXAMPLE ANALYSIS 16

Consider yet again the bracket of the foregoing examples, and as shown again here in Figure a. Let the thickness h of the horizontal portion of the bracket be: 3/16 inch. Let the other dimensions and distances be the same as in Example Analysis 12. That is,

$$\begin{gathered} a = 2 \text{ in}, \quad b = 12 \text{ in}, \quad c = 4 \text{ in}, \quad d = 8 \text{ in} \\ e = 4 \text{ in}, \quad h = 3/16 \text{ in}, \quad 2r = 1/4 \text{ in} \end{gathered} \tag{a}$$

Let the depth, or width, t of the bracket be: 1 inch.

Let the weight load W be 100 pounds.

Determine the tensile stress σ_Q at point Q. Suggest design alternatives to reduce σ_Q.

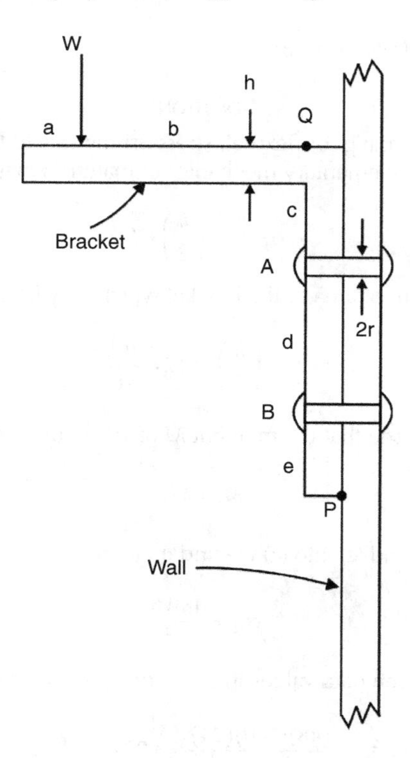

FIGURE A A rivet supported bracket.

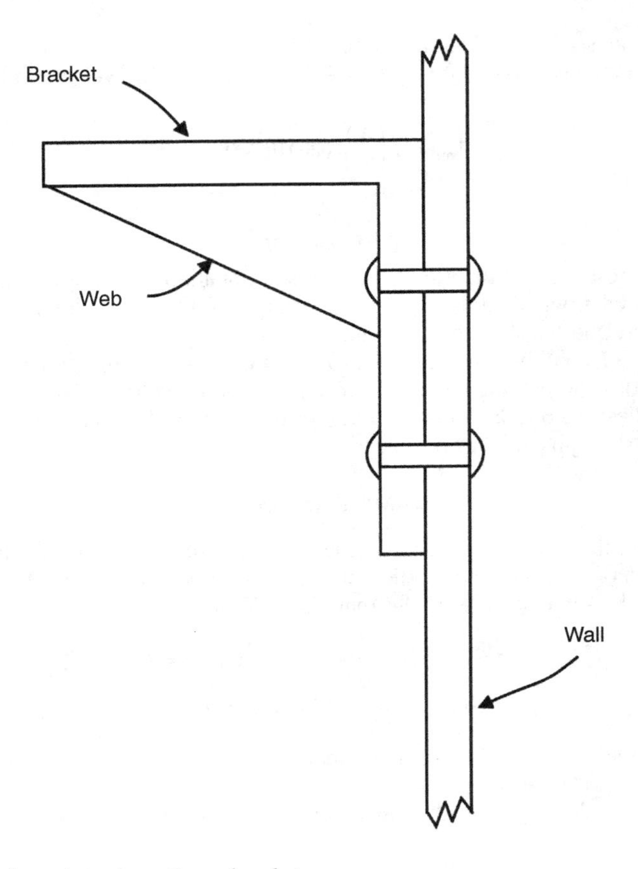

FIGURE B Use of a web to strengthen a bracket.

Solution

The stress at Q is simply a bending, or flexural, stress arising due to the moment M of the weight W about Q. Recall then from elementary mechanics of materials that σ_Q may be expressed as:

$$\sigma_Q = M\left(\frac{h}{2}\right) \div I \tag{a}$$

where I is the second moment of area of the bracket which may be expressed as:

$$I = t\left(\frac{h}{2}\right)^3 \div 12 = \frac{th^3}{96} \tag{b}$$

By inspection of Figure a we see that the moment M of W about Q is simply:

$$M = Wb \tag{c}$$

By substituting from Eqs. (b) and (c) into (a) we find σ_Q to be:

$$\sigma_Q = \frac{48Wb}{th^2} \tag{d}$$

Finally, by substituting the given data values into Eq. (d), σ_Q becomes:

$$\sigma_Q = \frac{(48)(100)(24)}{(1)(3/16)^2} = 3.277 \times 10^6 \text{ psi} \tag{e}$$

Design Alternatives

The stress given by Eq. (e) is at least two orders of magnitude larger than usual acceptable stress levels. There are several ways to reduce the stress:

1. Reduce the allowable weight load.
2. Increase the height h of the horizontal portion of the bracket.
3. Brace the bracket from below—say with a web.

The first alternative is unacceptable since the purpose of the bracket is to support weight loads.

The second alternative reduces the stress by the square of the height h. But this is an inefficient approach since it requires a major height increase to reduce the stress. For example, if h is increased to ½ inch the stress is still over 46,000 psi. Also, this does not address potential high stresses in the vertical portion of the bracket.

The third alternative provides a greater range of possibilities for reducing the stress. Figure b illustrates the general geometry.

Example Analysis 17 provides a brief discussion of the effectiveness of this design alternative.

EXAMPLE ANALYSIS 17

Review Example Analysis 16. Consider now a bracket with a supporting web as in Figure a.

To explore the usefulness of the web in reducing cantilever support bending stress, let the dimensions of the web be 28 inches long and 12 inches high as in Figure b. Let the web thickness be 1/8 inch.

a. Compute the centroidal second moment of area of the vertical end of the web against the wall portion of the bracket.

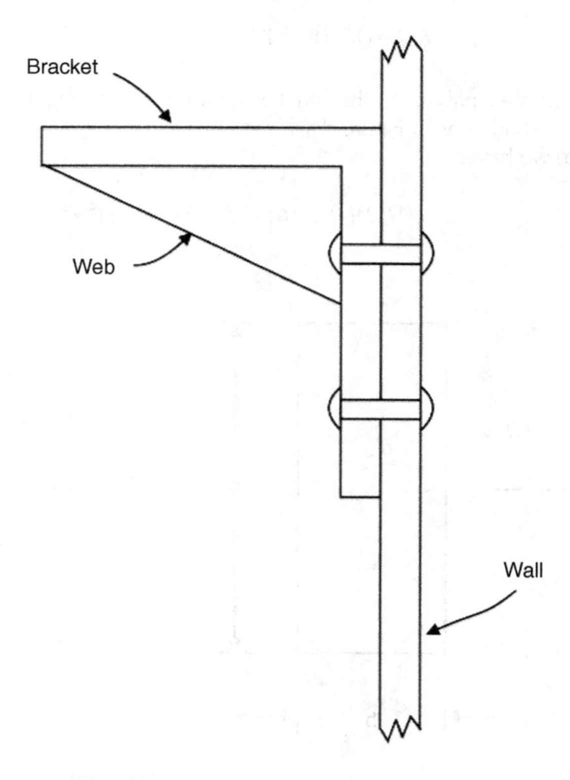

FIGURE A A web supported bracket.

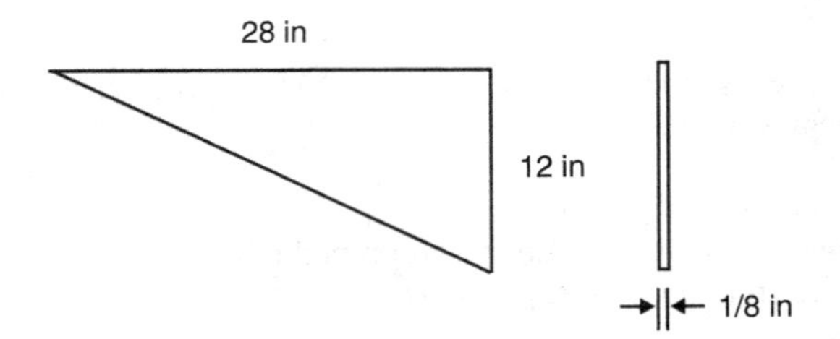

FIGURE B Web dimensions.

b. Compute the corresponding centroidal second moment of area of the end of the horizontal portion of the bracket (the "shelf") using the dimensions of Example Analysis 16.
c. Compare the results of a) and b). Discuss the implication in bracket design and in mechanical design in general.

Solution

a. Recall from elementary beam theory that the second moment of area *I* relative to the centroidal bending axis of a rectangular cross-section is simply:

$$I = \frac{bh^3}{12} \tag{a}$$

where here *b* is the width of the base of the cross-section and *h* is the height as in Figure c.

By using the web dimensions of Figure b we have:

$$I_{web} = (1/12)(1/8)(12)^3 = 18 \text{ in}^4 \tag{b}$$

b. Recall from Example Analysis 15 the bracket shelf base and height of the bracket shelf are 1 inch and 3/16 inch as in Figure d.

Using Eq. (a) we have:

$$I_{shelf} = (1/12)(1)(3/16)^3 = 5.495\left(10^{-4}\right) \text{in}^4 \tag{c}$$

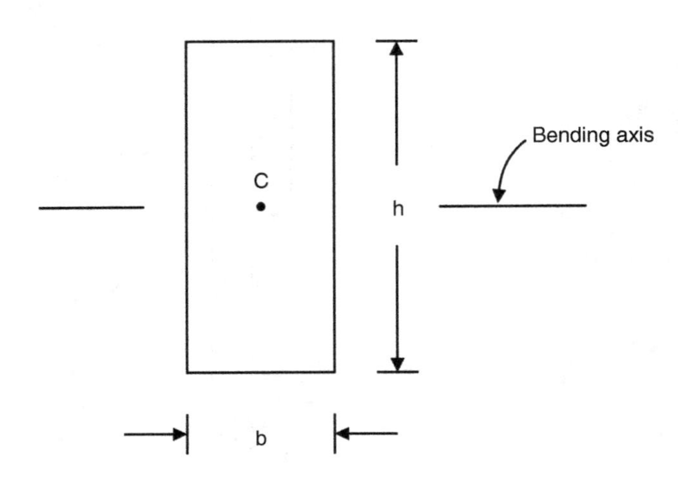

FIGURE C A rectangular cross-section, centroid C, and bending axis.

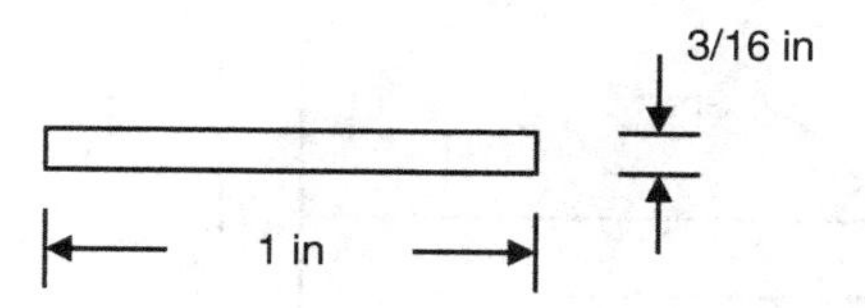

FIGURE D Bracket shelf cross-section.

c. Comparing the results of a) and b) we see that the second moment of area of the wall end of the web, I_{web}, is more than 30,000 larger than that of the shelf cross-section.

Since flexural (bending) stress is inversely proportional to the second moment of area, the presence of the web greatly reduces the bending stress.

To be more specific, consider the familiar bending stress formula:

$$\sigma = \frac{Mc}{I} \tag{d}$$

where c is now the distance from the bending axis to the extreme point of the cross-section. For a rectangular cross-section, c is simply half the height of the section. That is:

$$c = \frac{h}{2} \tag{e}$$

By substituting from Eqs. (a) and (e) into (d), the stress becomes:

$$\sigma = \frac{6M}{bh^2} \tag{f}$$

By substituting the given data values for the cross-section dimensions into Eq. (f) we have:

$$\sigma_{shelf} = 170.7 \text{ M} \quad \text{and} \quad \sigma_{web} = 0.333 \text{ M} \tag{g}$$

The ratio: $\sigma_{shelf}/\sigma_{web}$ exceeds: 500. That is the web reduces the stresses by more than a factor of 500.

Finally, observe in Eq. (f), the cross-section h appears squared in the denominator of the stress expression. This observation, and the results of Eqs. (b), (c), and (f), demonstrate the fundamental design principle:

Whenever a dimension variable in an expression is raised to a power, the value of the expression is changed exponentially by changes in the dimension variable.

EXAMPLE ANALYSIS 18

Consider again the bracket of Example Analysis 11 as shown in Figure a. As before, let the bracket be a relatively rigid "L-shape" member supported by rivets on its vertical side, and acting as a shelf on its horizontal portion. Assume further that the vertical support is rigid.

Next, suppose a box B with weight W is to be placed on the shelf portion of the bracket as presented in Figure b. Let B be modeled as a homogeneous block with mass m. Suppose further that as B is being placed on the shelf it is dropped onto the shelf while it is only a small distance above the shelf as represented in Figure c. The dropped box causes a sudden, or "dynamic," loading on the bracket.

Determine the stresses in the rivets due to the dynamic loading.

Solution

Observe in Figure a, and in the problem statement, that with the bracket and vertical support being rigid, the only elastic members are the rivets. If we assume further that the box is considerably heavier than the bracket, but not so heavy that the rivets are stretched beyond their elastic limits, then we can model the entire structure (the box, the bracket, the rivets, and the vertical support) as a linear mass-spring system.

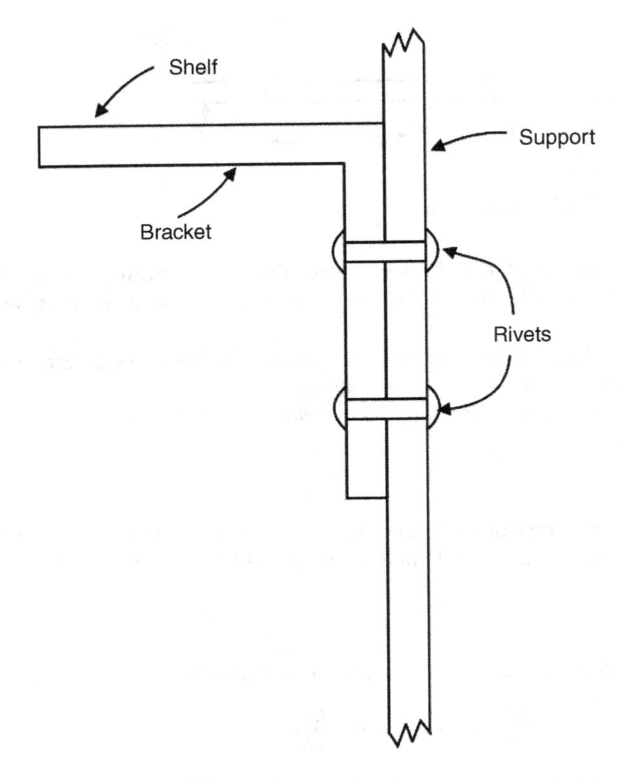

FIGURE A A bracket riveted to a vertical support.

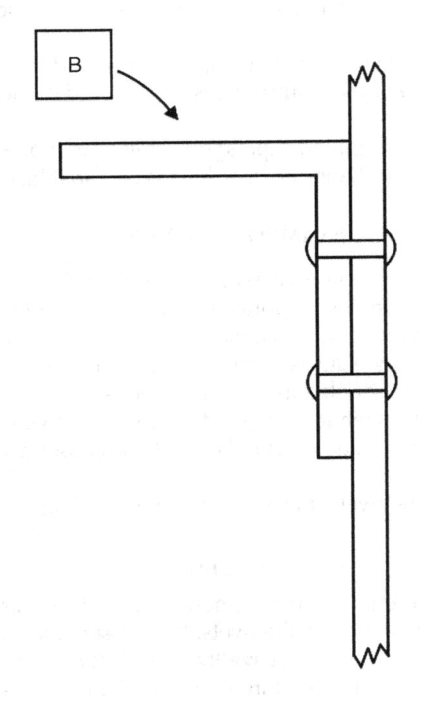

FIGURE B A block being placed on the bracket shelf.

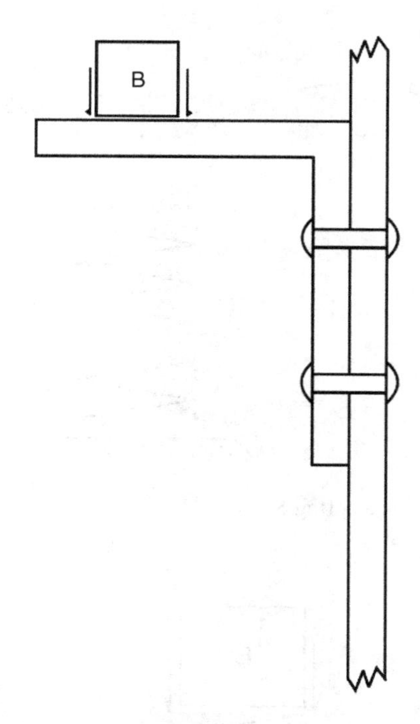

FIGURE C A block being dropped onto the bracket shelf.

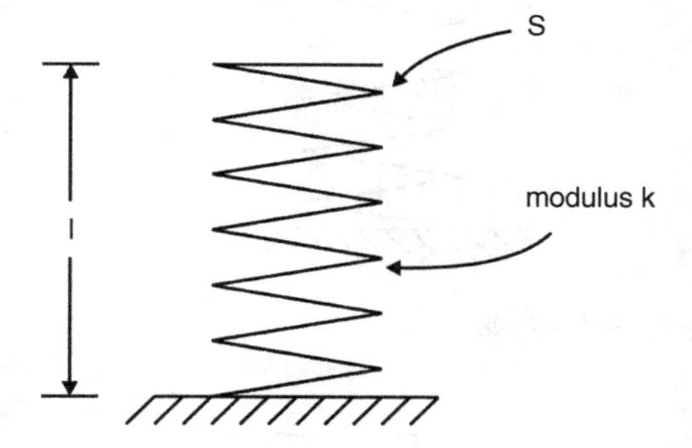

FIGURE D A linear spring S with modulus k and length ℓ.

If we model the system as a linear mass-spring system, we can readily determine the requested rivet stresses, due to the dynamic loading, by using the elementary work-energy method. To see this, consider a massless linear spring *S* with length ℓ and modulus *k* resting on a horizontal surface as represented in Figure d.

Next, imagine a block *B* with weight *W* and mass *m* (equal to *W*/*g*) to be placed atop *S* and then released when it is just a small distance above the spring as represented in Figures e and f.

As the block *B* continues to fall, and as δ increases, the resistive force *F* exerted by *S* on *B* also increases according to the well-known relation:

$$F = k\delta \tag{a}$$

As δ increases, the resistive force *F* of *S* also increases until it eventually becomes large enough to stop the fall of *B*. Let δ_{max} then be the maximum compression of *S*—that is, when *S* arrests the downward movement of *B*. Then from Eq. (a) the resistive force of *S*, when stopping *B*, reaches its maximum value F_{max} given by:

$$F_{max} = k\delta_{max} \tag{b}$$

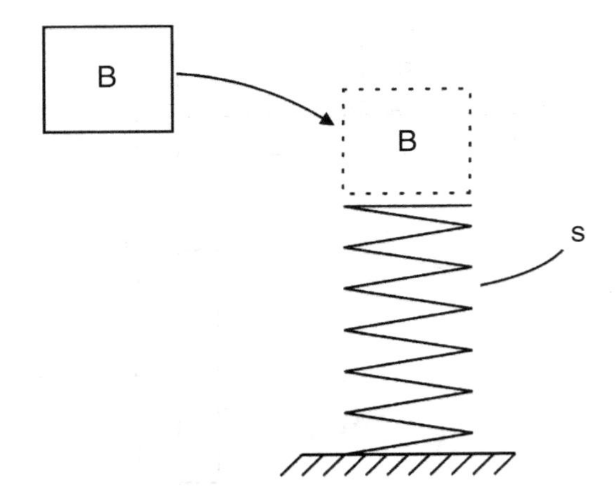

FIGURE E A block B placed atop a spring S.

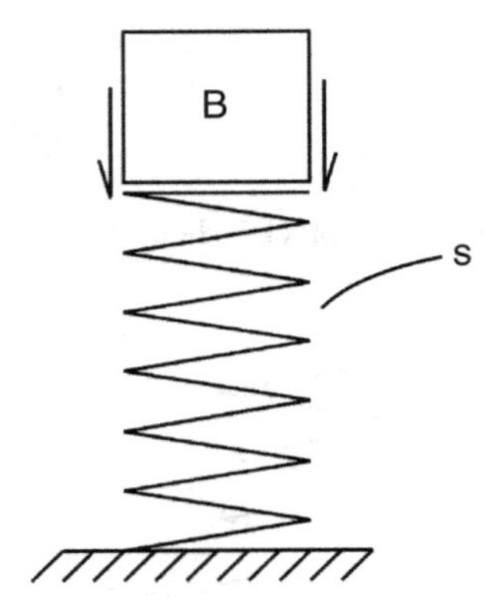

FIGURE F Block B released atop spring S.

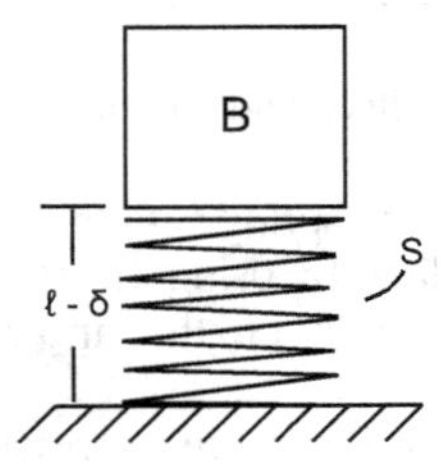

FIGURE G Block B compressing spring S by amount δ.

Recall now the work-energy principle which states that the work done on a system, or a part of a system, between two different configurations (or "states") is equal to the change in kinetic energies of the system between these two states. Analytically, the work-energy principle may be written as:

$$\text{Work} = \Delta KE \tag{c}$$

where *KE* represents the kinetic energy.

By applying this principle to the block B we see that the work done on B between the time it is released atop S until its fall is stopped by S, is due to two forces: gravity (or weight W) and the spring force F. That is:

$$\text{Work} = \text{Work}_{gravity} + \text{Work}_{spring} \tag{d}$$

The work done on B by gravity is simply the weight force: W (= mg) multiplied by the fall distance: δ_{max}. That is:

$$\text{Work}_{gravity} = W\delta_{max} \tag{e}$$

The work done on B by the spring force F (Work_{spring}) is a bit more detailled since F is proportional to the downward displacement δ as in Eq. (a). Therefore, Work_{spring} is the sum of the incremental work of F during the fall. That is:

$$\text{Work}_{spring} = -\int_0^{\delta_{max}} F\,d\delta = -\int_0^{\delta_{max}} k\,\delta d\delta$$

or

$$\text{Work}_{spring} = -(1/2)k\delta_{max}^2 \tag{f}$$

where the negative sign arises since the spring force F on B is directed upward and thus opposite to the downward movement of B.

Just before B is released from rest atop S its kinetic energy is zero (B is at rest). Also, when the downward movement of B is arrested by S, its kinetic energy is zero. Therefore, the change in kinetic energies between the two states ΔKE is also zero. That is,

$$\Delta KE = 0 \tag{g}$$

By substituting from Eqs. (d), (e), (f), and (g) into Eq. (c) we have:

$$\text{Work} = 0 \tag{h}$$

or

$$\text{Work}_{gravity} + \text{Work}_{spring} = 0 \tag{i}$$

or

$$W\delta_{max} - (1/2)k\delta_{max}^2 = 0 \tag{j}$$

or

$$k\delta_{max} = 2W \tag{k}$$

But from Eq. (b) we see that $k\delta_{max}$ is simply the maximum force F_{max} in the spring. Therefore, F_{max} is simply:

$$F_{max} = 2W \tag{l}$$

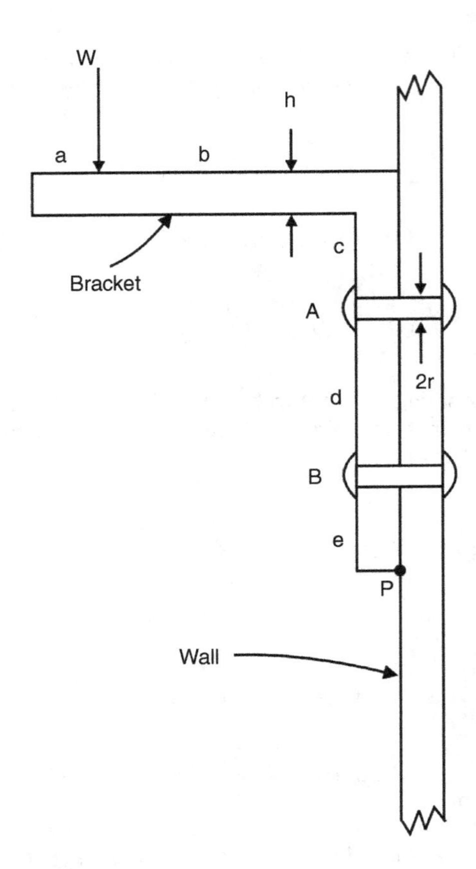

FIGURE H A rivet supported bracket.

Equation (l) shows that the sudden application of the weight load has the same effect as a static load with *twice* the magnitude! This is a well-known result and widely used "rule of thumb" in structural dynamics.

In view of the result of Eq. (l) we can immediately determine the requested rivet stresses due to the dynamic loading. Recall from Example Analysis 11 that for a static weight load *W*, the stresses in the rivets are given by Eqs. (k) and (l) of that example as:

$$\sigma_A = Wb(d+e)/(\pi r^2)\left[(d+e)^2+e^2\right] \quad \text{(m)}$$

and

$$\sigma_B = Wbe/(\pi r^2)\left[(d+e)^2+e^2\right] \quad \text{(n)}$$

where *b*, *d*, *e*, and *r* are dimensions of the bracket as shown in Figure h.

Therefore, replacing *W* by 2*W* in Eqs. (m) and (n) the requested dynamic rivet stresses are:

$$\sigma_A = 2Wb(d+e)/(\pi r^2)\left[(d+e)^2+e^2\right] \quad \text{(n)}$$

and

$$\sigma_B = 2Wbe/(\pi r^2)\left[(d+e)^2+e^2\right] \quad \text{(o)}$$

Comment

Observe in obtaining the results in Eqs. (n) and (o) we did not need to obtain the deformations of the rivets, but instead we simply used the results of Example Analysis 24 for the static loading.

EXAMPLE ANALYSIS 19

Review Example Analysis 18. In the Comment at the end of the example we observed that we were able to obtain the rivet stresses without needing to determine the rivet elongations. Instead, in finding the rivet stresses we simply used the results in Example Analysis 11 where the rivet elongations were determined and used.

Suppose now, with dynamic loading, we are interested in knowing the extent of the downward displacement of the block after it is suddenly placed on the bracket shelf. To this end, suppose that with all geometric and physical conditions for the bracket, block, rivets, and wall support being the same, determine the maximum downward displacement of the block.

Solution

With the assumptions of bracket and support rigidity, the rivets are the only elastic (or deformable) parts of the structural joint. Therefore, as the loading from the weight of the block is applied to the bracket shelf, the rivets will be stretched and the bracket will rotate counterclockwise as represented in Figure a (in exaggerated form).

Let θ represent the bracket rotation and let δ_A and δ_B be the elongations of rivets A and B, as in Figure b.

Figure c shows the overall dimensions of the bracket-joint, which are the same as in the previous example analyses. Figure d in turn presents a line drawing showing the rivet deformations (in exaggerated form) and two relevant dimensions, where the vertical and inclined lines represent the inside (or "interface") surfaces of the bracket and the wall support.

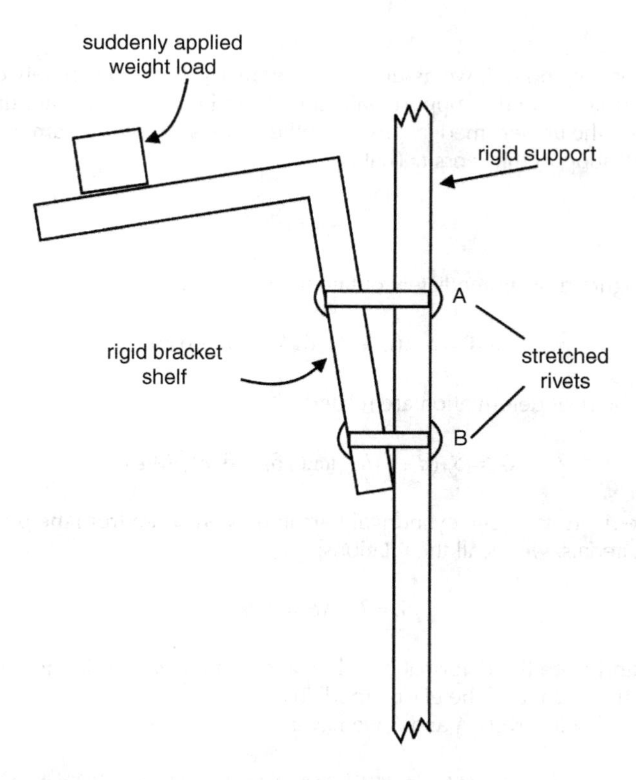

FIGURE A Stretched rivets and rotated bracket (exaggerated for illustration).

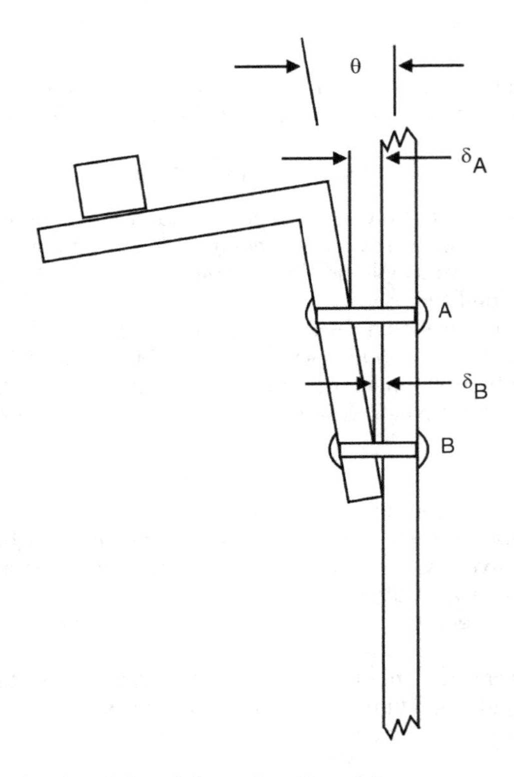

FIGURE B Bracket rotation θ and rivet deformations δ_A and δ_B.

Observe in Figures c and d, if we assume ideal geometry, with no gap between the unloaded interface bracket surface and the support wall surface, and that the rivets are undeformed in the unloaded state, then the undeformed length P of either rivet is simply the sum of the bracket thickness h and the wall support thickenss t. That is,

$$\ell = h + t \tag{a}$$

By inspection of Figure d we immediately obtain the relations:

$$\tan\theta = \delta_A/(d+e) = \delta_B/e = (\delta_A - \delta_B)/d \tag{b}$$

Then we see that the rivet deformation are related as:

$$\delta_A = \delta_B(d+e)/e \quad \text{and} \quad \delta_B = \delta_A e/(d+e) \tag{c}$$

We can model the rivets as elastic cylindrical bars in tension. Then from the principle of elementary strength of materials, we recall the relations:

$$\delta = F\ell/AE = \sigma\ell/E \tag{d}$$

where δ, F, ℓ, A, and E are the deformation (elongation), the axial tension force, the rivet length, the rivet cross-section area, and the elastic modulus.

By applying Eq. (d) with rivets A and B we have:

$$\delta_A = F_A\ell/AE = \sigma_A\ell/E \quad \text{and} \quad \delta_B = F_B\ell/AE = \sigma_B\ell/E \tag{e}$$

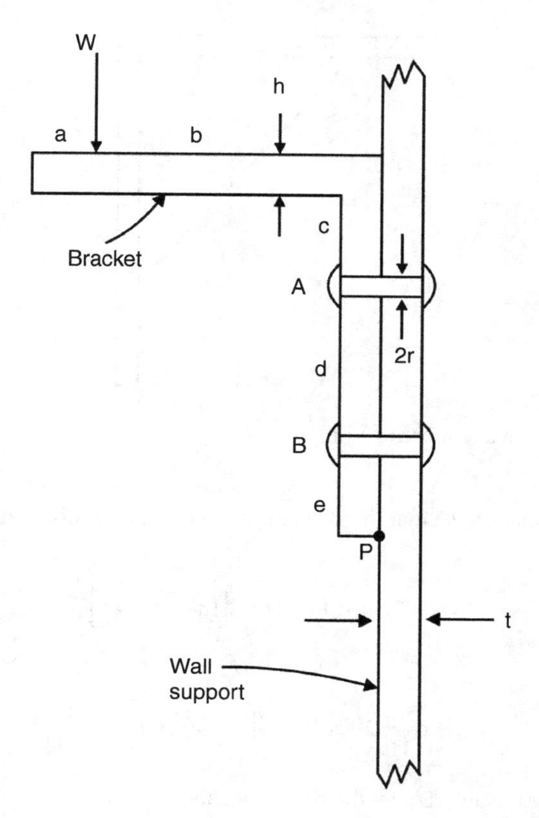

FIGURE C Bracket and support dimensions.

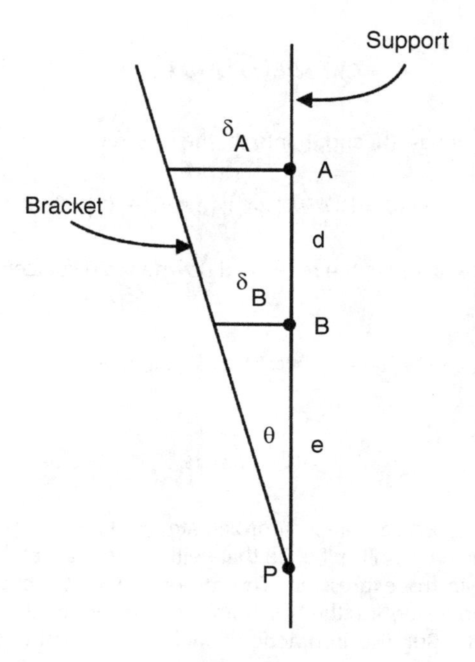

FIGURE D Rivet deformation geometry (exaggerated).

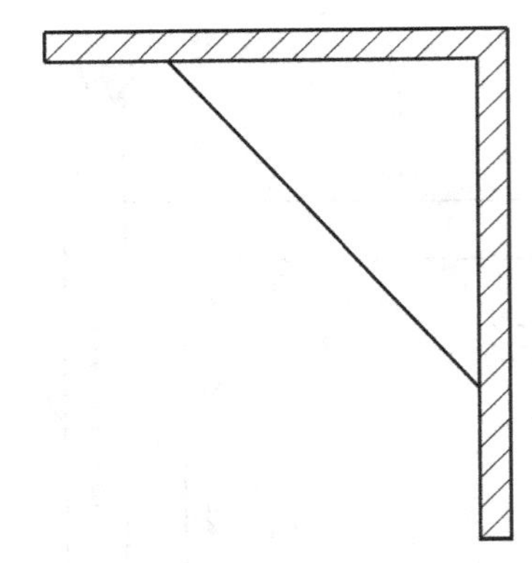

FIGURE E A bracket with a reinforcing web.

Recall from Example Analysis 19 that the stresses in rivets A and B due to the sudden loading by the block are:

$$\sigma_A = 2Wb(d+e)/(\pi r^2)\left[(d+e)^2+e^2\right] \tag{f}$$

or

$$\sigma_B = 2Wbe/(\pi r^2)\left[(d+e)^2+e^2\right] \tag{g}$$

By substituting these results into Eq. (e) the rivet elongations become:

$$\delta_A = 2Wb(d+e)\ell/E(\pi r^2)\left[(d+e)^2+e^2\right] \tag{h}$$

and

$$\delta_B = 2Wbe\ell/E(\pi r^2)\left[(d+e)^2+e^2\right] \tag{i}$$

By substituting the results of Eqs. (h) and (i) into (b) the bracket rotation angle θ is given by:

$$\tan\theta = 2Wb\ell/E(\pi r^2)\left[(d+e)^2+e^2\right] \approx \theta \tag{j}$$

Finally, by reviewing Figures a and b, the requested downward displacement δ of the block is seen to be:

$$\delta = b\theta = 2Wb^2\ell/E(\pi r^2)\left[(d+e)^2+e^2\right] \tag{k}$$

Comments

1. For a small rotation θ, we can safely approximate $\tan\theta$ by θ.
2. Observe in the principal result of Eq. (k) that neither the bracket thickness h nor the wall thickness t, appear in the expression. The reason is that the bracket is assumed to be rigid. In practice, however, if either (or both) of h or t is small, the rigidity assumption may be questionable. But also in practice, brackets are usually reinforced by a web as represented in Figure e.

EXAMPLE ANALYSIS 20

Consider again the dynamically loaded bracket of Example Analyses 18 and 19 and as shown again here in Figure a. As before, let the bracket and wall support by rigid.

Let the dimensions of the bracket, the rivets, and the wall thickness be as follows:

a = 6 in
b = 18 in
c = 6 in
d = 12 in
e = 5 in
h = 0.375 in
r = 0.1875 in
t = 0.625 in

Let the elastic modulus E of the rivets be:

$$E = 30 \times 10^6 \text{ lb/in}^2 \tag{b}$$

Finally, let the weight W of the suddenly applied block B be:

$$W = 500 \text{ lb} \tag{c}$$

Determine the maximum stresses in the rivets and the maximum downward displacement of the bracket of the point of application of the block.

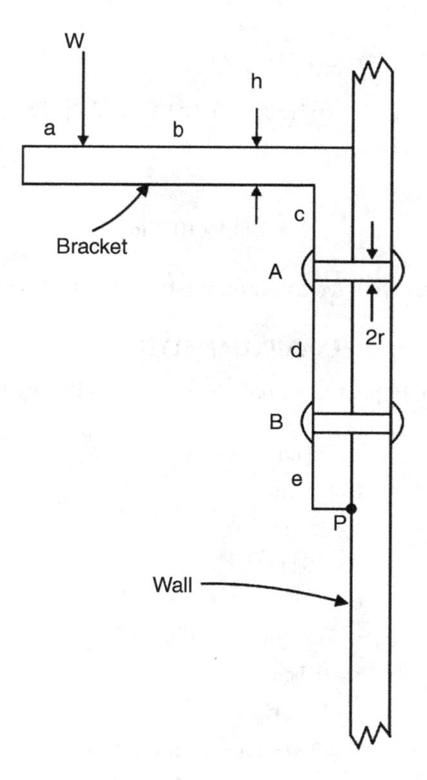

FIGURE A A rivet supported bracket.

Solution

From the results of Example Analysis 37 we have these expressions for the dynamic stresses:

$$\sigma_A = 2Wb(d+e)/(\pi r^2)\left[(d+e)^2+e^2\right] \quad \text{(a)}$$

$$\sigma_B = 2Wbe/(\pi r^2)\left[(d+e)^2+e^2\right] \quad \text{(b)}$$

Also, from the result of Example Analysis 19, the maximum downward displacement δ, due to the suddenly applied load is:

$$\delta = 2Wb^2\ell/E(\pi r^2)\left[(d+e)^2+e^2\right] \quad \text{(c)}$$

By direct substitution of the given data we immediately obtain the results:

$$\sigma_A = (2)(500)(18)(12+5)/(\pi)(0.1875)^2\left[(12+5)^2+(5)^2\right]$$

or

$$\sigma_A = 8823 \text{ lb/in}^2 \text{ (psi)} \quad \text{(d)}$$

and

$$\sigma_B = (2)(500)(18)(5)/(\pi)(0.1875)^2\left[(12+5)^2+(5)^2\right]$$

or

$$\sigma_B = 2595 \text{ lb/in}^2 \quad \text{(e)}$$

and

$$\delta = (2)(500)(18)^2(1.0)/(30)(10)^2(\pi)(0.1875)^2\left[(12+5)^2+(5)^2\right]$$

or

$$\delta = 3.114\times10^{-4} \text{ in} \quad \text{(f)}$$

where in the last computation we have assigned the rivet length P as $h+t$.

EXAMPLE ANALYSIS 21

Review Example Analysis 20. Repeat that analysis with the following data:

$$\begin{aligned}
&\text{a} = 15 \text{ cm}\\
&\text{b} = 45 \text{ cm}\\
&\text{c} = 15 \text{ cm}\\
&\text{d} = 30 \text{ cm}\\
&\text{e} = 13 \text{ cm}\\
&\text{h} = 1.0 \text{ cm}\\
&\text{r} = 0.5 \text{ cm}\\
&\text{t} = 1.5 \text{ cm}\\
&\text{E} = 200 \text{ GPa} = 2.0\times10^{11}\text{N/m}^2\\
&W = 250 \text{ kg (mass)}
\end{aligned} \quad \text{(a)}$$

Solution

Equations (a), (b), and (c) of Example Analysis 20 provide expressions for the stresses σ_A and σ_B in the rivets and for the downward displacement δ of the suddenly applied weight. For convenience we restate those expressions here:

$$\sigma_A = 2Wb(d+e)/(\pi r^2)\left[(d+e)^2+e^2\right] \quad \text{(b)}$$

$$\sigma_B = 2Wbe/(\pi r^2)\left[(d+e)^2+e^2\right] \quad \text{(c)}$$

and

$$\delta = 2Wb^2\ell/E(\pi r^2)\left[(d+e)^2+e^2\right] \quad \text{(d)}$$

By substituting the data of Eq. (a) into Eqs. (b), (c), and (d) we obtain the results:

$$\sigma_A = (2)\left[(250)(9.8)\right](0.45)(0.3+0.13)/(\pi)(0.5)^2(10^{-4})\left[(0.3+0.13)^2+(0.13)^2\right]$$

or

$$\sigma_A = 59.82 \text{ MPa} \quad \text{(e)}$$

and

$$\sigma_B = (2)\left[(500)(9.8)\right](0.45)(0.13)/(\pi)(0.5)^2(10^{-4})\left[(0.3+0.13)^2+(0.13)^2\right]$$

or

$$\sigma_B = 18.086 \text{ MPa} \quad \text{(f)}$$

and

$$\delta = (2)\left[(250)(9.8)\right](0.45)^2(1.0+1.5)(10^{-2})/(2.0)(10^{11})(\pi)(0.5)^2(10^{-4})\left[(0.3+0.13)^2+(0.13)^2\right]$$

or

$$\delta = 7.85\times10^{-6} \text{ m} \quad \text{(g)}$$

Comment

The multiplying factor 9.8 converts the mass in kilograms to the weight force in Newtons.

REFERENCES

1. Waters, E. O., D. B. Westrom and F. S. G. Williams. 1934. Design of bolted flanged connections. *Mechanical Engineering*.
2. Roberts, I. 1950. Gaskets and bolted joints. *ASME Journal of Applied Mechanics*, 1950.
3. Osman, M. O. M., W. M. Mansour and R. V. Dukkipati. 1976. On the design of bolted connections with gaskets subjected to fatigue loading. *ASME Journal of Engineering for Industry*, October.
4. Waters, E. O. and F. S. G. Williams. 1952. Stress conditions in flanged joints for low-pressure service. *ASME Transactions*, Paper No. 51-SA-4.
5. Timoshenko, S. 1956. *Strength of Materials*. New York: D. Van Nostrand.
6. Wahl, A. M. 1929. Stresses in heavy closely coiled helical springs. *ASME Transactions*, 51.

7. Sass, F., C. Bouche und A. Leitner. 1966. *Dubbels Taschenbuch fur den Maschinenbau*. Berlin: Springer-Verlag.
8. Holmbert, E. O. and K. Axelson. 1931. Analysis of stresses in circular plates and rings. *ASME Transactions*, Paper APM-54-2.
9. Waters, E. O. and J. H. Taylor. 1927. The strength of pipe flanges. *Mechanical Engineering*, 49.
10. Blake, A. 1982. *Practical Stress Analysis in Engineering Design*. New York and Basel: Marcel Dekker, Inc.
11. Harvey, J. F. 1967. *Pressure Vessel Design: Nuclear and Chemical Applications*. Princeton, NJ: D. Van Nostrand Company, Inc.
12. Bernhard, H. J. 1963–64. Flange theory and the revised standard B.S. 10:1962—Flanges and bolting for pipes, valves and fittings. *Proceedings of the Institution of Mechanical Engineers* (Lond.), 178(5), pt. 1.
13. Harvey, J. and J. P. Duncan. 1963. The rigidity of rib-reinforced cover plates. *Proceedings of the Institution of Mechanical Engineers* (Lond.), 177(5).
14. Frocht, M. M. and H. N. Hill. 1940. Stress concentration factors around a central circular hole in a plate loaded through pin in the hole. *Journal of Applied Mechanics* 7.
15. Webjörn, J. and R. W. Schneider. 1980. Functional test of a vessel with compact flanges in metal-to-metal contact. *Welding Research Council, Bulletin* 262, October.
16. Webjörn, J. 1967. *Flange Design in Sweden*, ASME Paper 67-PET-20..
17. Schwaigerer, S. 1961. *Festigkeitsberechnung von Bauelementen des Dampfkessel, Behälter und Rohrleitungsbaues*. Berlin: Springer-Verlag.
18. Webjrn, J. 1983. The effect of thermal shock in pipe connections with compact flanges. ASME Paper 83-WA/PVP-6.

SYMBOLS

a	Radius of plate; also moment arm
a_0	Mean length of flange sector
A	Cross-section
A_b	Bolt cross-section
A_r	Depth of tapered rib
b	Width of stiffener; also moment arm
B	General symbol for width
B_r	Depth of uniform rib
C	Coefficient of joint stiffness
d	Nominal bolt diameter
D	Flexural rigidity of plate
D_b	Bolt spacing
D_p	Experimental stress factor
e	Load eccentricity
E	Modulus of elasticity
E_b	Modulus of elasticity of bolt material
E_g	Modulus of elasticity of gasket
F	General symbol for external force
h	Depth of pipe stiffener
H	General symbol for depth; also thickness of flange
H_e	Effective depth of flange
H_0	Edge distance from hole
I	Second moment of area
k	Ratio of flange radii
K_b	Spring constant of bolt
K_g	Spring constant of gasket
L	General symbol for length

L_b Length of bolt
L_g Thickness of gasket
L_e Effective length
L_r Actual length of gusset
m Mean radius to thickness ratio
m_i Inner radius to thickness ratio
m_0 Gasket factor
M_0 Discontinuity moment
M_t Torsional moment
n Ratio of flange depth to thickness of pipe
N Number of bolts
P Uniform pressure
q Unit pressure
Q_0 Discontinuity shear
r Radius of pin or bolt shank
R Radius of bolt circle; also mean radius of vessel with stiffeners
R_i Inner radius of pipe
R_0 Outer radius of flange
S General symbol for stress
S_b Bending stress
S_F Flange stress due to dishing
S_g Gasket stress
S_{gi} Initial gasket stress
S_r Radial stress
S_t Flange stress due to twist
S_{tr} Total stress in rib
S_y Yield strength
ΔS Change in stress
t Thickness of plate
t_0 Auxiliary dimension
T Thickness of pipe or vessel
T_0 Thickness of back-up ring
T_r Thickness of rib
V Design parameter
V_p Volume of cover plate
W Total load applied to bolted joint
W_i Load per unit length of flange
W_{pl} Plastic load
Y Deflection
α Angle of gusset plate
β Shell parameter
δ_p Change in thickness
ε Ratio of stiffener depth to thickness of vessel
ε_p Plastic strain
θ Angle of twist
λ Spacing between rib centers
ϕ Ratio of stiffener width to thickness of vessel
ν Poisson's ratio

7 Clamps and Pipe Joints

7.1 INTRODUCTION

The existence of the many formulas and design aids that are utilized for the design of mechanical joints show that a great variety of components can be held together by using rivets, bolts, welds, and adhesives. More than six billion bolts are manufactured every year and more than two million rivets holding the Eiffel Tower together are impressive numbers.

The proper operation of all machines and structures depend on the integrity of their clamps and joints. Designing joints via the analysis of single fasteners alone, however, can lead to deleterious effects for a system as a whole. For example, it is risky to design a split hub or a pipe flange on the basis of a single bolt calculation alone. Indeed, every consideration of a mechanical fastener requires knowledge of geometry, materials, and fabrication techniques.

Unfortunately, no single formula can satisfy all design requirements, and for this reason, we continually try to consider a variety of options and trade-offs in our joint designs.

It is difficult to classify all mechanical joints under a single category given that there are virtually billions of machine parts forming either sliding or fixed connections. For example, links, connecting rods, pins, shaft bearings, or toothed wheels may belong to the sliding type of a joint family. The fixed joints can include boiler plates, flanges, vessels, pipes, and similar mechanical systems.

At times, joints are represented only by a fixed assembly of the components. Under the term of joints, this book also compiles examples of fixed as well as detachable elements, such as pipe supports, couplings, clamps, and split hubs. In the case of pipe couplings used in the oil-field explorations, the detachable or nondetachable characteristics of the joint will depend on a number of design and fabrication variables. Detachable joints may involve threads, cotter pins, or splines. Their choice can be determined by design or operational requirements. This is especially important because the majority of machine failures occur at the joints.

7.2 INTERFACES, OR SHRINK, FITS

The special case of a permanent-type mechanical joint is concerned with the concept of structural interference, which can be exemplified by the difference between the actual diameters of the shaft and the hole measured before the assembly. Such joints are important in wheels or railway rolling-stock, wormgear rims, pressure vessels, or bushing type of applications. These joints can be assembled by a mechanical force or by a shrink-fit process involving heating and cooling of mating component parts. One of the more recent approaches to achieve a shrink for is to use high-pressure oil as a wedge between the two parts of cylindrical assembly.

The advantage here is that the oil does not impair the condition of the surfaces in contact [1]. For example, Figure 7.1 shows that the mechanical press-fit depends to a large degree on the solution of a taper. The interacting surfaces can be lubricated with linseed oil or a similar coating to reduce the press-fit load and to minimize galling. The existence of galling and abrasions has the effect of reducing the interference pressure between the two cylindrical surfaces.

However, the press-fit carried out by a heating and cooling process of the respective parts is to be preferred. The internal cylinder can be cooled, for instance, by a liquid gas or dry ice. The mechanical joint, assembled by means of a temperature gradient, appears to be significantly stronger than the regular joint formed by an axial thrust due to a pressing operation.

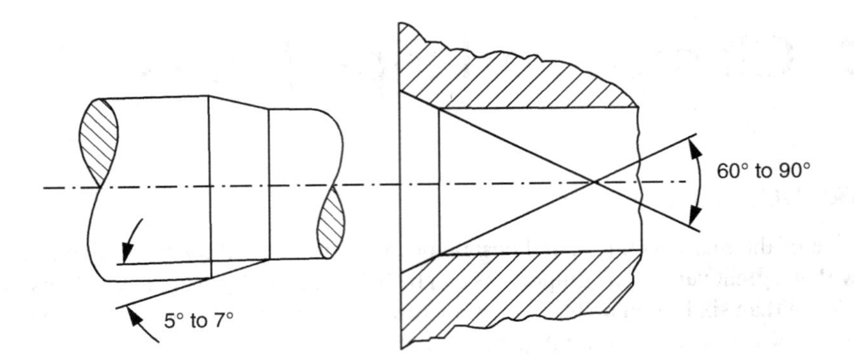

FIGURE 7.1 Example of press-fit tapers.

The majority of shrink-fit assemblies involve metals of comparable modulus of elasticity and the Poisson's ratio. Under these conditions, the assembly pressure p between the two cylinders can be given as:

$$p = \frac{E\delta\left(b^2 - a^2\right)\left(c^2 - b^2\right)}{2b^3\left(c^2 - a^2\right)} \tag{7.1}$$

where δ denotes the difference between the external radius of the inner cylinder in the unstressed condition and the internal radius of the outer cylinder. In a sense, δ defines the amount of radial interference which, during the assembly, causes a pressure p, sometimes called shrink-fit pressure. Figure 7.2 provides other notation for Eq. (7.1)

Equation (7.1) is derived on the basis that the increase in the inner radius of the outer cylinder, plus the decrease in the outer radius of the inner cylinder, must be equal to radial interference δ. The modulus of elasticity of the joint components is E.

The corresponding tangential S_t and radial S_r stresses at the inner surface of the outer cylinder are:

$$S_t = \frac{P\left(b^2\right) + c^2}{c^2 - b^2} \tag{7.2}$$

and

$$S_r = -P \tag{7.3}$$

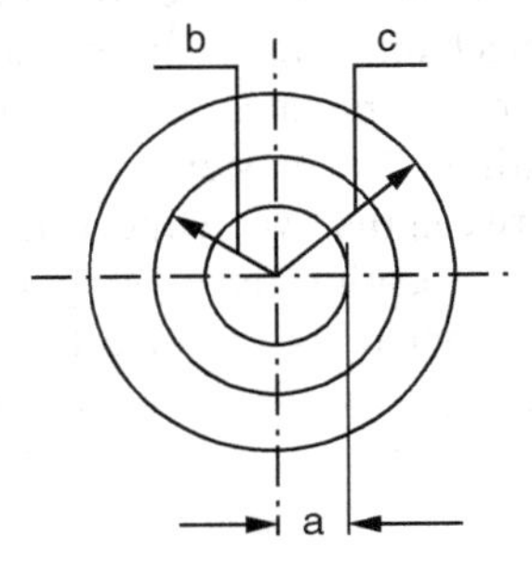

FIGURE 7.2 Symbols for shrink-fit assembly.

The maximum shearing stress at the same point can be expressed in terms of the tangential and radial stress components, acting as principal stresses. This yields:

$$\tau_{max} = \frac{E\delta c^2\left(b^2 - a^2\right)}{2b^3\left(c^2 - a^2\right)} \tag{7.4}$$

If the inner cylinder becomes a solid shaft, then for $a=0$, Eqs. (7.1) and (7.4) reduce to:

$$p = \frac{E\delta\left(c^2 - b^2\right)}{2bc^2} \tag{7.5}$$

and

$$\tau_{max} = \frac{E\delta}{2b} \tag{7.6}$$

In the foregoing analysis of a mechanical joint obtained by a shrink-fit process, the length of the two cylindrical components is assumed to be the same. In the case of a hub and shaft combination, that is at $a = 0$, the projecting portions of the shaft can react in such a manner that there will be some increase in the local compression as shown in Figure 7.3. In this illustration, we have a cross-sectional view, indicating the approximate distribution of the radial pressure.

If, instead of a shaft, the joint involves an inner cylinder subjected to internal pressure, the corresponding stresses can be superimposed on the shrink-fit stresses. The effect of shrink-fit is to produce a compressive tangential stress that decreases the maximum tangential stress in tension caused by the internal pressure.

This principle is utilized in mitigating the stresses in gun barrels by designing built up systems containing several cylindrical layers. By carrying this process a step further sufficiently high internal pressure can be applied resulting in a plastic flow of the inner cylinder. Next, after removing the initial internal load, a residual stress pattern will result due to the plastic deformation. Essentially, the inner part retains some residual compression, having a beneficial effect on the system when the working internal pressure is finally applied.

The method can be extended to several cylinders in a built up shrink-fit assembly by controlling the internal pressure until the entire wall of the composite cylinder has yielded. This process is

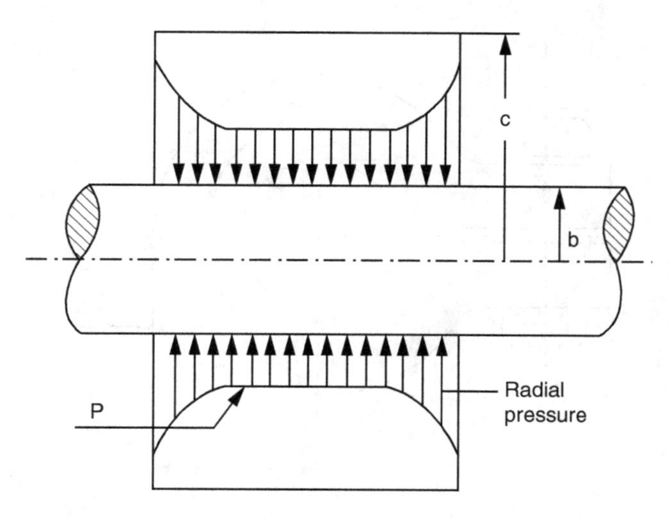

FIGURE 7.3 Shaft-hub interaction.

often called "auto-frettage" of thick cylinders [2] and it applies to design of hydraulic ram cylinders, accumulators, gun barrels, and similar high-pressure components.

Our interest here is restricted to the basic concept of a mechanical joint formed by the shrink-fit stresses with or without the auto-frettage.

7.3 SPLIT HUB CRITERIA

Figure 7.4 illustrates a mechanical joint often found in machinery and structures. The basic design problem is to determine the bolt tightening force required to resist the external torque M_t. The available torsional reaction, due to the contact friction between the hub and the shaft, is denoted by M_{tf}.

The particular case shown in Figures 7.4 and 7.5 is shown on a two-bolt clamping force. The response of the joint in question can be judged on the basis of a torque or axial resistance W. Since the exact distribution of unit pressure on the contact surface is generally unknown, it is necessary to make certain approximate calculations supported by tests. The force exerted on an individual bolt due to tightening torque is *V*. The total number of bolts in a general case can be denoted by *N*, so that the product *FN* signifies the shaft clamping force as shown in Figure 7.4. Hence, the coefficient of friction *f*, together with the assumed dimensions *x* and x_1, can be used to determine the approximate frictional torque available in the joint. Thus, the frictional couple resisting the external torque M_t is:

$$M_{tf} = FNfd \tag{7.7}$$

Taking moments about a point measured x_1 units from the center of the shaft and introducing the bolt preload *V* into Eq. (7.7), gives:

$$M_{tf} = \frac{VNfd(x + x_1)}{x_1} \tag{7.8}$$

In the case of axial equilibrium, implied by Figure 7.5, the approximate resistance *W* can roughly be determined as:

$$M = 2P_e Ldf \tag{7.9}$$

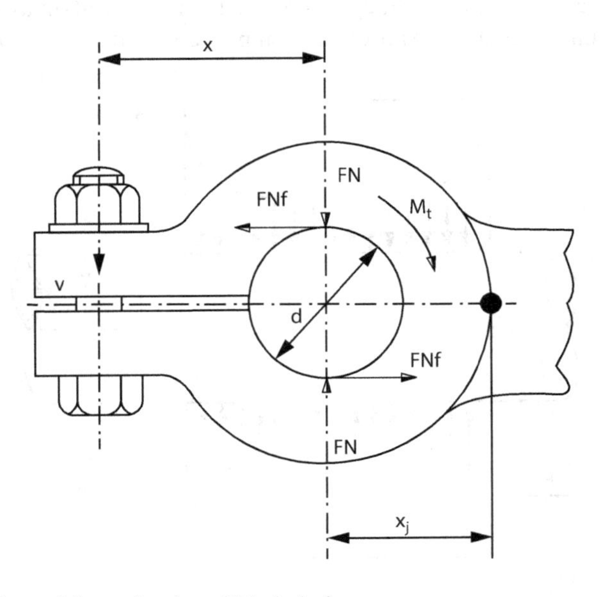

FIGURE 7.4 Equilibrium of forces for the split hub design.

where P_e defines an equivalent contact pressure in this joint, which when multiplied by the projected area of the shaft equal to (Ld) and the friction factor f, should give the axial resistance W as shown in Figure 7.5.

The distribution of contact pressure between the shaft and the hub is not uniform and it is likely to vary with the angle θ as indicated in Figure 7.6.

Assume, for instance, that the contact pressure can be represented by the following expression in terms of angle θ:

$$P = P_{max} \sin\theta \tag{7.10}$$

Then, the contact pressure can vary between 0 and P_{max} and the elementary torque resistance for one quadrant can be written as:

$$dM_{tf} = \frac{PLfd^2 d\theta}{4} \tag{7.11}$$

The total frictional torque may now be obtained by the summation of the resistance in the four quadrants:

$$M_{tf} = P_{max} Lfd^2 \int_0^{\pi/2} \sin\theta \, d\theta \tag{7.12}$$

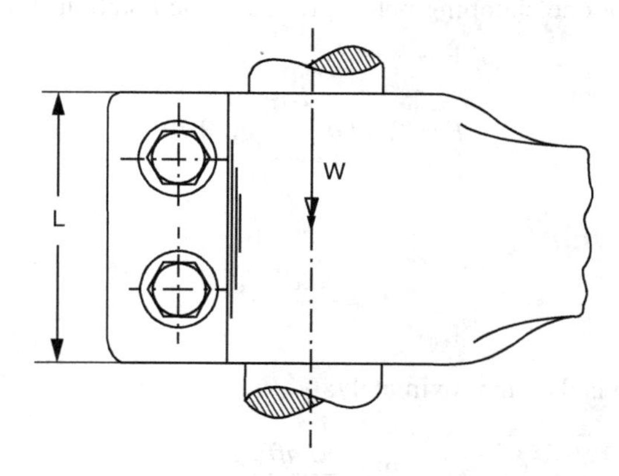

FIGURE 7.5 Plan view of split hub.

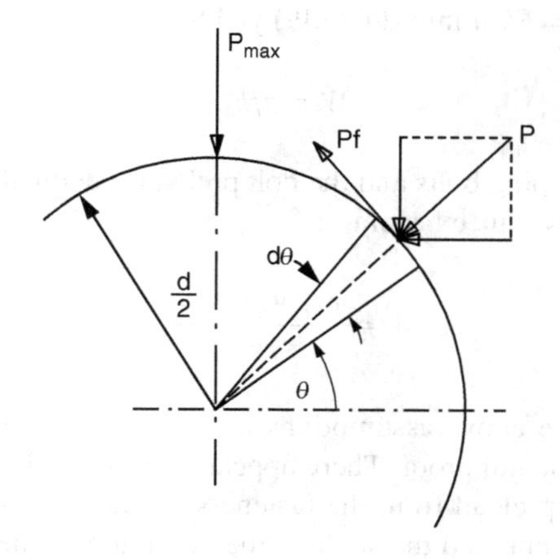

FIGURE 7.6 Symbols for definition of contact pressure.

This gives M_{tf} as:

$$M_{tf} = P_{max} Lfd^2 \tag{7.13}$$

Assuming again that the equivalent pressure is $(P_{max} / 2) = f / Ld$, we have P_{max} as:

$$P_{max} = \frac{2F}{Ld} \tag{7.14}$$

Equations (7.13) and (7.14) thus yield the following frictional torque for a clamping force of one bolt:

$$M_{tf} = 2Ffd \tag{7.15}$$

In terms of the bolt preload V and the number of bolts N, the final expression for M_{tf} becomes:

$$M_{tf} = \frac{2VfMd(x + x_1)}{x_1} \tag{7.16}$$

In considering the case of axial equilibrium, the problem can be analyzed by assuming that the summation of all the vertical components of pressure over one-half of the hub contact area should give the applied load F for one clamping bolt. In terms of the notation shown in Figure 7.6, the following relation ensues:

$$F = P_{max} Ld \int_0^{\pi/2} \sin^2 \theta d\theta \tag{7.17}$$

This gives:

$$F = \frac{\pi LdP_{max}}{4} \tag{7.18}$$

The total axial resistance is then approximately:

$$W = \frac{\pi LdfP_{max}}{2} \tag{7.19}$$

Introducing $P_{max} = 2P_e$ and $P_e = F/Ld$ into Eq. (7.19) yields:

$$W = \pi fF \tag{7.20}$$

Finally, for the case of N clamping bolts and the bolt preload V defined previously, the final expression for the axial frictional resistance becomes:

$$W = \frac{\pi fVN(x + x_1)}{x_1} \tag{7.21}$$

Observe that a number of the above assumptions can be considered as arbitrary unless we can produce some sort of experimental proof. There appear to be several uncertainties in such areas as the nature of transfer of bolt preload from the fasteners to the shaft, sinusoidal distribution of the contact pressure between the hub and the shaft or the definition of the equivalent contact pressure for the simplicity of the analysis.

It seems to be evident then that even a relatively simple mechanical joint, such as that illustrated by Figures 7.4 and 7.5, involves a number of complex interfaces, which cannot be described by elementary equations without the benefits of well-designed experiments. We are faced again with the deceiving simplicity of a mechanical joint.

7.4 THE SINGLE-PIN CLEVIS JOINT

Figure 7.7 depicts another example of a fundamental mechanical joint. It consists of two parallel cantilevered beams whose ends are pulled toward each other by a bolt, exerting a load w. It is known as a "clevis joint."

The clamping action illustrated in Figure 7.7 has numerous practical applications. Each arm of the joint can be modeled with the aid of a beam with sinking supports, on the assumption that the clamping bolt maintains the surfaces in contact through the provision of a suitable rigid insert.

From classical beam theory, the maximum bending stress S_b for the *two* beams, using the notation of Figure 7.7 is:

$$S_b = \frac{3WL}{BH^2} \tag{7.22}$$

The corresponding deflection Y is:

$$Y = \frac{WL^3}{EBH^3} \tag{7.23}$$

By setting the yield stress S_y equal to the maximum bending stress, and by eliminating w between Eqs. (7.22) and (7.23) Y becomes:

$$Y = \frac{S_y L^2}{3HE} \tag{7.24}$$

This is the basic design formula for a single-pin clevis joint, which defines the allowable deflection of $2Y$ for a specified value of the yield strength S_y, or a fraction thereof for a design factor of safety. It should be noted that the width B does not enter into the calculations using Eq. (7.24).

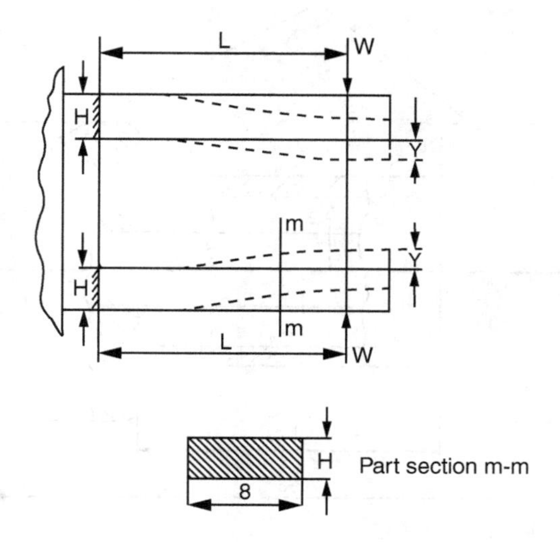

FIGURE 7.7 Clevis joint.

The required bolt load in terms of the yield strength of the clevis material then becomes:

$$W = \frac{BH^2 S_y}{3L} \tag{7.25}$$

In this case, the bolt load is governed only by the strength of the clevis.

7.5 A RIGID CLAMP APPLICATION

Figure 7.8 portrays a relatively rigid system used to clamp a softer component. The essential elements necessary for design include elastic moduli, nominal bolt diameter, bolt length, and the corresponding dimensions of the clamped tubular component.

In this case, we can assume, for example, that an aluminum tube is being compressed by two steel blocks by a standard bolt. The notation is shown in Figure 7.8. The tensile load in the bolt W_b will depend on the spring constants of the components bolted together, the external load W applied to the rigid blocks and on the initial bolt preload W_i. It has been shown [3] that the bolt load in this particular joint is:

$$W_b = \frac{W}{1+\phi} + W_i \tag{7.26}$$

where ϕ is:

$$\phi = \frac{4\left(b^2 - a^2 E_a L_b\right)}{d^2 E L_a} \tag{7.27}$$

If, instead of the bolt preload W_i, we specify the initial torque M_t, then according to the approximate formula, Eq. (2.17), the design expression becomes:

$$W_b = \frac{W}{1+\phi} + \frac{5M_t}{d} \tag{7.28}$$

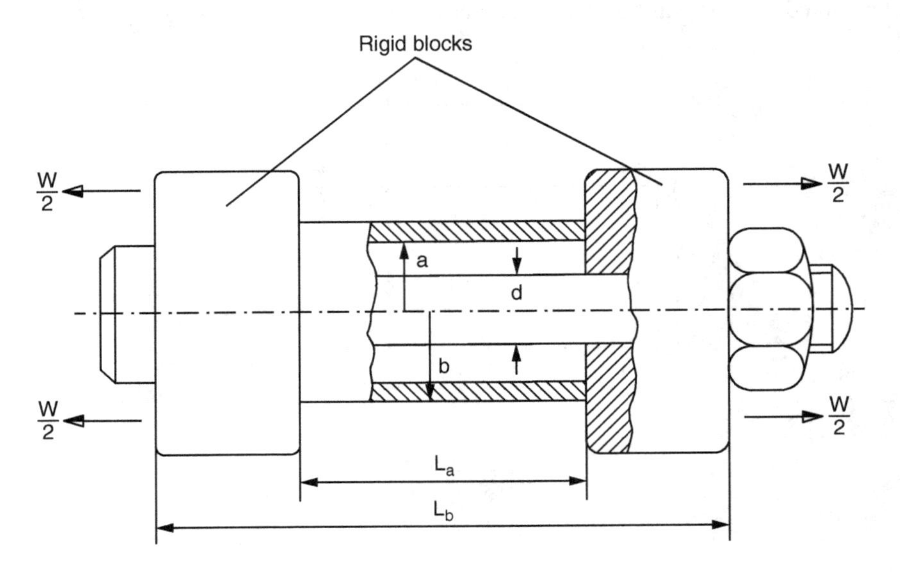

FIGURE 7.8 An example of a rigid clamp.

When the ratios E_a/E and L_b/L_a are found to be small, the parameter ϕ tends to zero and the actual bolt load reduces to:

$$W_b = W + W_i \tag{7.29}$$

This relationship was expressed in Chapter 5, regarding riveted and bolted joints, with particular reference to the case of a soft system clamped by a rigid bolt. In deriving Eq. (7.26) the tube is first assumed to decrease in length under the initial bolt load. This decrease is next reduced through the application of the external load W shown in Figure 7.8. These changes can be calculated using Hooke's law as:

$$\Delta L_a = \frac{W_i L_a}{A_t E_a} \tag{7.30}$$

The design method based on Eqs. (7.26) and (7.27) is recommended for fastener analysis by the Industrial Fastener Institute [4]. In its original form, applied to our case of a rigid clamp compressing the soft tube, the formula for the bolted load can be stated as:

$$W_b = \frac{K_b W}{K_b + K_t} + W_i \tag{7.31}$$

where K_b and K_t are:

$$K_b = \frac{\pi d^2 E}{4 L_b} \tag{7.32}$$

and

$$K_t = \frac{\pi\left(b^2 - a^2\right)E_a}{L_a} \tag{7.33}$$

The terms K_b and K_t represent the spring constants of the bolt and the tube, respectively. Hence, the parameter ϕ is the ratio of the two spring constants, K_t/K_b, as shown by Eq. (7.27).

7.6 DESIGN OF A C-CLAMP

Quite often, in standard mechanical applications, we are called upon to deal with a temporary joint required to hold a given structural member or an assembly of parts by means of a C-clamp such as that shown in Figure 7.9.

The strength of the mechanical joint in this case depends on the flexural rigidity of the clamp, on the assumption that the external load W is limited by the maximum allowable stress.

The total spread Y of the C-clamp legs under the load W can be approximated by the expression [5]:

$$Y = \frac{WR}{AE}\left[\left(2.43 + 0.71\frac{R}{n}\right) + 3.93\varepsilon\right] \tag{7.34}$$

The corresponding stress S, obtained as the sum of bending and tension at the inner surface of the clamp, can be expressed as:

$$S = \phi_0\left[\frac{W}{A} + \frac{(R+e)W(R-R_i)}{I}\right] \tag{7.35}$$

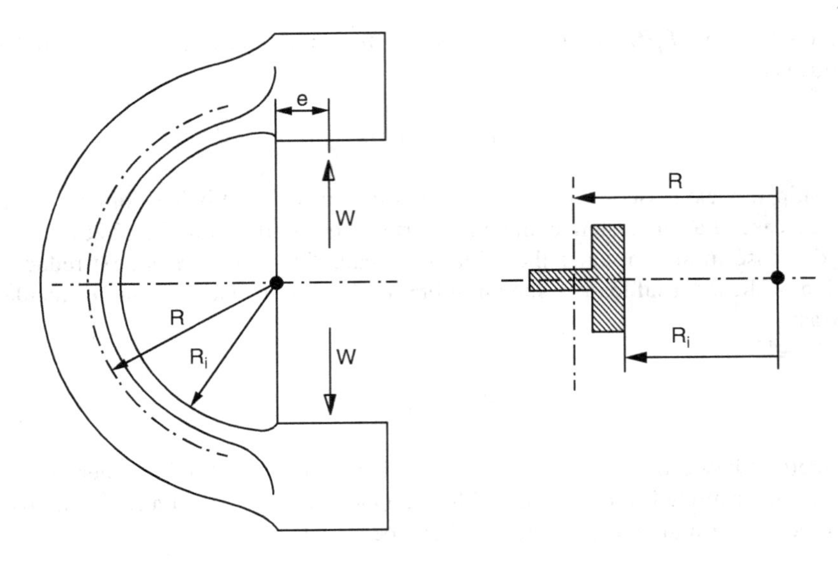

FIGURE 7.9 Example of a C-clamp.

In Eqs. (7.34) and (7.35), A is the area of the clamp cross-section, n is the distance between the neutral axis and the central axis, R is the radius of curvature about the centroid of the cross-section ε is the shear distribution factor, and ϕ_0 is the stress concentration factor for a thick curved member.

When the clamp becomes relatively thin the distance e tends to zero and then the foregoing equations can be reduced to the design formulas:

$$Y = 0.71\frac{WR^3}{EI} \tag{7.36}$$

and

$$S = \frac{W(R - R_i)R}{I} \tag{7.37}$$

By eliminating the load W between Eqs. (7.36) and (7.37), the deflection (clamp spread) Y becomes:

$$Y = \frac{0.71SR^2}{(R - R_i)E} \tag{7.38}$$

All the above formulas are applicable to clamps with a circular curvature and the type of cross-sections for which the value of the approximate parameter ϕ_0 is sufficiently accurate for most practical purposes. The method of the solution involves the assumption of a deflection Y and the calculation of the stress for a given shape and dimensions of the clamp cross-section.

For more details on design of thick, curved members for clamp application, the reader is referred to machine design literature as in References [4] and [5] and similar publications. When using Eq. (7.35), the approximate curved beam parameter ϕ_0 can also be obtained using Figure 7.10.

In the foregoing example, the role of the fastener is represented by the rigidity and spring-back action of the clamp itself. This is in contrast with the clevis type joint illustrated in Figure 7.7. The case of a split-hub design can be considered as some combination of that shown in Figure 7.4, and the C-clamp concept given in Figure 7.9. For practical purposes, however, the individual analyses of

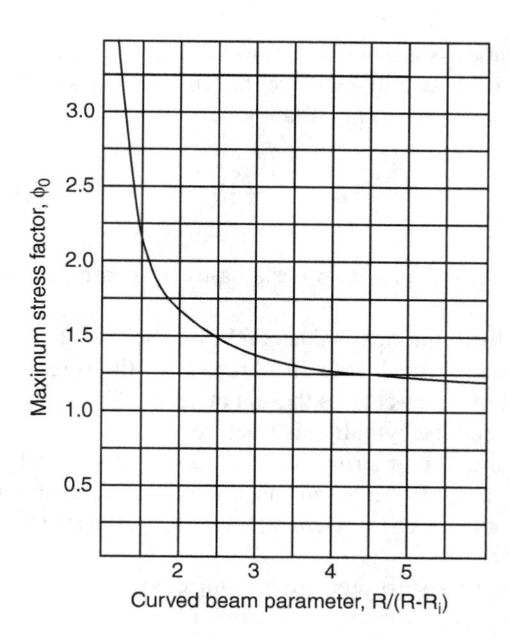

FIGURE 7.10 Approximate correction factor for curved beam design.

split hub, clevis joint, and C-clamp configurations should be sufficient during the preliminary stages of design of clamping devices defined here as detachable connections.

7.7 VESSEL AND PIPE SUPPORTS

This area of design, related to mechanical joint technology, is important but highly complex [2,6].

Large pressure vessels can be supported by skirts or symmetrically placed brackets. In the support skirt design, radial allowance is made to account for the pressure and temperature effects. Bracket supports can accommodate the motions by the process of sliding or rolling. When local bending effects in the vessel walls are encountered, the theory of beams on elastic foundation can be utilized in the calculations.

For an external support system consisting of a narrow ring fastened tightly around the outside of a vessel, or a pipe, the deflection caused by the circumferential ring loading P_0, the radial displacement Y_r may be calculated using the expression:

$$Y_r = 0.64\frac{P_0}{E}\left(\frac{r}{t}\right)^{3/2} \tag{7.39}$$

Where P_0 is expressed in pounds per circumferential inch, E in pounds per square inch, and Y_r is in inches; and where r is the mean radius of the vessel, or pipe, and t is its thickness.

It is assumed here that the ring is placed sufficiently far from the vessel closures or pipe flanges. Also, in the development of Eq. (7.39) it is assumed that there is no significant internal pressure. Thus, the local deformation depends upon the cylindrical stiffness.

When, however, the vessel, or pipe, has an internal pressure P_i, the corresponding ring load P_0 resisting dilatation is [2]:

$$P_0 = \frac{0.85}{0.64}\frac{A_r P_i (rt)^{1/2}}{A_r + t(rt)^{1/2}} \tag{7.40}$$

Equation (7.40) is based on the assumption that, initially, there is zero clearance between the outside diameter of the vessel and the inside diameter of the ring. The maximum bending stress produced in the vessel, or pipe wall, due to the radial dilation is:

$$S_b = \frac{1.17P_0(r)^{1/2}}{t(t)^{1/2}} \tag{7.41}$$

Figure 7.11 provides notation for these two basic cases of loading involving ring fastening for a cylindrical vessel or pipe.

It should be noted that the numerical values of P_0 for the two cases may or may not be equal, although the dimensions are identical. In both illustrations, the ring provides a means of support and restraint without the need for a weld or adhesion material. The entire connection depends on the elastic response of the ring and the cylindrical structure.

The choice of a skirt, a bracket, or a ring-type support is influenced by the environment and the method of assembly for a land- or sea-based installation. A problem may arise due to a thermal stress induced by a temperature gradient. Generally speaking, the steeper the temperature gradient, the higher the stress.

A typical case may involve a skirt attachment subjected to thermal strains and possible crack development in the welded joints as shown in Figure 7.12.

The situation in Figure 7.12 may arise in the skirt attachment of a coking drum [2], which can be subject to frequent heating and cooling gradients of significant magnitudes. In such a case, fatigue cracks can develop on the tensile face of the weld and possibly spread into the drum wall.

Due to their complexity and resulting difficulties in their analysis, the majority of piping and vessel restraints are conservative.

One of the more difficult topics addressed today involves the mechanism of interaction between pipes and the pipe clamps. The clamps are often used to provide the attachment points that can support and restrain a given piping system.

The split hub design as illustrated in Figures 7.4 and 7.5 provides a rather simplified view of the clamping action. But in recent years more success in modeling the clamping action has been attained by using finite element techniques considering both linear and nonlinear behavior and a variety of boundary conditions.

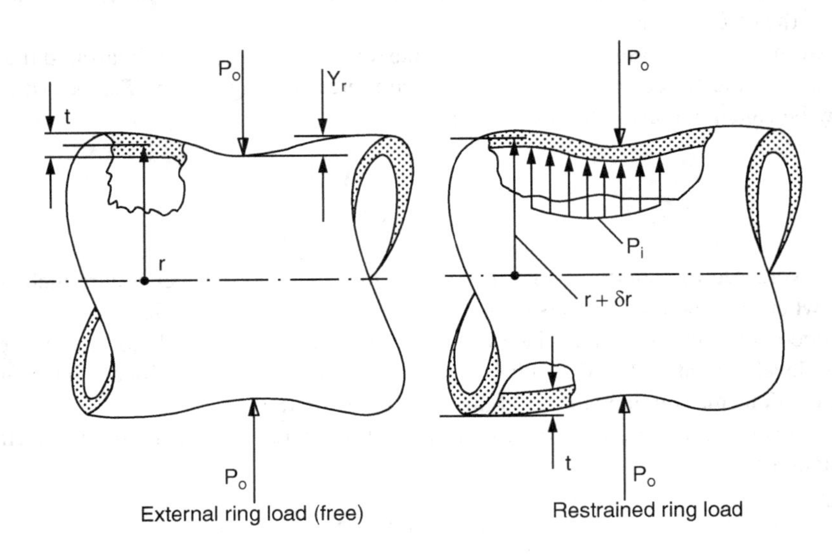

FIGURE 7.11 Notation for analysis of ring loading.

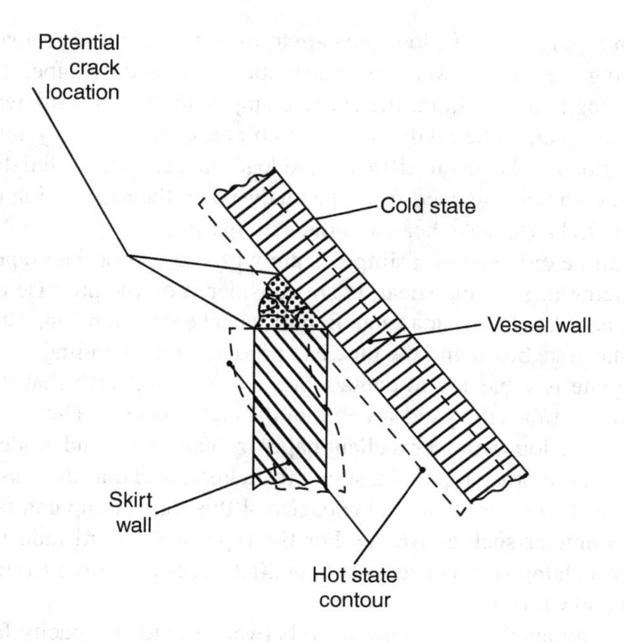

FIGURE 7.12 Areas of potential weld cracking.

The stiffness of the pipe branches and the supports can be extremely important. Figure 7.13 illustrates a typical clamp configuration for a pipe.

In this case, a deep-section curved beam, known as the "yoke beam," delivers a clamping pressure to the pipe by means of a round curved member known as the "load bar," and a "load distribution plate," suitably grooved to receive the load bar. This bar is bolted to the yoke beam with the two

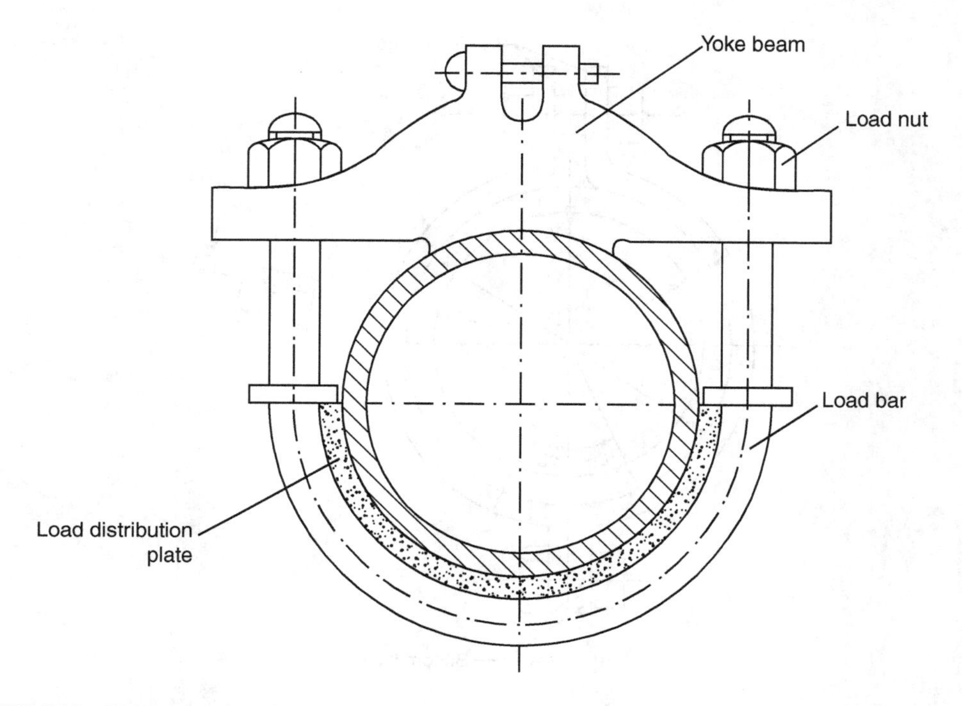

FIGURE 7.13 Yoke-type pipe clamp.

load nuts as shown in Figure 7.13. The load nuts are torqued to the initial preload, thereby providing the necessary clamping pressure between the distribution plate and the pipe.

For a pipe containing a hot medium, the entire clamp assembly is insulated with the exception of the load distribution plate. The analysis of this complex mechanical joint is usually made in terms of the clamp preload, yoke beam stiffness, and load-bar rate. The radial deformation and ovalization of the pipe wall can be excluded from the analysis, on the assumption that the most critical components are likely to be the yoke bean and the load bar [6].

The yoke beam can be modeled as a simple beam with the appropriate taper. Normally, out-of-plane-loading, producing torsion and shear, is not considered on the premise that the relevant pipe deformation can be neglected. The load bar is only checked for tension, although a differential expansion between the yoke beam and the pipe can produce some bending.

The distribution plate is subjected to contact pressure selected such that the allowable bearing stress does not exceed 1.5 times the yield strength of the plate material. Details of the recommended design calculations can be found in an excellent paper by Van Meter and Anderson [6].

Figure 7.14 illustrates another type of a standard industrial clamp used as a one-way restraint for piping. Experience shows here that load capacity of this pipe clamp can be related to a convenient dimensional parameter such as BH^2/D. For the type of standard industrial clamp shown in Figure 7.14, the highest clamp stress occurs at about 20 degrees measured from the plane of split, as calculated by Bilgin and Chow [6].

There are also indications that the relationship between the load capacity factor and the dimensional parameter BH^2/D is bilinear. The load capacity factor for this case may be defined as W/S, where W is the external load on the clamp parallel to the plane of split and S is the allowable clamp stress measured in the hoop direction.

It should be noted that the distribution of interface pressure and of the resulting hoop stresses in the clamp are highly complex and nonlinear.

Similar analyses can be made for other types of loading and configurations of clamping devices designed for pipe restraint.

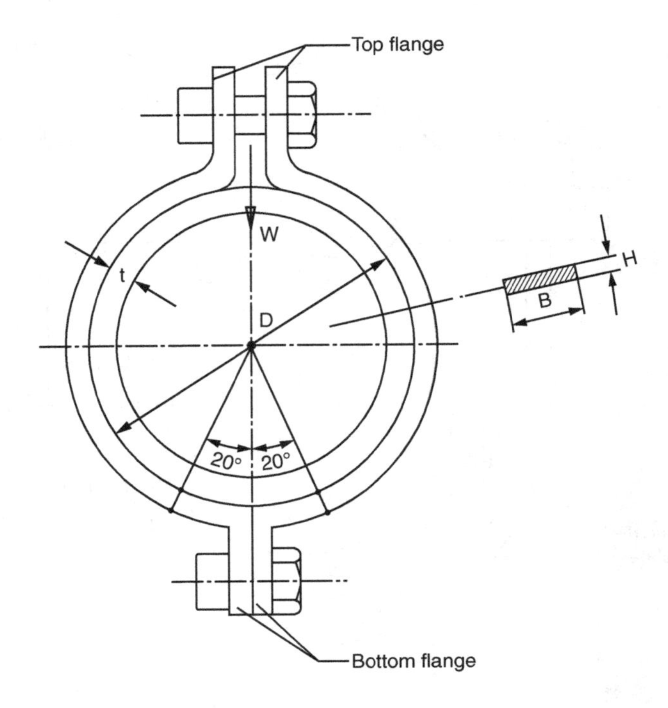

FIGURE 7.14 Standard pipe clamp.

Sophisticated analytical methods and finite element methods are still being used to further clarify the complicated interactions between the clamps and the piping.

The use of a clamp amounts to creating a mechanical joint, which causes localized stress distributions resulting from external support loads, internal pressure, and the thermal gradients.

One of the special problems in the design of pipe clamps is concerned with determining clamp stiffness that depends on a number of component stiffnesses arranged in series. As long as the relationship between the clamp load and the deflection is reasonably linear, the clamp stiffness for the case illustrated in Figure 7.14 can be approximated as:

$$K_C = 83\frac{Et^3}{(D-t)^2} \tag{7.42}$$

Equation (7.42) is based on the work of Chang and Line [6] who performed finite element analyses for pipe sizes ranging from 3 to 12 in in diameter. However, it should be recognized that the numerical term in Eq. (7.42) may change with changes in the clamp geometry. The clamp stiffness formula discussed here is based on the premise that the pipe is relatively thick, so that the interaction effects between the clamp and the pipe can be neglected.

The stiffness constant K_C units in Eq. (7.42) are given in N/cm. In English units, K_C is 47.4 lb/in.

In contrast with clamp design for thick pipes, the question of clamps operating on thin-walled piping subjected to high temperatures common to nuclear applications, demands that the structural interactions between the clamp and the pipe do not result in excessive stresses in the pipe wall. This becomes especially important in the applications of large diameter piping.

Pipe thinness and high thermal gradients require careful analysis involving insulating materials, clamp flexibility, and localized stresses.

One of the special findings applicable to thin-walled piping is concerned with the generally accepted criteria of cosine load distribution of contact pressure between the pipe and the clamp. Figure 7.15 presents a typical comparison between the cosine distribution model (CDM) and the finite element modeling (FEM).

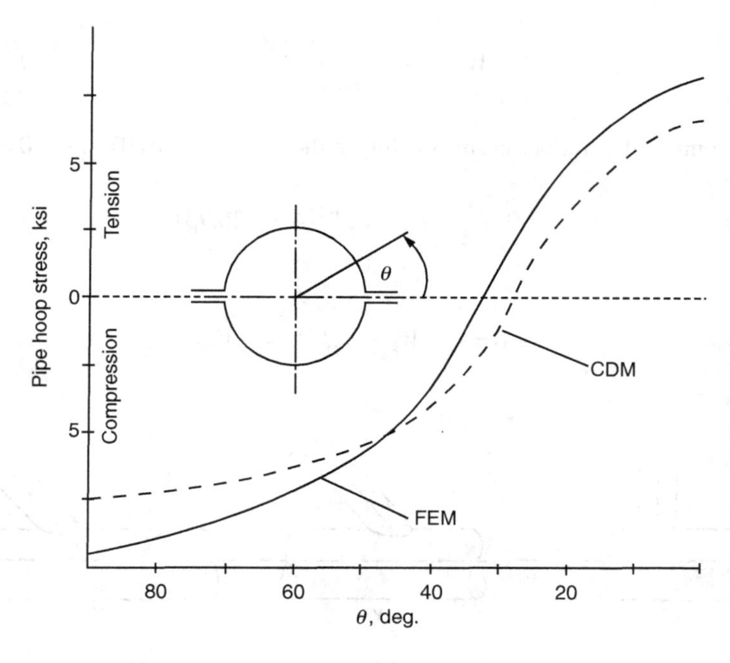

FIGURE 7.15 Approximate comparison of hoop stresses for clamp preload.

According to Anderson, Hyde, Wagner, and Severud [6] the cosine distribution of contact pressure does not provide a conservative estimate of hoop stress in a thin-walled, high temperature pipe at the clamp interface.

7.8 TUBULAR JOINTS

Circular tubular joints are used in many applications where the specific mechanical connections are made by full penetration welding. In essence, this area of technology covers the design of branched pipe joints known in industry as "T," "Y," "K" and other configurations.

Such joints can be subjected to axial tension, axial compression, in-plane as well as out-of-plane bending. The failure criteria may be defined in terms of the maximum load capacity, excessive deformation, or crack growth.

Since the theoretical prediction of a tubular branch capacity is very difficult, various industries have always relied on the development of empirical design formulas. A survey of this important area of joint mechanics [7] has resulted in a statistical study aimed at assuring a higher confidence level while maintaining a relatively simple format of the equations involved. The survey by Yura, Zettlemoyer, and Edwards includes the results of numerous investigations in the United States and elsewhere, representing professional associations, regulatory agencies, research institutes, and similar organization [7].

Figure 7.16 illustrates some of the typical joints. The capital letters: T, Y, and K denote the respective designs.

Current design formulas, recommended for the prediction of the ultimate capacity of the various tubular joints, may essentially involve some function of the d/D ratio denoted in this chapter by β.

The dependence of the ultimate axial or bending loading applied to the branch-part of the joint can be expressed in terms of sinθ, where θ denotes the acute angle between the branch and the main pipe body as shown in Figure 7.16. Figure 7.17 depicts the basic notation for a T-joint and Figure 7.18 shows basic loading conditions on a Y-joint.

For the case of axial tension of compression directed along the axis of the branch, the ultimate loading on the joint W_u can be calculated from the expression:

$$W_u = \frac{S_y T^2 (3.4 + 19\beta)}{\sin\theta} \tag{7.43}$$

For the specific joint configurations corresponding to the branch angles θ=π/4 or θ=π/2, we obtain:

$$\text{for } \theta = \frac{\pi}{4} \quad W_u = S_y T^2 (4.8 + 26.9\beta) \tag{7.44}$$

and

$$\text{for } \theta = \frac{\pi}{2} \quad W_u = S_y T^2 (3.4 + 19\beta) \tag{7.45}$$

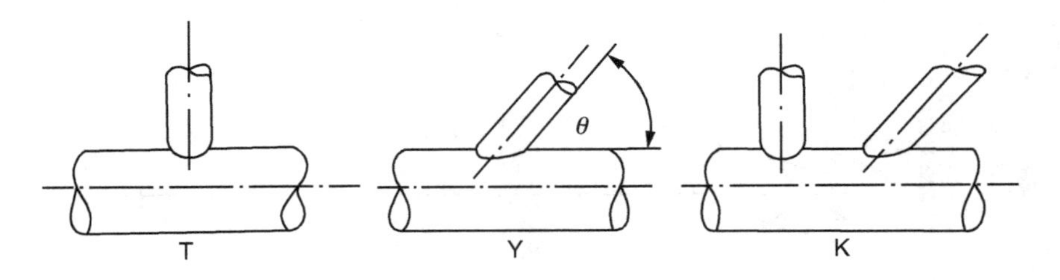

FIGURE 7.16 Typical tubular joints.

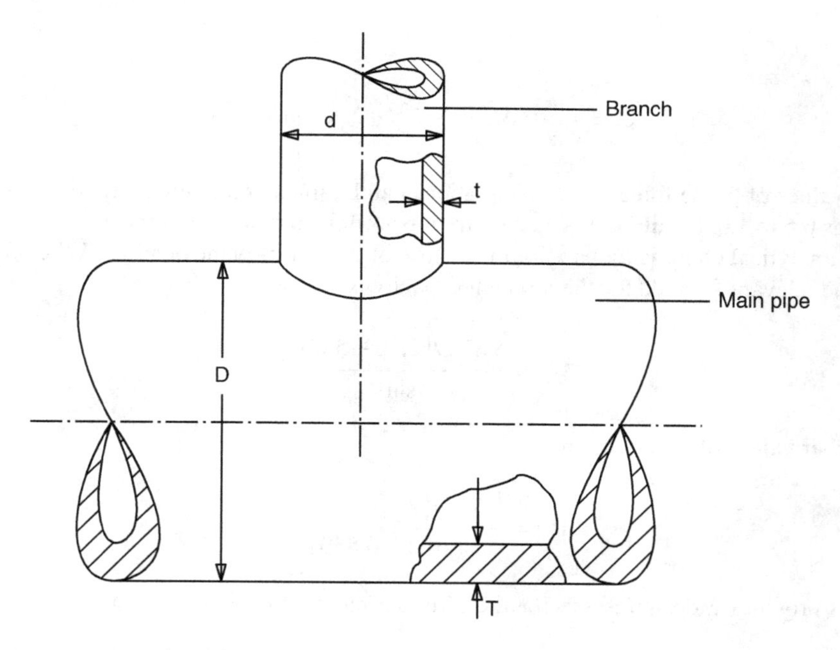

FIGURE 7.17 Basic notation for a T-joint.

It is noted that in the design for axial loading, the ultimate capacity is linear with respect to β (where $\beta = d/D$). This parameter can vary between 0.2 and 0.9. While the theoretical limits are 0 and 1.0, the majority of industrial configurations, designed for all kinds of loading, never reach the extreme proportions.

When the external bending moment is applied in the plane of the joint, the ultimate moment capacity for the two branch angles, θ=π/4 and π/2, respectively, can be calculated as:

$$\text{for } \theta = \frac{\pi}{4} \quad M_u = S_y T^2 D\left(3.8\beta + 21.5\beta^2\right) \tag{7.46}$$

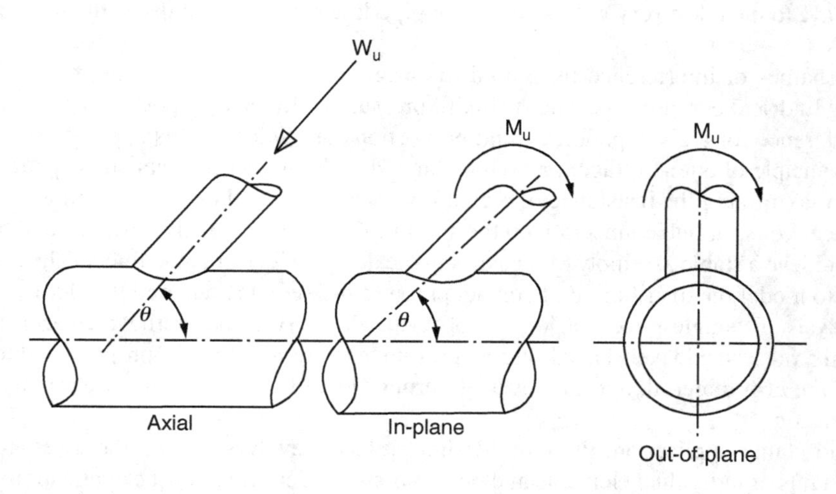

FIGURE 7.18 Basic loading conditions for a Y-joint.

and

$$\text{for } \theta = \frac{\pi}{2} \quad M_u = S_y T^2 D\left(2.7\beta + 15.2\beta^2\right) \tag{7.47}$$

For small values of β, the foregoing two equations can be linearized. However, with higher values of β, such as 0.6 to 1.0, the ultimate capacity trends become distinctly nonlinear.

Certain analytical complications arise in the case of the out-of-plane bending. When β is smaller than 0.6, the relevant formula for the moment capacity is:

$$M_u = \frac{S_y T^2 D\left(2.7\beta + 5.6\beta^2\right)}{\sin\theta} \tag{7.48}$$

For the higher values of β, we obtain

$$M_u = \frac{S_y T^2 D\left(2.7\beta + 5.6\beta^2\right)}{\sin\theta\left(1 - 0.83\beta\right)} \tag{7.49}$$

Figure 7.19 provides a design guide for the ultimate moment capacity based upon the foregoing equations.

It is normally assumed that the characteristic D/T ratios intended for the foregoing equations should vary between 20 and 100. Yura et al. [7] suggest that the effect of this ratio on the ultimate joint capacity in tubular design is rather small. The foregoing equations represent a lower bound capacity for the majority of joint configurations and types of loading. By a comparison with the API standards [8], the design formulas given in this section [7] should consistently predict lower joint capacities.

The question of the appropriate safety factors, however, particularly under the combined branch loading, is difficult to resolve because of the complications involved in predicting the reserve of joint strength in the plastic region.

7.9 SUMMARY

Countless joints found in industrial applications are held together by means of conventional rivets, bolts, welds, and adhesives. The joints can be permanently fixed or detachable. In the latter category, we have to include a very wide selection of pipe-line supports, clamps, split hubs, and similar connections.

The mechanics of interference fit is used to design and to assemble strong mechanical joints involving cylindrical components. The shrink-fit pressure is directly proportional to the amount of radial interference for a given geometry and proportions of the joint. This application leads to the so-called principle of auto-frettage, involving planned yielding and residual stress patterns for the purpose of maximizing the resistance of pressure vessels to internal pressure loading.

The concept of split hub connection on the shaft utilizes the frictional resistance and bolt torque control to achieve a stable assembly of the parts in the design of certain machinery. The main design problem is to model the distribution of contact pressure between the hub and the shaft.

The analysis of a single-pin clevis joint involves the elasticity of the cantilevered arms and a bolt, torqued to the prescribed level. The allowable pull-up deflection is directly proportional to the yield strength of the arm material, and the inversely proportional to the modulus of elasticity for given clevis dimensions.

In a rigid clamp application, the basic design problem revolves around the assessment of the spring constants of individual elements and bolt response under initial torque application. The main parameter is the ratio of spring constant of the bolt and that for the clamped component.

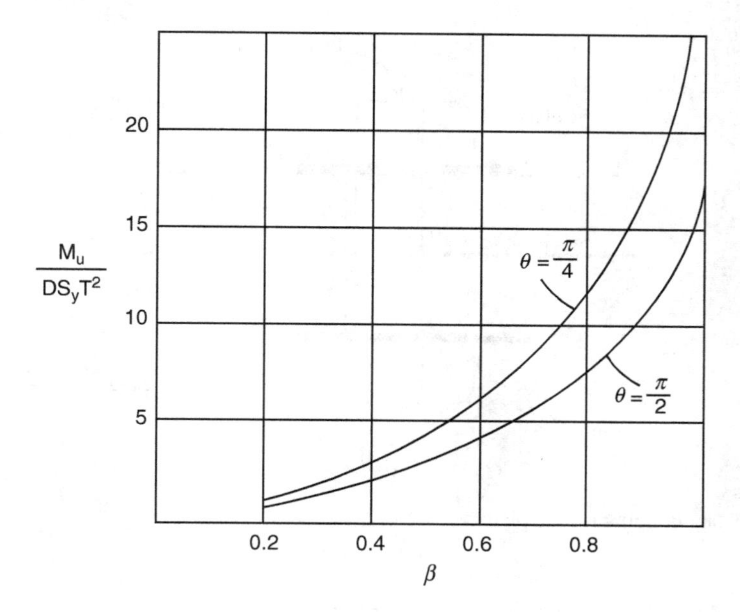

FIGURE 7.19 Moment capacity for out-of-plane loading.

A C-clamp can be analyzed as a curved structural member utilizing the conventional sectional properties and a correction for the displacement of neutral axis. The stresses can also be approximated using a correction factor, based on the curved beam tables available in the design literature.

A special area of mechanical joint technology involves the design and testing of pipe supporting brackets. The response of piping to localized bracket loading can be analyzed on the basis of external ring load under conditions of zero internal pressure or full internal pressure in the pipe.

In most cases, the mechanical joint, created by the bracket and the pipe wall, depends on the elastic response of the ring and the cylinder in contact, without the need for a weld or adhesive material.

The design of piping brackets and the restraint mechanisms, however, is a very complex area of engineering requiring modeling of linear and nonlinear characteristics of the clamps by means of finite element techniques.

The majority of pipe clamps are of the yoke beam or ring type involving highly nonlinear distribution of interface pressure and the resulting hoop stresses. The forces developed in the pipe clamps are caused by the combined effect of external loads, internal systems pressure, and thermal gradients. The conventional idea of the cosine distribution of contact pressure, although highly convenient, may not provide a conservative estimate of hoop stresses in a thin-walled piping.

The design of tubular joints, popular in many branches of industry, is concerned with the failure criteria of T, Y, K, and other geometrical configurations under in-plane and out-of-plane loading conditions. The design formulas developed by regulatory agencies and industry are relatively simple and represent lower-bound envelopes in order to guard against fabrication and operational uncertainties. The major design problem, however, is the selection of factors of safety and the assessment of plastic strength under the combined loading conditions.

7.10 DESIGN ANALYSES

EXAMPLE ANALYSIS 1

Two plastic plates are being clamped together by a metal bolt and washers as represented in Figures a and b.

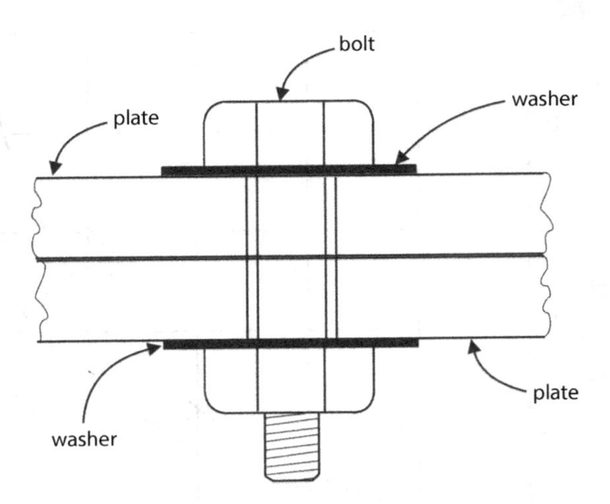

FIGURE A Bolt clamping plastic plates.

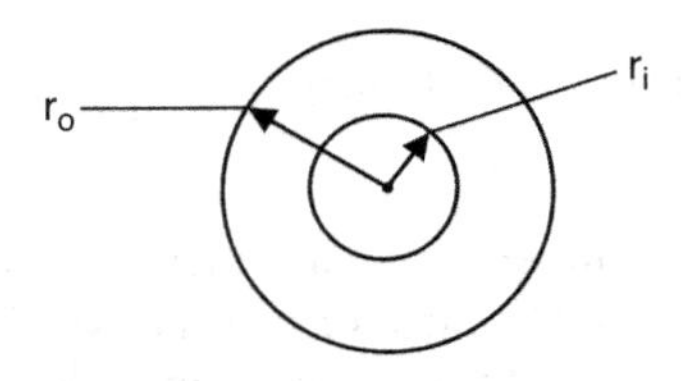

FIGURE B Washer profile and dimensions.

Suppose that the bolt shank diameter is 1.0 cm, and that the washer radii are: $r_o = 1.25$ cm and $r_i = 0.625$ cm. Let the maximum stress allowed on the plastic be: $\sigma_{max} = 7$ MPa.

Find the maximum tensile force in the bolt shank and correspondingly the maximum stress in the bolt shank.

Solution

Imagine a free-body diagram of a washer. The force F exerted on the washer by the bolt head or bolt nut will be balanced by a corresponding force F exerted by a plate on the washer. Consequently, the contact stress σ induced by the washer on the place is simply:

$$\sigma = \frac{F}{A_w} \tag{a}$$

where A_w is the washer area.

From Figure b and the given dimension for the washer A_w is:

$$A_w = \pi\left(r_o^2 - r_i^2\right) = \pi\left[(1.25)^2 - (0.625)^2\right] = 3.68 \text{ cm}^2 = 3.68 \times 10^{-4} \text{m}^2 \tag{b}$$

With the maximum stress σ_{max} on the plastic being 7 MPa, Eq. (a) yields the bolt force F as:

$$\begin{aligned} F &= \sigma_{\max} A_w = \left(7 \times 10^6\right)\left(3.68 \times 10^{-4}\right) \\ &= 2.576 \times 10^3 \text{ Pa} = 2.576 \text{ kN} \end{aligned} \tag{c}$$

Finally, the bolt shank stress σ_b is simply:

$$\sigma_b = \frac{F}{A_b} \tag{d}$$

where A_b is the cross-section area of the bolt shank. From the given bolt diameter d_b dimension A_b is:

$$\begin{aligned} A_b &= \left(\frac{\pi}{4}\right)d_b^2 = \left(\frac{\pi}{4}\right)(1.0)^2 \\ &= 0.7853 \text{ cm}^2 = 7.853\times10^{-5} \text{ m}^2 \end{aligned} \tag{e}$$

Therefore, by substituting the results of Eqs. (c) and (e) into (d) we find the bolt stress σ_b to be:

$$\begin{aligned} \sigma_b &= \frac{\left(2.576\times10^3\right)}{\left(7.853\times10^{-5}\right)} \text{ Pa} \\ &= 32.8\times10^6 \text{ Pa} = 32.8 \text{ MPa} \end{aligned} \tag{f}$$

EXAMPLE ANALYSIS 2

See the problem statement and the solution to Example Analysis 1. Suppose that the bolt head in contact with the washer has an average radius of 1.2 cm. Determine the stress on the washer.

Solution

The stress on the washer is simply:

$$\sigma_w = \frac{F}{A_{wb}} \tag{a}$$

where A_{wb} is the area of the washer in contact with the bolt head. Figure a shows a profile of the washer in contact with the bolt head, where r_{ob} is the average bolt head radius.

From the given data r_i is 0.625 cm and r_{ob} is 1.2 cm. Therefore, the washer/bolt contact area A_{wb} is:

$$A_{wb} = \pi\left(r_{ob}^2 - r_i^2\right) = \pi\left[(1.2)^2 - (0.625)^2\right]$$

or

$$A_{wb} = 3.297 \text{ cm}^2 = 3.297\times10^{-4} \text{ m}^2 \tag{b}$$

From the solution of Example Analysis 10, F is seen to be:

$$F = 2.576 \text{ kN} \tag{c}$$

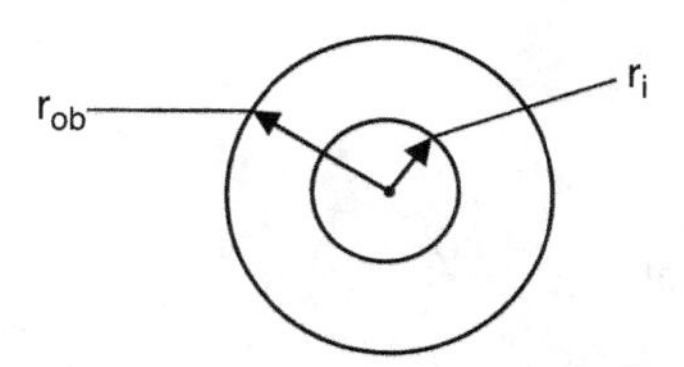

FIGURE A Washer profile and dimensions in contact with the bolt head.

Finally, by substituting from Eqs. (b) and (c) into (a) the washer stress σ_w is:

$$\sigma_w = \frac{2.576}{\left(3.297 \times 10^{-4}\right)} = 7.81 \text{ kPa} \tag{d}$$

REFERENCES

1. Dobrovolsky, V., K. Zablonsky, S. Make, et al. 1968. *Machine Elements.* Moscow: Mir Publishers.
2. Harvey, J. F. 1967. *Pressure Vessel Design: Nuclear and Chemical Applications.* Princeton, NJ: D. Van Nostrand Company, Inc.
3. Huston, R. L. and H. Josephs. 2009. *Practical Stress Analysis in Engineering Design*, 3rd Ed. Boca Raton, FL: CRC Press.
4. Shigley, J. E. 1963. *Mechanical Engineering Design.* New York: McGraw-Hill.
5. Blake, A. 1979. *Design of Curved Members for Machines.* Huntington, NY: Robert E. Krieger.
6. The American Society of Mechanical Engineers. 1980. *Effects of Piping Restraints on Piping Integrity*, PUP-40. New York: The American Society of Mechanical Engineers.
7. Yura, J. A., N. Zettlemoyer and I. F. Edwards. 1980. Ultimate capacity equations for tubular joints. OTC 3690, Offshore Technology Conference. Houston, TX.
8. American Petroleum Institute. 1977. *Recommended Practice for Planning, Designing and Constructing Fixed Offshore Platforms.* RP2A, 4th Ed., 1973, 9th Ed., 1977.

SYMBOLS

A	Cross-sectional area
A_b	Nominal cross-section of bolt
A_t	Cross-sectional area of tube
Ar	Area of ring cross-section
a	Inner radius
B	Width of section
b	Outer radius of tube; contact radius
c	Auxiliary symbol for a radius
D	Outer diameter of main pipe
d	General symbol for diameter
E	Modulus of elasticity
E_a	Modulus of elasticity for aluminum
e	Auxiliary dimension
F	Contact force
f	Coefficient of friction
H	Depth of section
I	Second moment of area
K_b	Spring constant for bolt
K_c	Clamp stiffness
K_t	Spring constant for tube
L	Length
L_a	Length of tube
L_b	Length of bolt
M_t	External torque
M_{tf}	Frictional torque
M_u	Ultimate bending moment
N	Number of bolts
n	Displacement of neutral axis

P	Shrink fit pressure
P_e	Equivalent contact pressure
P_i	Internal pressure
P_{max}	Maximum contact pressure
P_o	Circumferential loading
R	Radius to center of gravity
R_i	Inner radius of clamp
r	Mean radius of tube
S	General symbol for stress
S_b	Bending stress
S_r	Radial stress
S_t	Tangential stress
S_y	Yield strength
T	Thickness of main pipe
t	Wall thickness of tubing
V	Bolt force
W	External load
W_b	Actual bolt load
W_i	Bolt preload
W_u	Ultimate axial load
$x_1 x_2$	Arbitrary distances
Y	Deflection
Y_r	Radial deflection
$B = d/D$	Diameter ratio
δ	Radial interference
ΔL_a	Change in tube length
ε	Shear distribution factor
θ	Arbitrary angle
$\phi = K_t/K_b$	Ratio of spring constants
ϕ_o	Stress correction factor
τ_{max}	Maximum shear stress

8 Joint Connections
Pins, Couplings, and Other Joint Fittings

8.1 INTRODUCTION

There are a number of industrially important joint fittings wherein a primary consideration for successful operation involves a detailed study of the various and often not obvious stresses applied. Some of the couplings are almost ubiquitous in their application with years of existing experimental data and standards applications for use in machine design and practical applications. Others of the topics discussed address a specific application or industrial use but nevertheless whose industrial importance requires analysis and discussion. Included in this chapter of apparently disparate topics are:

Cotter Pin Joints
Key Connections
Pipe Couplings
Abutment Failure
Eyebar or Knuckle Joints
Structural Pins
Chain Drives and Coupling Links
Wire Rope Fittings
High-Pressure Threaded Configurations

8.2 DESIGN OF COTTER PIN JOINTS

Joints with cotter pins have, over the years, served in applications where simplicity, rapid assembly, and disassembly were of importance. Typical areas of use in the past included piston rod connections, engine crossheads for steam and internal combustion engines, as well as special elements of tool fixtures. Only such features as through-holes, grooves, and locking devices prevented cotter joints from being used on a very extensive scale. Nevertheless, the design characteristics of cotter joints are applicable to numerous practical situations and justify our inclusion of these configurations in the general area of mechanical joint design.

Figure 8.1 depicts a typical assembly of the unstrained components of a cotter joint resisting loads of constant magnitude. The shank contains a through type opening with rounded ends to accept the key of a similar cross section.

In considering the essential features of a cottered joint, we find that the bearing surfaces of the interacting elements of the joint have to be rounded off to reduce abrupt geometrical transitions and to improve stress patterns.

The cotter keys may involve single or double taper, although preference is given to single-tapered keys for manufacturing reasons. The customary taper lies within the values of 0.01–0.20. For the design of more frequently disassembled joints, some past practices indicate a range of taper of 0.1–0.2. On the other hand, 0.01–0.05 taper limits may be more suitable for more permanent cottered

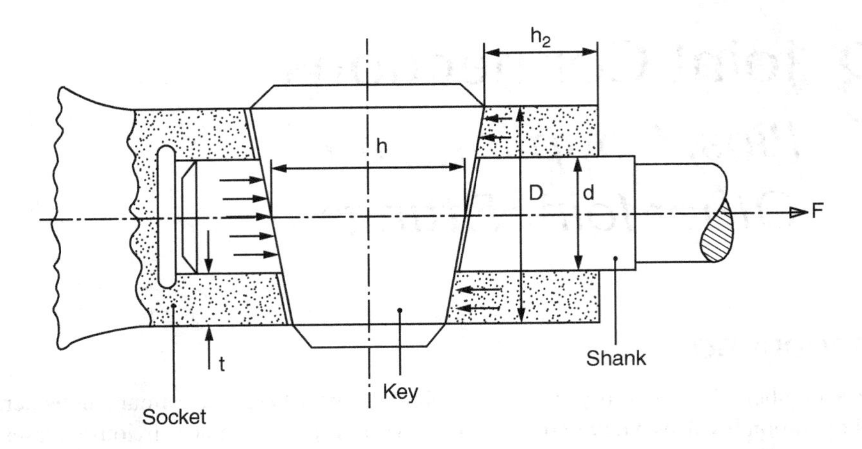

FIGURE 8.1 Side view of a cotter pin joint.

fastenings. The final choice of the design range of the taper will depend on the frictional characteristics of the joint material and the desired frictional removal resistance.

In general, the cottered fastenings can be assembled for either unstrained or prestressed conditions, depending on the magnitude and the manner of load application.

Figure 8.2 provides a view of the shank and its opening for the pin.

An external force F, applied along the shank axis, is resisted by the compression of the key and the extension of the openings in the socket.

The nominal tensile force F on the shank can be expressed as:

$$F = d\left(\frac{\pi d}{4} - b_0\right)S_t \tag{8.1}$$

where S_t is the tensile stress and where Figure 8.2 shows the diameter d and the slot width b_0. The compressive stress S_C applied to the cotter key is then:

$$S_C = \frac{F}{b_0 d} \tag{8.2}$$

Assuming that $S_C = 1.5S_t$ for the usual metallic materials, Eqs. (8.1) and (8.2) show that the ratio of b_0/d should be on the order of 0.3. The allowable shear stress for the shank can be taken as $S_t/\sqrt{3}$, using the rule of thumb that shear strength ≅ tensile strength/$\sqrt{3}$. Since there are two shear planes resisting the external load F, we obtain:

$$F = 2h_i d\,\frac{S_t}{\sqrt{3}} \tag{8.3}$$

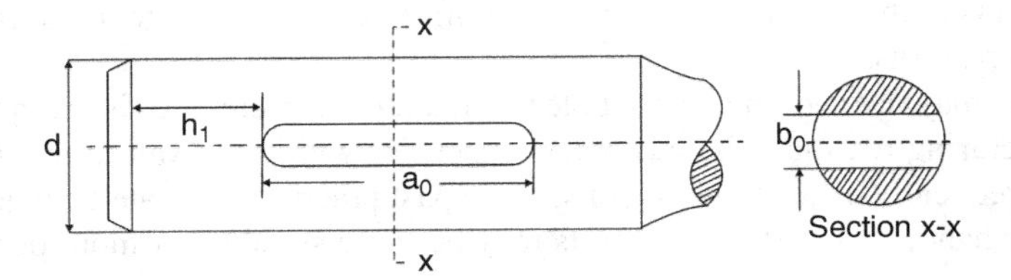

FIGURE 8.2 The shank and pin channel.

where Figure 8.2 shows the dimension h_1. Note this expression does not account for the slot a_0b_0. Hence the *allowable* shear stress should be lower.

For $S_C = 1.5S_t$ and $b_0 = 0.3d$, Eqs. (8.2) and (8.3) yield $h_1 = 0.4d$. The usual ratio h_1/a_0 should be about 0.5–1.0, where again Figure 8.2 shows all the relevant dimensions.

To avoid crushing the bearing surfaces of the female socket, the following design relation should be considered:

$$F = 2b_0tS_C \tag{8.4}$$

Observe that t defines the thickness of the socket wall which is equal to $(D-d)/2$. Hence, the external socket diameter can be defined with the aid of Eq. (8.4) as:

$$D = d + \frac{F}{b_0S_C} \tag{8.5}$$

When the material for the shank and socket is the same Eqs. (8.2) and (8.5) give $D = 2d$.

The strength of the socket in tension is governed by the relation:

$$S_t = \frac{4F}{\pi\left(D^2 - d^2\right) - 8b_0t} \tag{8.6}$$

By setting $D = 2d$ and $b_0 = 0.3d$, Eq. (8.6) reduces to:

$$S_t = \frac{F}{2.4d^2 - 0.6dt} \tag{8.7}$$

The allowable design load under these conditions is:

$$F = 2.4d(d - 0.25t)S_t \tag{8.8}$$

If the design stress calculated from Eq. (8.7) is significantly lower than the allowable tensile stress S_t, then the socket diameter can be decreased locally without drastically affecting the stress distribution in compression due to the key effect. The limiting dimension in this case is h_2, as shown in Figure 8.1.

The shear-out strength of the socket edges can be expressed here as:

$$F = 2.3th_2S_t \tag{8.9}$$

It is often customary to make $h_1 = h_2$ for this type of cotter joint, so that Eq. (8.9) can be restated as:

$$F = 0.92dtS_t \tag{8.10}$$

The cotter key depth h can be calculated on the assumption of beam bending, provided the ratio of h/D is smaller than about 0.5. It is also reasonable to assume that the contact pressures between the key, shank, and the socket are relatively uniform. Figure 8.3 illustrates the design model for this stress condition.

The bending moment at the center of the cotter key is:

$$M = \frac{F}{8}(d + 2t) \tag{8.11}$$

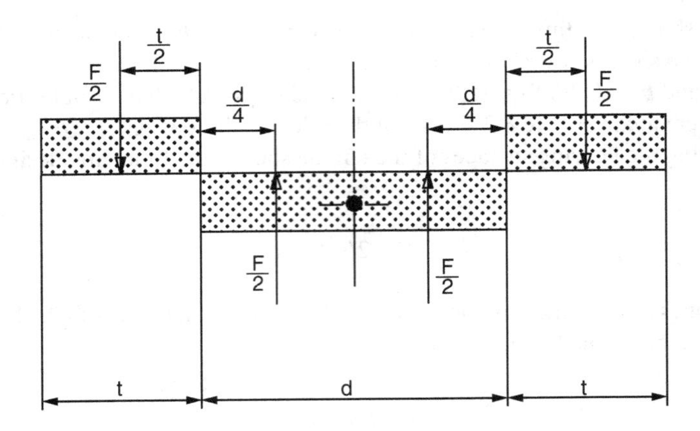

FIGURE 8.3 Load diagram for cotter key.

Since, according to the classical theory of flexure, we can define:

$$S_b = \frac{M}{Z} \tag{8.12}$$

where

$$Z = b_0 \frac{h^2}{6} \tag{8.13}$$

then, combining Eqs. (8.11) through (8.13), and observing that for the socket geometry $2t = D{-}d$ yields:

$$h = 0.5\left[\frac{3FD}{b_0 S_b}\right]^{0.5} \tag{8.14}$$

Introducing, for example, $b_0 = t = 0.3d$ into Eq. (8.14), gives:

$$h = 2\left(\frac{F}{S_b}\right)^{1/2} \tag{8.15}$$

Cotter keys designed in this manner are usually made from steel. Figure 8.1 shows the corresponding dimension h. Again, the assembly in Figure 8.1 is assumed to be unstrained and it is intended for resisting loads of constant magnitude.

When, however, the joint is prestressed, the cotter key is assembled in such a manner as to cause some initial stresses before the external load is applied. These residual stresses are needed for the dependable operation of the cottered fastening. Since the problem of the residual effects is difficult to solve, it is generally acceptable to size the joint components according to the basic rules of the unstrained cotter joints, on the premise that the design load can be made 25% higher than the maximum expected external load [1].

Figure 8.4 shows an alternative to a flat cotter key by employing a cylindrical component with or without a taper.

If instead of a flat cotter key, the joint employs a cylindrical component with or without a taper, various designs are possible, such as those indicated in Figure 8.4. When taper is used, the

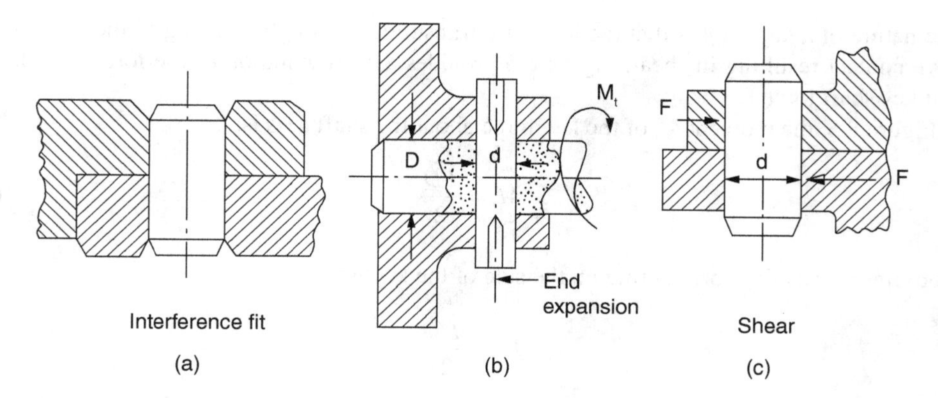

FIGURE 8.4 Cylindrical pin applications.

customary ratio is about 1/50 allowing better disassembly. Cylindrical pins can be held by interference and frictional effects as indicated in Figure 8.4a. Other applications may feature cylindrical or taper pins with grooves formed by cutting operations.

Yet another design, such as that shown in Figure 8.4b, can be locked into a given position by the forced opening of the pin ends thereby expanding the pin diameter at its ends.

For the case of shear, as in Figure 8.4c, the pin diameter can be given by the expression:

$$d = 1.13\left(\frac{F}{\tau}\right)^{1/2} \tag{8.16}$$

where τ represents the allowable shear strength of the pin material which can be approximated as $S_t/\sqrt{3}$, where S_t is the corresponding strength in tension.

When the pin joint transmits the torque M_t (where $M_t = FD$), as shown in Figure 8.4b, then the pin diameter can be calculated from (8.16) as:

$$d = 1.13\left(\frac{M_t}{D\tau}\right)^{1/2} \tag{8.17}$$

where D represents shaft diameter transmitting torque M_t. Cylindrical and tapered pins are sometimes made from a high-strength tool steel. Typical applications may be found in the design of slipping clutches, intended as control mechanisms for mechanical overloads. This type of a control is possible if the pin can be designed to fail in a brittle manner at a prescribed shear load. Here we have the special case of materials application where a controlled brittle failure is used as a safety device.

8.3 KEY CONNECTIONS

Many machines involve a combination of the shaft and wheel-type assemblies in such applications as gears, pulleys, discs, wheels, and hubs. The problem of joining such members can be reliably solved by the use of a key joint designed to transmit the torque. Such a joint should be relatively inexpensive and suitable for relative ease of assembly and disassembly.

Key joints, in general, have evolved into splined connections with both types being subject to complicated load and stress conditions. Although further developments of this group of mechanical joints have led to the appearance of keyless connections, there are still many applications where the conventional key and spline designs are optimal [2].

The nature of a key joint is that the loads are transmitted through crushing of the sides in compressive contact resulting in shear. Figure 8.5 provides a representation of the forces exerted in a typical key/shaft joint.

In Figure 8.5, the moment M_i of the key force about the shaft axis is:

$$M_t = \frac{FD}{2} \tag{8.18}$$

The maximum crushing force acting on the side of the key is:

$$F = \frac{hLS_C}{2} \tag{8.19}$$

where L is the length of the key and S_C is the design value of the compressive stress allowed in this particular design.

Combining Eqs. (8.18) and (8.19) gives a convenient formula for the design length L of the key.

$$L = \frac{4M_t}{hDS_C} \tag{8.20}$$

Alternatively, the condition of a shear transfer through the key given the design length as:

$$L = \frac{2M_t}{b_0 D_\tau} \tag{8.21}$$

From a design perspective, the larger of the two values of L, in Eqs. (8.20) and (8.21), should be selected, provided, of course, that L is not greater than the hub length.

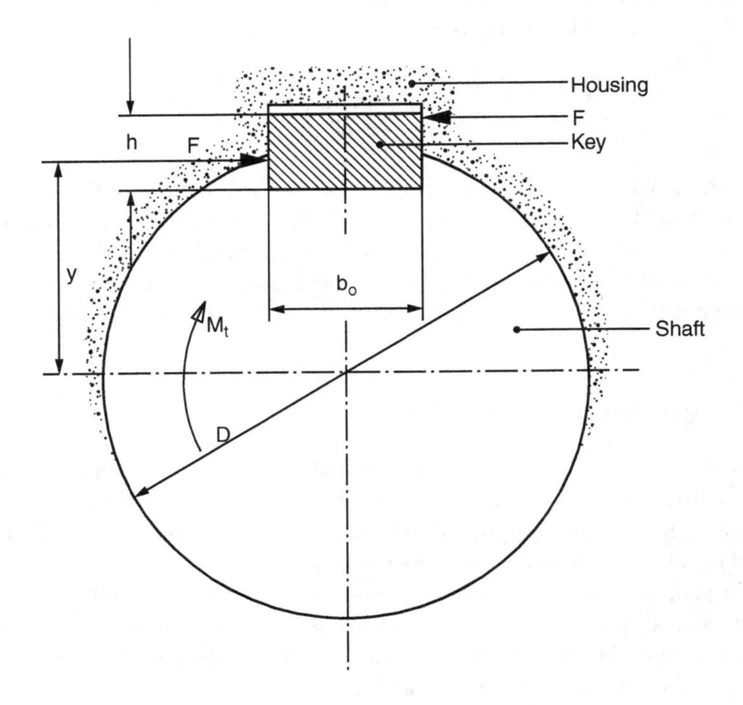

FIGURE 8.5 Forces in a key joint.

It may also be noted that for $S_C = 1.5S_t$ and $\tau = S_t/\sqrt{3}$ the foregoing expressions yield $b_0 = 1.3h$. When the length of the hub is insufficient, as judged on the basis of the selected value of L, two keys should be employed and placed 180° apart.

The key joint of Figure 8.5 can be used as a convenient model for explaining the basic mechanism of load transfer from the shaft to the hub. The relevant joint configuration may depend on a number of requirements, including the direction of the applied torque. This is clear from the observation of Figure 8.6, where one of the designs is reversible. Sunken and flat keys used in this case are generally of a rectangular cross-section.

Other key applications can be characterized by a taper ratio of about 1/100. They can also act in pairs as wedges which, however, can cause significant contact pressures on the bearing surfaces of the keyway and the key.

The reversible key-joint design, of Figure 8.6, can be based on the 90°–120° offset of the shaft bearing surfaces. The length of the keys in general should be matched to the length of the hub as closely as possible to avoid any hub misalignment. This appears to be particularly important when relatively short hubs have to be fastened by means of taper keys. Greater accuracy can normally be achieved through the use of a straight key, as in Figure 8.5, or through the application of an unstrained joint based on the so-called Woodruff key, as illustrated in Figure 8.7.

The compressive strength criterion for the Woodruff key can be stated as:

$$L = \frac{2M_t}{D(m + h - D)S_C} \tag{8.22}$$

Figure 8.8 illustrates the principle of a strained-key joint design. Since this connection is formed by means of a taper key, a radial thrust is created through the interaction between the hub keyway and the shaft.

In the joint assembly the normal pressure is created by driving the key with a force P. The magnitude of the normal force, which determines the extent of the contact pressure on the key, can be calculated using the expression:

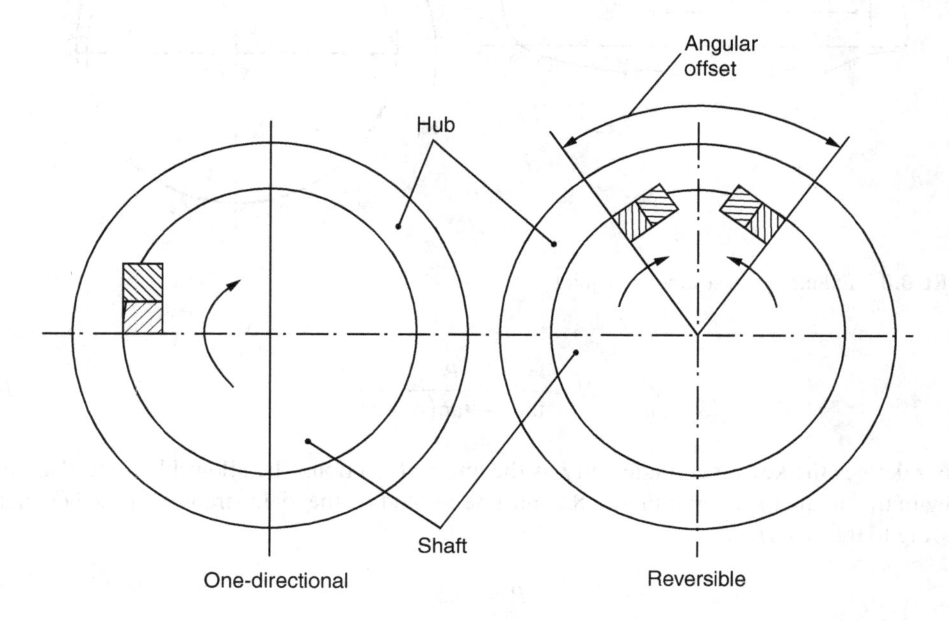

FIGURE 8.6 Tangential key design.

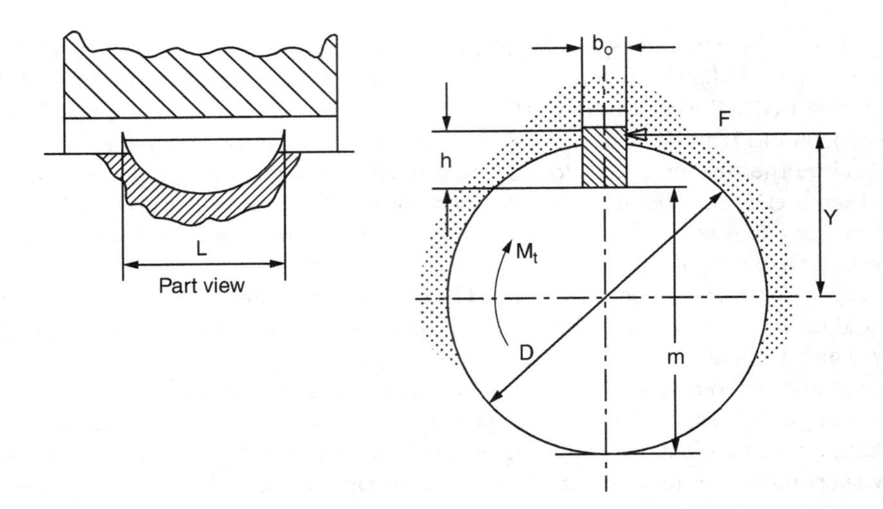

FIGURE 8.7 Woodruff key design.

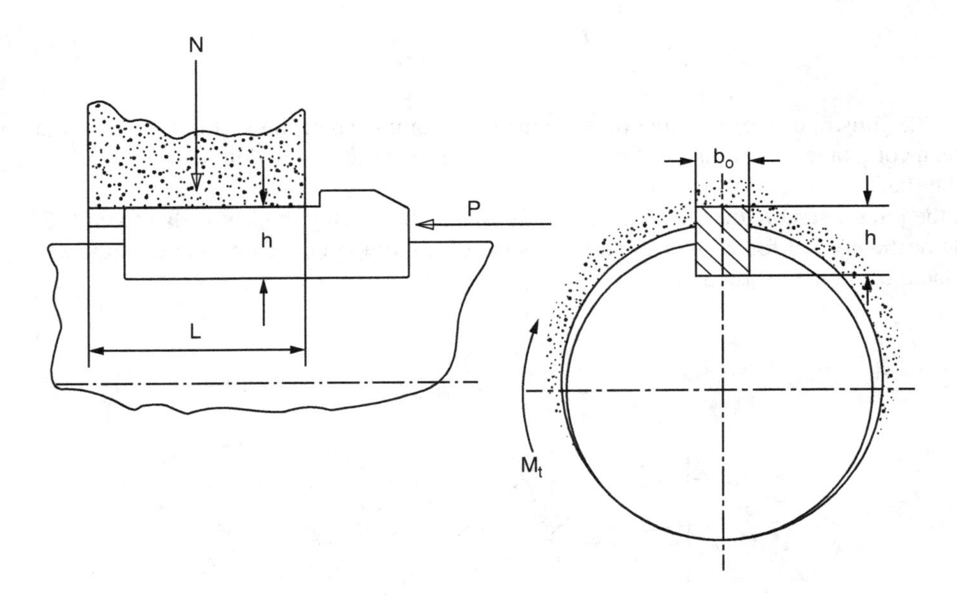

FIGURE 8.8 Example of a strained key joint.

$$N = \frac{P}{\tan\rho + \tan(\alpha + \rho)} \tag{8.23}$$

where α defines the key taper angle and ρ is the angle of friction. The allowable force P applied to the key in the manner shown in Figure 8.8 must be limited by the yield strength of the key material according to the criterion:

$$P = b_0 h S_y \tag{8.24}$$

To analyze the joint shown in Figure 8.8, consider, for example, the equilibrium of an element of the key subjected to the forces of the taper as postulated in Figure 8.9. Taking moments about the center of the key cross-section leads to the expression:

$$N\varepsilon - fNh = 0 \tag{8.25}$$

Solution of Eq. (8.25) gives $\varepsilon = fh$, where f is the coefficient of friction. The reactive forces and moments resulting from the applied moment and the assembly load P can be shown to act on the shaft in the manner as illustrated in Figure 8.10.

It is convenient here to assume that the pressure distribution on the lower half of the shaft surface varies as the cosine of the angle measured from the vertical diameter as shown in Figure 8.10. This gives:

$$q = q_{max} \cos\theta \tag{8.26}$$

At $\theta = 0$, $q = q_{max}$, and at $\theta = \pi/2$, $q = 0$ which agrees with the assumed load distribution at the end points. Then the relation between the external load N and the maximum unit pressure q_{max} can be obtained from the integration:

$$N = q_{max} \int_0^{\pi/2} DL \cos^2\theta \, d\theta \tag{8.27}$$

Solution of Eq. (8.27), yields:

$$N = \frac{\pi LD}{4} q_{max} \tag{8.28}$$

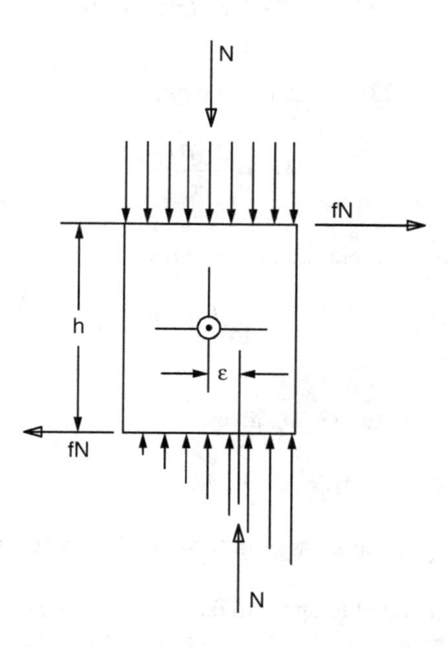

FIGURE 8.9 Elemental model of key equilibrium.

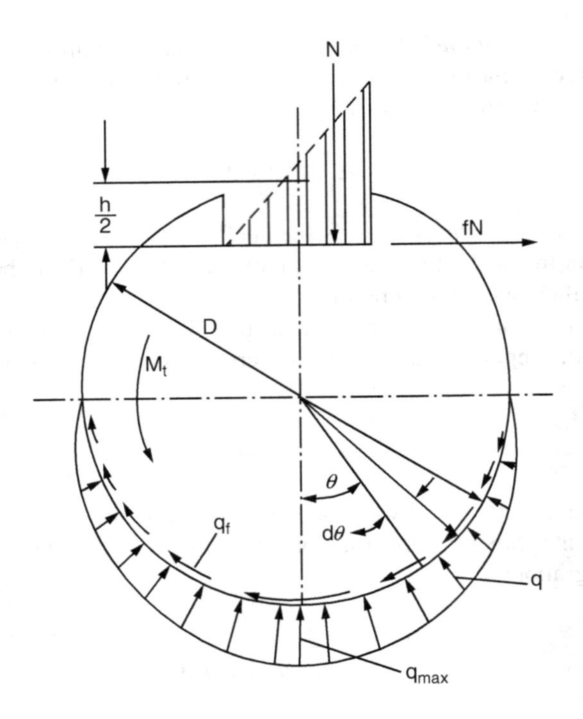

FIGURE 8.10 Equilibrium of forces on the shaft surface.

Since the elementary frictional force is now $q_f = qf$, where f is the coefficient of friction, the total frictional moment M_f can be defined as:

$$M_f = \frac{q_{max}}{2}\int_0^{\pi/2} D^2 Lf \cos\theta \, d\theta \tag{8.29}$$

This integration and use of Eq. (8.28) results in the expression:

$$M_f = \frac{2NfD}{\pi} \tag{8.30}$$

The sum of all the moments taken about the center of the shaft is:

$$M_t = M_f + fN\frac{(D-h)}{2} + N\varepsilon \tag{8.31}$$

Recalling that $\varepsilon = fh$, and utilizing Eq. (8.30), gives:

$$M_t = fN(1.14D + 0.5h) \tag{8.32}$$

To protect against the crushing of the keyway in the hub, the appropriate limit should be assumed for the normal force N.

The foregoing analysis is intended to again make the point that even the simplest joint configuration cannot be represented by a single formula or a simple theory. The calculations did not even begin to address the complex problem of the torsionally induced shaft stresses due to the presence of a keyway [3].

8.4 SPLINES

The analysis of an infinitely long groove in a circular shaft was first attempted more than 100 years ago. Several experimental attempts have been made over the years using the two-dimensional models which, however, could not accurately describe the real stress distribution. It was not until 1975, when this problem was addressed with the aid of a three-dimensional photoelastic model. The experimental approach still persists because of insurmountable mathematical difficulties in obtaining a reliable solution to the boundary value problem posed by a keyway of finite length.

The failures traced to the presence of a keyway seem to be due to a circumferential crack; subsurface peeling-type of a fracture on a cylindrical surface; or a 45° helix crack propagation observed in certain spline connections.

Multiple splines fall essentially in the same area of joint mechanics as that dealing with the keyed joints. These can be of a sliding or a fixed type. The action of a fixed spline is similar to that of a taper key. Multiple splines can be found in automobiles, tractors, and other machines. The method of assembly depends on the required accuracy with which the hub is to be fitted onto the shaft.

Better centering of the spline should, of course, permit a better distribution of forces. Spline tooth cross-section types can include straight sides, and triangular patterns or involute curves. In general, the involute splines are superior for the following reasons:

- Involute shape avoids abrupt changes in geometry and assures lower stress concentration factors.
- Production methods of involute shape are well developed and more economical.
- Involute splines permit better centering of the assembled parts.

The limit of the transmitted spline moment is often based on the crushing strength of the surfaces in contact as:

$$M_t = R_m LAbnS_C \tag{8.33}$$

where R_m denotes the spline mean radius, A is the area of bearing surface per unit length and n defines the number of splines. The effective length of the spline L can be taken equal to the length of the hub. However, the correct value of the compressive strength S_C should be found experimentally for a particular spline design.

8.5 PIPE COUPLINGS

Current deep underground and deep sea explorations require long, reliable piping strings. These strings, often two to four miles long, are held together by threaded connections. It is important that such connections represent an acceptable sealing mechanism and permissible design stresses. The stresses can be caused by tensile, burst, or collapse loading.

There are indications that 50%–80% of all the string failures can be traced back to some aspect of a pipe connection [4]. The pipe coupling requirements in modern industry may include leak proof performance at gas pressures exceeding 10,000 psi. It is also reported that the make-up torque is not the only variable in the mechanism of leakage control. It is important to meet the connection make-up requirements and to control the make-up stresses. Much effort, therefore, has gone into the development of devices suitable for field control of the make-up variables.

There are at least six types of threaded connections employed by the drilling and oil industries. Figure 8.11 provides a partial view of a typical tubular connection.

The thread specifications of the American Petroleum Institute range from round to buttress type thread forms. In round thread applications the clearance between the crests and roots of the threads can be on the order of 0.003 in. Soft tin or zinc plating of coupling threads should provide additional

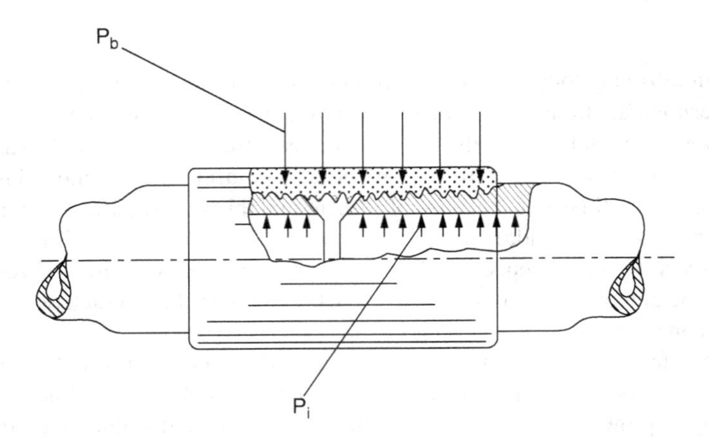

FIGURE 8.11 Example of tubular coupling.

filler for load distribution and some protection against thread galling. Figure 8.12 illustrates a typical buttress form for this purpose.

Figure 8.13 provides a sketch of a special thread form used in a pipe-coupling connection known as the "8-ACME" thread.

It appears that essentially all thread forms such as round, buttress, and ACME type, require a substantial bearing pressure between the contact surfaces to assure a good mechanical joint.

The sealing is normally attained through a metal-to-metal contact. When large clearances are employed, the threads are not designed to function as seals and separate secondary sealing provisions are utilized such as resilient Teflon rings. These rings can also form a barrier against corrosion. The rings are inserted in grooves machined into the threaded surface of the coupling.

The analysis of stresses induced in the connection during make-up and in service can be made on the basis of shrink-fit and taper-wedge theories of design. Figure 8.14 illustrates some of the details of the shrink fit joint assembly.

We briefly discussed the theory of shrink fit as it related to the pressing of bushings, developing interference strains. The amount of the radial interference δ caused by the wedging effect can be estimated as:

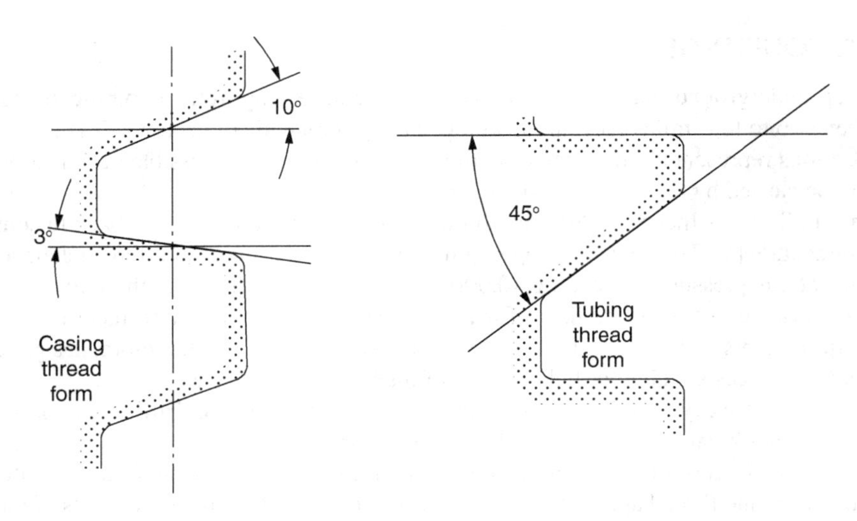

FIGURE 8.12 Buttress thread designs.

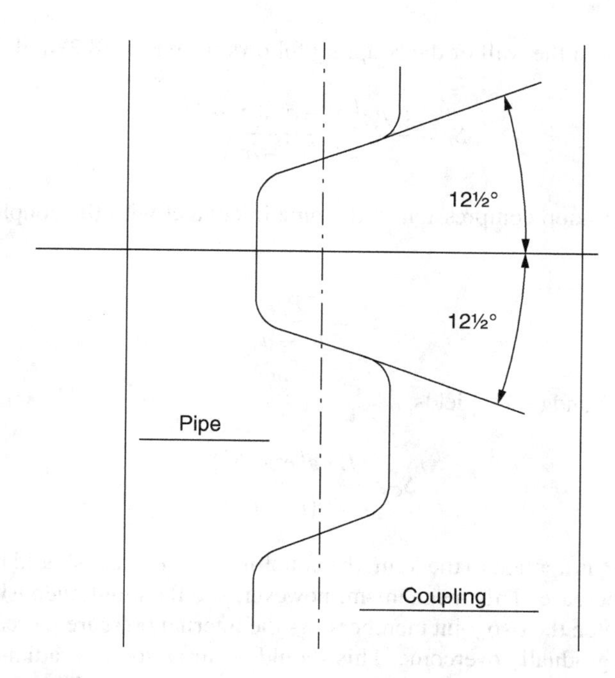

FIGURE 8.13 8-ACME thread type.

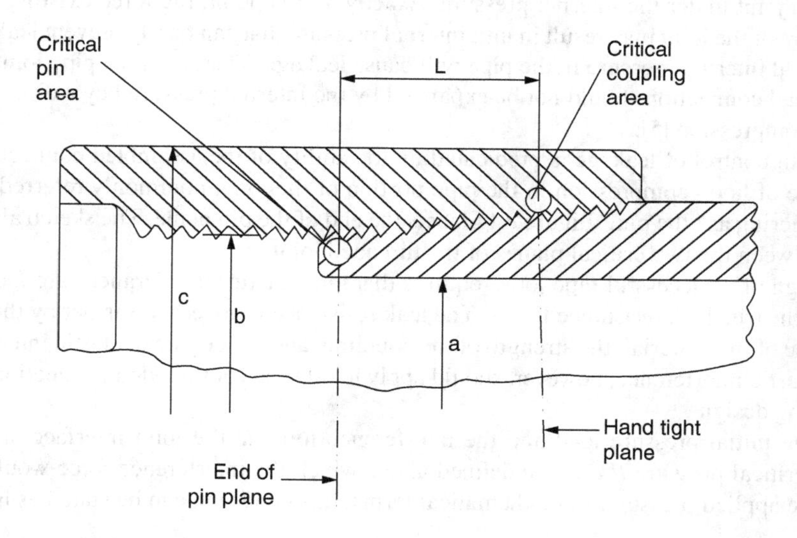

FIGURE 8.14 Location of critical stress areas.

$$\delta = \frac{ing}{2} \tag{8.34}$$

where, i denotes the taper, n is the number of completed turns during the joint make-up and g defines the pitch of the mating threads.

We can calculate the interference pressure p_b in the joint using Eqs. (7.1) and (8.34) as:

$$P_b = \frac{Eing\left(b^2 - a^2\right)\left(c^2 - b^2\right)}{4b^3\left(c^2 - a^2\right)} \tag{8.35}$$

The tensile hoop stress in the wall of the coupling follows from Eqs. (8.2) and (8.35) as:

$$S_h = \frac{Eing\left(b^2 - a^2\right)\left(c^2 + b^2\right)}{4b^3\left(c^2 - a^2\right)} \tag{8.36}$$

Similarly, the value of hoop compression in the pipe in contact with the coupling can be obtained from the expression:

$$S_C = \frac{2P_b b^2}{b^2 - a^2} \tag{8.37}$$

Combining Eqs. (8.35) and (8.37) yields:

$$S_C = \frac{Eing\left(c^2 - b^2\right)}{2b\left(c^2 - a^2\right)} \tag{8.38}$$

As internal pressure P_i is applied to the joint, the coupling tensile stress should increase and the pipe compressive stress decrease. This mechanism, however, should simultaneously increase the interference pressure between the two joint members. As the internal pressure increases further, the pipe compressive stress is gradually overcome. This should occur as soon as radial deformation, caused by the make-up operation of the joint, becomes equivalent to the radial expansion of the pipe wall or the entire joint under the internal pressure. Exactly at this point, the force existing between the two members of the joint may result in unit internal pressure that can barely prevent leakage. In fact, any additional internal pressure in the pipe will cause leakage. Therefore, the pipe joint involving a tapered thread connection should not be expanded by the internal pressure beyond the condition of zero hoop compression [5].

Successful control of leakage is predicated on the ability of the assembled connection to retain some degree of hoop compression in the pipe portion of the joint, commonly referred to in petroleum engineering as "the pin." Figure 8.14 labels the end of the pin plane. The sketch also shows the length L between the two critical planes of the threaded joint.

The design of a successful pipe joint requires that manufacturing tolerances are included in the analysis of the true leak resistance limits. The leak resistance is governed further by the joint diameter, the type of the material, the strength of the coupling, and the degree of elastic interference. The theory of elastic interference, however, should apply to all types of threaded connections utilized in pipe coupling design.

When the initial pressure load and the interference force at the joint interface are essentially equal, the critical pressure P_C can be defined above which the interference force would have to be less than the applied pressure. In mathematical terms, this condition can be stated as [6]:

$$P_C = \frac{E\delta\left(c^2 - b^2\right)}{2bc^2} \tag{8.39}$$

Combining Eqs. (8.34) and (8.39) also gives:

$$P_C = \frac{Eing\left(c^2 - b^2\right)}{4bc^2} \tag{8.40}$$

Equations (8.39) and (8.40) indicate that the critical internal pressure that can be applied to a joint is independent of the pipe weight, but it is influenced by the coupling dimensions and the extent of radial interference. The foregoing expressions then represent the fundamental principle which must be considered in the design of a leak-resistant pipe joint.

The joint make-up techniques in the field varied in the past with the type of joint application. For example, buttress and ACME connections used a make-up mark on the pin member, while others employed applied torque or the number of turns to control the interference pressure.

Over the past 70 years, metal-to-metal seals have proven to be reliable and durable. Particularly with the development of modern surface finishes, the metal-to-metal seals have become almost indestructible even when subjected to wear or galling.

The number of make-up turns in a threaded connection is related to the thread taper, thread pitch, and thread depth. The external shoulder makes an excellent visual method for controlling the joint make-up. Any additional applied torque to shoulder the joint is not harmful.

In general, new technological advances involving metal-to-metal conical seals, with controlled surface finishes and external shoulders, can be employed in practically all types of demanding oil and gas explorations.

8.6 ABUTMENT FAILURE

Some of the preliminary considerations of the rivet and bolt loading in single and double shear joints have been included in Chapter 5, dealing with the design strength of a single fastener.

We define the abutment portion of a joint as that part of a plate member involving the hole for a single bolt.

In this section, we consider the behavior of the abutment portion of the fitting when a single bolt or a pin joins three plate-like members together as shown in Figure 8.15. This particular design involves separate bushings for all the plate members. There are three primary abutment failure modes: Tensile; tear-out shear; and local compression failure modes.

Figure 8.16 illustrates, in an elementary way, the tensile failure of the plate member. In this case, the lug plate is assumed to be pulled apart due to the tension along the centerline.

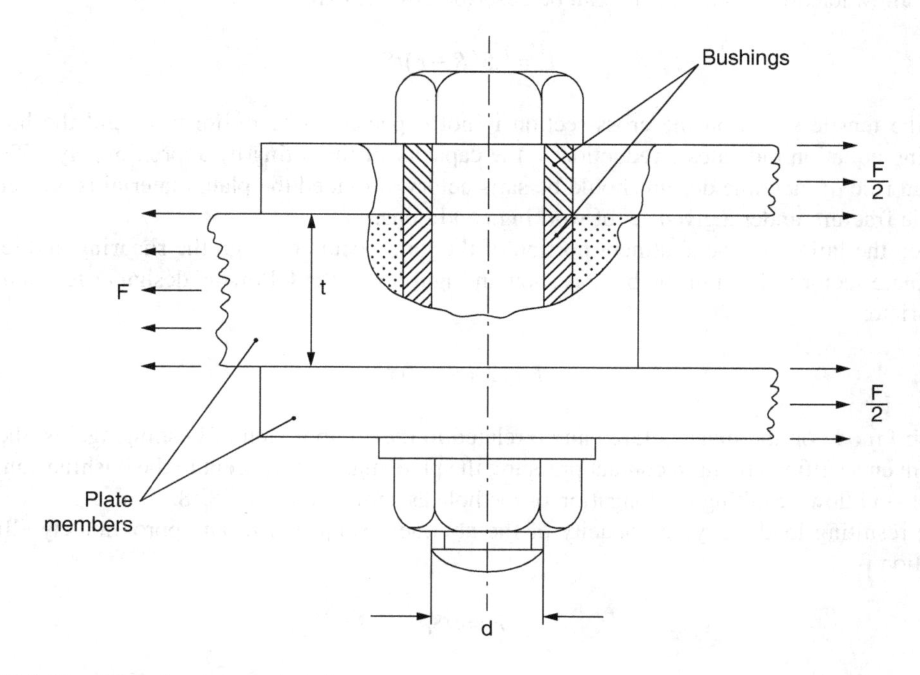

FIGURE 8.15 Example of single-bolt fitting.

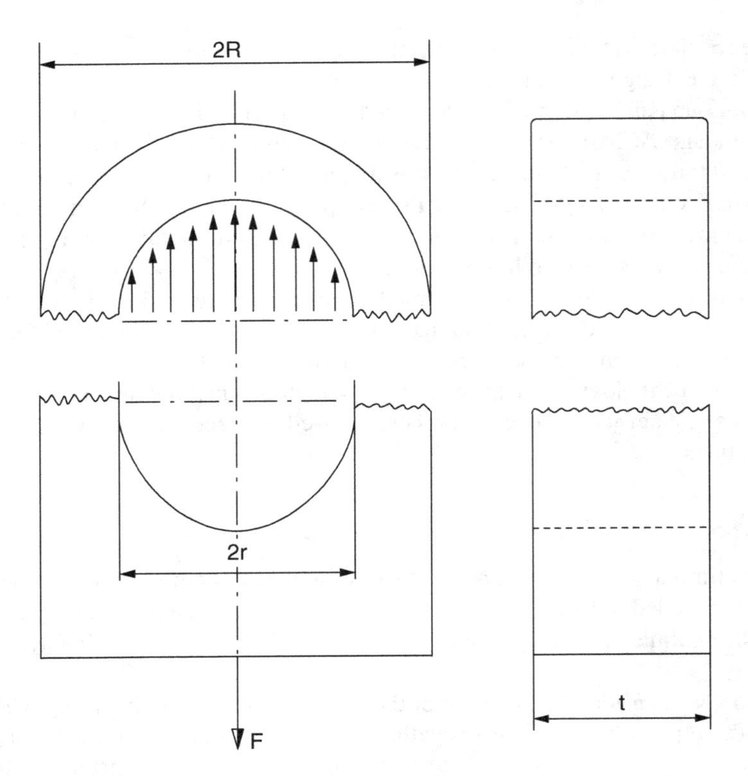

FIGURE 8.16 Tensile failure of the abutment plate.

The allowable load on the fitting can be described by the expression:

$$F = 1.5(R - r)tS_u \tag{8.41}$$

Since the tensile stress on the cross-section is not expected to be uniform around the hole, the foregoing equation indicates a reduction in the capacity of the fitting by approximately 25%. This practical rule of machine design should be satisfactory, provided the plate material is not sensitive to brittle fracture under a given set of working conditions.

When the failure of the abutment portion of the joint occurs through the shearing and tear-out of the plate sector in front of the bolt, as show in Figure 8.17, the following design criterion may be appropriate:

$$F = 2t(R - r)\tau \tag{8.42}$$

The third mode of abutment failure can be related to the compression of bushing against the plate wall. Given a sufficiently high contact pressure, the plate material adjacent to the bushing can begin to crush and flow, resulting in elongation of the hole as shown in Figure 8.18.

The resulting load-carrying capacity of the abutment subjected to an approximately elliptical elongation is:

$$F = rtS_C \tag{8.43}$$

The formula given by Eq. (8.43) suggests a safety margin of approximately 50% to prevent hole elongation under load. It is also a good practice to check the bearing pressure between the bolt and

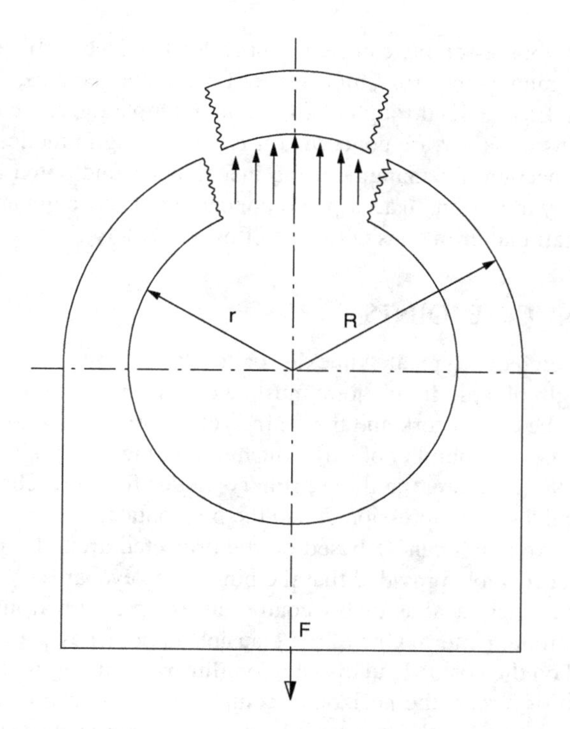

FIGURE 8.17 Tear-Out of the abutment plate.

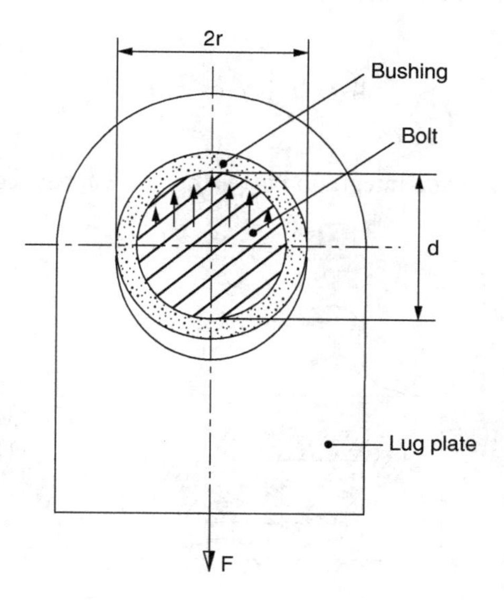

FIGURE 8.18 Elongation of abutment plate.

the bushing. Hence, for a nominal bolt diameter d and a 50% margin of safety, the allowable fitting load becomes:

$$F = 0.5dtS_C \tag{8.44}$$

The common practice here is to consider the bolt as a removable part requiring some finite manufacturing tolerances between the bolt and the bushing. Since the fitting may also be subject to reversible loads, any looseness in the bolt assembly can create shock-load and wear effects such as those found in the landing gears, gun mounts, hoisting, mooring, or towing connections.

It should be noted that Eqs. (8.41) through (8.44), in their simple form, are only approximate. As far as the various margins of safety are concerned it is good design practice to employ relatively high factors of safety to account for manufacturing inaccuracies and potential tendency to brittle behavior. This is especially important in aerospace applications. The weight and cost of such fittings should be a relatively small matter in considerations of overall safety.

8.7 EYEBAR OR KNUCKLE JOINTS

Figure 8.19 provides a sketch of a typical "knuckle" or "eyebar" joint.

While the basic strength of a pin in this joint must be of concern, the remaining two other components of the joint, i.e., the eyebar fork and the main eyebar, behave essentially as eyebars which, at least theoretically, can fail in a number of ways. Such failure modes are similar to those indicated in Figs. 8.16 through 8.19. Therefore, the three primary modes for the eyebar joint should include tension, tear-out shear and local compression due to the pin contact.

The average compressive stress can be based on the projected area of contact. This procedure appears to be generally acceptable provided that the pin fits the eyebar with essentially zero clearance. However, even under such ideal assembly conditions, the pressure around the pin is expected to vary according to a definite pattern. One of the plausible approaches to evaluating the compressive stress may be based on the cosine load distribution illustrated in Figure 8.20.

From the equilibrium of forces, the horizontal components of q, shown in Figure 8.20, are in balance with each other. Assuming that the total load on the main eyebar is W, the summation of vertical components of q can be obtained as:

$$W = 2r \int_{0}^{\pi/2} q \cos\theta \, d\theta \tag{8.45}$$

By substituting from Eq. (8.26) and integrating gives the relation between the maximum unit pressure and the total load W as:

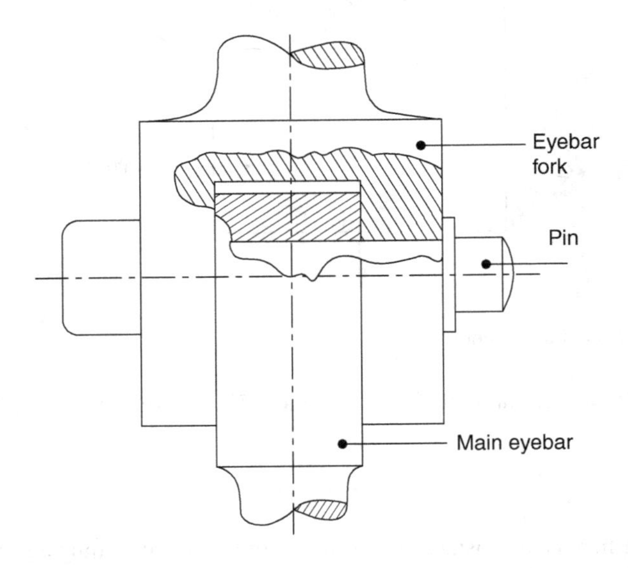

FIGURE 8.19 Knuckle joint.

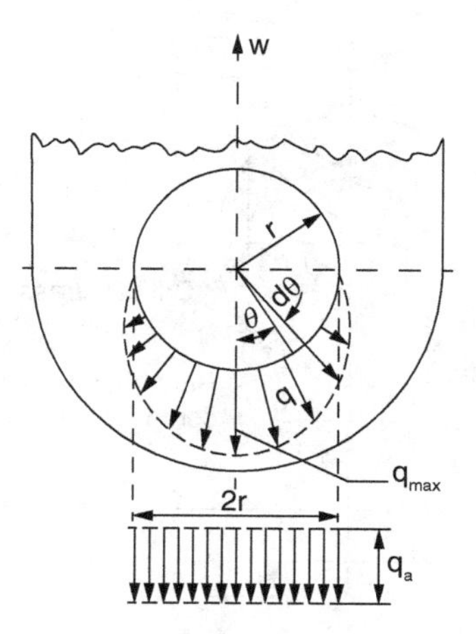

FIGURE 8.20 Cosine loading on eyebar.

$$q_{max} = \frac{2W}{\pi r} \tag{8.46}$$

In Eq. (8.46) q_{max} denotes the maximum unit load, and r is the inner radius of the eyebar which is assumed to be essentially equal to the pin radius. Dividing q_{max} by the width of the eyebar cross-section, we obtain the maximum compressive stress based on the cosine model. Recalling that the average projected pressure is proportional to q_a, as in Figure 8.20, we find q_a to be:

$$q_a = \frac{W}{2r} \tag{8.47}$$

Therefore, according to Eqs. (8.46) and (8.47), the ratio of the maximum to average contact pressure on the basis of the cosine model is: 1.27.

The magnitude of the actual clearance, and its effect on the eyebar joint design, appears to be rather poorly defined in design literature because of the theoretical complexity of the problem involved. However, the assumption of the cosine load distribution is probably a reasonable approximation of the load transfer under zero clearance, while a concentrated load model naturally corresponds to a very large clearance between the pin and the eyebar.

Figure 8.21 illustrates the typical eyebar geometry considered in this chapter.

For eyebar design analysis, using thick-ring theory, the first step is to express the bending moment in terms of θ (see Figure 8.21), the ring geometry, and the load W [7]. The maximum stresses obtained in this manner are significantly higher than those corresponding to the eyebar response under the conditions of zero clearance. For a typical ratio of R/r equal to 2, the relevant stress ratio may be on the order of 1.65.

The extent of pin clearance and the manner of load distribution in an eyebar joint will affect the levels of the calculated critical stresses. Therefore, eyebar design requires the use of relatively high factors of safety. The reason for this is that the conventional criteria obtained using traditional mechanics of materials procedures may not be adequate for the case where the materials fail in a brittle manner.

As we discussed in Chapter 3, the problem of "ductile-to-brittle transition" in metals was painfully recognized on many occasions since the early 1900s. If the nominal stress is of a tensile nature, such as that expected in a critical region noted in Figure 8.21, all conditions are ripe for catastrophic crack propagation through a brittle matrix of the material.

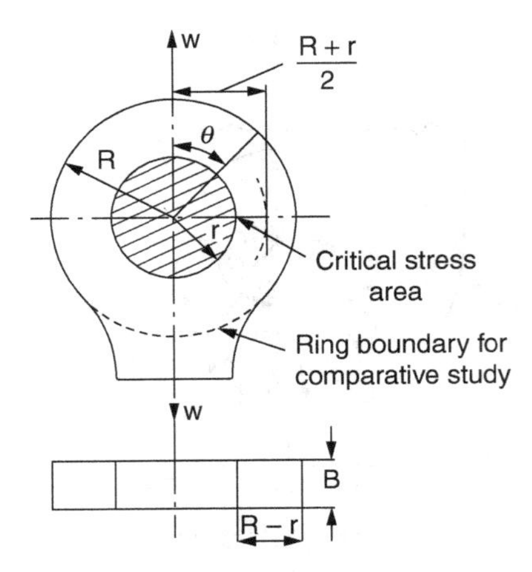

FIGURE 8.21 Eyebar geometry.

If we denote the eyebar curvature ratio by $\lambda = R/r$, and a stress correction factor for clearance by ψ, then it can be seen that the critical tensile stress in the eyebar is [7]:

$$S_{max} = \frac{W_{\psi}}{\lambda r B} \tag{8.48}$$

where Figure 8.22 provides ψ as a function of the clearance ratio $\eta = e/r$. If we now express the maximum eyebar stress in terms of the primary tensile stress on the net cross-sectional area, the following simple formula results:

$$S_{max} = \frac{2\psi S_t(\lambda - 1)}{\lambda} \tag{8.49}$$

The degree of design conservatism and the rationale for selecting the appropriate factors of safety must be tempered with considerable engineering judgment.

The parameter e, defined as the radial clearance between the pin and the eyebar, is not easy to control or measure. It is also clear from Figure 8.22 that the stress factor ψ tends to a constant value

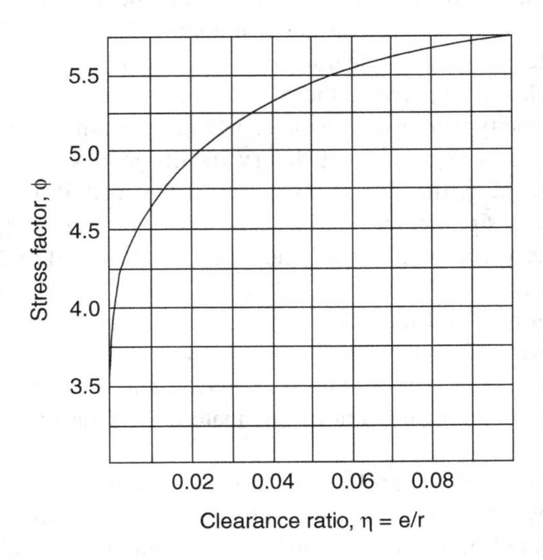

FIGURE 8.22 Stress factor for eyebar joint.

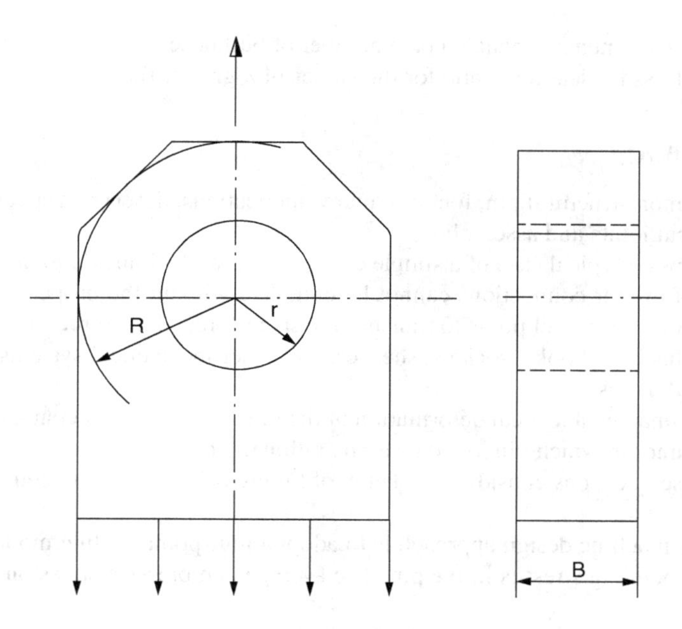

FIGURE 8.23 "Bull nose" eyebar configuration.

for relatively low numbers of n. For all practical purposes, this condition is reached for $n \simeq 0.1$, giving the maximum theoretical value for ψ equal to about 5.7. For example, taking $\lambda = 2$, Eq. (8.49) shows that $S_{max} = 5.7S_t$, which is consistent with the thick-ring theory based on the concentrated load criterion [7].

When the shape of the eyebar corresponds to the so-called bull nose configuration shown in Figure 8.23, experiments have suggested a reduction in design conservatism associated with the use of a stress correction factor for the conventional eyebars [8].

Figure 8.24 provides design curves for bull nose joints for typical values of λ, useful in the design of heavy equipment—such as used in construction work.

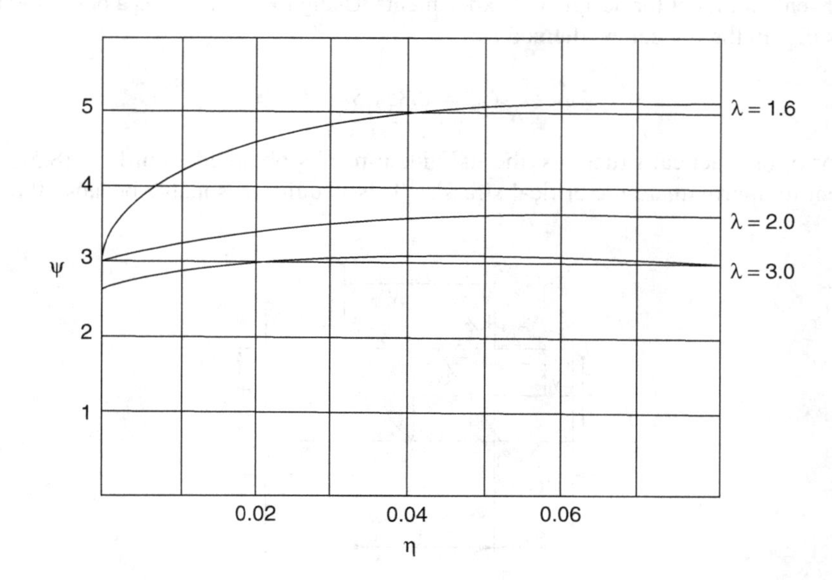

FIGURE 8.24 Stress factors for "Bull nose" eyebars.

The results of experiments conducted on a number of bull nose eyebars show relative insensitivity of the eyebar stress to clearance ratio for the values of λ greater than 2.0.

8.8 STRUCTURAL

The role of a common structural pin, found in many applications, deserves a special mention in the study of mechanical joints and assemblies.

Figure 8.19 shows a typical case of a simple cylindrical pin of a knuckle joint.

The analysis of eyebar connections cannot be complete without the proper sizing of pins. The primary function of a structural pin is to transmit the shear force. Such forces are common in swivels, clevis links, shackles, hooks, sockets, sheaves, and other mechanical systems and components performing critical duties.

Because of the unavoidable local deformation of the pin and the parts it contacts, questions often arise as to the manner in which pin loading can be rationally defined.

To address those questions consider the sketch of Figure 8.25 illustrating equilibrium of the pin forces.

The customary machine design approach is to adopt a four-point loading model for the purpose of calculating the bending stresses in the pin. The average compressive stress on the projected pin area is:

$$S_c = \frac{W}{Ld} \tag{8.50}$$

The bending stress S_b expressed in terms of the compressive stress S_c of Eq. (*8.50) is then

$$S_b = 1.27 S_c \frac{(k+2)k_1^2}{k} \tag{8.51}$$

It is important to note here that the dimensionless ratios k and k_1 refer directly to Figure 8.25, where $k = L/a_0$ and $k_1 = L/d$.

The concept of an average compressive stress on the projected area is hard to justify. However, it is still a convenient model for design and experiments. Using the stress S_C as a basis, the maximum shear stress τ_{max} in the pin can be defined as:

$$\tau_{max} = 0.85 k_1 S_C \tag{8.52}$$

In the majority of practical situations, the individual results obtained from Eqs. (8.51) and (8.52) are sufficient to approximate the critical stresses. This is quite reasonable because the maximum

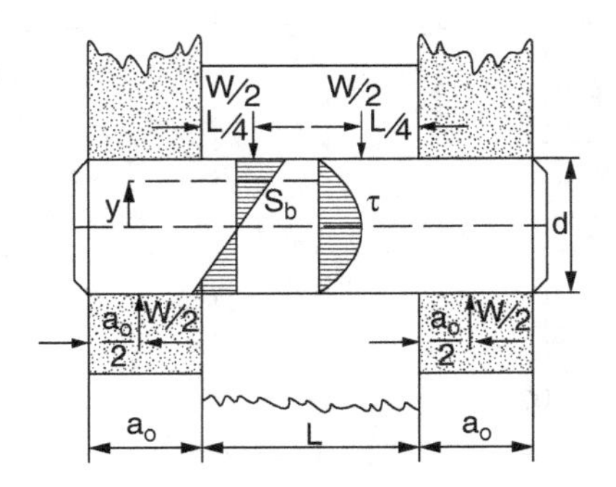

FIGURE 8.25 Equilibrium of pin forces.

bending stress resides at the extreme fiber while the maximum shear stress is found at the centerline of the pin.

It may be recalled from Chapter 3 that shear stresses should equilibrate on four sides of an element simultaneously. The transverse shear stress must also be equal to zero on a free surface where only one pair of shear stresses can exist and which cannot satisfy the conditions of equilibrium.

It may be necessary, at times, to estimate the magnitude of the combined stress at a point such as that defined by radius y in Figure 8.25. If we denote the ratio y/d by k_2, then the ratio of the principal stress S to the bending stress can be calculated from the expression:

$$\frac{S}{S_b} = k_2 + \frac{\left[9k_1^2k_2^2(k+2)^2 + 4k^2\left(1-4k_2^2\right)^2\right]^{0.5}}{3k_1(k+2)} \tag{8.53}$$

The parameter k_2 can vary between 0 and 0.5. When $k_2 = 0$, the maximum stress is equal to that given by Eq. (8.52). Also, when $k_2 = 0.5$, Eq. (8.53) reduces to Eq. (8.51).

The formula given by Eq. (8.53) indicates that as long as the pin length is equal to or greater than its diameter, combining the stresses does not cause the maximum principal stress to numerically exceed the conventional bending stress found at the outer fiber [7].

8.9 CHAIN DRIVES AND COUPLING LINKS

Coupling links, chain drives, and similar machine elements represent component parts or multiply connected systems. There is a popular saying that the chain is only as strong as its individual link, which, despite its relative geometrical simplicity, presents a significant design challenge. Each link is a statically indeterminate structure requiring the application of strain energy methods and experimental techniques in estimating the relevant structural integrity.

In many situations, we can procure conventional chain drives and other chain systems without the necessity of specifying the methods of design or testing because of the existing vast experience in industry related to these products.

The chain drive, consisting of chain links and sprockets, can be used in power transmission with good efficiency and relatively high peripheral velocity. Chain drives can be found in agricultural machinery, bicycles, motorcycles, rolling mill mechanisms, conveyors, coal-cutters, or lathes, to mention a few.

The disadvantage of coupling links include high production costs, need for careful maintenance and the dependence of the chain drive on the constant velocity requirements. The common shape of a hinge-connected chain link resembles an eyebar design, as shown in Figure 8.26.

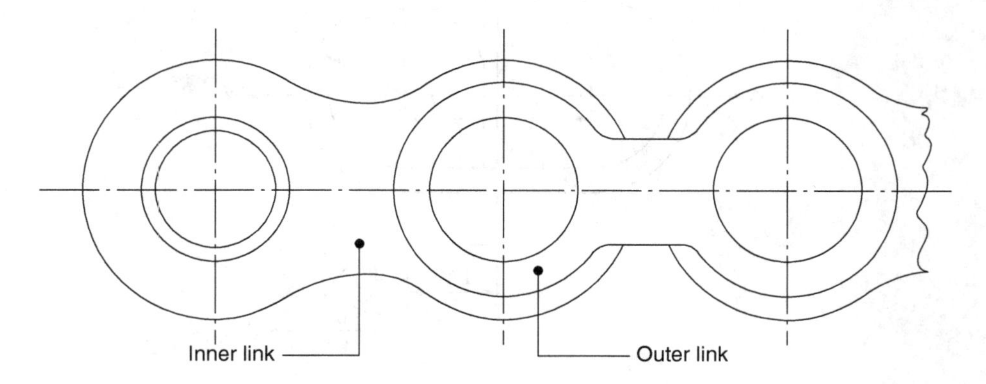

FIGURE 8.26 Hinge-connected chain links.

This particular chain consists essentially of alternating inner and outer links hinged together by means of bushings and pins. When rollers are used instead, the chain can be protected against any undue wear caused by the sprocket teeth. The conventional design methods, given for lugs and eyebars discussed in this chapter, can be extended to sizing of outer and inner link plates in a hinge-connected chain.

The pitch, defined as the distance between the pin on the roller centers, and the overall width, representing the length of pins and rollers between the link plates, constitute the main geometrical characteristics of the hinged-link assemblies. The final strength and the operating features of the particular chain are usually obtained experimentally.

In the case of shackles, railway couplings, and more standard chains, the design model of a symmetrical link can be used in the evaluation of critical stresses based on the theory of curved beams of circular cross-section.

Figure 8.27 illustrates such as design model.

Since the common chain link is a statically indeterminate structure, the fixing moment M_0 cannot be found from the conventional method of the equilibrium of forces. The solution, however, is obtained rather easily from the expression:

$$\int_0^{\pi/2} M_1 \frac{\partial M_1}{\partial M_0} R\,d\theta + \int_0^{L/2} M_2 \frac{\partial M_2}{\partial M_0} dx = 0 \tag{8.54}$$

where

$$M_1 = M_0 - PR\sin\theta \tag{8.55}$$

and

$$M_2 = M_0 - PR \tag{8.56}$$

Integrating Eq. (8.54) with the aid of Eqs. (8.55) and (8.56), and noting that $P = W/2$, yields:

$$M_0 = \frac{WR(2R+L)}{2(\pi R+L)} \tag{8.57}$$

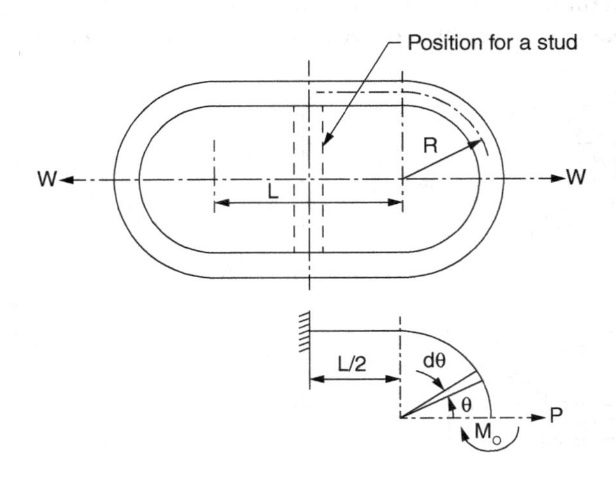

FIGURE 8.27 Design model for a chain link.

The maximum bending moment in this type of a chain link operating without the central support is equal to M_0. The usual analysis of curved beams and hooks shows that the maximum stress appears at the inner surface of the component. However, link configurations with even and uneven ends can, at times, be subject to maximum stresses at the outer surfaces. This is more likely to happen with the higher ratios of L/R.

If by r we denote the radius of a circular cross-section, the majority of r/R ratios for all links in practice fall between 0.2 and 0.5. The more conventional ratios of L/R will be found to vary between 0 and 4. Note that $L/R = 0$ corresponds to the geometry of a plane circular ring.

The dashed lines in Figure 8.27 show how the strength of a typical chain link can be increased by inserting a central reinforcement. The design analysis problem, however, becomes more involved, requiring the calculation of the two redundant quantities such as the fixing moment M_0 and the axial compressive force in the reinforcing stud. Such an analysis leads to somewhat lengthy expressions in terms of L/R ratio [7].

8.10 WIRE ROPE FITTINGS

One of the more important mechanical connections is concerned with the provision of a suitable end fitting for wire rope applications.

The mechanics of a tensile joint, almost ubiquitous in the real world to nature, receives surprisingly little attention in the majority of design texts. This trend is also evident in British, German, and Russian publications dealing with the more conventional topics of strength of materials.

The designer of wire rope fittings must resort to laboratory tests and engineering judgment if appropriate practical design formulas are not readily available. In this section, we attempt to illustrate the general nature of the tensile joint design problem in terms of more elementary concepts of joint mechanics. A rather idealized model of an end fitting will be studied, which can be extended to ropes, cables, tie-bars, and similar tensile components.

Pulling or jacking heavy loads suspended from cables often require unique gripping systems or swage fittings found in conventional and offshore applications. Imagine, for example, forces involved in controlling the tension in a suspension cable on the Golden Gate Bridge in San Francisco or on an offshore rig requiring 3.5 in.-diameter wire rope and 1400 tons load capability.

Sophisticated pulling equipment is now available in industry for performing various tensile tests on the premise that the wire rope can be threaded through a puller and grip system providing the same loading during the operation. Such systems are often designed to hold the rope in suitable wedges at any point along the rope's length, until a permanent end fitting can be secured for the specified loading conditions.

Galvanized steel wire strand in rope design is generally used for guying poles, overhead structural members, railroad hardware, and similar structures. For more critical overhead duties involving trolley systems and special anchors, improved grades of steel are required to assure greater strength and toughness.

For the production of splicing and fittings, the wire rope is cut and properly seized to assure uniform tension in the socket after the pouring of molten zinc and to achieve the full strength of the rope.

Figure 8.28 illustrates some of the typical end fittings.

Type A, shown in Figure 8.28, is referred to as a "swaged socket," formed by applying external pressure or other metal working technique. Sketch C illustrates the basic elements of a pin-type shackle, the design of which involves curved beam theory and pin analysis. The strength characteristics of the U-shape portion of the shackle are similar to those of the chain link analyzed in the preceding section. The sizing of the pin can be performed with the aid of the conventional theory of beams.

Type B, shown in Figure 8.28, can be designed on the basis of the available taper and the frictional adhesion between the body of the fitting and the appropriate filler material of metallic or nonmetallic nature.

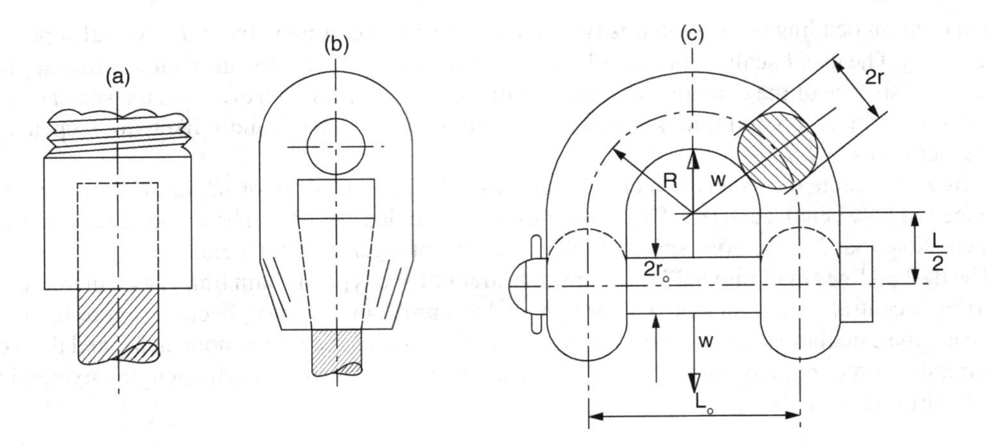

FIGURE 8.28 Wire rope fittings.

Experience shows that end-fittings for wire rope can be expected to have high efficiency. It appears, for example, that swaged and zinc-filled sockets can be rated as high as the rope for which the fitting is designed. Other configurations, such as the shackle or thimble type, are likely to have 80% efficiency in relation to the rated capacity of the rope.

Suppose we consider the design analysis of the shackle-type fitting shown in Figure 8.28. To assure equal-strength design of the pin and the U-shape part on the basis of bending, the assumption can be made that the pin behaves as a simply supported beam of length L_0 subjected to a central load W. For the notation indicated in Figure 8.28c, the section modulus of the pin is: $\pi r_0^3/4$. The relevant bending moment at the center of the beam for $L_0 = 2R$, is: $WR/2$. Hence, the bending stress S_b in the pin is:

$$S_b = \frac{2WR}{\pi r_0^3} \tag{8.58}$$

The maximum bending moment for the U-shape portion of the fitting can be described with the aid of Eq. (8.57). For the section modulus of $\pi r^3/4$, the critical bending stress is then:

$$S_b = \frac{2WR(2R+L)}{\pi(\pi R+L)r^3} \tag{8.59}$$

For the condition of equal bending strength, Eqs. (8.58) and (8.59) yield:

$$\frac{r_0}{r} = \left(\frac{\pi R+L}{2R+L}\right)^{1/3} \tag{8.60}$$

Introducing $\beta = L/R$, Eq. (8.60) transforms into:

$$\frac{r_0}{r} = \left(\frac{\pi+\beta}{2+\beta}\right)^{1/3} \tag{8.61}$$

Figure 8.29 provides a graph of the function of Eq. (8.61) for values of β between 0 and 1. It is clear from the plot that the effect of U-shape length on the shackle-pin size is, for all practical purposes, rather small. Hence, the parameter β can be selected here on a basis other than strength.

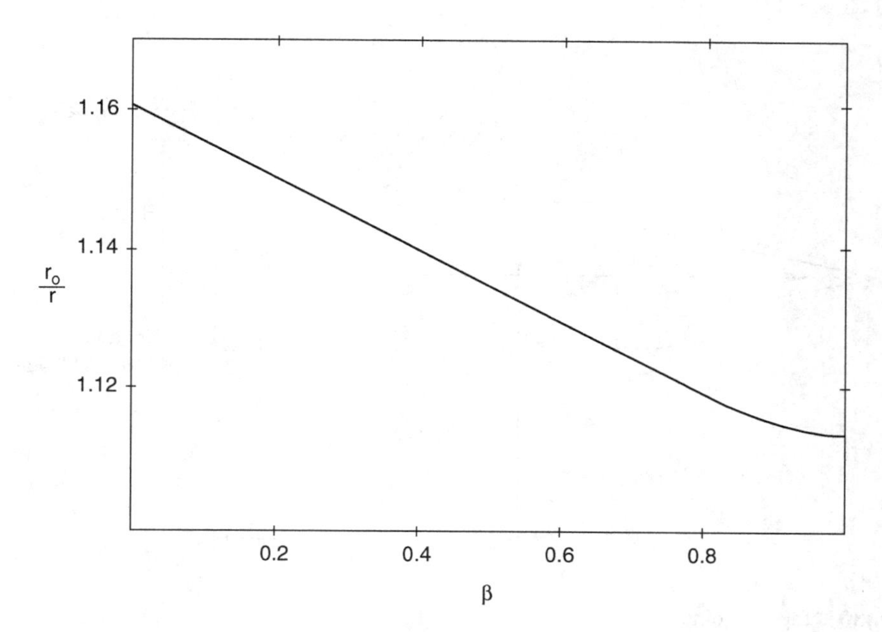

FIGURE 8.29 Size effect of U-shape member.

The wire rope socket, shown in Figure 8.28b, involves a tapered interface which provides a degree of resistance to pull-out of the rope, cable, or a rod depending on the cone angle and some form of frictional adhesion factor. The normal pressure acting on the interface can be very high when the cone angle and the frictional adhesion strength are both low. We discuss the mechanics of adhesion in some detail in Chapter 11.

As a conservative assumption for the preliminary calculations, the conventional friction factor between the body of the socket and the filler material can be used. This type of analysis can, at best, be approximate and contingent upon the resistance of individual strands to pull-out if the mechanical joint consists of the conventional socket and the wire rope.

This fact is well known in industry and the manufacturer of rope and cable terminations must rely on qualification testing. The design analysis which follows should only be looked upon as an example of a conservative boundary compatible with a mechanical model which may be used in experimental analysis. However, even the best theoretical predictions of the strength of a wire rope fitting should not be used in lieu of the performance test.

Figure 8.30 provides a relatively crude mechanical model of the end fitting. The half-angle of the cone is exaggerated for the purpose of the illustration and the filler-end does not show any specific details of strands or the filler material. The term N is the equivalent total interface force reacting with the filler. The corresponding frictional resistance is assumed to be directly proportional to the friction factor f.

Considering the equilibrium of the forces represented in Figure 8.30, we obtain:

$$W = N(\sin\alpha + f\cos\alpha) \tag{8.62}$$

When $\alpha = 0$, $W = fN$, representing only the frictional effect if the joint is assembled with some residual compression between the wire rope matrix and the socket wall.

In reality, the external load has to overcome additional adhesion forces. As the cone angle α increases to the limit, the end fitting forms a shoulder and the effect of interface friction, according to Eq. (8.62), must approach zero. This is, indeed, a gross-over-simplification and a trivial solution

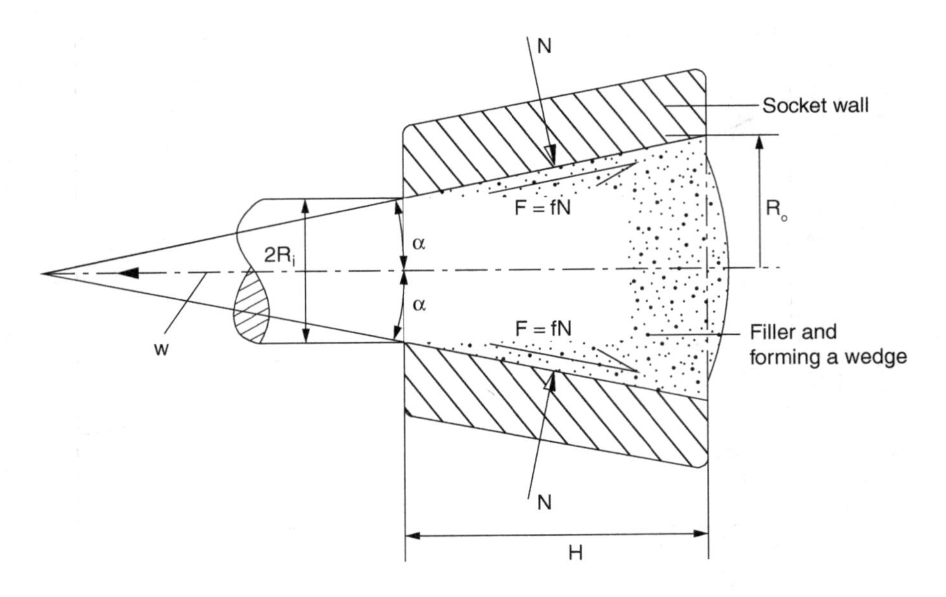

FIGURE 8.30 Tapered joint.

stating that $W = N$. Hence, any reasonable interpretation of Eq. (8.62) should be restricted to the values of α, sufficiently removed from the two extremes defined as 0 and $\pi/2$.

The contact pressure q between the socket and the filler can be estimated as follows: From the geometry depicted in Figure 8.30, the contact area is:

$$A = \pi(R_0 + R_i)\left[(R_0 - R_i)^2 + H^2\right]^{1/2} \tag{8.63}$$

Hence, postulating that $q = N/A$, and using Eqs. (8.62) and (8.63) gives:

$$q = \frac{W}{\pi(\sin\alpha + f\cos\alpha)(R_0 + R_i)\left[(R_0 - R_i)^2 + H^2\right]^{1/2}} \tag{8.64}$$

Observing that the foregoing trigonometric functions can be expressed in terms of R_0, R_i, and H, Eq. (8.64) simplifies to:

$$q = \frac{W}{\pi(R_0 + R_i)(R_0 - R_i + fH)} \tag{8.65}$$

The contact pressure q can now be used to check the stresses in the socket on the assumption that the bond between the socket and the filler can be modeled with the aid of the friction factor f.

Although the foregoing calculation appears to be sufficiently conservative, the representative magnitude of the parameter f can only be established experimentally.

8.11 HIGH-PRESSURE THREADED CONFIGURATIONS

The design of high-pressure closures and connections for 10,000–100,000 psi systems in commercial and research applications cannot be complete without due considerations of thread characteristics. This is particularly important in situations where pressure cycling and repetitive use of equipment cause us to examine the fatigue resistance of a threaded configuration subjected to stress concentration.

There appears to be a definite trend in industry towards the use of high-pressure components as well as hydrostatic methods of material forming, which puts more emphasis on design, in view of the general awareness of safety.

Joint technology in high-pressure closures usually involves bolts holding down the closure or a conventional threaded nut acting as the closure itself. In either case, the basic problem here is to resist the end load created by the internal pressure. A detailed configuration is most likely to depend on the closure design based on the internal thread, which satisfies an optimum combination of a minimum component size and reasonable access to the vessel interior.

The use of a thread connection, however, presents the designer with a number of questions concerning the transfer of loads and stresses. This is a typical problem of a mechanical threaded fastener joint interface. The form of the thread, as well as the extent of the undercut below the bottom thread in the vessel, can determine the structural integrity of the system. The classical analysis or finite element modeling may then be used to assess the superposition of the bending and longitudinal stresses in the undercut area.

The other major design consideration is related to the stresses in the threads. The conventional method in this case is based on the calculation of average shear stresses, where the shear diameter and the overall length of the threaded portion under load are the principal variables. Unfortunately, the distribution of thread loading and shear stresses is far from uniform, as proven by experience and analysis. For example, the thread load distribution curve for a large screw closure shown in Figure 8.31, indicates that for a conventional thread design, the first thread may carry two to four times the theoretical average load. Similar behavior was noted in thread stresses illustrated in Figure 3.10 and as shown again here as Figure 8.31.

Recognition of this problem in closure design has led to a number of investigations of thread forms with respect to flank angle, start-up friction, thread thickness, corner radii, and the mechanical compliance to assure acceptable structural integrity and a more uniform load transfer.

The two main candidates for the best thread form for a high-pressure closure design are still Acme and buttress threads. The various advantages and shortcomings of these two designs have been debated for a long time now without a clear resolution as to which form has proved to be distinctly better. Figure 8.33 shows a modified profile of both threads, superimposed upon one another. This modification consists essentially of providing a more generous root radius to reduce the stress concentration.

In the case of the buttress thread, the maximum bending stress is also smaller because of the greater thickness at the root. The thread thickness through the pitch diameter, however, is

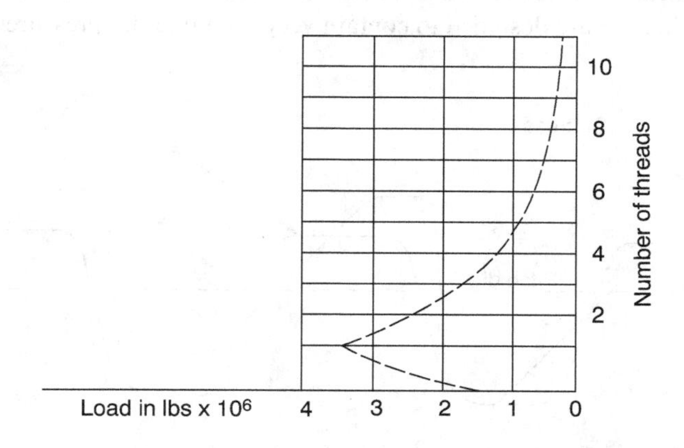

FIGURE 8.31 Load distribution curve for a conventional thread form. (Courtesy of Autoclave Engineers, Inc.)

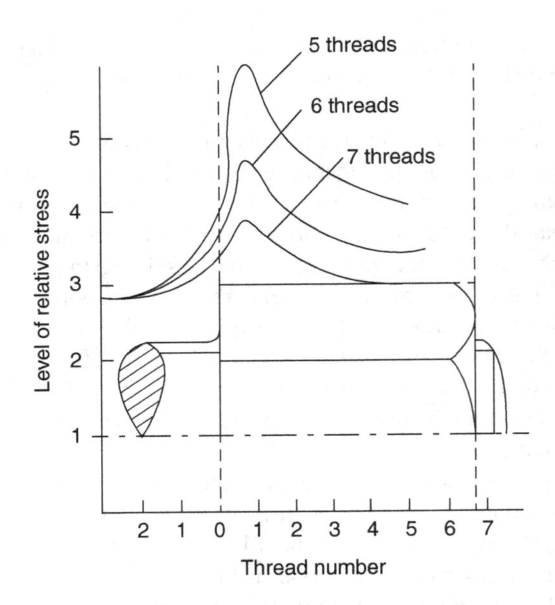

FIGURE 8.32 Variation of peak stresses with number of threads in a conventional design.

approximately the same for both designs. Although the load-side flank angle on the buttress thread is smaller than that for the Acme contour, the theoretical difference in the radial force transmitted to the cylindrical wall of the closure is not significant.

Since the choice between the two thread forms shown in Figure 8.33 is still largely subjective, a new contour has been developed known as the Gasche Resilient Thread intended for large vessel closures. This thread form is based on the idea of machining a semi-circular groove on the nut and the corresponding semi-circular groove in the vessel wall. The closure joint is obtained by providing a specially designed, solid-wound spring fitted into the nut groove. The resiliency of the spring allows a certain amount of equalization of the thread loading. When this joint is refined further through the introduction of a small amount of taper, as illustrated in Figure 8.34, further equalization of the thread loading can be achieved.

Figure 8.34 shows how the resilience of the spring, combined with the effect of the taper, provides a satisfactory load distribution, for the three thread forms discussed here.

While the modified resilient thread concept is not particularly simple, this may be a relatively small price to pay for the benefits derived from the more uniform load and stress distribution in the case of a large threaded closure designed to contain very high internal pressures.

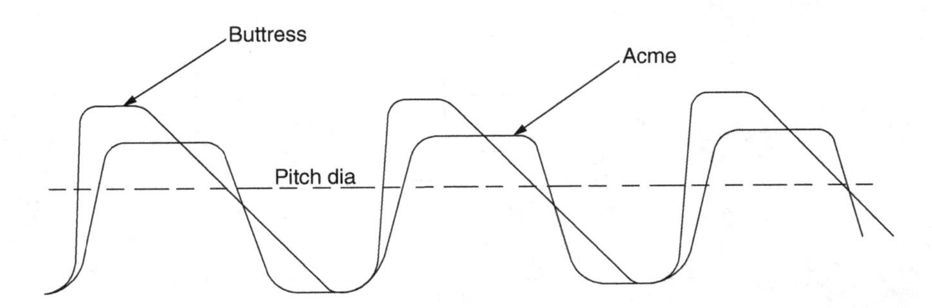

FIGURE 8.33 Comparison of modified thread forms (Courtesy of Autoclave Engineers, Inc.).

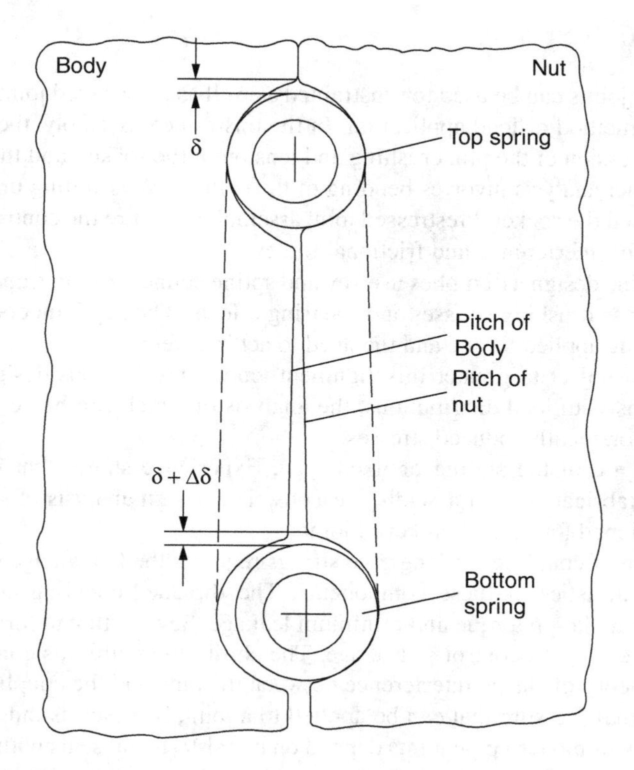

FIGURE 8.34 Gasche resilient thread design for large closures. (Courtesy of Autoclave Engineers, Inc.)

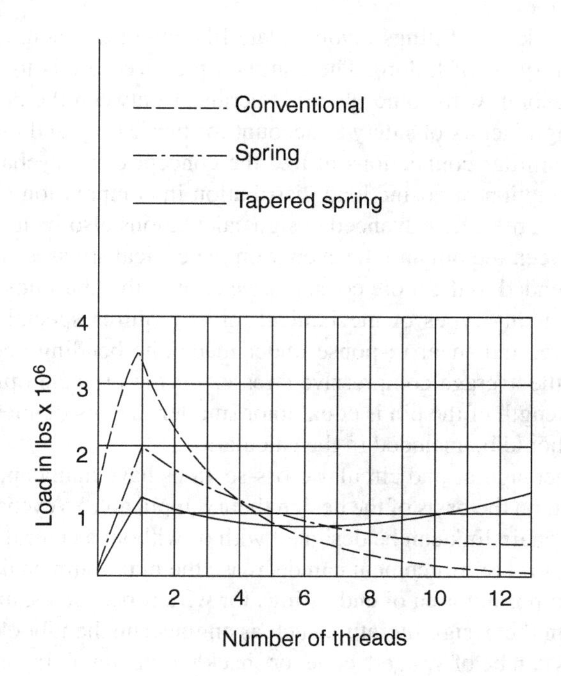

FIGURE 8.35 Comparison of load distribution for the various thread designs. (Courtesy of Autoclave Engineers, Inc.)

8.12 SUMMARY

The design of cotter joints can be used for unstrained as well as prestressed joints, depending on the magnitude and the method of load application. In the unstrained assembly, the design is based on the shear and compression of the pin, crushing and tension of the socket, and the tensile strength of the shank. More exact analysis involves bending of the cotter key, assuming uniform contact pressures on the shank and the socket. Prestressed joint assemblies require the control of residual effects and careful design for interference and frictional forces.

Numerous machine design assemblies use key and spline connections to transmit the torque. The basic forces here cause crushing stresses and shearing effects. The key joint configuration depends on the direction of the applied torque and the need to act in reverse.

Shorter keys are subject to greater misalignment tendencies. Various designs have evolved to address the problems with load distributions, the analysis of which can be very complex, particularly in the case of torsionally induced stresses.

Multiple splines are of the sliding or fixed type. Experience shows that involute splines are superior for stress, fabrication, and assembly reasons. The design analysis of splines is essentially similar to that developed for individual keyed joints.

The extensive use of couplings in long pipe strings requires the knowledge of the tensile, burst, and collapse characteristics of these components. The threaded-coupling connections must be designed for correct make-up torque and minimum leakage. Several thread forms have been developed for the purpose of good control of leakage. The ability to assure this control depends on the correct use of the theory of elastic interference between the pipe and the coupling.

The critical internal pressure that can be applied to a joint, however, is independent of the pipe weight. The majority of modern pipe joints depend on metal-to-metal seals with strict control of the number of make-up turns. Design theory and extensive field experience since the early 1900s have been combined to assure satisfactory pipe coupling performance for the most demanding applications in the oil and gas industries.

The conventional single-bolt fittings involve plate-like abutment members subjected to tensile, tear-out, and elongation types of failure. The common practice here is to employ a bolt or pin as a removable part of the joint, with some specific tolerances between the bolt and the bushing. It is recommended to use high factors of safety to account for fabricating and materials variables.

Knuckle joints and similar connections utilize the concept of an eyebar. The design approach often involves the assumption of cosine load distribution in compression and the combined effect of tension and bending in the eye. Advanced design calculations also include the assessment of the effect of clearance between the pin and the eyebar on the critical stresses. The conventional eyebar calculations can be extended to the more conservative case of the "bull nose" eyebar joint.

The role of pins in many types of mechanical joints requires special care in design, related to bending, compression, and shear response under load. The bending and shear stresses can be expressed in terms of the average compressive stresses on the projected pin area. For special pin proportions where the length of the pin is equal to or smaller than its diameter, the combined effect of bending and shear should be included in the calculations.

Coupling links of rectangular and circular cross-sections have many applications. The majority of these can be analyzed on the basis of the design theory applicable to eyebars or chain links in tension. The conventional chain link can be designed with or without a central support called a "stud." The provision of the central reinforcement can decrease the maximum chain stress by about 20%.

The extensive and important area of end fittings for wire rope, cables, and similar tensile members is seldom treated in the design literature such as engineering handbooks and texts on machine design. The end fittings can be of swaged, cone, or shackle type, involving cylindrical, conical, eyebar, U-shape, and pin configurations. While experience shows that the efficiency of modern fittings of this type can be relatively high, the design analysis involves a number of simplifying assumptions which therefore requires considerable experimental support.

Since no single model or knowledge of a mechanical property will suffice in the design of the majority of tensile end-fittings, it is strongly recommended to conduct qualification tests on all new joint designs.

The development of high-pressure closures in modern industry has involved significant research into the distribution of loads and stresses in large threaded connections. The two best thread forms currently recognized for this application historically include the Acme and buttress configurations. A relatively new contour, known in industry as the Gasche Resilient Thread, allows a more uniform load transfer. This design, combined with the taper, offers superior load distribution in large threaded closures.

8.13 DESIGN ANALYSES

EXAMPLE ANALYSIS 1

Figure a contains an overhead and side view line drawing of a typical chain link. Suppose the links and connecting pins are metal (for example, steel) with tensile strength: $\sigma_{max} = 30$ ksi and shear strength: $\tau_{max} = 10$ ksi.

Let the pin radius r be 0.25 in. Let the links have a rectangular cross-section with height $h =$ 0.5 in. and thickness $t = 0.125$ in.

Determine the maximum force F_{max} that can be sustained by the chain-link system.

SOLUTION

Assume ideal geometry. The system will then either fail due to excess shear at the pin or due to excess tension in the connecting links.

Consider first the pin shear: the cross-section area of the pin A_{pin} is:

$$A_{pin} = \pi r^2 = \pi(0.25)^2 = 0.196 \text{ in}^2 \tag{a}$$

The shear stress τ on the pin is then

$$\tau = \frac{(F/2)}{A_{pin}} \tag{b}$$

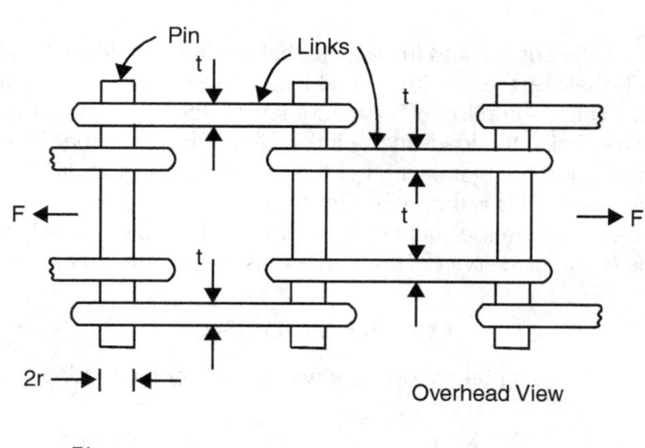

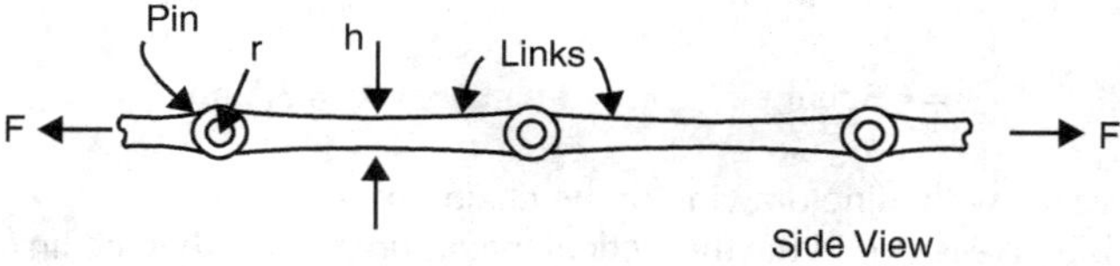

FIGURE A Chain-link joint.

Note that with two parallel links each pin shear surface need support only half the tension load.

By substituting from Eqs. (a) into (b) we have:

$$\tau = 2.546F \text{ psi} \tag{c}$$

where F is expressed in pounds.

In view of Eq. (c), the maximum pull force F_{max} will occur when the shear stress reaches its limit τ_{max} (10 ksi). By substituting 10 ksi for τ in Eq. (c), F_{max} is:

$$F_{max} = \frac{10,000}{2.546} = 3927 \text{ lb} \tag{d}$$

Consider next the link tension: the cross-section area of a link A_{link} is simply:

$$A_{link} = ht = (0.5)(0.125) = 0.0625 \text{ in}^2 \tag{e}$$

The tensile stress σ in the links is then:

$$\sigma = \frac{(F/2)}{A_{link}} \tag{f}$$

Note as with the shear on the pins, with two parallel links, each link need only support half the tension load.

By substituting from Eqs. (e) into (f) we have:

$$\sigma = 8.0F \text{ psi} \tag{g}$$

where, as before, F is expressed in pounds.

In view of Eq. (g), the maximum pull force F_{max} will occur when the tensile stress reaches its limit σ_{max} (30 ksi). By substituting σ_{max} for σ in Eq. (g), F_{max} is:

$$F_{max} = \frac{30000}{8} = 3750 \text{ lb} \tag{h}$$

By comparing the results of Eqs. (d) and (h), we see that the links are likely to fail in tension before the pin fail in shear. Therefore, the maximum pull load on the chain-link system is: 3,750 lb.

A word of caution. The assumption of ideal geometry is too optimistic. If there is even a slight difference in the parallel links, the loads in the links will not be the same. This means that one of the links will need to support a larger portion of the load than the other link. The value for F_{max} in Eq. (h) will then be too high. This is the reason for using a safety factor.

Note also that the contact stresses between the links and the pins is likely to be high, particularly as the chain-link is put into service. This then is reason for further increasing the safety factor.

EXAMPLE ANALYSIS 2

Repeat Example Analysis 1 using the following physical and geometric data:

$$\sigma_{max} = 200 \text{ MPa} \qquad \tau_{max} = 70 \text{ MPa}$$

$$r = 6.3 \text{ mm} \qquad h = 13 \text{ mm} \qquad \tau = 3.5 \text{ mm} \tag{a}$$

For convenience we show the line drawing of the chain link again here:

As before the objective is to find the theoretical maximum force F that the link joint can sustain. We can obtain this objective by following the solution procedure of Example 1.

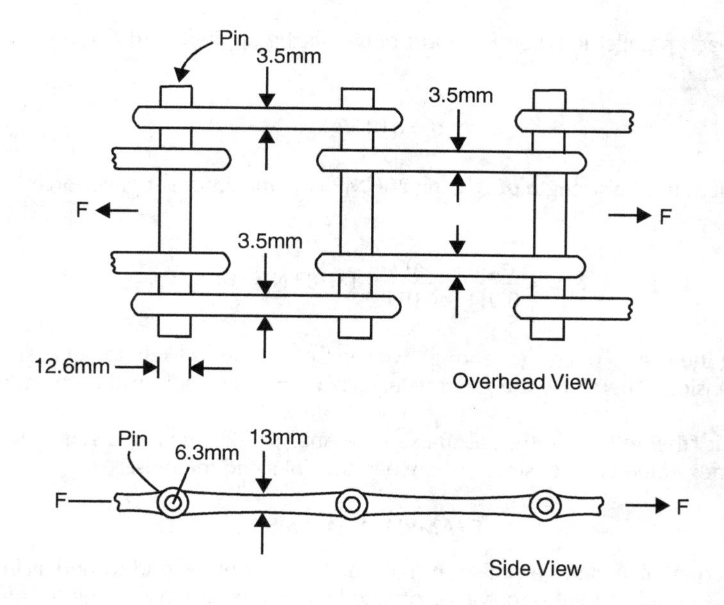

FIGURE A Chain-link joint.

To this end, in considering pin shear, the cross-section area of the pin, with the new given data is:

$$A_{pin} = \pi r^2 = \pi\left(6.3\times10^{-3}\right)^2 = 124.7\times10^{-6}\ \text{m}^2 \tag{a}$$

The shear stress τ on the pin is then:

$$\tau = \frac{(F/2)}{A_{pin}} \tag{b}$$

where, as before, with ideal geometry, each end of the pin need support only half the applied load F.

From the result of Eq. (a), τ is:

$$\tau = 4.01F\ \text{kPa} \tag{c}$$

With a maximum shear stress τ_{max} being 70 MPa, the maximum applied force F_{max} before pin shear failure is:

$$F_{max} = \frac{\tau_{max}}{4.01} = \frac{70,000}{4.01} = 17.45\ \text{kN} \tag{d}$$

Next, for the tension on the link, the cross-section area of the link with the new data is:

$$A_{link} = (3.5)(13)\times10^{-6} = 45.5\times10^{-6}\ \text{m}^2 \tag{e}$$

As before, the tensile stress σ in the links is:

$$\sigma = \frac{(F/2)}{A_{link}} \tag{f}$$

where again each parallel link need support only half the applied load F. From the result of Eq. (e) σ is:

$$\sigma = 0.011F \text{ kPa}$$

With a maximum tensile strength σ_{max} being 200 MPa the maximum applied force F_{max} before link tension failure is:

$$F_{max} = \frac{\sigma_{max}}{0.011} = \frac{200}{0.011} = 18{,}182 \text{ N} = 18.182 \text{ kN} \tag{g}$$

By comparing the results of Eqs. (d) and (g), we see that the pin is likely to fail in shear before the links fail in tension. Therefore, the theoretical maximum pull load on the chain-link system is: 17.45 kN.

This result is "theoretical" in that, as in Example Analysis 12, we have assumed ideal geometry, and we have neglected contact stresses between the links and the pins.

EXAMPLE ANALYSIS 3

Two circular, equal diameter, transmission shafts are to be coupled end-to-end as in Figure a.

Suppose the coupling joint is to consist of a collar and pins as illustrated in Figure b.

Let the shafts each have radius R and let the pin radii be r, as in Figure c.

Let an axial moment (or "torque") T be transmitted by the shafts through the coupling joint. Assuming ideal geometry, determine the stress exerted on the pins.

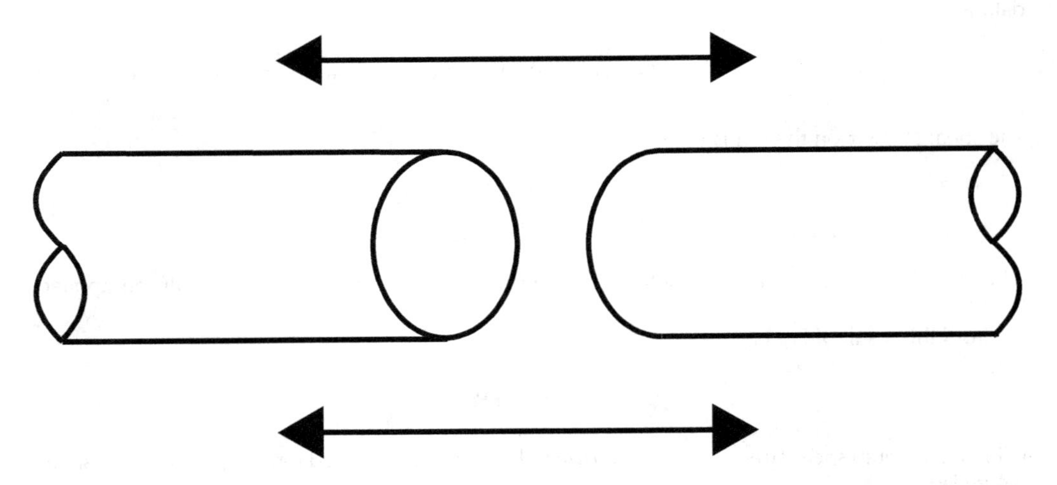

FIGURE A Transmission shafts to be coupled together.

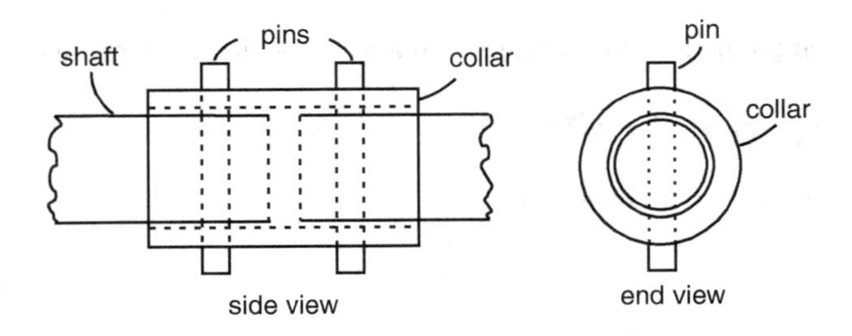

FIGURE B Collar/Pin coupling joint.

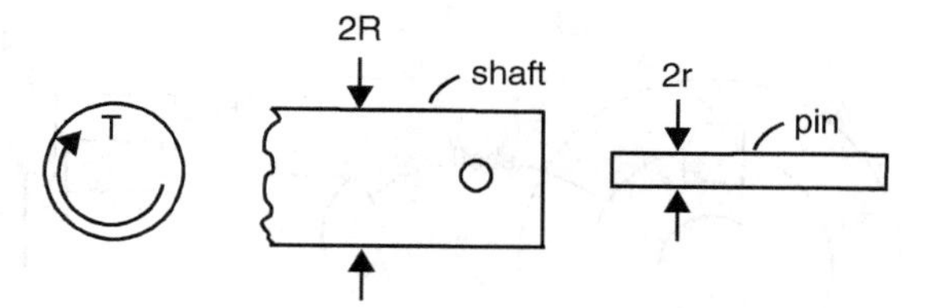

FIGURE C Transmission shaft and coupling pin.

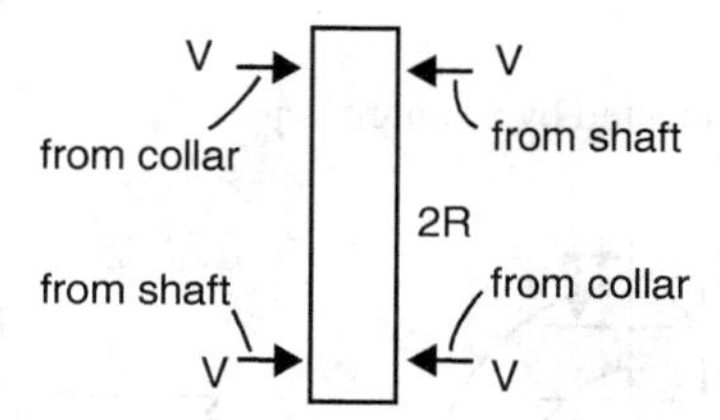

FIGURE D Shear forces are the shaft/coupling pin.

Solution

From the end view in Figure b, we see that during the torque transmission the shaft and collar will exert shear forces at the ends of the pins as in Figure d. Assuming ideal geometry, and then considering a moment balance about the shaft axis, we see that the shear force magnitude V is related to the transmission torque T by the simple expression:

$$2RV = T \quad \text{or} \quad V = \frac{T}{2R} \tag{a}$$

Finally, the shear stress τ on the pin is simply the shearing force divided by the pin cross-section area A. That is:

$$\tau = \frac{V}{A} = \frac{T}{2\pi r^2 R} \tag{b}$$

Comment

Equation (b) presents a design formula for the pin radius r given a transmission torque, shaft radius, and shear stress strength. Since in practice the coupling geometry is likely to be less than ideal, a safer design formula is obtained by simply assuming all the shear force concentration at one end of the pin. Under this assumption Eq. (b) is replaced by:

$$\tau = \frac{T}{\pi r^2 R} \quad \text{or} \quad r = \left[\frac{T}{\pi r \tau}\right]^{1/2} \tag{c}$$

EXAMPLE ANALYSIS 4

Square keys are commonly used in shaft/hub connections as represented in Figure a. When a moment T is applied to the shaft, the key is subjected to shear forces.

We can keep the shear forces from creating excess shear stresses by either: (i) increasing the key cross-section area; or (ii) increasing the length of the key.

Suppose a shaft with radius r is transmitting a moment T through the key to the hub. Let the key cross-section side be: a and length be: ℓ as in Figure b.

Determine an expression for the shear stress τ in the key in terms of the shaft moment and the shaft and key dimensions.

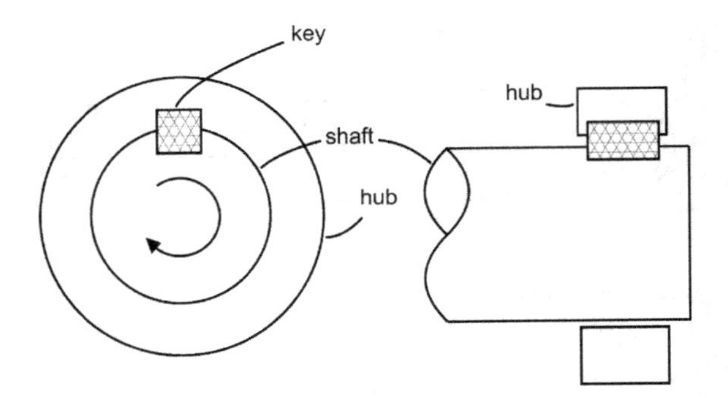

FIGURE A A shaft and hub connected by a square key.

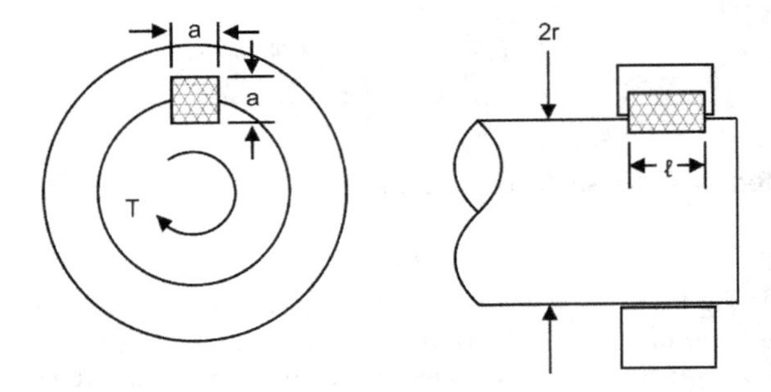

FIGURE B Shaft and key dimensions.

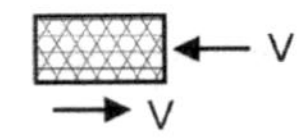

FIGURE C Equivalent shear forces on the upper half of the key.

Solution

Envision a free-body diagram of say the upper half of the key, housed in the hub. Figure c depicts the equivalent shearing force components V on the half key.

By inspection and comparison of Figures b and c we immediately see that V is:

$$V = \frac{T}{r} \tag{a}$$

The shear stress τ is simply the shear force V divided by the sheared stress area A. Therefore, τ is:

$$\tau = \frac{V}{A} = \frac{\left(\frac{T}{r}\right)}{(a\ell)}$$

or simply

$$\tau = \frac{T}{ra\ell} \tag{b}$$

EXAMPLE ANALYSIS 5

Review Example Analysis 4. Suppose we have a 2 in. diameter shaft with a ½ in. square key. Suppose the key length is 1½ in. Suppose further that the maximum shear stress τ_{max} of the key material is 30,000 psi. Determine the maximum shaft moment T_{max} which can be transmitted through the joint.

SOLUTION

From Eq. (b) of the solution of Example Analysis 34 we have the relation between the shaft moment T and the shear stress τ:

$$\tau = \frac{T}{ra\ell} \quad \text{or} \quad T = ra\ell\tau \tag{a}$$

where as before r, a, and ℓ are the shaft radius, key side width, and key length, respectively.

By substituting the values of the given data into Eq. (a) we obtain the maximum shaft moment T_{max} as:

$$T_{max} = (1)(0.5)(1.5)(30)(10)^3 = 22{,}500 \text{ in lb}$$

or

$$T_{max} = 1875 \text{ ft lb} \tag{b}$$

COMMENT

The result of Eq. (b) is a theoretical maximum. The shear stress maximum is an ideal material value and the formula of Eq. (a) is developed assuming ideal geometry. Therefore, for actual designs, we should limit the shaft moment transmission to only a fraction of the value given by Eq. (b).

EXAMPLE ANALYSIS 6

Repeat Example Analysis 5 using the following data:

Shaft diameter: $2r = 50$ mm
Square key side: $a = 12.5$ mm
Key length: $\ell = 37.5$ mm
Maximum key shear stress: $\tau_{max} = 200$ MPa

SOLUTION

From Eq. (b) of the solution of Example Analysis 4 the shaft moment T is:

$$T = ra\ell\tau \tag{a}$$

By substituting the given data values, T_{max} is:

$$T_{max} = \left[(25)(10^{-3})\right]\left[(12.5)(10^{-3})\right]\left[(37.5)(10^{-3})\right]\left[(200)(10^{6})\right] \text{ Nm}$$

or

$$T_{max} = 2.344 \text{ kN m} \tag{b}$$

REFERENCES

1. Sass, F., Ch. Bouche and A. Leitner. 1966. *Dubbels Taschenbuch fur den Maschinenbau*. Berlin: Springer Verlag.
2. Dobrovolsky, V., K. Zablonsky, S. Make, et al. 1968. *Machine Elements*. Moscow: Mir.
3. Orthwein, W. C. 1979. A new key and keyway design. *Journal of Mechanical Design*, Vol. 101, American Society of Mechanical Engineers, Paper No. 78-WA/DE-7.
4. Weiner, P. D. and F. D. Sewell. 1967. New technology for improved tubular connection performance. *Journal of Petroleum Technology*. 19(3): doi:10.2118/1601-PA
5. Blose, T. L. 1970. *Leak Resistance Limit-Tubular Products*. Paper 70-Pet-9, New York: American Society of Mechanical Engineers (ASME).
6. Blose, T. L. and R. L. Vingoe. 1972. *New Developments in High Pressure Oil Well Tubular Products*. Paper 72-Pet-27, New York: American Society of Mechanical Engineers (ASME).
7. Blake, A. 1982. *Practical Stress Analysis in Engineering Design*. New York and Basel: Marcel Dekker, Inc.
8. Scott, R. G. and J. C. Stone. 1981. *The Effects of Design Variables on the Critical Stresses of Eye Bars under Load: An Evaluation by Photoelastic Modeling*. UCRL-85805. Livermore, CA: Lawrence Livermore National Laboratory.

SYMBOLS

A	Area of contact
a	Radius of cylinder
a_0	Length of groove or pin support
B	Width of eyebar
b	Radius of cylinder
b_0	Width of groove or key
c	Radius of cylinder
D	Diameter of shaft or socket
d	Diameter of bolt or shank
E	Modulus of elasticity
e	Radial clearance
F	Force
f	Coefficient of friction
g	Pitch of thread
H	Depth of socket
h	Depth of key
h_1	Distance to end of shank
h_2	Limit distance
i	Taper ratio
$k = L/a_0$	Length ratio for eyebar
$k_1 = L/d$	Length to diameter ratio for pin
k_2	Auxiliary parameter
L	General symbol for length
L_0	Shackle distance
M	General symbol for moment
M_f	Frictional moment
M_0	Fixing moment
M_t	Torsional moment
M_1	Bending moment in curved part
M_2	Bending moment in straight part
m	Distance to groove

N	Normal force
n	Arbitrary number
P	Concentrated load
P_b	Bearing pressure
P_c	Critical pressure
P_i	Internal pressure
q	Contact pressure
q_a	Average contact pressure
q_f	Frictional drag
q_{max}	Maximum contact pressure
R	Radius of eyebar or curved beam
R_i	Small radius of cone
R_m	Mean radius of spline
R_0	Large radius of cone
r	Radius of circular cross-section
r_0	Radius of pin in shackle
S	General symbol for stress
S_b	Bending stress
S_c	Compressive stress
S_h	Hoop stress
S_t	Tensile stress
S_u	Tensile strength (ultimate)
S_y	Yield strength
S_{max}	Maximum combined stress
t	Thickness
W	External load
x	Arbitrary distance
y	Arbitrary distance
Z	Section modulus
α	Angle of taper
$\beta = L/R$	Length ratio in shackle
δ	Radial interference; clearance
ε	Distance to center
$\eta = e/r$	Clearance ratio
θ	Arbitrary angle
$\lambda = R/r$	Curvature ratio
ρ	Angle of friction
ψ	Stress correction factor
τ	Shear stress
τ_{max}	Maximum shear stress

9 Design of Welded Joints

9.1 INTRODUCTION

The basic function of a welded joint is to transfer the stress across a mechanical boundary and to maintain a geometrical relationship between the various components comprising a particular system. Thus, the role of a weld is similar to that of any other fastener and demands similar care in design.

Often a welded connection is generally simpler and more compact than a conventional joint involving bolts or rivets. Additionally, various manufacturing complications can be avoided such as drilled holes, framing elements, or gusset plates when designing for welding. Other desirable features of a welded joint include smaller weight and easier maintenance.

In general, the successful performance of every structure depends upon the integrity of the joint and the connected members. The connections that are not adequately designed invite local overstress and eventual structural failure.

By analogy to conventional mechanical fasteners, the design of a welded joint should include the considerations of strength, stiffness, deformation, load capability, and economy. These considerations require experience and design knowledge related to the types of welds, allowable stresses, working equations, material limitations, and special features of weld behavior—particularly in the presence of incipient cracks and the heat affected zone (HAZ).

This chapter, as well as other portions of the book, continues to emphasize the basic philosophy that the structural response of any mechanical interface is sufficiently complex so that a single design formula cannot reliably model the behavior of the entire joint.

While it is exceedingly important to select a rational and well-proven method of design for a particular welded connection, it should be realized that the performance of the joint is affected by many variables related to countless details of the welding process.

The general term "welding" has, over the centuries, denoted the process of connecting metal parts under the application of heat and pressure. The term welding then evolved into essentially describing the method of joining the metals by means of an electric arc, applied for the first time over 100 years ago [1]. Current arc welding methods include the applications of submerged arc, shielded metal arc, and gas metal arc technology to the fabrication of welded components and entire systems. At least 40 welding processes are now classified by the American Welding Society (AWS) [2].

Control of heat is perhaps the most important variable affecting weld integrity. The thermal input creates a number of physical changes leading to plastic deformations and residual stresses. The increase in temperature at and around the weld causes a decrease in both yield point and the elastic modulus of the material. Experience shows, however, that residual stresses cannot be formed until the temperature of the welded region approaches 1400EF when the material, such as steel, loses the capacity to resist loading caused by differential movements.

Residual stresses induced by welding can approach the yield strength of the weld material. These stresses can be kept at a minimum by applying strict controls of fabrication, specifying the appropriate preheating cycles, and avoiding weld intersections. Similar care is required in controlling distortions, so that tedious and costly mechanical straightening can be avoided at a later state of the construction.

The presence of a highly localized residual stress region in the weld or HAZ, extending to about one to two weld widths, may precipitate fracture due to a relatively small crack. Here, therefore, lies the main disadvantage of welding: A brittle, fracture-resistant structure cannot be assured by only

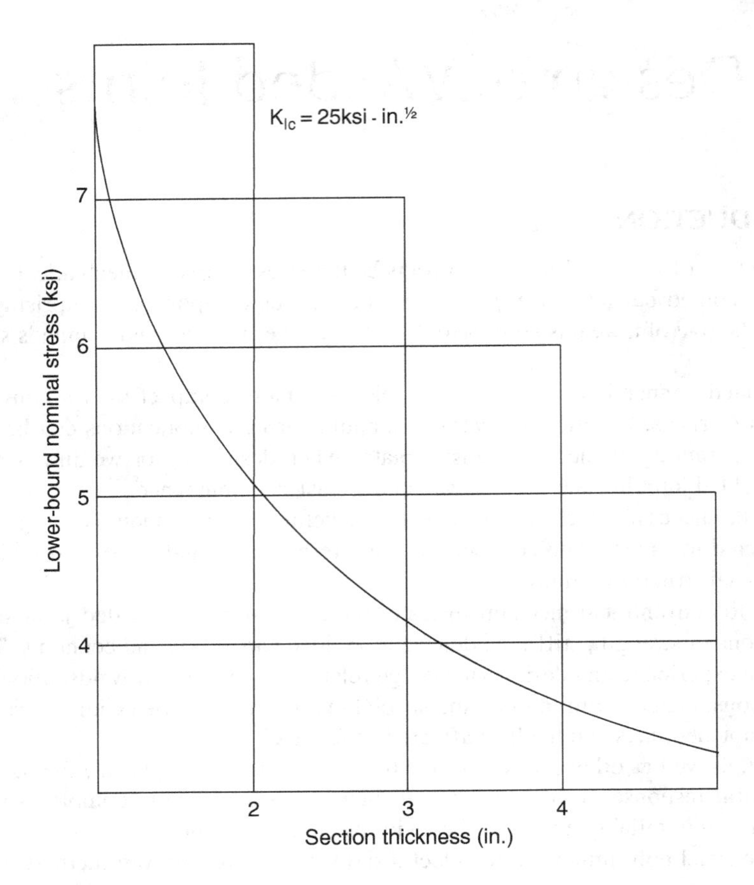

FIGURE 9.1 Lower bound design stress for fracture control for steel.

the selection of a working stress satisfying a conventional design criterion. Good practice dictates that the probability of a brittle failure can be minimized if we can adhere to the following principles:

- Avoid stress gradients
- Provide crack arresters
- Design for redundant stress paths
- Avoid highly restrained configurations
- Specify preheat and postheat criteria
- Specify low-hydrogen electrodes and the submerged-arc process
- Avoid arc strikes and dynamic abuse of welds

When it is difficult to qualify the material, weld conditions and design methodology for a particular structure, it may be necessary to invoke a highly conservative process of structural review based on the absolute levels of safe design stress, such as that given for steel and aluminum in chart form by Figures 3.21 and 3.22 and shown again here for convenience in Figures 9.1 and 9.2.

9.2 TYPICAL WELDED JOINTS

There are two basic types of welds involving either fillet or groove configurations. These can be further grouped into joints and weld classifications. The joints may be characterized as tee, lap, butt, corner, or edge type as shown in Figure 9.3. The weld type classification requires machining or flame-cutting preparation of the mating parts into the proper shapes for good welding quality.

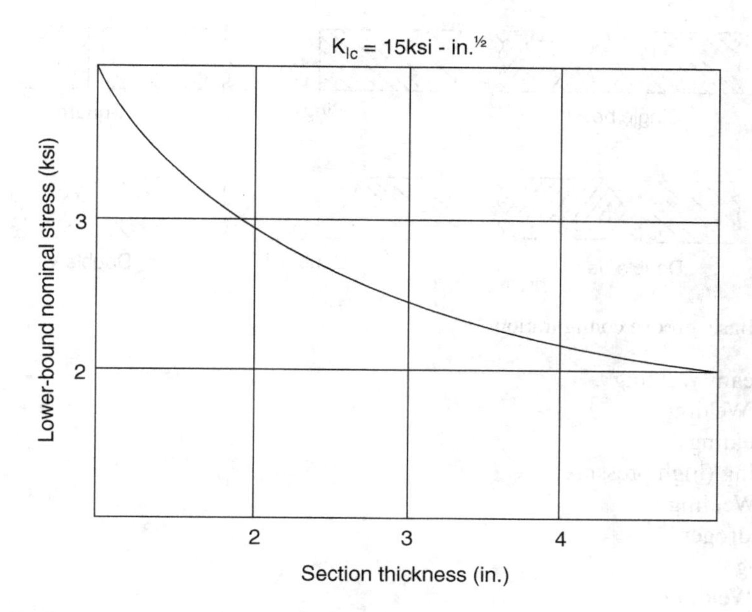

FIGURE 9.2 Lower bound design stress for aluminum.

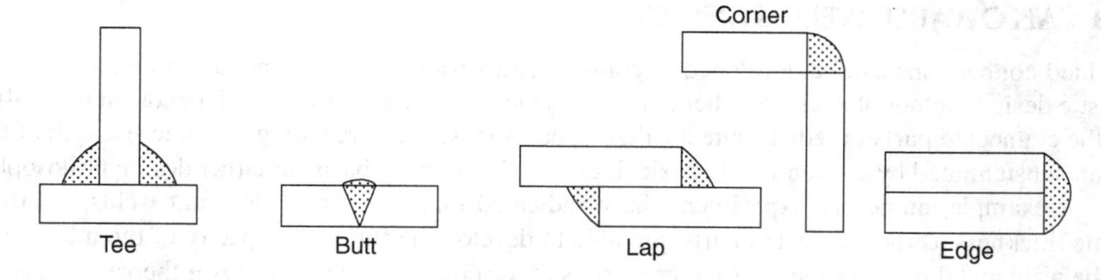

FIGURE 9.3 Basic weld joint configurations.

Figure 9.4 illustrates this concept involving various groove configurations. The primary difference between the two weld designs is the manner in which the forces are transferred from one member to another. The joint-type welds shown in Figure 9.3 are likely to be subjected to all combinations of stresses, since the weld material is placed on the edges of the base metal. On the other hand, full penetration welds in the grooves shown in Figure 9.4 behave essentially like the connected components because the joint forces are transmitted by direct shear, compression, or tension.

In general, groove welds can be recognized according to their characteristic form such as square, bevel, J or V. Several of these may be used as either single or double welds. Experience also shows that groove welds tend to be stronger than the corresponding fillet weld designs although many structural joints are made by applying fillet welding.

In practice, the two basic types of welds may be combined to form a particular connection.

There are numerous processes available for making these welds. They include:

- Fusion Welding
  - Oxyacetylene Flame
  - Electric Arc
- Resistance Welding
  - Spot Welding
  - Seam Welding

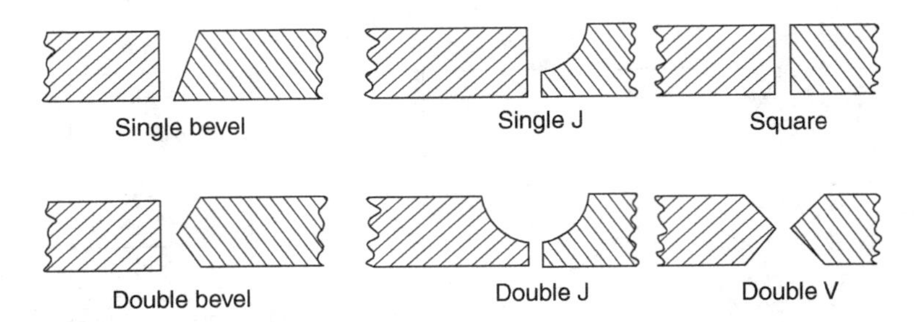

FIGURE 9.4 Basic groove configurations.

Electron Beam Welding
Ultrasonic Welding
Friction Welding
Cold Welding (high pressure)
Explosive Welding
Atomic Hydrogen
Hard Facing
Lap Seam Welding
Diffusion Welding

9.3 ALLOWABLE WELD STRESSES

Welded connections can be developed to comply with various requirements using either elastic or plastic design methodologies. In either case, a satisfactory joint performance depends on the ability of the connected parts to redistribute localized peak stresses. The resulting ultimate strength of the joint, substantiated by conventional physical tests, is the primary basis for either design philosophy.

For example, numerous experiments have indicated that full penetration butt welds, with the same thickness as the connected parts, are able to develop the full load capacity of the main parts if the weld metal matches the physical properties of the connected parts [1]. Over the years, various tests have been made on single welds and complete structural joints to determine the maximum weld stresses that can be attained.

Table 9.1 shows the elastic design allowables recommended by the AWS [3]. Recommendations for matching of the base and weld metals have been developed by the American Institute of Steel Construction (AISC). Table 9.2 [4] provides the relevant material selections.

Welded connections, designed with the aid of a plastic theory, generally operate at higher stress levels than those sized on the basis of elastic methodology. For plastic design, full penetration groove welds are assumed to be capable of developing stresses equal to the tensile yield of the weaker of the base or weld metal [5]. Fillets can be made to develop stresses equal to the shear yield of the weaker of the base or weld metal. Conservative plastic design allowables may be obtained by multiplying the corresponding elastic allowables of Table 9.1 by: 1.67.

9.4 CRACK INITIATION AND FATIGUE

The process of weld forming in a joint can be compared to the casting of steel. Accordingly, the internal structure of a weld can be expected to develop poor fusion, porosity, inclusions, and shrinkage cracks. These effects, together with the metallurgical transformations, are produced due to the internal heterogeneities and the presence of HAZ.

The HAZ is that volume of the adjacent base metal that has not been melted, but in which the microstructure and the mechanical properties have been changed by the process of welding. The most significant result of HAZ is that the metal adjacent to the weld is coarsened. These effects,

TABLE 9.1
AWS Elastic Design Allowable Weld Stresses

Type of Weld and Stress	Permissible Stress	Required Strength Level
Complete penetration groove welds		
Tension normal to the effective throat.	Same as base metal.	Matching weld metal must be used.
Compression normal to the effective throat.	Same as base metal.	Weld metal with a strength level equal to or less than matching weld metal may be used.
Tension or compression parallel to the axis of the weld.	Same as base metal.	
Shear on the effective throat.	.30 × nominal tensile strength of weld metal (ksi) except stress on base metal shall not exceed .40 × yield stress of base metal.	
Partial penetration groove welds		
Compression normal to effective throat.	Same as base metal.	Weld metal with a strength level equal to or less than matching weld metal may be used.
Tension or compression parallel to axis of the weld.	Same as base metal.	
Shear parallel to axis of weld.	.30 × nominal tensile strength of weld metal (ksi) except stress on base metal shall not exceed .40 × yield stress of base metal.	
Tension normal to effective throat.	.30 × nominal tensile strength of weld metal (ksi) except stress on base metal shall not exceed .60 × yield stress of base metal.	
Fillet welds		
Stress on effective throat, regardless of direction of application of load.	.30 × nominal tensile strength of weld metal (ksi) except stress on base metal shall not exceed .40 × yield stress of base metal.	Weld metal with a strength level equal to or less than matching weld metal may be used.
Tension or compression parallel to the axis of weld.	Same as base metal.	
Plug and slot welds		
Shear parallel to faying surfaces.	.30 × nominal tensile strength of weld metal (ksi) except stress on base metal shall not exceed .40 × yield stress of base metal.	Weld metal with a strength level equal to or less than matching weld metal may be used.

together with the geometrical stress risers, surface ripples, and voids, can increase the chance of crack initiation with the subsequent reduction in strength and fatigue resistance.

Because of their unique geometry, some of the weldment configurations can act as built-in stress risers. Sharp corners and notches, for example, can be found in welded areas where weld penetration is incomplete. Figures 9.5 and 9.6 depict typical examples of such design induced discontinuities.

In the case of a schematic view of a single-fillet tee joint, Figure 9.5, there is an obvious eccentric load which creates uneven stress distribution. There is an overall bending of the weldment as shown by the beam-type distribution of normal stresses. The tensile load component causes uneven tensile stresses because of the local stress riser.

The complete double-fillet tee joint of Figure 9.6, has a built-in crack of length a and corner radius equal to about $e/2$. For relatively small values of e, this configurational crack can be even sharper, indicating a high stress concentration factor such as one might expect to exist at the bottom of a sharp notch.

Since the tee joint is often used in a tensile mode, such as that shown in Figure 9.6, the designer should be aware of this detrimental feature of the joint wherever a critical tensile load path is

TABLE 9.2
Matching Weld and Base Metal

Weld Metal	60[a] (or 70)	70	80	90	100	110
Type of Steel	A36 A53 Gr B A106 Gr B A131 A139 Gr B A375 A381 Gr Y35 A500 A501 A516 Gr 55, 60 A524 A529 A570 Gr D,E A573 Gr 65 API 5L Gr B ABS Gr A,B,C,CS,D,E,R	A242 A441 A537 Class 1 A516 Gr 65, 70 A572 Gr 42–60 A588 A618 API 5LX Gr 42 ABS Gr AH,DH,EH	A537 Class 2 A572 Gr 65		A514 A517 2½″ & over	A514 A517

[a] Strength level in ksi

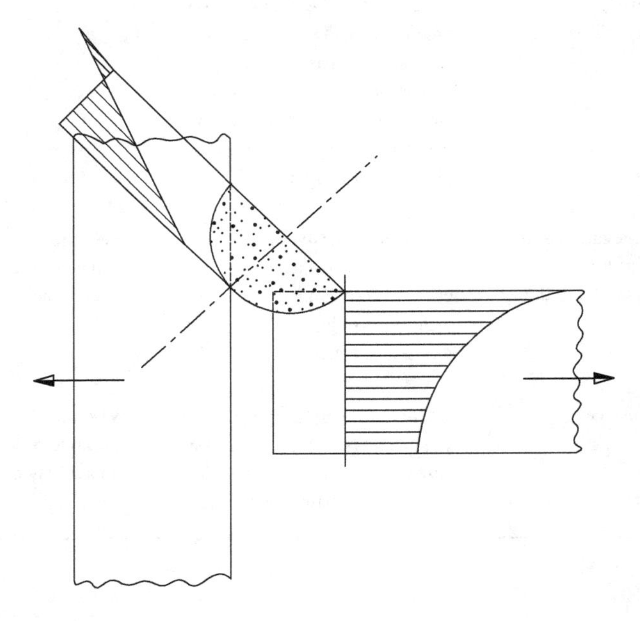

FIGURE 9.5 Single-fillet tee joint.

involved. Although the foregoing illustrations involved tee configurations, other designs are possible where the geometry, accentuated by the lack of penetration, could become the source of initiation and propagation of a structural crack.

According to the current state of knowledge of weld zone cracking, there can be various metallurgic and process causes of crack development. These can be summed up here as follows:

- Solidification cracks due to hot tearing of weld metal
- Liquation cracks of HAZ
- Cold cracking in HAZ
- Lamellar tearing of base metal
- Stress-relief cracking of HAZ
- Hydrogen-assisted cracking of HAZ and weld metal
- Restraint cracking of HAZ and weld metal

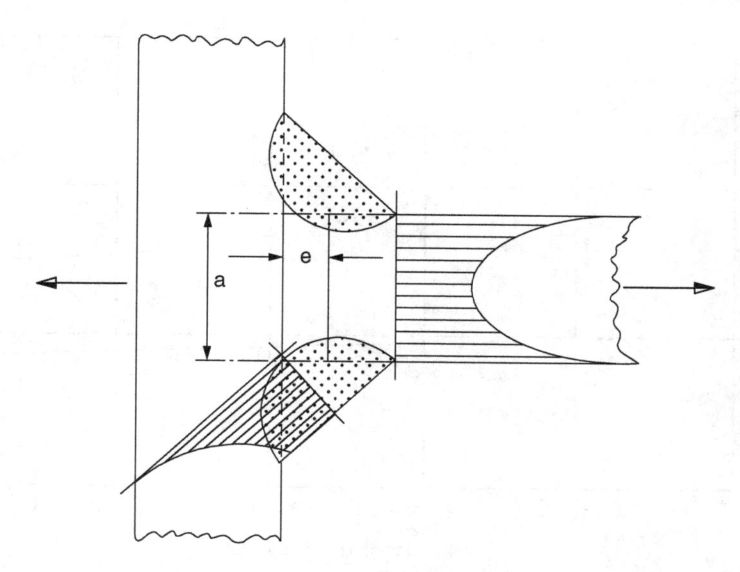

FIGURE 9.6 Double-fillet tee joint.

Since the weld regions can provide a trigger for initiating the cracks, special controls must be put on the welding processes, particularly when welding materials which are sensitive to temperature and strength transitions. This element of quality control and careful analysis aimed at avoiding undue stress concentrations are necessary to establish good practices in weldment design and manufacture.

The fatigue life of a structural weld consists of three phases:

- Initiation of a microscopic crack
- Crack growth to a critical size
- Exceeding the strength of the cracked part

It is difficult to quantify the effect of incipient cracks on the fatigue strength of a weld because of a lack of uniform criteria for experimental analysis. Lack of weld penetration, such as that shown in Figures 9.5 and 9.6, lowers the fatigue strength of a transverse weld significantly. Porosity and inclusions decrease fatigue strength in proportion to the decrease in effective weld area, while the presence of severe notches decreases the fatigue resistance, regardless of tensile strength. These effects arise primarily in the crack initiation phase. It follows then, that any improvement in fatigue life due to an enhanced tensile strength will be realized only when notch severity is reduced.

When a structural steel member involves a welded butt joint, the resultant fatigue life is expected to be lower if the joint is oriented transverse to the load direction. There is virtually no effect of longitudinal butt welds on the fatigue strength of carbon steel. Figure 9.7 provides, in graphical form, an envelope of the maximum stress in terms of the number of fatigue producing cycles for a welded butt joint [11].

It appears that the maximum stress for this case is about constant for the number of fatigue cycles greater than about one million. Figure 9.8 [1] provides an approximate envelope of test results for the fatigue life of transverse and intermittent longitudinal fillet welds in carbon steel joints.

At longer lives, say over one million of cycles, the behavior of the welded constructional alloys and carbon steels is quite similar. However, the alloy steels are generally more sensitive to notch effects. A general guide is that intermittent longitudinal welds and the transverse welds are expected to lower the fatigue resistance of a structural joint. In addition, experience shows that the majority of weld failures in practice are initiated at the weld discontinuities.

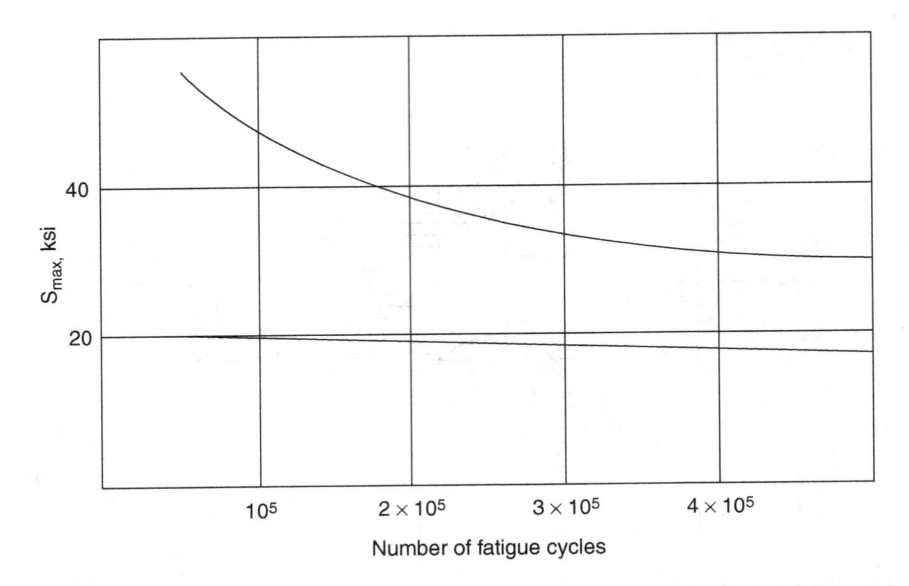

FIGURE 9.7 Fatigue envelope for carbon steel transverse butt welds.

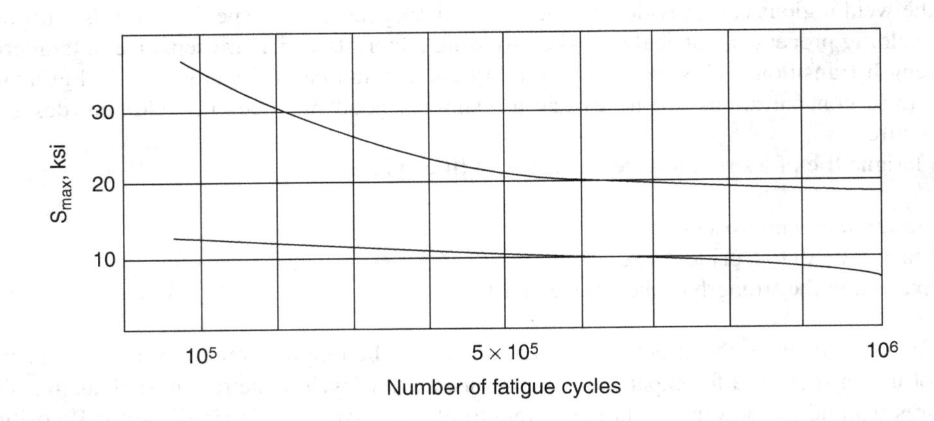

FIGURE 9.8 Fatigue envelope for carbon steel fillet welds [1].

In the case of stiffeners welded to webs, fatigue failure starts at the termination of the web to stiffener fillet weld. When a stiffener is welded to the flange, the tensile failure is usually observed at the toe of the fillet weld on the flange.

9.5 DESIGN ASSUMPTIONS

As stated previously, there are two basic types of weld configurations known as "butt" and "fillet" welds. With the former type, the strength is found by multiplying the cross-sectional area of the thinnest connected plate by the allowable tensile or compressive stress for the weld material. The design of a fillet weld, however, is still largely empirical and it is related to the allowable shear strength across the smaller dimension defined as the theoretical weld throat. For equal weld legs of the fillet, this definition leads to the familiar multiplier of $\sqrt{2}/2$, as in Figure 9.9.

According to the AWS, the strength of a fillet weld, regardless of the direction of the applied force, is simply the throat cross-sectional area multiplied by the allowable shearing stress for the weld metal. This is a practical rule of thumb for initial design purposes, however, the actual stress

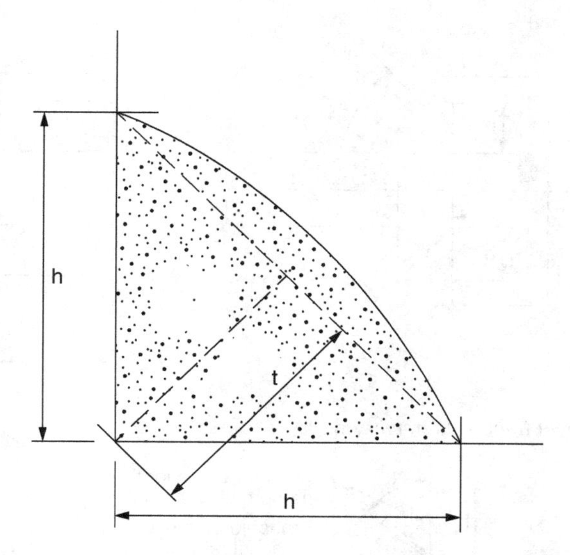

FIGURE 9.9 Theoretical throat of fillet weld.

pattern is much more complicated and certain local areas of a particular joint configuration may bear only a limited resemblance to the average shear strength of the weld throat.

In line with the concept of average shear stress, the total shear load is divided by the total available cross-sectional area which may or may not coincide with the plane of rupture.

When welds are sized according to the elastic design method, the conventional beam theory can be used. The relevant design equations can be summed up as follows:

$$\text{Axial stress} \quad S_a = \frac{W}{A_W} \tag{9.1}$$

$$\text{Bending stress} \quad S_b = \frac{Mc}{I} \tag{9.2}$$

$$\text{Torsional stress} \quad \tau = \frac{M_t r}{I_p} \tag{9.3}$$

$$\text{Horizontal shear} \quad \tau = \frac{VQ}{Iw} \tag{9.4}$$

Since the transverse forces applied to a beam cause bending, normal and shear stress components can result in a weld connection. The effect of torsional loads can be reflected in varied stress distributions for different cross-sectional shapes, where the shearing stress can be assumed to vary linearly from the center of the weld joint by analogy to the shear stress distribution in a shaft subjected to twist.

For a symmetrical fillet weld in tension, as illustrated in Figure 9.10, experiments indicate that it is safe to assume that the stress at the throat section is essentially tensile. On this basis, it is customary to use the design formula:

$$S_a = \frac{\sqrt{2}W}{2bh} \tag{9.5}$$

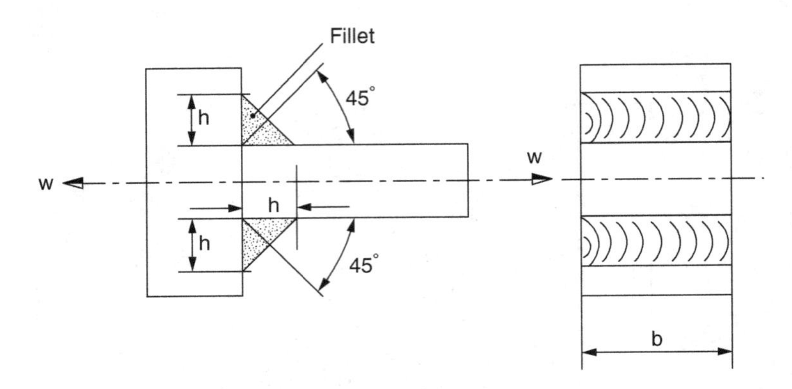

FIGURE 9.10 Symmetrical fillet weld in tension.

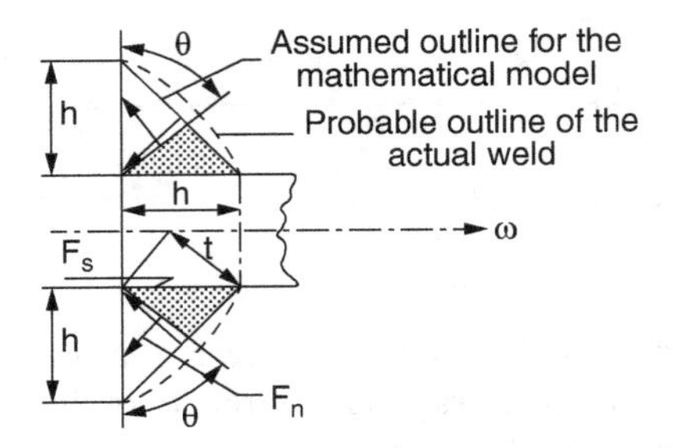

FIGURE 9.11 Free-body diagram of fillet weld.

Equation (9.5) should be considered to be largely conservative because one half of the total external load W was assumed to act on the theoretical throat cross-section equal to $hb\sqrt{2}/2$.

9.6 ANALYSIS OF FILLET WELDS

Figure 9.11 [6] presents a free-body diagram of a double-fillet weld. In the figure, we assume a weld seam of unit length and an arbitrary section be defined by θ. The normal and shear components are denoted by F_n and F_s, respectively. The throat dimension measured at angle θ is taken to be equal to t. In terms of the specified quantities, this dimension can be expressed as:

$$t = \frac{h}{\sin\theta + \cos\theta} \tag{9.6}$$

Typically, we assume $\theta = 45°$, so that Eq. (9.6) gives $t = h/\sqrt{2}$. In the derivation that follows, θ is considered to be a variable quantity for the purpose of the mathematical model. Summation of the forces along the line of W then gives:

$$2F_s \sin\theta + 2F_n \cos\theta = W \tag{9.7}$$

Since the components F_s and F_n must be balanced in order to not produce any horizontal reaction on the system, we obtain:

$$F_s \cos\theta - F_n \sin\theta = 0 \tag{9.8}$$

Solving Eqs. (9.7) and (9.8) for the shear component F_s yields:

$$F_s = \frac{W \sin\theta}{2} \tag{9.9}$$

For a length of weld seam equal to b, the shear stress can now be calculated using Eqs. (9.6) and (9.9). This gives:

$$\tau = \frac{W \sin\theta(\sin\theta + \cos\theta)}{2bh} \tag{9.10}$$

Similarly, the normal stress is:

$$S_n = \frac{W \cos\theta(\sin\theta + \cos\theta)}{2bh} \tag{9.11}$$

An examination of Eq. (9.11) shows that at no point of the weld does the theoretical stress exceed the value computed from Eq. (9.5).

Also, for practical reasons, when fillet welds are placed on both sides of a plate, as in Figure 9.10, the maximum size of the weld leg should be limited to about 75% of the thickness of the thinner plate.

9.7 WELD LINE FORMULAS FOR DESIGN

The evaluation of the maximum resultant weld stress and the design of a welded connection can be simplified if we consider the weld as a line without any cross-sectional area. It can be shown, for example, that such a geometrical property as section modulus of any thin area is equal to the property of a line configuration multiplied by its thickness.

Consider the case of a thin ring with thickness t and average radius r, as illustrated by Figure 9.12. The second moment of area and the section modulus are: πr^3 and πr^2, respectively. By multiplying these expressions by the thickness t of the ring, we obtain well-known formulas for a relatively thin ring. It is easy to show that the error made by such approximations is rather small, provided the thickness of the ring is less than 10% of the mean radius.

The great majority of practical cases the weld size is small compared to the overall dimensions of a welded joint. Therefore, the method of treating the weld as a line is sufficiently accurate and very simple to use [7]. On this basis, the original design formulas, Eqs. (9.1) through (9.4), can be restated in terms of the unit forces as:

$$\text{Axial shear force} \quad q_a = \frac{W}{L_w} \tag{9.12}$$

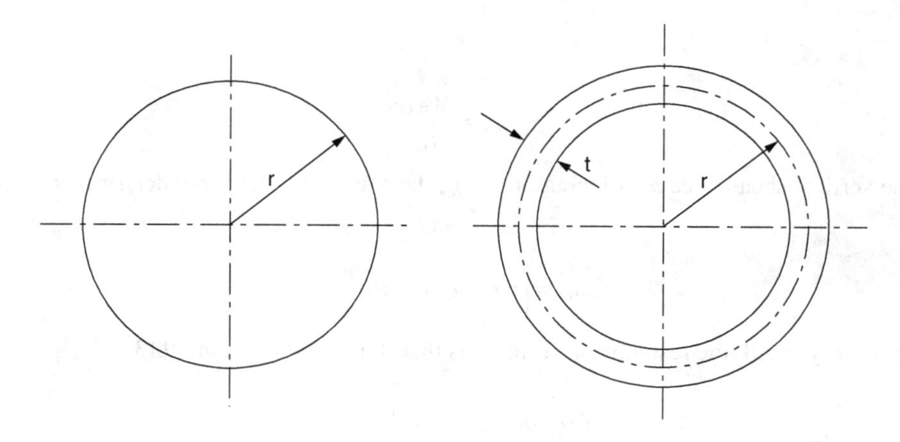

FIGURE 9.12 A circle model of a thin ring.

$$\text{Bending shear force} \quad q_b = \frac{M}{Z_w} \tag{9.13}$$

$$\text{Torsional shear force} \quad q_t = \frac{M_t r}{I_w} \tag{9.14}$$

$$\text{Horizontal shear force} \quad q_v = \frac{VQ}{I} \tag{9.15}$$

The weld forces, defined in Eqs. (9.12) through (9.15), have units of force divided by length. Hence, Z_w and I_w should also have the dimensions of in^2 and in^3, respectively. The geometric properties of weld connections treated as lines are shown in Table 9.3.

The following steps can be used to evaluate the stresses for a welded connection:

- Select one or more locations in the welded joint where the combined shear forces are likely to be of interest.
- Calculate weld line properties with the aid of Table 9.3 or a similar summary of formulas.
- Calculate the appropriate shear force components on the basis of formulas such as those given by Eqs. (9.12) through (9.15).
- Determine the resultant shear vector by combining the relevant shear force components.
- The resultant weld stress can be obtained by dividing the resultant shear force by the actual throat of the fillet.

To illustrate some of the procedural steps given above, consider a welded bracket carrying an eccentric load W shown in Figure 9.13.

It is evident from Figure 9.13 that the particular weld joint is subjected to a transverse load W and a twisting moment We. The transverse shear force follows directly from Eq. (9.12):

$$q_a = \frac{W}{2b} \tag{9.16}$$

In considering the total shear effect at a distance r from the center of gravity, such as that shown in Figure 9.13, it may be first convenient to define the torsional shear force components on the basis of Eq. (9.15). This gives:

$$q_x = \frac{Wey}{I_w} \tag{9.17}$$

and

$$q_y = \frac{Wex}{I_w} \tag{9.18}$$

Since the vertical shear force q_y is increased by q_a, the resultant shear force determined vectorially becomes:

$$q_r = \left[\left(q_a + q_y\right)^2 + q_x^2\right]^{1/2} \tag{9.19}$$

When q_a is very small, the resultant shear force is that illustrated in Figure 9.13.

$$q_r = \left(q_y^2 + q_x^2\right)^{1/2} \tag{9.20}$$

TABLE 9.3
Properties of Various Welds Treated as Line

b = width d = depth	Bending about Horizontal Axis (*)	Torsion
	$Z_w = \dfrac{d^2}{6}$ in.^{2}	$I_w = \dfrac{d^3}{12}$ in.^{3}
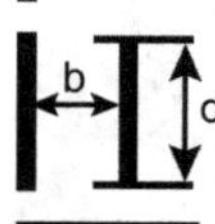	$Z_w = \dfrac{d^2}{3}$	$I_w = \dfrac{d(3b^2 + d^2)}{6}$
	$Z_w = bd$	$I_w = \dfrac{b^3 + 3bd^2}{6}$
$x = \dfrac{d^2}{2(b+d)}$ (**) $y = \dfrac{b^2}{2b+d}$	$Z_w = \dfrac{4bd + d^2}{6}$ $Z_w = \dfrac{d^2(4b+d)}{b(2b+d)}$ at bottom	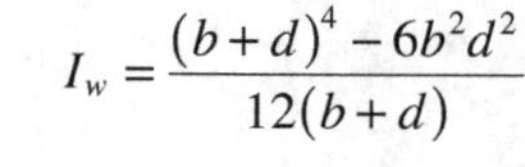 $I_w = \dfrac{(b+d)^4 - 6b^2d^2}{12(b+d)}$
 $x = \dfrac{b^2}{2b+d}$	$Z_w = bd + \dfrac{d^2}{6}$	$I_w = \dfrac{(2b+d)^3}{12} - \dfrac{b^2(b+d)^2}{2b+d}$
$y = \dfrac{d^2}{b+2d}$	$Z_w = \dfrac{d^2(2b+d)}{3(b+d)}$ at bottom	$I_w = \dfrac{(b+2d)^3}{12} - \dfrac{d^2(b+d)^2}{b+2d}$
	$Z_w = bd + \dfrac{d^2}{3}$	$I_w = \dfrac{(b+d)^3}{6}$
$y = \dfrac{d^2}{b+2d}$	$Z_w = \dfrac{d^2(2b+d)}{3(b+d)}$ at bottom	$I_w = \dfrac{(b+2d)^3}{12} - \dfrac{d^2(b+d)^2}{b+2d}$
	$Z_w = bd + \dfrac{d^2}{3}$	$I_w = \dfrac{b^3 + 3bd^2 + d^3}{6}$
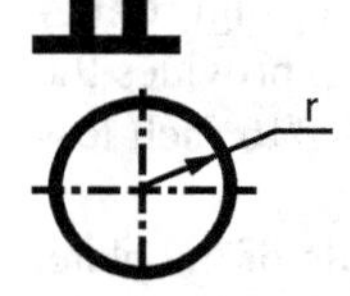	$Z_w = \pi r^2$	$I_w = 2\pi r^3$

(*) For the asymmetrical sections shown, the maximum bending stress will be found at the bottom of the section
(**) The values of x and y coordinated locate the center of gravity of the section

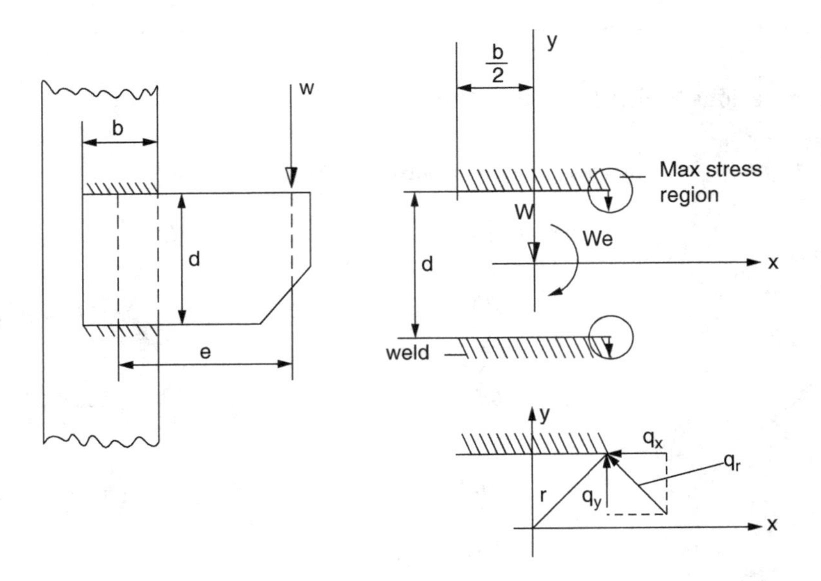

FIGURE 9.13 Example of welded bracket.

If the allowable weld stress in shear is denoted by r_a, the size of the throat t can be calculated as:

$$t = \frac{q_r}{\tau_a} \tag{9.21}$$

The corresponding size of the weld leg h becomes:

$$h = \frac{\sqrt{2}q_r}{\tau_a} \tag{9.22}$$

The simplest process here is to substitute all numerical values and to use the torsional line property from Table 9.3.

9.8 WELDED LAP JOINTS

Welded structural connections can be termed "rigid," "semi-rigid," or of "simple construction," depending on the degree of rigidity between the connected members. Rigid construction is found in many frames and it is probably the most advantageous application of welding technology. Rigid joints assure structural continuity that generally allows a more efficient use of materials. The systems based on intermediate rigidity or semi-rigid design are used if the moment-rotation characteristics are known. In simpler configurations, the connections consist of either web angles or beam seats. Such structural systems are expected to allow free rotation of the joint under load.

If a joint transmits only axial load, it is one of the simplest connections to design. Figure 9.14 illustrates the two most common lap joints. In either of these configurations, Eq. (9.12) provides the method of design analysis. Standard practice, however, limits the fillet weld size to 1/16 inch less than the dimension of the thinnest connected part.

The connection in Figure 9.14(a) should have the weld length the same on each side of the plate. The AWS welding code also recommends that the length of fillet d, along each side, should be greater than or equal to the perpendicular distance between them denoted by b. This insures full development of the plate, despite the so-called shear lag.

Shear lag results from the incipient deformations caused by shear stresses. This phenomenon can be better described by assuming that the relevant weld length under the load is actually getting

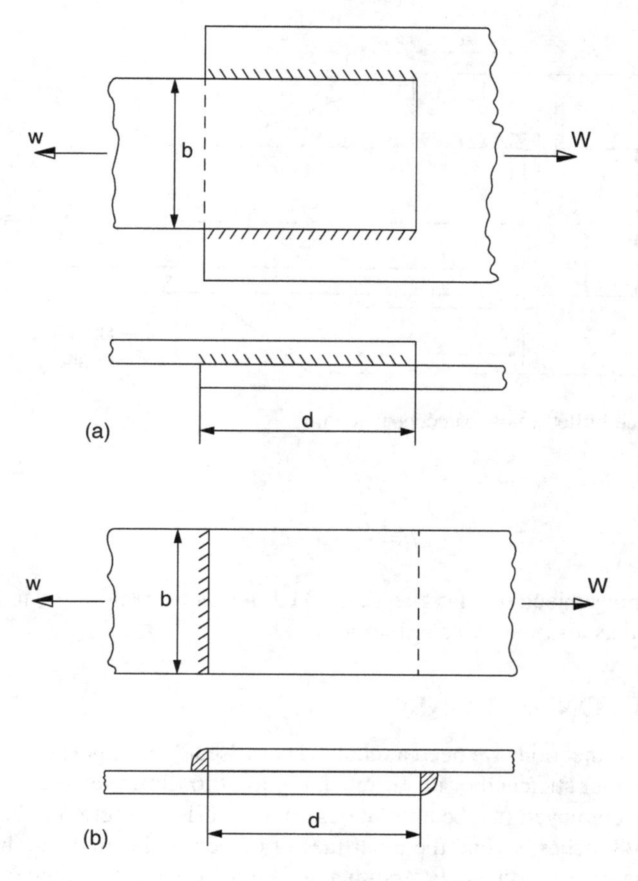

FIGURE 9.14 Standard lap joint.

smaller. As this occurs, the effective width of the plate also tends to decrease since the central portion of the plate, as in Figure 8.14a, carries less load. The AISC welding code recommends the minimum "d" for the connection shown in Figure 9.14b, to be five times the plate thickness.

Occasionally, a situation develops where a single- or double-angle member is subjected to repeated eccentric loads as shown by Figure 9.15. When this occurs, the weld joint should be designed to reduce fatigue-related weld cracking problems. This can be accomplished by balancing the shear forces in the welds about the central axis of the member.

Equilibrium of forces and moments acting on the connections leads to the specific relationships if welds *a*, *b*, and *c* are of the same size. In algebraic terms, this can be defined as:

$$L_w = a + b + c \tag{9.23}$$

Equilibrium of moments of weld lines about the central axis gives:

$$a(c-e)+\frac{1}{2}(c-e)^2 = be+\frac{e^2}{2} \tag{9.24}$$

Combining Eqs. (9.23) and (9.14) results in:

$$a = \frac{eL_w}{c} - \frac{c}{2} \tag{9.25}$$

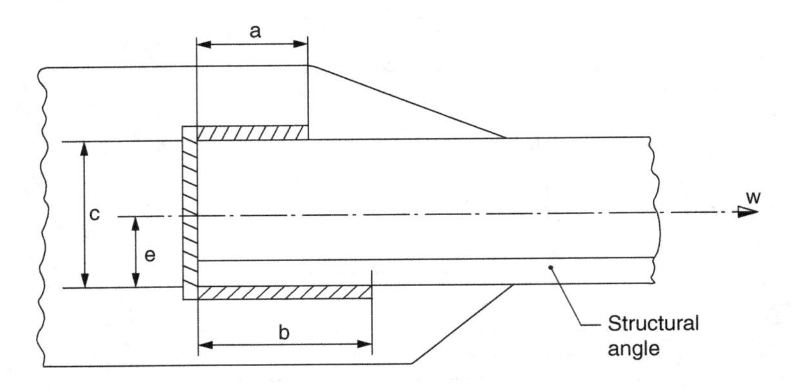

FIGURE 9.15 Balanced fillet welds to eccentric loading.

and

$$b = \frac{L_w}{c}(c - e) - \frac{c}{2} \quad (9.26)$$

If the connection is not subjected to fatigue, the weld dimensions do not need to be balanced and the joint can be designed as a simple lap configuration.

9.9 BEAM SEAT CONNECTIONS

In beam-to-column connections we need a reliable beam "seat" to support the beam. The beam seat (or support) can be either stiffened or unstiffened. Figure 9.16 illustrates the unstiffened version.

The typical angle employed in a beam seat design has a critical section in bending which is normally found about 3/8 inches beyond the inner face of the leg of the angle as shown in Figure 9.16.

The location of beam reaction can be assumed to be uniformly distributed over a length N of the beam to just satisfy web buckling requirements.

The flexible beam seat usually consists of an unreinforced structural angle. The resultant shear force on the weld attaching the flexible seat to the column is obtained by the vector summation of the axial shear force and bending shear force components. The calculations determine the weld size as a function of leg length d_o, shown in Figure 9.16. In most cases, the length is assumed first followed by sizing of the weld.

The recommended procedure then starts with the location of the point where the beam reaction W is applied so that the eccentricity of the load e_o can be defined. The next step is to select the required thickness of the angle t_o in order to assure sufficient bending stiffness. The last portion of the calculation is concerned with the size of the fillet weld designed to fasten the seat angle to the column.

An example of a practical formula for sizing of the weld leg is:

$$h = \frac{W\left(d_o^2 + 20.25e_f^2\right)^{1/2}}{19.2d_0^2} \quad (9.27)$$

Here, the eccentricity of the load can be determined as:

$$e_f = a_0 + 0.5N \quad (9.28)$$

The required length of the beam seat may be approximated from the empirical expression:

$$b_0 = \frac{W\left(e_f - t_0 - 0.375\right)}{4t_0^2} \quad (9.29)$$

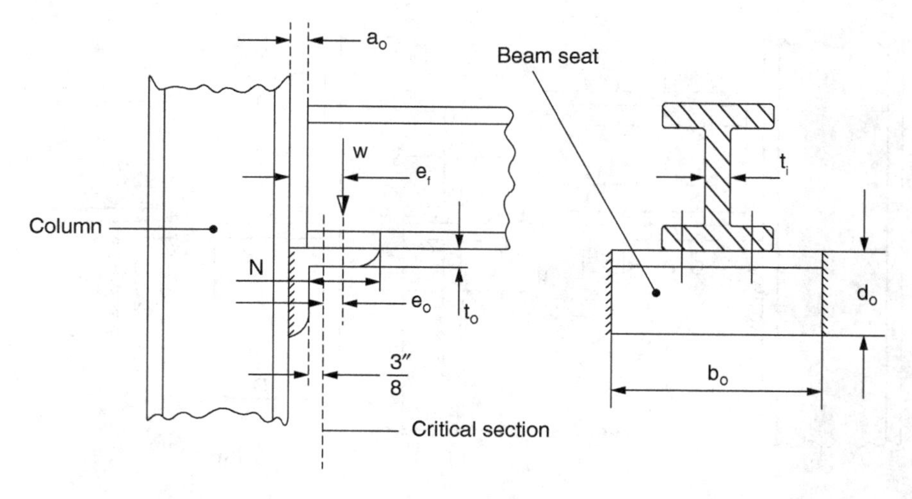

FIGURE 9.16 Flexible beam seat.

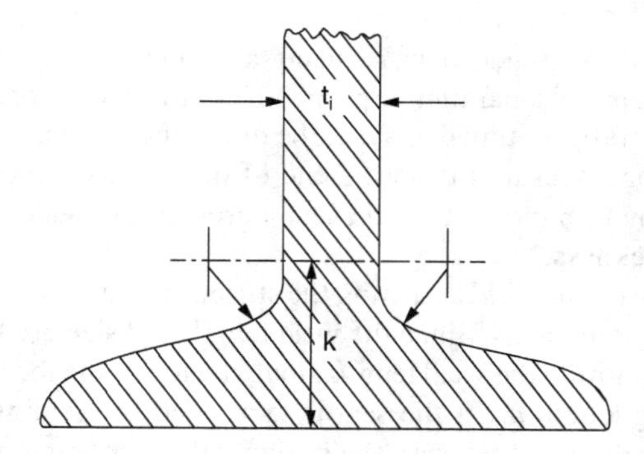

FIGURE 9.17 Part section of the beam.

To select the critical dimension N, welding practice [4] provides the following aid for the calculations:

$$N = \frac{W}{24t_1} - k \tag{9.30}$$

where k is a dimensional quantity which can be obtained from handbooks of rolled steel sections. Essentially, k denotes the distance from the bottom of the beam to the top of the fillet of the section as shown in Figure 9.17. The values of k and t_1 are therefore available for the calculation of the contact length N.

The clearance a_0, indicated in Figure 9.16 and measured from the column to the end of the beam, is often on the order of 0.5 in. The foregoing dimensions and the knowledge of the seat load W can be utilized in Eq. (9.29) in determining the appropriate relationship between the critical dimensions b_0 and t_0 of the seat angle.

Assuming that for a good design $h = 0.8t_0$, the solution involves Eqs. (9.27) and (9.29) containing three unknowns h, d_0, and b_0. Hence, with one trial value, the resulting equations can be solved. Note, that in the foregoing empirical formulas, the beam seat load W is given in kips while all other dimensions are in inches. All the relevant dimensional symbols are given in Figures 9.16 and 9.17. This procedure is based on the premise that the maximum bending stress in the seat angle does not exceed 24,000 psi.

Large beam reactions may require the thickness of the seat angle to exceed that of the available sections. If this occurs, a stiffened seat, as in Figure 9.18, can be used. This connection can be

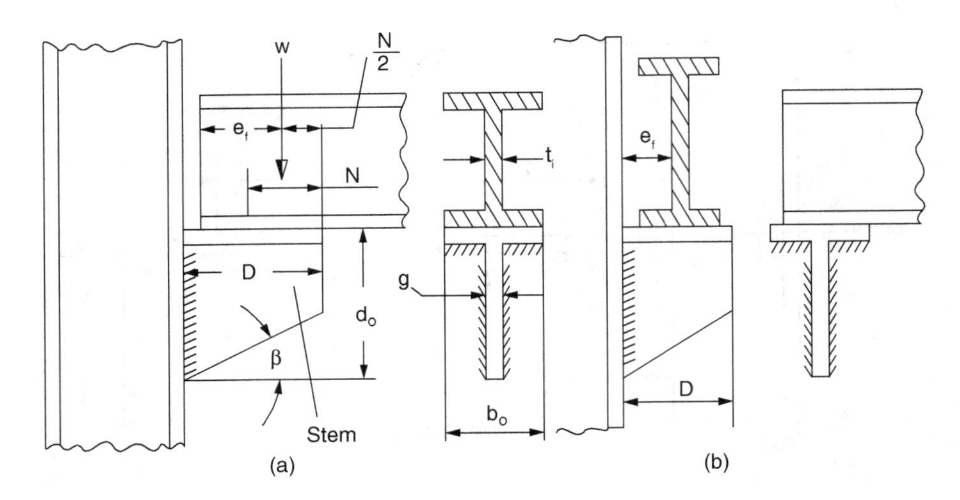

FIGURE 9.18 Stiffened beam seat.

fabricated by welding two plates together or by using a standard structural tee. The stiffened seat can support beams with their webs parallel or perpendicular to the seat stiffness. When the web is parallel, the reaction is generally assumed to act at the midpoint of the greater of the length required for bearing or web buckling, measured from the end of the seat as shown in Figure 9.18(a). Local buckling of the stem does not, in most cases, present a problem, provided the stem thickness is not less than the beam web thickness.

The resultant shear force on the weld attaching the stiffened seat to the column is again the vector sum of axial and bending components. Since the weld length and size are both unknown, the design effort can be reduced by assuming the weld length along the top of the tee (dimension "b_0 " in Figure 9.18) to be about 0.4 of leg length d_0. If the seat is fabricated by welding two plates together, the connecting weld should be designed to carry the horizontal shear being developed. The horizontal shear force can be modeled with the aid of Eq. (9.17).

The thickness of the stiffened seat can be selected on the basis of an approximate method [7] from the expression:

$$g = \frac{(3e_f - D)W}{10D^2 \sin^2 \beta} \tag{9.31}$$

Equation (9.31) applies when the beam rests on the bracket in such a manner that its web is at right angles to the stiffener of the bracket, and when the maximum bending stress in the seat does not exceed 20,000 psi. The bearing length N can still be found from Eq. (9.30), while the vertical length d_0 follows from the empirical expression:

$$\frac{W}{h} = \frac{23d_0^2}{\left(d_0^2 + 16e_f^2\right)^{1/2}} \tag{9.32}$$

In selecting the final proportions of the beam seat shown in Figure 9.18, the thickness of the bracket web g should not be less than the thickness of the beam web t_1. Also, the width of the top bracket plate D generally extends beyond the length of bearing denoted by N. The choice of the depth of the stiffener d_0 depends on the weld leg h as shown by Eq. (9.32). In this design, good practice demands that $h < 0.67t_1$. The eccentricity e_f is often found to be equal to about 80% of the dimension D. Also, the thickness of the bracket web g can be taken equal to 1.5h.

Finally, if the seat is made up of plates, the welds connecting the top plate to the web should be as strong as the horizontal welds on the column-to-bracket joint. All the relevant symbols are given in Figure 9.18.

9.10 COLUMN BASE PLATE DESIGN

The column base plate of a structural joint presents a number of unique design problems. For example, the base plates are required to distribute the compressive column while the overhanging portion behaves as a cantilever beam. Figure 9.19 illustrates the key dimensions of a typical base plate.

Table 9.4 lists some typical allowable bearing pressures.

TABLE 9.4
A. I. S. C. Bearing Pressure Allowables in Pounds per Square Inch

Granite	800
Sandstone and limestone	400
Portland cement	600
Hard brick in cement and mortar	250

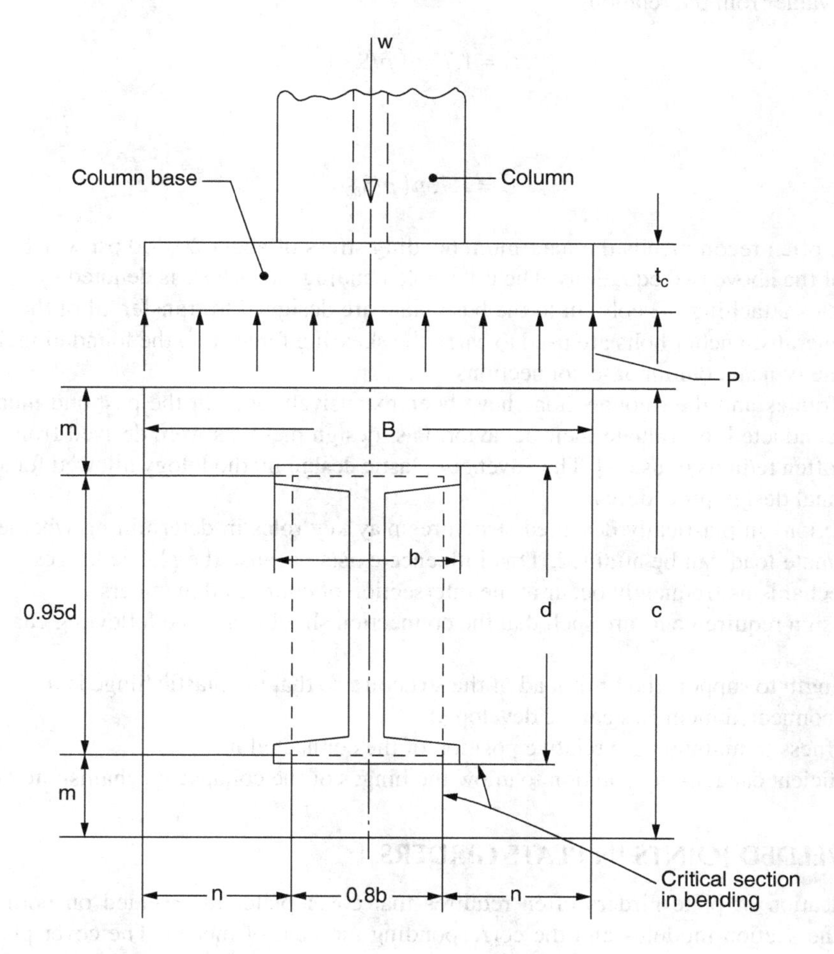

FIGURE 9.19 Example of a column base plate.

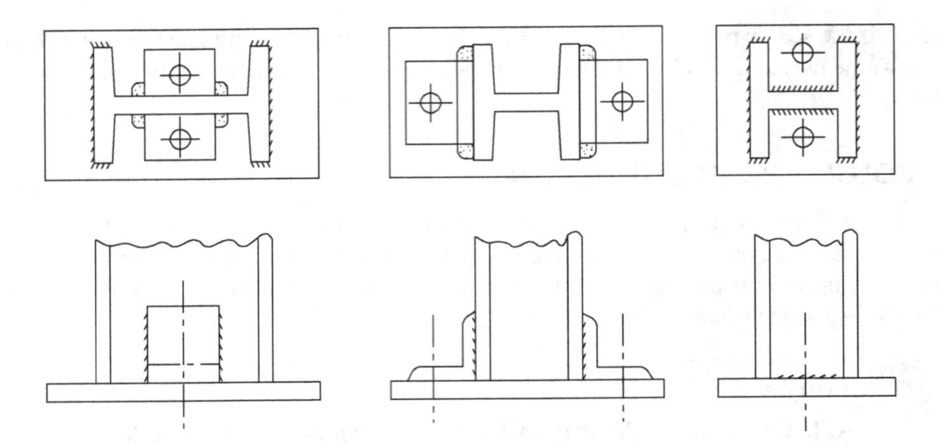

FIGURE 9.20 Typical column base connections.

A procedure for this design case is to start with the base area equal to W/p and determine the control dimensions m and n. The next step is to calculate the thickness of the base plate t_c by using the larger value from the relations:

$$t_c = 1.73m\left(p/S_{ab}\right)^{1/2} \tag{9.33}$$

or

$$t_c = 1.73n\left(p/S_{ab}\right)^{1/2} \tag{9.34}$$

The AISC often recommends the maximum bending stress of about 20,000 psi which can be used in either of the above two equations. The allowable bending stress here is denoted by S_{ab}.

The welds attaching the column to the base plate are designed to transfer all of the column end forces. Generally, anchor bolts are used to carry the shearing forces into the foundation. Figure 9.20 shows some typical column base connections.

Rigid frames and their connections have been extensively used in the past and much research has been conducted to evaluate their behavior. Past design methods were derived from experience and were often tedious to use [1]. The advent of plastic design methodology allowed for simpler and more rational design procedures.

Connections in plastically designed structures play key roles in determining whether the computed ultimate load can be attained. This influence exists because the plastic hinges that form the failure mechanisms frequently occur at the intersection of connected members.

The design requirements are such that the connection should have the following characteristics:

- Strength to support the limit load of the structure so that the plastic hinge is the weaker of the connected members can be developed.
- Stiffness to maintain the relative position of the connected members.
- Sufficient capacity for rotation to allow the hinges of the collapse mechanism to develop.

9.11 WELDED JOINTS IN PLATE GIRDERS

The application of plate girders often requires that cover plates are welded on both flanges, to increase the section modulus and the corresponding moment of inertia. The cover plates may be either continuous or partial, depending on a particular requirement. For example, in the case of stress control, the cover plates may be welded at locations of high bending moment. The welds can

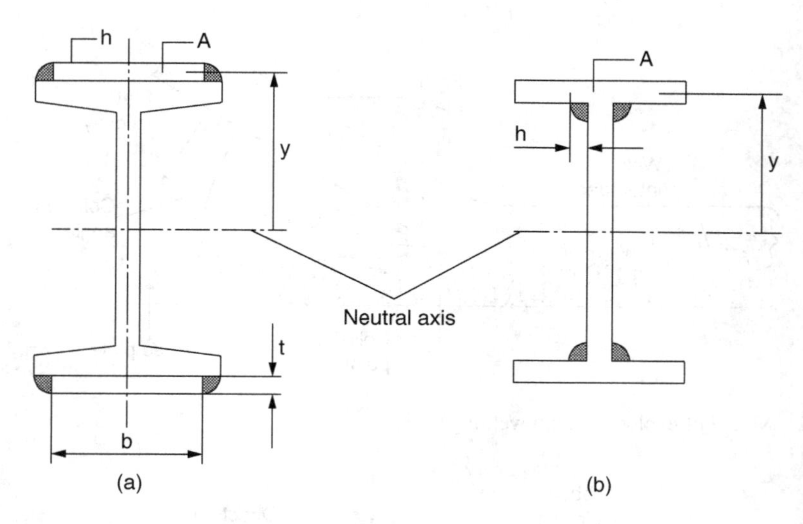

FIGURE 9.21 Weld location in plate girders.

be used to join the cover plate to the flange or the flange to the web of a plate girder as shown in Figure 9.21.

The size of the fillet weld in this application can be determined from the approximate equation [7]:

$$h = 0.05\frac{VAy}{I} \tag{9.35}$$

Equation (9.35) is based on the AISC standards. The same specification limits the unsupported width of the cover plates between the nearest line of welds, not to exceed 40 times the thickness ($b/t < 40$).

Figure 9.21 presents the relevant notation. The parameter V represents the vertical shear on the beam at the appropriate section in kips, while A is the area of the plate held by the welds. The second moment of area refers to the whole section, involving the rolled beam and the cover plates in the case of Figure 9.21(a) or the web and the flange plates shown in Figure 9.21(b).

In the case of flange type girders, the design premise is that the flanges carry the bending moment and the web resists the shear stresses. It is noteworthy that, although the welds joining the flange to the web of a beam are stressed in horizontal shear, the actual values of these stresses are relatively low. This is quite different from the case of a connection which acts as a lifting lug and where the weld must carry the same load as the plate. The maximum shear stress, in the case of girder sections illustrated in Figure 9.21, is found along the neutral axis. The parameters h, A, y, and I in Eq. (9.35) are in inches to the appropriate power.

9.12 SPECIAL WELDED JOINTS

It was shown so far that welding methodology has evolved in concert with design theory and regulatory specifications. The demand for remote metal joining processes in hazardous situations, has also led to the development of explosive seam welding techniques that often produce superior joints. These joints can be better than the conventional connections utilizing mechanical fasteners, swaging, fusion welding, soldering, or adhesives [8].

The basic principle of explosive welding process rests in utilizing a high-velocity, angular collision of metal plates, which leads under the appropriate surface conditions to interatomic link ups. Figure 9.22 illustrates a mechanical model of explosive seam welding.

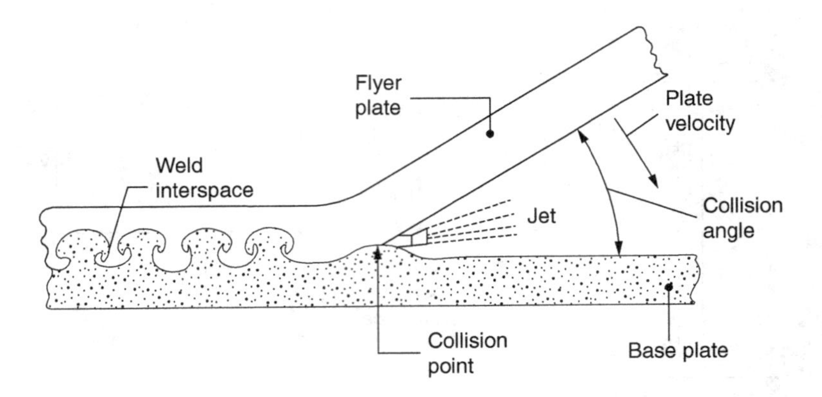

FIGURE 9.22 Model of explosive seam welding.

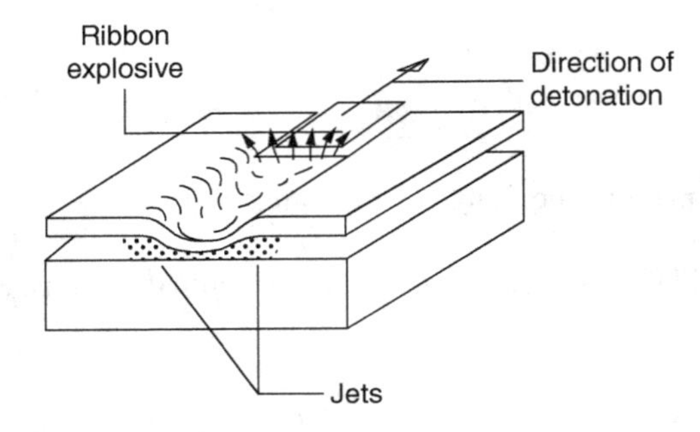

FIGURE 9.23 Deformation pattern in explosive seam welding.

The process of jet collision involves a velocity of impact of several thousand feet per second and creates contact pressures on the order of several million pounds per square inch. These impact conditions result in high temperatures on the material's surface and subsequent deformations. The closest analogy to this process can be found in vacuum bonding where surface oxides are removed to allow interatomic link ups.

Current applications involve explosive cladding or seam welding. In the case of cladding, the compressive input travels at velocities of 4,000 to 10,000 feet per second. This process, however, is limited to joining plates of less than 10 feet in length. The explosive seam welding requires more powerful explosive materials, and develops a velocity of propagation on the order of 26,000 feet per second. Figure 9.23 graphically depicts the mechanism of deformation in seam welding. This is a very efficient process.

The major variables involved in design of explosive welding include:

- Material properties
- Plate thickness and size
- Amount of explosive material
- Plate separation
- Surface conditions
- Magnitude of explosive shock wave

Explosive seam welding has evolved into four types of lap and tube joints. Figure 9.24 provides end views of joined plates via explosive welding. Bement [8] lists some 18 features and advantages of explosive seam welding procedures, ranging from the process, to the design, reliability, and safety characteristics which have contributed to the popularity of this method.

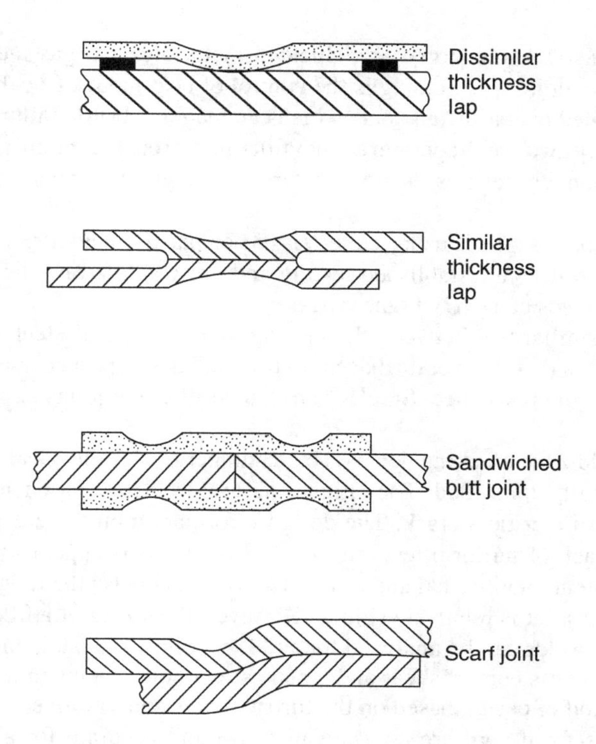

FIGURE 9.24 Types of explosive seam joints.

One of the special uses of explosive seam welding involves the repair of fuel channels in nuclear reactors. The use of this process, instead of a robotic fusion welder, reduces personnel radiation exposure, and reactor downtime. Other areas of interest may include pipelines, large space structures, or vessels containing hazardous materials.

It is evident, from a brief review of explosive welding technology that the particular weld designs still have to depend on empirical data and general experience with the mechanical joints made by explosive techniques.

Another important area of modern welding technology is based on the process of diffusion. For example, aluminum alloys can be diffusion-welded throughout a wide range of temperatures and pressures. Often, the welding operation is carried out in a vacuum or inert gas environment. The only limitation in this case is the melting point of the alloy under consideration. Sometimes, the aluminum alloys are brazed using a copper interlayer, resulting in good quality joints which can withstand significant temperatures during heat treatment.

In the case of steel, more conventional brazing and fusion welding methods are used. However, diffusion welding can be applied to plain carbon steels in a protective atmosphere or in air in special situations involving, for example, large, flat surfaces. With special surface preparation and atmosphere control, stainless steels can also be diffusion-welded.

Finally, we should single out the role of diffusion welding in joining many dissimilar metals even when the melting points of the two metals and their metallurgical structures are quite different. Potential problems with this process can be mitigated by the selection of an appropriate interlayer compound.

9.13 SUMMARY

Welded connections offer various advantages over many of the mechanical fastener approaches. Major factors to consider include: strength, stiffness, load capability, and economics. Modern methods of welding involve submerged arc, shielded metal arc, and gas metal arc technologies. Perhaps, the most difficult task in welding is the control of heat inputs. Localized residual stresses are found in the so-called heat affected zone, which can lead to a brittle failure.

There are essentially two weld configurations; fillet and groove configurations. The difference is largely due to the manner the stresses are transferred. The groove or butt-type welds are usually stronger.

Welded joints can be designed, using either elastic or plastic methods. The design allowables have been developed from long established testing and are compiled by the AWS. Plastic design often assures that the welds can carry higher stresses.

There are certain similarities between the casting and welding of steel. The most significant problem with the HAZ is that the metal adjacent to the weld develops a coarse microstructure leading to reduction in fatigue resistance. Single- and double-fillet tee joints can act as built-in stress risers.

The problem of weld zone cracking depends on a number of metallurgical and process variables which should be carefully controlled. The fatigue life of a welded component is influenced by the initiation and growth of a critical crack. The design information on the crack effect is difficult to establish because of lack of uniform test criteria. The alloy steels appear to be more sensitive to notch effects. Intermittent longitudinal and transverse welds decrease the fatigue strength of a joint.

The design of a butt joint is relatively simple. However, the sizing of a fillet weld remains to be largely empirical. The welds can be analyzed for axial, bending, torsional, and transverse loading.

The mathematical assessment of shear and tensile stress components in a fillet weld shows that the conventional method of design based on the throat area is conservative.

Weld line formulas for design are convenient to use and accurate for all practical purposes. Simple expressions for finding weld line properties are available in standard handbooks. The appropriate shear force components acting on welded brackets can be obtained from the equations of static equilibrium.

The design analysis of welded lap joints depends on the degree of rigidity between the connected members. Shear lag results from the incipient deformations caused by shear stresses. Various regulatory bodies recommend the appropriate proportions for lap joint design, in order to assure good fatigue resistance.

Beam-to-column connections can be of stiffened or unstiffened variety. Simplified formulas are available for weld sizing and past design experience offers certain guidelines for selecting the optimum dimensions of the joint components. A number of practical design rules are based on the recommended maximum stresses and the most efficient configurations.

The common problem of a weld design for column base plates involves bearing pressure allowables and plate bending criteria. In addition to empirical design methods, plastic hinge methodology can be employed.

In various practical situations, cover plates can be welded to form the entire girders or to increase the strength of the existing structural shapes. Identical design formulas can be used for both cases.

Developments in welding technology have been recently extended to explosive seam welding, explosive cladding, diffusion welding, and similar techniques found to be superior to a number of conventional joining processes. In this category of welded joint, the designs should be based on empirical data.

9.14 DESIGN ANALYSES

EXAMPLE ANALYSIS 1

Two identical steel plates are welded together with a butt weld as represented in Figure a. The plates are 12 inches (30.48 cm) long, 10 inches (24.0 cm) wide, and 3/8 inches (0.953 cm) thick. The weld width is ¼ inches (0.635 cm).

Although not perfect, the weld is estimated to be at least 85% efficient.

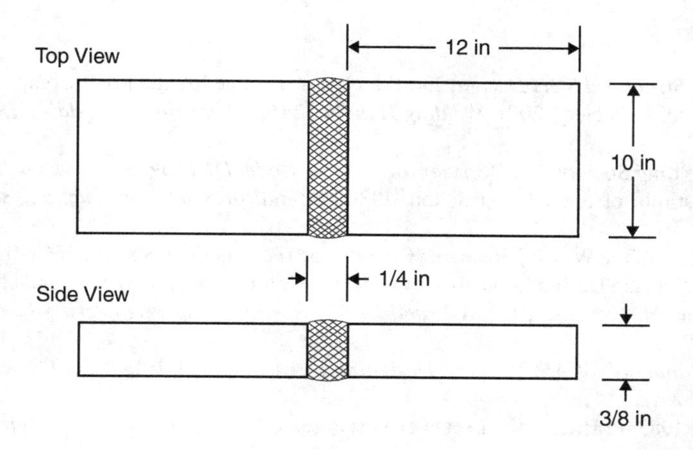

FIGURE A Top and side views of welded plates.

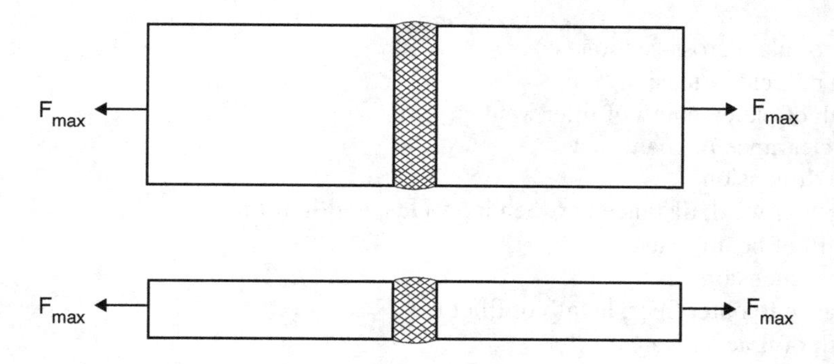

FIGURE B In-plane loading of welded plates.

The yield stress, S_{yp} of the plates is approximately 36 ksi (248 MPa) and the yield strength, S_{yw} of the weld is 40 ksi (276 MPa).

With a safety factor (SF) of 2, determine the maximum safe in-plane load F_{max}, as in Figure b, which can be carried by the welded plates.

Solution

Since the 85% efficient weld strength is 34 ksi (234 MPa) which is smaller than the 36 ksi strength of the plates, a failure from overload is expected to occur at the weld. Therefore, the maximum safe load F_{max} may be expressed as:

$$F_{max} = \frac{\eta S_{yw} A}{SF} \tag{a}$$

where η is the weld efficiency and A is the loaded area of the weld. From the given geometric data, A is

$$A = (10)(3/8) = 3.75 \text{ in}^2 \left(24.19 \text{ cm}^2\right) \tag{b}$$

Finally by substituting the data into Eq. (a) F_{max} becomes

$$F_{max} = \frac{(0.85)(40)(10^3)(3.75)}{2} = 63.750\, kip (283.7 \text{ kN}) \tag{c}$$

REFERENCES

1. Tall, L. 1974. *Structural Steel Design*, 2nd Ed. New York: The Ronald Press Company.
2. American Welding Society. 2001. *Welding Handbook (Fundamentals of Welding)*, 9th Ed. New York: AWS.
3. American Welding Society. 1977. *Structural Welding Code, D1.1–69*, 3rd Ed. New York: AWS.
4. American Institute of Steel Construction. 1980. *Manual of Steel Construction*, 8th Ed. New York: AISC.
5. Joint Committee of the Welding Research Council and the American Society of Civil Engineers (ASCE-WRC). 1971. "Plastic Design in Steel—A Guide and Commentary," ASCE Manual No. 41, 2nd Ed.
6. Huston, R. and H. Josephs. 2009. *A Practical Stress Analysis in Engineering Design*, 3rd Ed. Boca Raton: CRC Press.
7. *Procedure Handbook of Arc Welding Design and Practice*, 14th Ed. 2000. Cleveland: The Lincoln Electric Company.
8. Bement, L. J. 1983. Practical small-scale explosive seam welding. *Mechanical Engineering*. 105:53–59.

SYMBOLS

A Area of plate cross-section
A_w Area of weld in tension
a Width of plate; length of fillet weld
a_0 End clearance in beam seat
B Plate dimension
b Length of weld; distance between lap welds; width of I beam
b_0 Length of beam seat
C Plate dimension
c Distance to outer fiber; length of fillet weld
D Width of bracket plate
d Depth of section; length of lap weld
d_0 Length of beam seat leg
e Clearance in Tee joint; eccentricity
e_0 Distance to critical section
e_f Maximum offset
F_n Normal force component
F_s Shear force component
g Thickness of stiffened seat
h Weld leg
I Moment of inertia
I_p Polar moment of inertia
I_w Polar moment of weld line
k Dimensional parameter
L_w Cumulative length of weld
M General symbol for bending moment
M_t Torsional moment
m Auxiliary dimension for base plate
N Critical distance in web crippling
n Auxiliary dimension for base plate
P Bearing pressure
Q First moment of area
q_a Unit axial shear
q_b Unit bend force
q_r Resultant shear force

q_t	Unit torsional force
q_v	Transverse unit shear
q_x	Horizontal component of shear force
q_y	Vertical component of shear force
r	Radial distance; also ring radius
S_a	Axial stress
S_{ab}	Allowable bending stress
S_b	Bending stress
S_{max}	Maximum stress in fatigue
S_n	Normal stress
t	Weld throat; also ring or plate thickness
t_c	Thickness of column base plate
t_0	Thickness of beam seat angle
t_1	Thickness of beam web
V	Transverse shear force
W	External load
w	Plate thickness, Eq. (9.4)
x	Horizontal distance to weld; also distance from neutral axis to c.g. or cover plate
y	Vertical distance to weld
Z_w	Section modulus of weld line
β	Taper angle of stiffener
θ	Angle in weld analysis
τ	Shear stress
τ_a	Allowable weld stress in shear

10 Membrane Joints

10.1 INTRODUCTION

In this relatively brief chapter, we consider some rather special joints and connections used primarily for affixing membranes, diaphragms, and barriers to their supports. We also describe an experiment for measuring diaphragm displacements.

We begin with a consideration of vacuum barrier fixation.

10.2 VACUUM BARRIER TECHNOLOGY

Vacuum barrier applications in research and diagnostic systems may involve such configurations as flat plates, cylinders, conical surfaces, and shallow spherical caps. These structural elements and systems often act as thin membranes attached to relatively rigid housings and similar machine parts. Hence, vacuum barriers constitute a class of important mechanical joints designed to resist a maximum pressure differential of one atmosphere although it is feasible that similar barrier systems can also be developed for larger pressure differentials.

Since mechanical joint integrity is an important consideration in vacuum barrier technology, the theoretical design predictions often require support by full-scale tests to destruction. The term "barrier" is defined here as a window, quill, or a similar mechanical interface having circular, elliptical, or other shape. The structural response may involve bending, tension, creep, or buckling characteristics depending on the particular geometry, type of loading, and materials involved. The relevant barriers acting as mechanical joints can involve bolted, welded, or adhesive elements designed to resist complex stress interactions often caused by large deformations and membrane-type forces in barrier walls.

The choice of membrane materials includes a wide range of properties with polymers (plastics) supplementing those requirements where a viscoelastic behavior may be of special importance. The more conventional designs utilize yield, ultimate strength, elongation, and the elastic modulus data for the theoretical predictions and experimental analysis.

In many applications of vacuum barrier technology, Mylar, and other similar materials are used in making window-like elements of the particular equipment or experimental rigs. Mylar is a well-known polyester introduced by DuPont some 75 years ago and has since undergone numerous evaluations of performance under specific conditions of loading and environment. In a chemical sense, Mylar is obtained from the polymer formed by the condensation reaction between ethylene glycol and terephthalic acid. The material is used successfully as a tough, clear film in a number of applications including packaging, photographic industry, and diagnostics research.

Adhesives are available for laminating Mylar components to various materials, while heavier gages of this material can be shaped by stamping and forming fabrication techniques. Because of its dimensional stability and high mechanical strength, Mylar is particularly appropriate for vacuum barriers. In the forthcoming review of some of the practical aspects of barriers acting as mechanical interfaces, a number of metallic windows and shell-like elements will also be included.

10.3 DESIGN OF A CIRCULAR MEMBRANE

In the design of a thin mechanical barrier across an opening to withstand a given pressure differential, we need to consider a number of theoretical and experimental issues. For example, consider the system illustrated in Figure 10.1 where a circular membrane is clamped between two rigid rings.

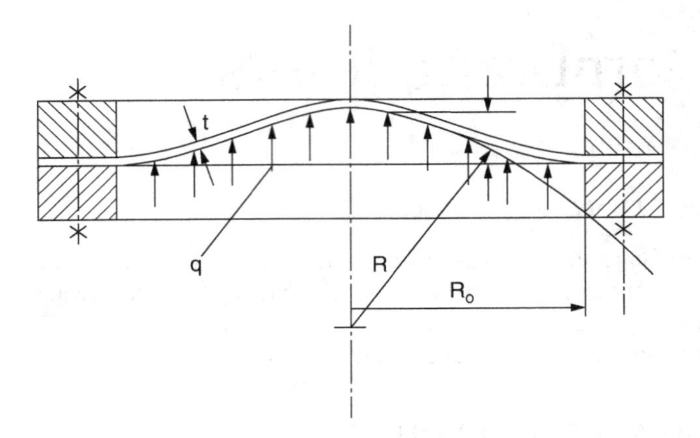

FIGURE 10.1 Typical membrane configuration and symbols used.

The barrier has an initial thickness t, sufficient to support the pressure differential q. As the central deflection δ develops, the barrier adopts an approximately spherical curvature represented by radius R shown in the sketch of Figure 10.1.

The manner in which this process takes place must depend on the ratio R_0/t. For instance, for a relatively small value of R_0/t the barrier behaves more like a classical circular plate clamped at the edges. It may also be stated, however, as a general practical guide that when the deflection becomes larger than about one-half the plate thickness, the middle surface undergoes significant straining so that the corresponding membrane stress should not be ignored. This mechanism allows the plate to carry part of the transverse load by means of a diaphragm tension. The tension, in turn, is resisted by the clamping force of the mechanical joint represented by the two rings shown in Figure 10.1.

The edge of the plate is prevented from moving by the joint forces that can be the result of friction, welding, adhesion, or bolting, depending on a particular requirement and design. The barrier, together with the clamping rings, constitutes a complex system that cannot easily be modeled with a single design formula accounting for all the variable parameters involved.

The behavior of a more conventional thin plate in the regime of large deflections is nonlinear. It is governed by the δ/t ratio and the clamping conditions at the support. The relevant deflection and stress formulas for the barriers, acting as circular plates subjected to a uniform pressure differential q can be expressed as:

$$\frac{qR_o^4}{Et^4} = H(\delta,t) \tag{10.1}$$

and

$$\frac{SR_0^2}{Et^2} = F(\delta,t) \tag{10.2}$$

Figures 10.2 and 10.3 provide dimensionless graphs for the deflection H and stress F factors of Eqs. (10.1) and (10.2) for various support and loading configurations.

Specifically for the three curves of Figure 10.2, we have:

H_1 Simple support
H_2 Edge restrained in a vertical direction; zero tension in the plane of the plate
H_3 Fixed support at the periphery with full diaphragm tension

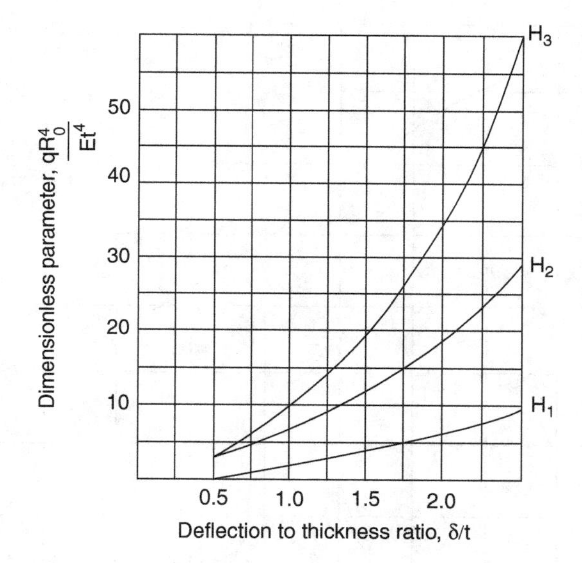

FIGURE 10.2 Design factors for large deflection of circular plates.

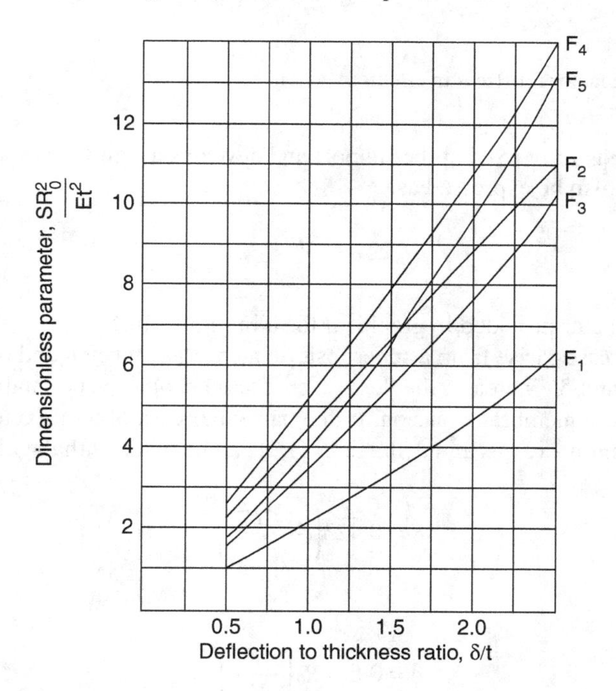

FIGURE 10.3 Stress factors for large deflection of circular plates.

Similarly, the conditions for the five curves of Figure 10.3 are:

F_1 Center stress under zero constraint at the support
F_2 Edge stress for vertical restraint at the support
F_3 Center stress for vertical restraint at the support
F_4 Edge stress for fixed support and full diaphragm tension
F_5 Center stress for fixed support and full diaphragm tension

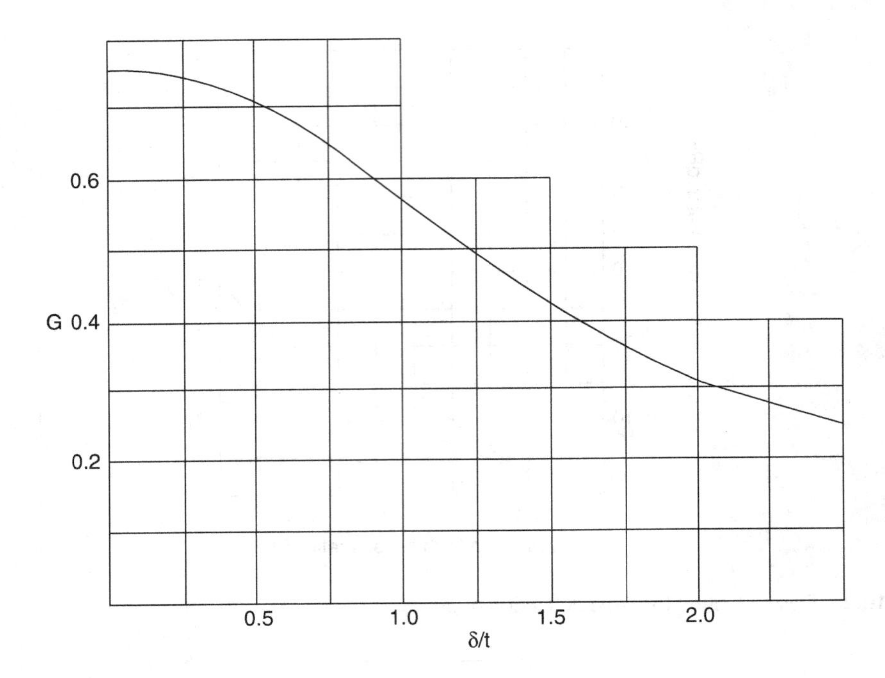

FIGURE 10.4 Design factor for stress in circular vacuum barriers.

When a circular plate is fully fixed at the support and develops a significant membrane tension, the maximum stress can also be expressed as:

$$S = \frac{qR_0^2 G}{t^2} \tag{10.3}$$

where G is a design factor provided in graphical form in Figure 10.4.

The diaphragm stresses arise from a superposition membrane tension and bending (flexure).

As the ratios R_0/t and δ/t increase, the flexural stiffness becomes very small and the entire loading is resisted by a pure membrane tension. In the particular case of a thin circular membrane, the analysis of the maximum stresses and deflections [1,2] can be made with the aid of the expressions:

$$S = 0.423\left(\frac{Eq^2R_0^2}{t^2}\right)^{1/3} \tag{10.4}$$

and

$$\delta = 0.662R_0\left(\frac{qR_0}{Et}\right)^{1/3} \tag{10.5}$$

It appears that the location of the critical stress in the circular vacuum barrier can be either at the center or at the edge, depending on the magnitude of the dimensionless ratio δ/t. The calculations based on the large deflection formulas show that, for the center and edge stresses to be approximately equal, the ratio δ/t should be about 3. For all δ/t values greater than 3, the maximum stress can be assumed to be at the center of the vacuum barrier.

Since the formulas given by Eqs. (10.4) and (10.5) have essentially been derived for very flexible and thin diaphragms, one-to-one correlation with the more conventional expression, Eqs. (10.1) and (10.2) may or may not be attainable. Different mathematical models and boundary conditions demand that each particular case of a circular barrier, as defined in this chapter, should be carefully examined and tested prior to adopting the appropriate design theory for a specific working environment.

10.4 BEHAVIOR OF MYLAR BARRIERS

Vacuum barrier designs found in various applications generally require experimental validation. Unfortunately, such validation is seldom available due to proprietary restrictions, as well as difficulty in experimental simulation techniques. The exact design formulas are not available and it is often necessary to resort to classical formulas, such as those quoted in this chapter, or to empirical correlations. Each particular case then should be considered on its own merits. This is especially true when the barrier materials can have mechanical properties ranging from viscoelastic to entirely brittle.

Figure 10.5 provides an approximate correlation of some earlier test results for a number of circular Mylar windows.

In this series of experiments, the barriers were between 0.0075 and 0.0150 in thick. While the results in Figure 10.5 did not account for the rate of strain, the trend appeared to be quite evident as to what degree the burst pressure was sensitive to the size of barrier diameter. This finding is in some agreement with the theoretical predictions based on Eqs. (10.4) and (10.5), indicating that for a given stress, q is proportional to $1/R_0$. Here, R_0 denotes one half of the window diameter.

While the trend shown in Figure 10.5 appears to be well defined, true Mylar barrier behavior as a mechanical interface is much more complex. The performance of this type of a mechanical joint

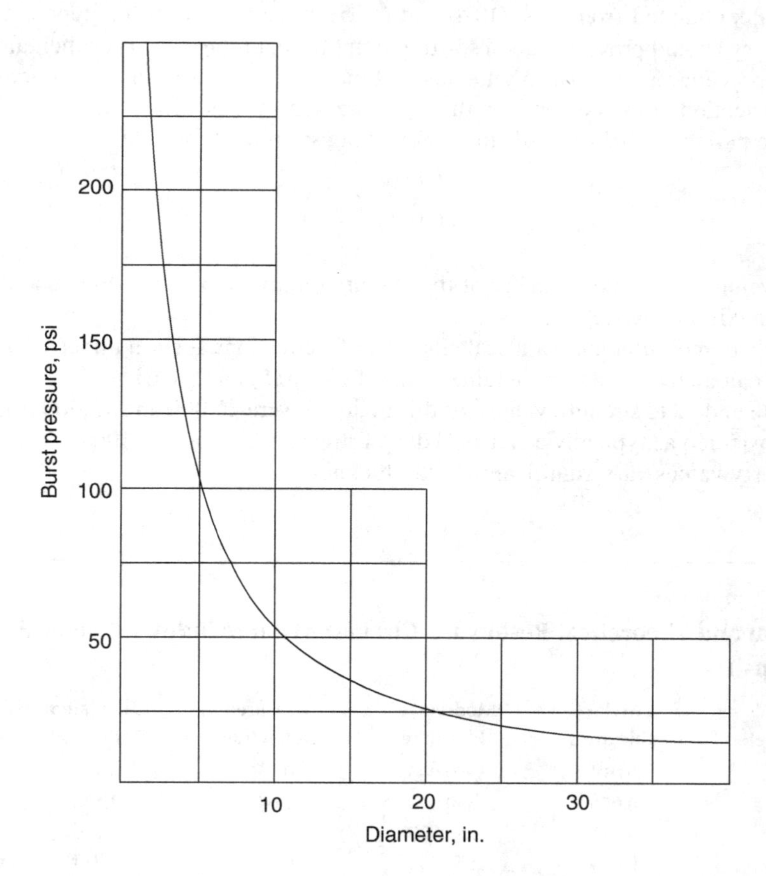

FIGURE 10.5 Approximate correlation of burst pressures for circular Mylar windows (0.0075 to 0.0150 in thick).

depends on the structural properties of the Mylar and geometry of the barrier, as well as on the clamping arrangement designed to hold the barrier in place. Often the clamping force is provided by a bolt force. The terms barrier, diaphragm, or window are often used in the literature interchangeably, depending on the emphasis on a particular feature of the system. Diaphragm, however, usually implies a thin membrane in tension acting across a given pressure drop.

Since Mylar has distinct characteristics of both an elastic solid and a viscous fluid, the working stress must be a function of both strain and time. Hence, the most significant measurement will depend on the rate of strain and loading history prior to cycling. The calculations are naturally much more complicated because of a nonlinear response of the material and large-deflection behavior of a membrane under stress.

Experimental work carried out in a laboratory environment on a 21.5 in diameter, 0.014 in thick circular Mylar window [3], has led to the following observations:

- For stress levels below the yield point of this type of a material, the diaphragm should support a load without additional deformation, with the exception of a small effect of creep when maintaining the work load over an extended period of time.
- At stress levels above the yield point, the material cannot support a constant load. With the load maintained at a constant level, the diaphragm should continue to deform until failure. Such a constant supply of energy is found in the case of a pressure drop across a vacuum barrier.

While the results obtained from Eqs. (10.4) and (10.5) may not be in full agreement with the test values, the theory should provide a good starting point for planning the experimental work.

One of the procedures for verifying the level of maximum stress in the diaphragm would be to monitor the deflection at the center. The theory expressed by Eqs. (10.4) and (10.5) indicates that eliminating the parameter (qR_0/t) leads to the following simplified formula:

$$S = 0.965E\left(\frac{\delta}{R_0}\right)^2 \tag{10.6}$$

Thus, for a given material and diameter of the diaphragm, the maximum stress should vary as the square of the maximum deflection.

Table 10.1 lists measurements and calculations of deflections for a number of circular Mylar windows. The calculated values are obtained using Eqs. (10.5) and (10.6).

The ultimate and yield strength values for this material were 25,000 and 12,000 psi, respectively, and can be considered as typically advertised data. Little reference is usually made as to whether or not such property values vary significantly with thickness.

TABLE 10.1
Experimental and Theoretical Results for Circular Mylar Windows (Dimensions are in inches and psi)

Window Diameter	Mylar Thickness	Modulus of Elasticity	Measured Deflection	Calculated Deflection	Calculated Stress
12.00	0.014	550,000	0.962	0.887	13,640
36.00	0.050	550,000	2.647	2.538	11,480
10.75	0.014	550,000	0.870	0.776	13,900
9.75	0.014	550,000	0.714	0.714	11,390
42.00	0.076	550,000	2.850	2.850	9,780

It is important to know, in special applications, whether a mechanical vacuum barrier will remain stable as a tight and a reliable joint after cycling. Current observations are that, when the maximum stress does not appreciably exceed the yield strength of the material, the deflection from cycle to cycle does not change.

One theoretical exception would be the case of a diaphragm, having very high radius to thickness ratio subjected to a permanent dead-weight load.

It should also be noted that tests of Mylar barriers to failure have shown that the material's thickness at the center could be significantly reduced. For the specific case of a 21.5 in diameter and 0.014 in thick circular diaphragm, the reductions are 57% and 36% for the center and edge regions, respectively. This appears to be consistent with the theory of the maximum stresses being at the center rather than at the edge for a very thin circular plate stretched above the elastic limit. This is also in contrast with the more conventional plate behavior, given by a chart in Figure 10.3, for the case of a significant edge restraint. It can be reasoned that such a trend could have been anticipated on the basis of δ/t ratio.

Since a system's reliability is always an important consideration, the application of Mylar as a seal of sufficient quality in the type of a joint described above must be subject to limitations of the membrane clamping method at the edge. This process should not result in an unduly high clamping force. The important criterion here is to allow for a substantial contact area, generous edge radii and a rigid clamping ring to distribute the bolt loading.

There are indications that a clamping pressure on the order of 700 psi may be sufficient to hold a relatively large diameter Mylar diaphragm in place. This would depend, of course, on the corner radii, control of friction and adhesion characteristics at the interface between Mylar and the appropriate metallic components. The final proof of the vacuum barrier integrity, however, must be determined from experimental results.

10.5 METALLIC DIAPHRAGM EXPERIMENT

The use of a circular diaphragm is not limited to the applications of Mylar or similar viscoelastic materials. In a series of tests [4], the calculated deflection response for tantalum and aluminum circular diaphragms was compared with that from tests. Figure 10.6 schematically illustrates how the deflection measurements were obtained with the aid of a linear transducer.

Table 10.2 lists the geometric and physical property of the materials used in the experiment.

The diaphragms fabricated to the above dimensions and properties were intended to be vacuum barriers. However, in order to determine the ultimate factors of safety for this application, the test pressures were gradually increased to the point of diaphragm failure. The relevant failure modes

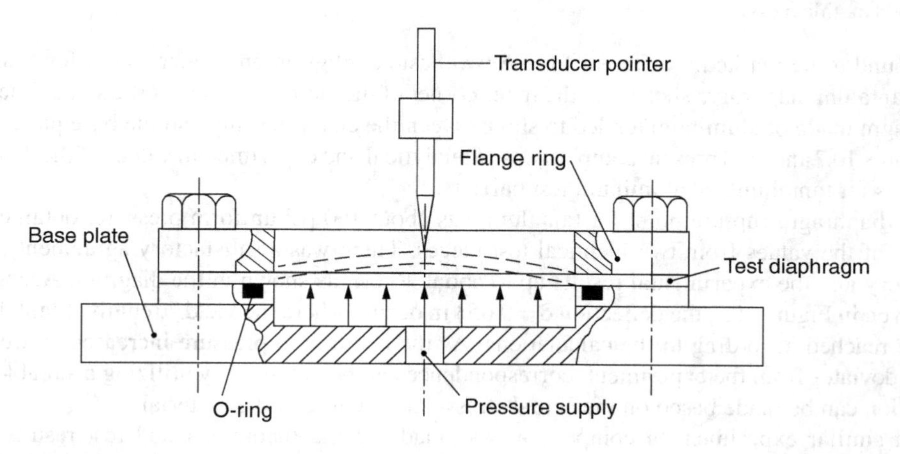

FIGURE 10.6 Experimental setup for diaphragm tests.

TABLE 10.2
Geometric and Physical Data for Test Diaphragms

Diameter of Diaphragm, in.	Metal Thickness, in.	Modulus of Elasticity, psi	Poisson's Ratio	Yield Strength, psi	Type of Material
2.5	0.004	27×10^6	0.35	47,700	Cold rolled tantalum
19.685	0.020	10×10^6	0.33	35,000	6061-T6 aluminum

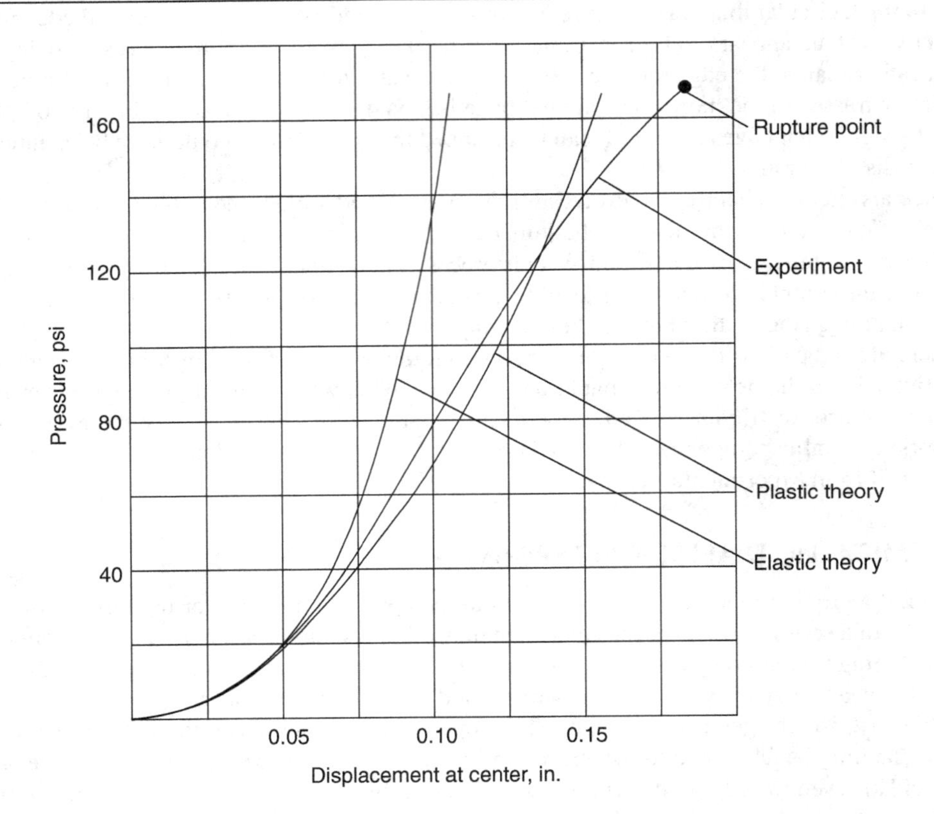

FIGURE 10.7 Comparison of analytical and experimental results for the tantalum diaphragm (2.5 in. diameter, 0.004 in. thickness).

were found to be markedly different for the two basic configurations under study; for example, a small tantalum diaphragm sheared at the inner corner of the clamping ring. At the same time a large diaphragm made of aluminum tended to slip between the clamping ring and the base plate.

Figures 10.7 and 10.8 present comparisons of analytical and experimental values of the diaphragm pressures for tantalum and aluminum test barriers.

The diaphragm rupture point for tantalum was about 160 psi uniform pressure, obtained as an average of the values from two identical test pieces. There was a satisfactory agreement between the theory and the experimental results up to about 40 psi, as shown in the diagram. According to data given in Figure 10.7, the deflection $\delta = 0.053$ in occurs where the yield strength of tantalum can also be reached according to the calculations. As the diaphragm pressure increases, however, the theory deviates from the experiment, correspondence can be obtained by utilizing a suitable plastic correction can be made based on the actual stress–strain curve of the material.

In a similar experiment, a comparison was made of the theoretical and test results for the large aluminum diaphragms. Figure 10.8 provides the relevant comparison suggesting that the

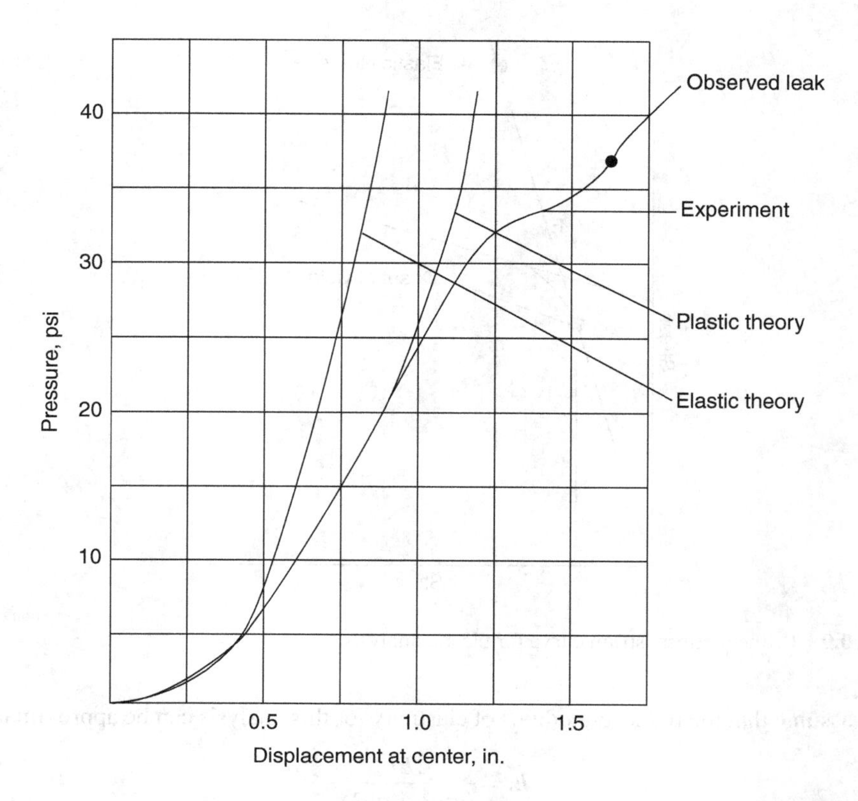

FIGURE 10.8 Comparison of analytical and experimental results for the aluminum diaphragm (19.685 in. diameter, 0.020 in. thickness).

design is suitable as a vacuum barrier prior to reaching the yield point of 6061-T6 aluminum. The interpretation of Figure 10.8 shows that the aluminum diaphragm can carry at least 15 psi of the differential pressure, which roughly corresponds to the calculated yield strength of the aluminum window used in the test. The diaphragm, however, was not loaded to destruction because the failure of the mechanical joint was caused by a premature slipping of the test piece between the clamping flange ring and the base plate as illustrated in Figure 10.6.

10.6 PLASTIC CORRECTION IN DIAPHRAGM DESIGN

The theory of large deflection, as in Eqs. (10.4), (10.5), and (10.6), is shown in Figures 10.7 and 10.8 as steeply rising curves for stress values exceeding the yield strength of the two materials. For the dimensions and material properties given in Table 10.2, the pressure versus the elastic deflection characteristics can be derived from the following expressions:

$$q = 0.00566E\delta^3 \quad \text{(Tantalum)} \tag{10.7}$$

and

$$q = 0.0000074E\delta^3 \quad \text{(Aluminum)} \tag{10.8}$$

When the diaphragm deflection is sufficiently high to cause stresses greater than the yield strength of the material, the deflection curve may be corrected using the actual stress–strain curve of the diaphragm material. To this end, suppose the actual stress–strain curve can be modeled with the aid of a bilinear relationship such as that illustrated in Figure 10.9.

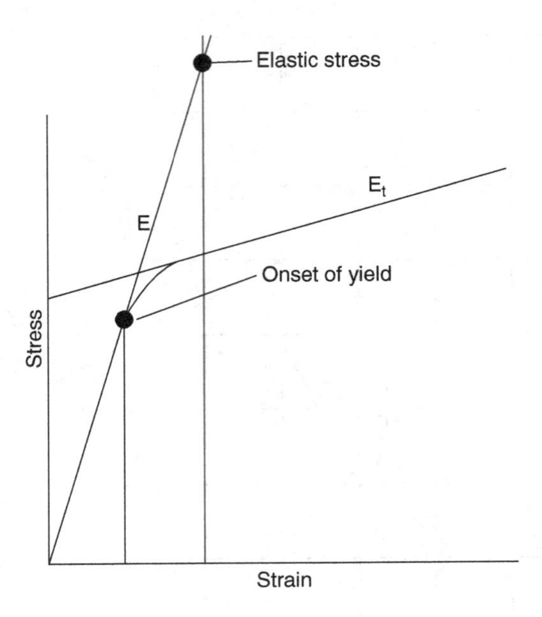

FIGURE 10.9 Bilinear stress–strain curve for plastic analysis.

Let us assume that the reduced modulus of elasticity for this analysis can be approximated as [5]:

$$E_r = \frac{4EE_t}{\left(E^{1/2} + E_t^{1/2}\right)^2} \tag{10.9}$$

Figure 10.9 defines the elastic and tangent moduli intended for use in Eq. (10.9).

The respective elastic and tangent moduli intended for the use in Eq. (10.9) are defined in Figure 10.9. For $E_t = 0.25E$, for example, the reduced modulus of elasticity is $E_r = 0.44E$. Hence, utilizing this plastic correction in Eqs. (10.7) and (10.8) gives the relevant pressure-deflection characteristics in Figures 10.7 and 10.8, which are in a much better agreement with the experiments.

Unfortunately, the failure mode of the aluminum barriers illustrated in Figure 10.8 becomes further complicated by the process of slipping of the test pieces between the flange-ring and the base plate causing a tear-out model of rupture. The experimental curve shown in Figure 10.8 also contains a depressed portion of the profile at larger diaphragm deflections. Because of this phenomenon, the maximum membrane tension in the diaphragm, represented by the curves in Figure 10.8, could not be attained in the experiment with aluminum windows. Further tests are required to draw more general conclusions.

10.7 SPHERICAL CAP BARRIERS

While vacuum barrier design, in general, has obvious pressure vessel characteristics, textbooks seldom use the barrier configurations as examples of illustrating the behavior of this special class of pressure vessel components. This is not surprising, however, when one considers the mission of the barrier and diaphragm systems providing a mechanical function of sealing a gas or a liquid medium across a specified pressure-drop boundary.

We stated at the beginning of this chapter that the materials used in these applications can range from very pliable and viscoelastic to highly rigid and brittle. An example of a brittle vacuum barrier employed in special diagnostic tests is a shallow spherical cap made of a high purity beryllium available in industry. Such a beryllium quality is often defined in terms of a specific amount of

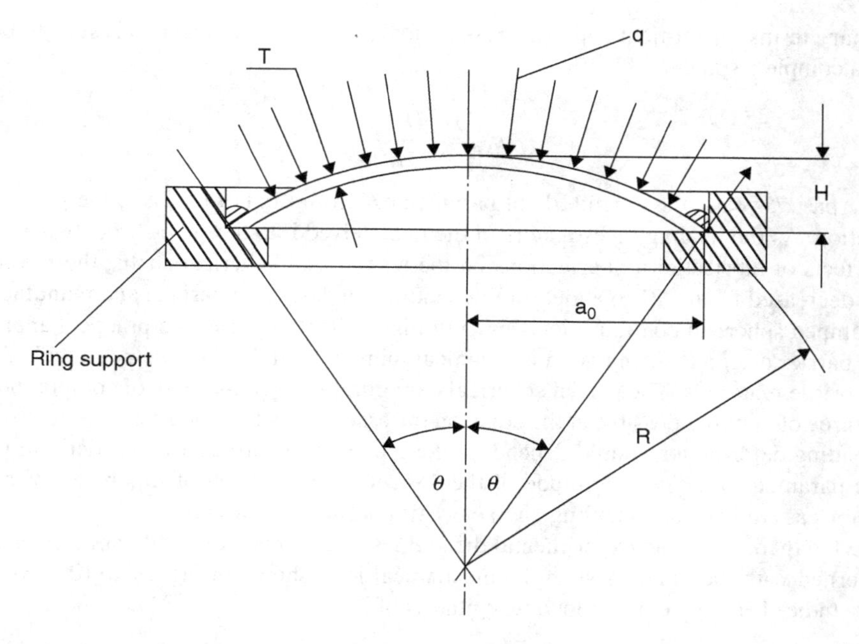

FIGURE 10.10 Notation for a spherical-cap barrier.

the impurities measured in PPM (parts per million). Figure 10.10 shows the pertinent notation and geometry of a spherical-cap vacuum barrier used in industrial experiments [6].

A uniform pressure q is applied to this window-type joint, resting on a ring support which is substantially more rigid than the barrier itself. This particular joint was made by a welding process with a portion of the base metal being melted. The geometry and the general proportions of this barrier suggest that the spherical part of the joint should have behaved as a spherical vessel under external pressure [7]. These tests were essentially designed for pressures to failure, to get a better idea of the ultimate factor of safety for design.

Due to a complex response of a spherical cap under external pressure, especially during the post-buckling process, a high degree of sensitivity to the manufacturing imperfections can be expected. Experience of numerous investigators in the field, between the years of 1940 and 1965, has shown a considerable experimental scatter, both with respect to the R/T ratio and the geometric parameter λ_0 defined as:

$$\lambda_0 = 2\left[3\left(1-\nu^2\right)\right]^{0.25}\left(\frac{H}{T}\right)^{0.5} \tag{10.10}$$

Assuming the Poisson's ratio ν for beryllium to be: 0.027, Eq. (10.10) reduces to:

$$\lambda_0 = 2.63\left(\frac{H}{T}\right)^{0.5} \tag{10.11}$$

It may be of interest to note that the radius of curvature R does not explicitly appear in Eq. (10.11). It can also be shown that the depth H for a relatively shallow cap is a function of radius R leading to two forms of Eq. (10.11), which essentially give identical results [8].

The observed discrepancies between the theoretical buckling loads and the test results have, over the years, led to the development of the formulas which could better predict the ultimate capacity of a shallow, spherical cap under external pressure. To this end, it became necessary to consider large deformations, the essential features of shallow configurations, and the basic parameters E, T, and R.

In elementary terms, the critical collapse pressure for a shallow, spherical cap is seen to be similar to that of a complete sphere:

$$q_{CR} = K\frac{ET^2}{R^2} \tag{10.12}$$

During the past 75 years, the magnitude of parameter K from Eq. (10.12) has undergone a remarkable evolution. As the leading investigators in the field delved into the snap-through characteristics and the effects of fabrication imperfections on the response of spherical shells, the K values have gradually decreased from 1.21 to about 0.13 depending on design, materials, and manufacture.

The clamped spherical configuration shown in Figure 10.10 represents a practical application to a vacuum barrier design assuring good mechanical joint and window geometry suitable for ductile as well as brittle materials. The partial spherical configuration is in the state of compression, with a limited degree of bending possible at the edge, on the assumption that the support is relatively rigid.

The bending deformation should depend on the type of a buckling response determined by the geometric parameter λ_0. The magnitude of the K factor is a function of this behavior and initial imperfections as well as post-buckling characteristics of the shallow cap.

Despite the theoretical and experimental difficulties, mathematicians and engineering practitioners concerned with the uniqueness of the mechanical joint shown in Figure 10.10, have persisted with their studies [7] to give the following synthesis of the problem:

- Differences in behavior of a complete sphere and a shallow cap are mainly caused by the degree of fixation at the cap boundary. The collapse pressure increases with the support rigidity.
- Snap-through characteristics of a complete sphere and a shallow cap are nearly the same.
- For a weak cap boundary, the critical buckling pressure of the cap is lower than that of the complete sphere.

For a relatively thin and shallow spherical cap, acting as a vacuum barrier similar to that illustrated in Figure 10.10, the approximate critical pressure can be calculated from the equation:

$$q_{CR} = \frac{(0.25 - 0.0026\theta)(1 - 0.000175m)E}{m^2} \tag{10.13}$$

where m denotes the ratio R/T and θ is the central half angle of a spherical cap in degrees.

Equation (10.13) is intended to be used between the practical limits of θ equal to 20 and 60 degrees, where θ is defined in Figure 10.10.

The formula given by Eq. (10.13) was originally derived for very thin caps having m ratios on the order of 400 or more. However, this model also appeared to apply to somewhat thicker caps made out of industrial grade of beryllium. Table 10.3 lists test and theoretical predictions for the three vacuum barriers, fabricated as spherical caps. All dimensions are given in English units.

TABLE 10.3
Summary for Spherical-Cap Barriers

T in.	R in.	A0 in.	θ deg.	E psi	H in.	q_{CR}, psi Experiment	q_{CR}, psi Calculated (Eq. 10.13)
0.010	2.40	0.781	19	42×10^6	0.131	140	140
0.006	2.40	0.781	19	42×10^6	0.131	32	49
0.020	6.00	2.837	28	42×10^6	0.719	44	78

Researchers have shown that the scatter found in the correlation of the theoretical and experimental collapse pressure for the thin vessels is generally quite high. The results summarized in Table 10.3 follow the same trend and care should be taken when applying these results to similar design configurations and brittle materials. The tabulated data also includes a_0/H ratios lower than 8.

A spherical cap can be regarded as shallow when the relevant ratio is higher than 8. Despite these limitations, the results should be of interest to designers concerned with the fabrication and testing of mechanical interfaces involving spherical caps.

10.8 ROLLING DIAPHRAGM INTERFACE

The role of a convoluted cylindrical diaphragm, such as that shown in Figure 10.11, is to provide a mechanical interface between two gasses or liquids under a given pressure drop ΔP. This system is sometimes referred to as a "positive expulsion device" where actuating pressure causes diaphragm inversion by bending the material continuously through a full angle of 180°.

We can establish an approximate calculation model assuming that the neutral axis of the diaphragm material coincides with its centerline and that the strain energy due to the axial and shear forces can be neglected as being small in relation to bending and hoop extension energy.

The strain energy of bending can be obtained as the product of the plastic moment and the full angle of bend expressed in radians. The unit energy due to the circumferential strain is the product of the plastic hoop stress and the corresponding total hoop strain.

Essentially, the diaphragm is considered to be held rigidly in the plane AA while a continuous plastic deformation under pressure drop Δp moves the diaphragm end from plane BB to CC. During this process, each element of the diaphragm undergoes longitudinal strains due to pure bending and hoop strains caused by the overall increase in the radius.

When we set the external work done on the diaphragm by the actuating pressure ΔP equal to the total internal energy of deformation, the design formula for the actuating pressure in a cylindrical diaphragm can be obtained as [8]:

$$\Delta P = \frac{(k+m)\left(0.55+4m^2+2m\right)Sy}{mk^3} \tag{10.14}$$

Where we can use Figure 10.11 to obtain the dimensionless parameters k and m, with k and m being R/t and r/t, respectively. Correspondingly, Eq. (10.14) can be simplified to:

$$\Delta P = \frac{\left(0.55k+4m^2+2m\right)Sy}{mk^2} \tag{10.15}$$

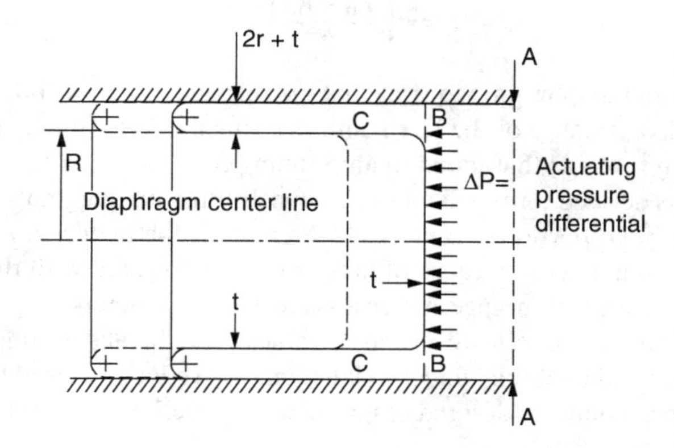

FIGURE 10.11 Rolling diaphragm.

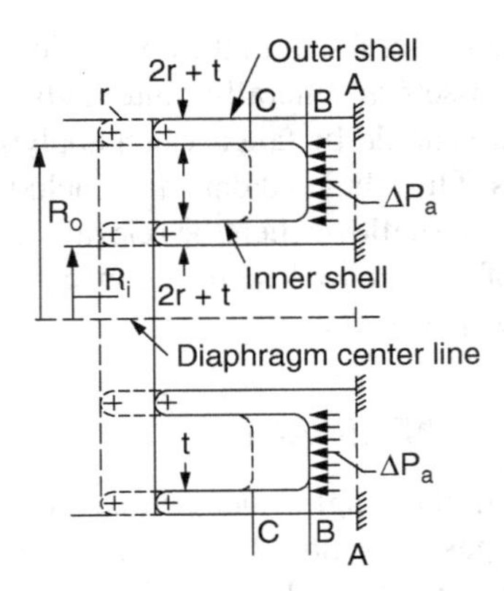

FIGURE 10.12 Annular version of a rolling diaphragm.

If it is necessary to increase the capacity of the diaphragm interface without damaging the mechanical joint, either at the *AA* plane or due to plastic deformation of the diaphragm wall at any point along the inversion path, we can select the annular geometry using Figure 10.12.

Utilizing assumptions similar to those applicable to the single cylinder design shown of Figure 10.11, the approximate expression can be derived:

$$\Delta P_a = \frac{\left[0.55(k_0 + n_0) + 4m(2m+1)\right] S_y}{m\left(k_0^2 + n_0^2\right)} \tag{10.16}$$

where, in this particular case, $k_0 = R_0/t$ and $n_0 = R_i/t$. The parameter m is the same for Eqs. (10.14) through (10.16). Also, the term S_y defines the yield strength of the diaphragm material for cylindrical and annular configurations.

We can make a comparison between the cylindrical and annular designs on the basis of equal volumes and equal pressures. For example, by denoting the equivalent thickness of the annular diaphragm by t_a, we obtain the expressions:

$$t_a = 0.87\left(\frac{R_i r \Delta P_a}{S_y}\right)^{1/2} \tag{10.17}$$

and

$$t_a = t\left(\frac{R_0 - R_i}{R}\right)^{1/2} \tag{10.18}$$

The mechanics of a rolling diaphragm requires that the relevant material should exhibit good elongation and ductility. Also, because of the large deformations involved, it may be desirable to select a nonheat-treatable alloy, such as that found in aluminum products.

The design equations derived for the diaphragms constituting unique joints show that the actuating pressures are directly proportional to the yield strength of the material. Also, from a practical point of view, it is well to note that because of large strains associated with the process of toroidal inversion of the cylinder wall, thin-gage materials and large diameters may be required in order to minimize the working stresses. However, large values of k, k_0, and n_0 imply a decrease in the resistance of the wall to local buckling, under axial compressive loads caused by the actuating pressure differentials. Hence, a delicate design compromise may well be involved in sizing a particular diaphragm system.

10.9 SUMMARY

The topic of membrane joint interfaces described in this chapter involves vacuum barrier configurations acting as membranes, thin shells, and plates. The relevant joints can include bolted, welded, or adhesion-type elements. In special barrier applications, the materials can range from viscoelastic to brittle. The mechanical properties of interest are usually defined as yield, ultimate strength, elongation, or the elastic modulus.

Thin circular membrane under pressure develops an approximate spherical curvature and the middle surface undergoes a significant straining. The edges are held by a combination of shear and friction. Large deflection theory of plates can be used to predict the membrane stresses and maximum deflections. The location of the critical stress depends on the boundary conditions and the ratio of deflection to thickness.

Many windows and barriers are made from viscoelastic materials such as Mylar. The designs using viscoelastic barriers require careful tests and analysis, and each particular case should be considered on its merits. The performance depends on the barrier material and the methods of clamping. The stress is a function of the strain and time. At stress levels above the yield, the barrier cannot support a constant load.

Tests on Mylar barriers show that the material's thickness at the center of the diaphragm can be reduced significantly. The experience shows that a clamping pressure of about 700 psi should be sufficient even for a relatively large diaphragm.

Test results are also available for tantalum and aluminum circular barriers. The mode of failure is different for large and small barriers. The pressures at failure, predicted on the basis of elastic theory, exceed the experimental values by a significant margin. The use of plastic correction indicates a much better correlation between the calculations and the experiments. Larger diaphragms are likely to fail by pull-out at the support prior to reaching the maximum membrane at the center.

Spherical caps used as windows are often made out of a brittle material such as industrial quality beryllium. When the pressure is applied to the convex side of the window, the factor of safety should be based on the buckling pressure. The experience shows a significant scatter in correlating theory and tests for all kinds of spherical caps and complete vessels. However, an empirical formula can be used for predicting the critical pressure for a relatively thin and shallow spherical cap.

A mechanical interface in a positive expulsion device is often based using the theory for a cylindrical diaphragm subjected to a continuous plastic inversion. The relevant analysis results in formulas for the actuating pressure for the cylindrical and annular configurations.

For a given geometry and proportions of a rolling diaphragm, the actuating pressure is directly proportional to the yield strength of the material. The relevant materials should exhibit good elongation and ductility. The best performance is found with thin-gage materials and large diameters, provided the diaphragm system has sufficient resistance to buckling due to axial compression.

10.10 DESIGN ANALYSES

EXAMPLE ANALYSIS 1

A floor mounted doorstop is an example of a mechanical joint used in virtually all office buildings and public facilities. Although doorstops are ubiquitous, they are hardly ever noticed or come to mind, unless the fasteners fail or come away from the floor.

Doorstops have various designs ranging from a wooden wedge on the floor to a hinge rotation restrictor, to a viscoelastic (damper) mechanism. Perhaps the most common design is a floor-mounted barrier arresting an opening door. Figure a provides a figurative drawing of this common device.

Suppose an engineer is asked to design a floor-mounted doorstop for arresting the movement of a heavy door in a highly populated facility. Suppose further that the doorstop design is to be used with each of the many widely used doors in the facility. Suppose even further that past

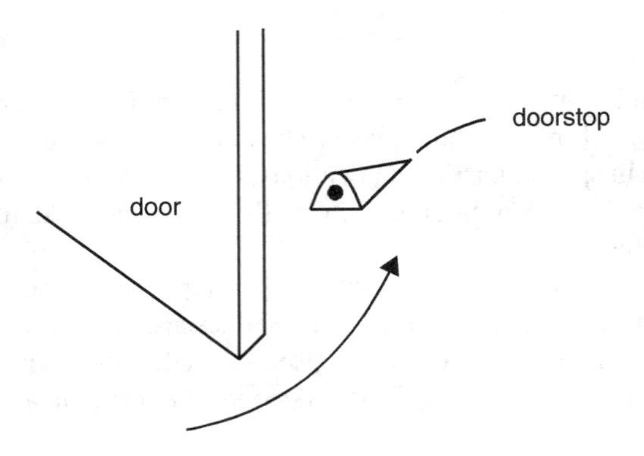

FIGURE A A doorstop for a swing door.

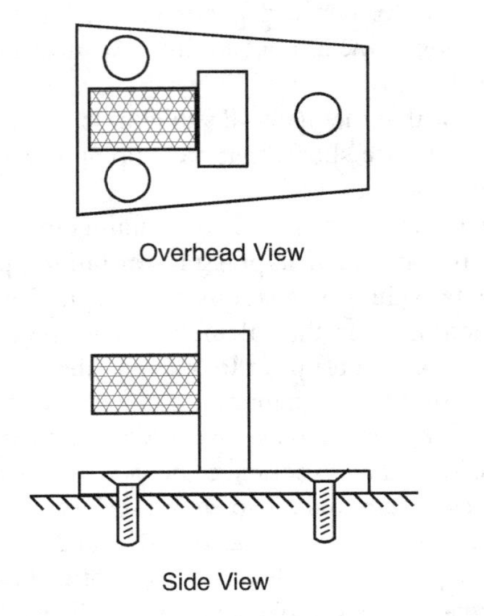

FIGURE B Model of a floor-mounted doorstop.

experience has shown that a common failure is the breaking, or yielding, of the fasteners attaching the doorstop to the floor.

Solution

In view of this assignment and of the past experiences, suppose for a beginning analysis we consider the shear stresses on the fasteners as the door impacts the doorstop.

To this end, let the doorstop be modeled as a rigid trapezoidal base plate supporting a rigid structure with a soft rubber-like projection positioned to encounter the lower corner of the swinging door, as represented in Figure b.

With these assumptions, the only elastic components of the doorstop model are the floor fasteners and the rubberized door-contact projection.

Rubber-like compounds and similar polymers are viscoelastic materials providing resistance to deformation proportional not only to the extent of the deformation, but also to the *rate* of the deformation. In the spirit of our preliminary analysis, for simplicity we will assume that the rubber-like stopper component can in turn be modeled as a linear spring.

Finally, for modeling simplicity we represent the swinging door as a moving block.

Figure c provides a side view of the resulting model where B is a moving block with mass m, modeling the swinging door; V is the speed of B; and k the linear spring modulus modeling the rubber-like stopper projection. The frame of the doorstop is assumed to be rigid. The fasteners

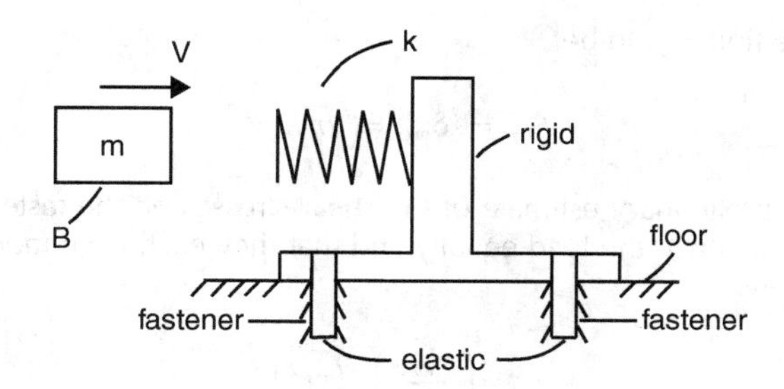

FIGURE C Resulting simplified doorstop model to estimate fastener shear stresses.

mounting the stopper frame to the floor are elastic but inserted and imbedded in the rigid stopper frame and rigid floor. Therefore, as the frame tends to slide along the floor, the fasteners are subjected to shear loading.

As block *B* contacts the spring and begins to compress it, the spring will respond with a resisting (arresting) force *F* on *B* where *F* is proportional to the deformation. The magnitude of this arresting force will increase until the movement of *B* is stopped. At the instant when *B* is stopped the magnitude of the resisting spring force will have a maximum value: F_{max}.

With the frame of the doorstop being rigid, the law of action-reaction, and/or a free-body diagram of the frame shows that when the spring force has its maximum value (F_{max}), this force will be transmitted to the fasteners.

By using the work-energy principle (as in Example Analysis 37) we can readily determine F_{max} in terms of the block mass *m* and its preimpact speed *V*. To this end, observe that the preimpact kinetic energy K_i of the block *B* is simply: $(1/2)mV^2$, and once the movement of *B* is arrested (zero velocity) the kinetic energy K_f of *B* is zero. Therefore, the change in kinetic energy ΔK, from impact to arrest is simply:

$$\Delta K = K_f - K_i = 0 - \left(\frac{1}{2}\right)mV^2 = -\left(\frac{1}{2}\right)mV^2 \tag{a}$$

Next, the work *W* done on *B* by the spring is simply the integrated spring force *F* through the spring deformation δ up to the maximum spring deformation δ_{max}, when *B* is arrested. That is,

$$W = \int_0^{\delta_{max}} (-F)d\delta = \int_0^{\delta_{max}} -k\,\delta d\delta = -\left(\frac{1}{2}\right)k\delta_{max}^2 \tag{b}$$

where the negative sign occurs since the force on *B* is opposite to the motion direction of *B*.

Finally, the work-energy principle states that:

$$W = \Delta K \tag{c}$$

By substituting from Eqs. (a) and (b) into (c) we have:

$$-\left(\frac{1}{2}\right)k\delta_{max}^2 = -\left(\frac{1}{2}\right)mV^2$$

or

$$k\delta_{max}^2 = mV^2$$

or

$$\delta_{max} = \left(\sqrt{m/k}\right)V \tag{d}$$

Since F_{max} is $k\delta_{max}$, we find F_{max} to be

$$F_{max} = k\delta_{max} = \left(\sqrt{mk}\right)V \tag{e}$$

We can now make a preliminary estimate of the shear stresses on the fasteners: Supposing that the three fasteners experience the load equally and that they each have radius *r*, the shear stress τ on a fastener is simply:

$$\tau = \left(\frac{F_{max}}{3}\right) \div \left(\pi r^2\right)$$

or

$$\tau = \frac{\left(\sqrt{mk}\right)}{3\pi r^2} \tag{f}$$

EXAMPLE ANALYSIS 2

Review Example Analysis 1. Make a listing of the principal assumptions and then discuss their reasonableness and how they could affect the validity of the results.

Solution

Aside from the assumption about the applicability of the usual principles of mechanics of mateirals, here are the principal assumptions in our preliminary analysis of the doorstop design:

1. The focus is on the shear stresses in the fasteners—neglecting the possible importance of tensile stresses and/or stresses in other parts of the structure.
2. Aside from the fasteners and the cap of the doorstop, the base and frame can be modeled as being rigid.
3. The stopper cap is modeled as a linear spring.
4. The swinging door can be modeled as a moving block with mass *m*.
5. All three fasteners absorb the shear force equally.

DISCUSSION

1. The focus upon fastener shear stresses is reasonable since: (i) materials fail in shear more readily than in tension or compression; (ii) the doorstop geometry being necessarily short, and contained to a region near the floor, results in shear being the principal loading force; and (iii) the fastener radius is likely to be the smallest of all the doorstop dimensions.
2. Assuming the doorstop frame to be rigid is also reasonable in that the longer geometrical dimensions of the frame will keep it from deforming to any significant extent and thus, in effect, be the same as rigid. The assumption of a rigid base, however, may not be as good in that the fastener holes may be a source of stress concentration and consequently some deformation which could affect the assumed equal loading on all of the fasteners (see 5. below).
3. A linear spring model of the stopper cap is not a good model of a rubber-like, polymer material. A polymer will act as a "shock absorber" where the resistant force will be dependent not only upon the deformation but also upon the *rate* of deformation. Thus, a linear spring model is not a good representation of the material response.

   That having been said, a linear spring model may still be of use if it is regarded as representing the entire structural response of an actual design incorporating the behavior of a non-rigid frame and base, and the deformation of the fastener.

   A question arising then is: What is the value of the spring constant? The answer is probably easiest to determine via an experiment. We thus may have an example of an evolving, iterative analytical/experimental design.

While such an iterative process may not be necessary for a doorstop where over-design is not an issue, such a process may well be economical for the design of more critical joint components.

4. Modeling the swinging door as a moving block is equivalent to thinking of the door as a particle with mass *m*. While this is not a good model, due to the neglect of the angular momentum of the door, the model may still be of use if the mass *m* is sufficiently large. If the dimensions and weight of the door are known, a separate dynamic analysis of the door may be useful in determining a value for *m*.
5. The assumption that all three fasteners absorb the shear force equally is the least good of all the assumptions. For equal load sharing there needs to be ideal geometry and a rigid base plate, neither of which can be attained. Therefore, from a joint design perspective, it is best to assume that one of the three fasteners will absorb the largest portion of the load and thus be the first to deform and fail. Equation (f) of Example Analysis 41 should thus be replaced by the expression:

$$\tau = \frac{\sqrt{mk}}{\pi r^2} \tag{a}$$

REFERENCES

1. Timoshenko, S. and S. Woinowsky-Krieger. 1959. *Theory of Plates and Shells.* New York: McGraw-Hill.
2. Roark, R. J., W. C. Young, R. G. Budynas, et al. 2012. *Roark's Formulas for Stress and Strain.* New York: McGraw-Hill.
3. Page, J. R. Aug. 3, 1970. Some Discussion on Large Diameter Mylar Vacuum Window Design. Lawrence Livermore National Laboratory, Livermore, CA, Internal Report ENN 70–54.
4. Peterman, K. A. Feb. 22, 1982. A Comparison of Large Deflection Plate Analysis vs. Experimental Response of Circular Flat Plate Diaphragms. Lawrence Livermore National Laboratory, Livermore, CA, Internal Report ENN 82-6.
5. Bleich, F. 1952. *Buckling Strength of Metal Structures.* New York: McGraw-Hill.
6. Koger, R. G. 1980. *Written Communication.* Livermore, CA.
7. Ramm, E. 1982. *Buckling of Shells.* Berlin, Heidelberg, New York: Springer-Verlag.
8. Huston, R. L. and H. Josephs. 2009. *Practical Stress Analysis in Engineering Design*, 3rd Ed. Boca Raton, FL: CRC Press.

SYMBOLS

a_0	Radius of cap base
E	Modulus of elasticity
E_r	Reduced modulus
E_t	Tangent modulus
F_i through F_5	Stress factors
G	Design factor
H	Depth of spherical cap
H_1 through H_3	Deflection factors
K	Collapse pressure parameter
$k = R/t$	Radius to thickness ratio
$k_0 = R_0/t$	Radius to thickness ratio
$\mathrm{m} = R/t$ or r/t	Radius to thickness ratios
$\mathrm{n}_0 = R_i/t$	Radius to thickness ratio
ΔP	Diaphragm actuating pressure
ΔP_a	Actuating pressure for annular diaphragm
q	Uniform pressure
q_{CR}	Critical pressure

R	Radius of curvature
R_o	Outer radius
R_i	Inner radius
r	Radius of bend
S	Stress
S_y	Yield strength of material
T	Thickness of spherical cap
t	General symbol for thickness
t_a	Equivalent thickness of diaphragm
δ	Maximum deflection
θ	Central half angle of cap
λ_0	Geometrical parameter
υ	Poisson's ratio

11 Design for Adhesion

11.1 INTRODUCTION

Major developments have taken place in the area of adhesive joints and adhesives since World War II. From the everyday household cements, up to the exotic heat-curing adhesives in aircraft applications, many ready-made formulations have been utilized in mechanical joint design and fabrication. The science of adhesive joints, however, was born in modern laboratories after the art of making adhesive bonds was tested by practice in the field. From field experience, various traditions and rules were developed which had to be critically examined by analyzing the forces existing between the adherends and the adhesives.

By definition the adherend describes a solid body which is held by a layer of an adhesive material in contact with another solid surface. The adhesive counteracts the effects of surface roughness, as well as any boundary layer found between the two solid surfaces.

The science of physics and chemistry of bonding attempts to provide the explanation of the molecular forces as well as the magnitude and the location of the breaking stress. However, our knowledge of the precise relation of the breaking stress to the various joint and materials parameters depends on a careful study of the intermolecular forces, adhesion boundary, and joint mechanics.

The variety and complexity of adhesive joints are astonishing and the past 50 years, in particular, have seen numerous research publications dealing with the growing body of the science of adhesion.

This chapter gives only a very brief overview of some of the basic characteristics of adhesives related to the mechanical strength and performance of adhesive connections. It is hoped that these topics are of interest to engineers concerned with the design of mechanical and structural joints. However, for a detailed study of the complete field of adhesion, the reader is directed to an excellent monograph by Bikerman [1].

11.2 CHARACTERISTICS OF SOLID SURFACES

The strength of adhesives can be applied to fractured, as well as new, fabricated surfaces. In general, a ruptured surface is rough and a number of irregularities are formed by a plastic action and stress relieving so that there is no perfect fit along the interface created by fracture. The function of an adhesive then is to mask the effects of a ruptured surface by filling the valleys and irregularities separating the two homogeneous phases.

Since many valleys and peaks represent the so-called surface roughness it is possible to estimate the actual area of interface by statistical means. Several methods are available for calculating the area ratio of the actual to the average geometrical surface. The degree of roughness affects also the magnitude of the conventional friction factor.

Some delicate measurements indicate that the average slope of the tallest hills on the surface can vary between 6E and 27E giving the respective tangents of 0.1 and 0.5. Experiments also indicate that in addition to roughness there is always a degree of waviness of the surface creating a variable clearance between the two solids available for penetration of the adhesive.

In the case of stainless steel, for example, subjected to a standard finish, the average roughness may be on the order of 2 to 5 microns. For a mirror polish this value may drop to about 0.02 microns. The American Standards Association defines 26° of roughness depending on the fabrication process [2].

In addition to roughness and waviness a solid surface may be porous. Materials such as leather, textiles, paper, or wood are porous throughout.

In some other situations only the solid surface is porous as, for example, in the case of anodized aluminum. The oxide film of this process may be on the order of 1 to 10 microns. In terms of this magnitude of roughness such a film should be regarded as thick. Standard oxide layers which form on almost all metals under suitable atmospheric conditions are thinner and less porous than the film on anodized aluminum. It may also be of interest to note that common glass in a humid environment develops a porous surface which may be verified by measuring the refractive index.

All metals are polycrystalline in a random fashion; however, the effects of machining create a preference in terms of crystallographic orientation. When more than one chemical phase is involved and exposed by machining or by means of internal damage, the electrical potential difference between the phases promotes corrosion. This process, together with impurities, is likely to influence the strength of an adhesive in a mechanical joint.

The chemical composition of a given metal varies across the surface and it is different from that of the bulk material. The existence of such a variation can be verified in a "microhardness" test which also shows that the surface stress can vary with the depth of probe indentation. In most cases, the surface hardness is higher than that of the underlying metal. For a mechanically polished aluminum the maximum difference in hardness may be on the order of 35% [11].

The process of migration of foreign atoms, known as adsorption, is important in understanding adhesion. The mechanism of adsorption is usually explained in terms of gases although the adhesives are applied as liquids. The atoms and molecules of an adhesive cling to the solid surface and the pores. When the solid is in a state of a fine powder this process can easily be achieved.

The problem of adsorption to metal plates, bars, and similar components, however, is more difficult to define and requires the determination of the actual area of interface by experimental and analytical techniques.

11.3 THE MECHANICS OF ADHESION

The assembly of mechanical joints by purely mechanical means using rivets or chemicals in a welding process is different from that of utilizing an adhesive to fasten the two solids together. In the so-called solid-to-solid adhesive there are several unique aspects of adhesion mechanics deserving our special attention.

It can be shown experimentally that two solids pressed against each other in a vacuum develop a mechanical bond. To break such a bond, a measurable force is required. It is quite reasonable to assume that the formation of the mechanical bond can be caused by forces of attraction between the individual atoms. The magnitudes of such forces, of course, must depend on the materials surface roughness and other physical irregularities.

In another instance of adhesion, such as that of a nail driven into wood or soft metal, the interaction pressure is derived from the elastic springback of the interface. Consequently, a definite amount of effort is required to extract the nail. However, the frictional resistance developed during the penetration and extraction process can be modified and reduced by a layer of lubricant.

When a rod is rotated inside a metal cylinder, under a given contact pressure, a considerable amount of surface "entanglement" can be created. The joint so formed can resist significant shear and tension stresses during the process of pulling the two pieces apart. This type of solid-to-solid adhesion is utilized in industry, utilized in such operations as bonding of copper and other metals by mechanical means at relatively low temperatures.

Although this art is rather ancient, the mechanism of this process has seldom been investigated theoretically. Perhaps the most likely explanation can be based on the theory of elastic response, where a protrusion of one surface is forced into a depression on the opposite surface. The process of locking the two surfaces together is similar to the action of a snap fastener. In the absence of a depression, the protrusion can, of course, act as a nail. If, in addition to a purely normal pressure, the surfaces can undergo some tangential displacement, a certain amount of "entanglement" or "plastic mixing" is developed. Shearing action is then required to separate the adhesive bonding to such surfaces. The rate of separation also becomes important and it introduces additional theoretical complexities in understanding the finer points of adhesion when the two solids behave in a tacky manner. Detailed evaluation of this problem is beyond the scope of this treatise, although it can be found in the literature [1].

When the adherence is created due to a special coating placed between the two solids, the mechanics of adhesion can be explained in terms of a hooking phenomenon, creating a strong bond. Such a bond is seldom broken. For example, postal stamps, glued leather strips, aluminum panels, or glass fiber composites usually fail without separating the adherends from the adhesives.

The best hooking process is obtained with porous and rough surfaces. The adhesive material enters the porous surface layer and it is intermixed to the degree that a mechanical separation of the two ingredients becomes virtually impossible. The depth of penetration of the adhesive material depends on the porosity and other microscopical features of the adherends affecting mechanical interlocking. The breaking stress of the joint increases with the increase of the depth of penetration and mixing.

The mechanics of adhesion and the absolute strength of a given bond are strongly influenced by the presence of the so-called weak boundary layer, consisting essentially of impurities, low strength components, and chemical films. The texture and chemistry of such layers may be difficult to locate and define prior to a cleaning operation. Chemical contaminants exhibiting a high physical strength need not be removed.

The mechanical methods of cleaning are best suited for carbon steels, while a chemical cleaning process appears to be better for stainless steels, titanium, or aluminum. In general, the breaking stress of a joint will increase with the degree of the appropriate cleaning procedure. In the case of aluminum, the breaking stress after cleaning was reported to be increased sevenfold [1].

Cleansing is often difficult due to the variety of contaminants and adhesion parameters involved in a single fabrication process. The most common contaminant, of course, is air, which should be displaced by the application of a suitable liquid adhesive.

Considerable theoretical work has been done, in this country and elsewhere, concerning the process of wetting and the associated liquid phenomena, including viscosity of adhesives. Unfortunately, it is difficult to compare the results of the various investigations because of the differences in materials, design, and test apparatus involved.

The process of air removal from the surfaces of adherends can only be successful when the capillary contact angle and the viscosity of the liquid adhesive are kept as low as possible. Furthermore, the solvent used with the adhesive should have a low surface tension. It should also be noted that the surface roughness per se has a limited influence on the process of air removal.

The breaking strength of an adhesive joint can be affected by the magnitude of contact pressure and the duration of the contact pressure during the fabrication. Although experimental data are available in some specific areas of adhesive joint technology, it is difficult to make any sweeping generalizations concerning such effects. It will suffice to note that the magnitude and the duration of the contact pressure enhance the strength of the adhesive butt joints. Since, however, the viscosity of Newtonian liquids is sensitive to temperature, it is also necessary to examine the effect of temperature on the time of duration of contact pressure required for the process of fabrication.

11.4 SETTING AND FLOW OF ADHESIVES

Although some commercial adhesives remain tacky, many are subject to what is known as a "cold flow" phenomenon. The process of setting can be accomplished by a chemical reaction, removal of solvent, or by cooling. The rate of setting can be controlled through purification and crystal sizing. The setting characteristics are also preferred to the use of solvents in adhesives and coatings.

Adhesives which solidify on cooling can be utilized on porous as well as nonporous materials. The two most important hot-melt adhesives include asphalts and solders. The applications of asphalts range from masonry to paper, cloth, and paperboard. In a more sophisticated application, polyvinyl butyral adhesive is used for the windshields of passenger automobiles. There is a great variety of industrial adhesives involving mixtures and chemical formulations too numerous to mention. Many common household and commercial adhesives solidify through the process of solvent removal, which can be accomplished by evaporation or by the use of porous adsorbents such as paper or wood. The evaporation method, however, is often too slow and can run well into months.

The accepted method of setting by a chemical reaction is not always satisfactory because the prediction of performance of a chemical mixture is very complex. Plaster of Paris is good for porous and nonporous surfaces and it acts as an adhesive. The well-known Portland cement, however, although similar in texture, does not quite behave as an adhesive.

The processes of solidification of adhesives have a number of undesirable features from the point of design of a mechanical joint, which has to rely on the final strength and volumetric changes. It is known, for example, that a number of latexes and cements are subject to shrinkage. Hence, if the adhesives cannot completely fill the pores and voids between the two rigid adherends, the strength of the bond must be somewhat lower. In the manufacture of a plastic joint, the problem of shrinkage is especially important.

Some of the adhesives classified as sealers also have high coefficients of shrinkage but they are still relatively pliable at the time when stress concentrations develop. Addition of inorganic substances to an adhesive product is often desirable in order to minimize the effect of shrinkage.

Volume increase during the process of cooling (for instance, water-to-ice transition) can expand the adhesive into the surface cavities and pores. The action creates the so-called snap-fastener effect with the corresponding increase in the strength of the bond. A number of solders, containing such components as lead, tin, or cadmium, tend to increase in volume during the transformation from liquid to solid. The length of time required for various transformations, including crystallization, is often critical and it is best determined from experience.

11.5 THE CONCEPT OF STRESS

The foregoing paragraphs outlined, a number of characteristics related to aging and setting phenomena of the adhesives. The final mechanical strength of the joint can be exceeded in the adherend, boundary layer, or the adhesive itself. Failure occurring exactly along the adherend-adhesive boundary is difficult to postulate. For example, a stratum of metal oxide can constitute such a boundary, although a sharp frontier between the metal and its oxide does not exist. It is more likely that the weakest strength area will be found in the bulk of crystals of the adhesive, leading to the so-called condition of failure in cohesion. Bikerman has shown how to prove in probabilistic terms [1] that no rupture can occur between the two adjoining phases.

Experimental analyses of brittle solids suggest that larger samples are generally weaker than the smaller test specimens. In the case of two filaments made of an identical material, the numerical value of the breaking stress of a shorter piece of a filament is greater. When the filament length is increased to a 100-fold, the corresponding value of the breaking stress can be decreased by as much as 50%. This analogy, when applied to the relative dimensions of the thickness of interface and the adhesive film, suggests again that the breaking stress of the interface must be higher.

The proof of this concept, of course, hinges on the definition of thickness of the interface. If we take the thickness to be roughly equal to the space between the adjacent rows of atoms, rather surprising results can be obtained if we assume that the bending stress varies inversely as the square of the thickness of the interface. For a typical household product, the thickness of the adhesive may be 10^6 times as great as the thickness of the interface.

In some of the old theories, the concepts of molecular attraction, free surface energy, and electrostatic attraction were suggested as the primary causes of the strength mechanism development between the adherend and the adhesive. Recent experiments, however, have shown repeatedly that a true adhesional failure between the two different materials cannot be supported by either theory or test [1]. This leads to an inevitable conclusion that the real breaking stress of the adhesive joint must be controlled by the strength of its weakest phase.

In terms of the given external force, producing a tensile or a shear failure, the nominal stress can be approximated by dividing the force by the appropriate area of the cross section. The difference between the actual and theoretical strength of the joint is governed by the geometrical stress concentrations, as well as by the presence of flaws. Again, the size effect must be considered because a larger specimen has a greater probability of containing an inclusion, crack, or other weakness detrimental to structural integrity. It appears that the probability theory of strength is seldom used, even in modern design practice, although the origin of this theory could be dated back over 100 years.

The effect of flaws on the structural integrity of a Hookean solid can be modeled with the aid of the stress concentration factors derived from theory and experiment [3,4]. While the crack along the line of pull has essentially no harmful effect, the flaw normal to the line of the external force must be detrimental. If the shape of the flaw can be approximated by an elliptical hole then the ratio of the major to minor axis determines the extent of stress concentration. Theoretically, this ratio can be very high, indicating unusually high numerical values of stress concentration factors. Fortunately, plastic deformation entirely alters the distribution and magnitude of the stress field and mitigates the extent of the stress gradient. This convenient mechanism applies as long as the adhesive does not become too brittle.

Molecular cohesion is usually expressed in stress units. This stress is assumed to act across a plane perpendicular to an external force. The result of this action is the increase of the distance between the atoms on both sides of the stress plane. The molecular cohesion is then a material property, depending on the composition and the atomic structure of the solid. As stated above, the real cohesion stress must be lower when corrected for flaws.

When the adhesive is allowed to cure in the absence of external, mechanical constraints, the calculation for a cohesion stress does not require a correction for residual stresses. That is, free shrinkage or expansion in itself should not result in a residual stress field. However, when residual stress of a tensile nature is present, the external tensile stress required to break the adhesive joint must be correspondingly smaller. If, in addition to the shrinkage stress, the adhesive is subject to relaxation, the prediction of the breaking stress becomes complicated and requires careful experimental verification.

Numerous studies have been made of the mechanics of shrinkage in terms of the conventional stress analysis and experiments. For a layer of the adhesive having thickness h_0 in a joint formed by two parallel plates, the adhesive material may be prevented from shrinking by design of the total composite system. If the same layer of the adhesive were allowed to set without external constraints, its thickness would have been a little smaller, say, h_1. The actual strain caused by the constraint of the shrinkage can then be:

$$\frac{\Delta L}{L} = \frac{h_0 - h_1}{h_0} \tag{11.1}$$

where, in conventional elasticity, ΔL represents a small change in any linear dimension L.

According to Hooke's law, the stress normal to the adherend plates must be:

$$S_a = \frac{(h_0 - h_1)E_a}{h_0} \tag{11.2}$$

where S_a is the shrinkage, or frozen, stress in the adhesive layer and E_a denotes the modulus of elasticity of the adhesive. Also by this definition, the tensile breaking stress of the adhesive joint has to be:

$$S_t = S_m - S_a \tag{11.3}$$

where S_m denotes the strength of the molecular type of cohesion of the adhesive material free of voids and imperfections.

In the foregoing discussion, the terms "residual" and "frozen," as well as "shrinkage," were used rather interchangeably, only to indicate that a natural tendency of the said adhesive to respond (in this case to shrinkage during setting) was prevented by some mechanical means. In the case of a conventional welded joint, for example, the term "residual stress" means the stress developed from "shrinkage" during cooling of the weld area. The term "frozen" in the science of adhesion reflects the existing, initial state of stress due to the external, rather than the internal, constraints.

Suppose an adherend plate of thickness H is elastically deformed by the action of a concentrated force F, derived from shrinkage of the adhesive layer having a thickness h, as shown in Figure 11.1. If the adherend plate is free to deform as indicated by the dotted line, then, by analogy to a beam bending theory, we can say that:

$$\frac{M}{EI} = \frac{2}{2r + H} \tag{11.4}$$

In this elementary model, r is the distance from the center of curvature to the plane common to the adherend and the adhesive, M is the equivalent external moment causing bending of the adherend plate and EI is the conventional flexural rigidity of the adherend.

The external bending moment in this case is assumed to be equal to:

$$M = \frac{FH}{2} \tag{11.5}$$

Also, the corresponding stress in the adhesive is:

$$S_a = \frac{F}{hb} \tag{11.6}$$

Combining Eqs. (10.4), (10.5), and (10.6) gives:

$$S_a = \frac{EH^2}{3(2r + H)h} \tag{11.7}$$

When the radius of curvature is large compared to the thickness of the adherend, Eq. (11.7) reduces to:

$$S_a = \frac{EH^2}{6hr} \tag{11.8}$$

Equation (11.8) is approximate at best. This and more elaborate theories, however, have been investigated experimentally indicating that in the two-thirds of the adherend nearer to the adhesive the stresses were actually of the compressive nature [1]. The model shown in Figure 11.1 assumes a symmetrical stress distribution, for a beam in pure bending, on which a direct compressive stress due to F can be superimposed.

The problem of a shrinkage stress in a cylindrical adhesive can be reviewed with the aid of the Lamé formulas of solid mechanics. The application of these formulas to the tubular lap joint, illustrated in Figure 11.2, is based on the premise that the adhesive and the adherend materials are essentially homogeneous, Hookean solids. This particular discussion follows from the considerations of the shrink-fit analogy where a straight cylindrical hub of outer diameter $2(a+h)$ and inner diameter $2a$ interacts with a solid circular shaft [5].

The process of shrinking the adhesive layer around the solid cylindrical adherend results in developing an interface pressure P which causes Lamé stresses. These can be of radial, tangential,

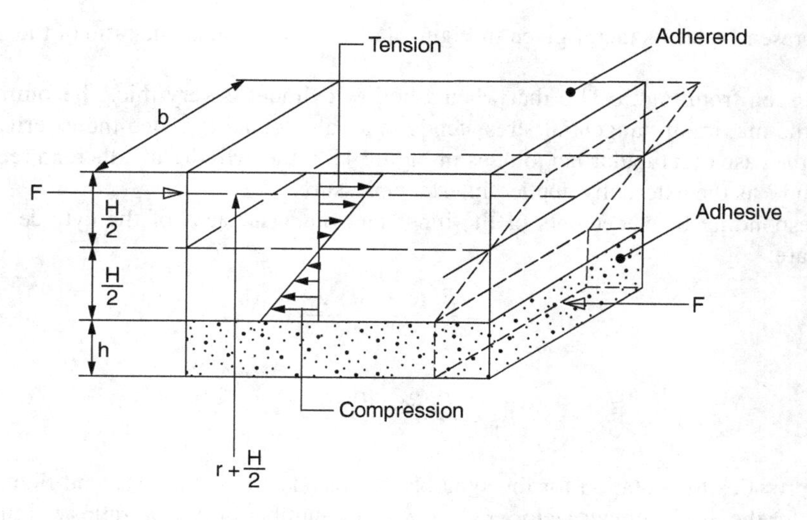

FIGURE 11.1 A plate joint.

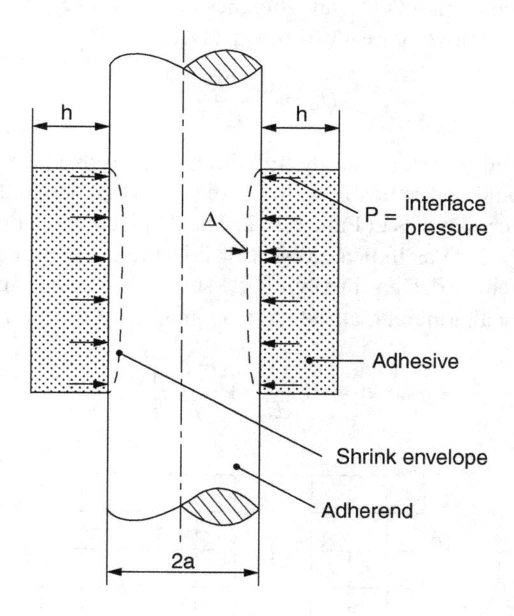

FIGURE 11.2 A tubular lap joint.

axial, and shear type. This specific model is concerned with the maximum radial stress which must be numerically equal to the interface pressure at the inner boundary of the adhesive defined by radius a.

The total amount of shrinkage can be defined as the absolute sum of the two individual, radial displacements of the inner and outer surfaces in contact. The inner displacement surface here is that of the adherend while the outer deformation refers to the adhesive.

When a cylindrical member, such as a pipe, vessel, or a sleeve, is subjected to uniform external pressure, the maximum tangential stress at the inner surface of the cylinder can be expressed as:

$$S_\theta = PK_4 \tag{11.9}$$

where K_4 represents a stress factor given in Figure 11.3 as a function of the ratio of the inner radius to thickness.

It may be seen from Figure 11.3 that when a hollow cylinder is very thick, becoming almost a solid shaft, the maximum tangential stress tends to a value equal to twice the externally applied pressure. In the case of a tubular lap joint as in Figure 11.2, the cylindrical adherend represents the solid shaft and P is the externally applied interface pressure.

The corresponding displacements of the inner and outer surfaces of the cylinder toward the central axis are:

$$U_i = \frac{P(R_0 - R_i)K_3}{E} \tag{11.10}$$

and

$$U_0 = \frac{P(R_0 - R_i)K_5}{E} \tag{11.11}$$

Figure 11.4 provides the notation for the symbols of Eqs. (11.10) and (11.11), and Figure 11.5 provides values for the displacement factors K_3 and K_5 for a number of typical ratios of inner radius to wall thickness. The wall thickness is simply: $R_0 - R_1$.

When the ratio of the inner radius to wall thickness becomes equal to zero, Figure 11.5 gives $K_5 = 0.7$, which leads to the following result from Eq. (11.11):

$$U_0 = \frac{0.7PR_0}{E} \tag{11.12}$$

Equation (11.12) can be used to determine the maximum radial displacement at the outer surface of a solid cylinder. This could be applicable to the case of a cylindrical adherend acted upon by the interface pressure due to shrinkage as in Figure 11.2. The displacement at the center of the adherend must, of course, be equal to zero as indicated by $K_3 = 0$ in Figure 11.5 and Eq. (11.10). In line with the shrink-fit theory of machine design, involving the sum of the radial displacements of the adhesive sleeve and the cylindrical adherend at radius a, an analytical formula can be stated as:

$$\Delta = P\left(\frac{hK_3}{E_a} + \frac{0.7a}{E}\right) \tag{11.13}$$

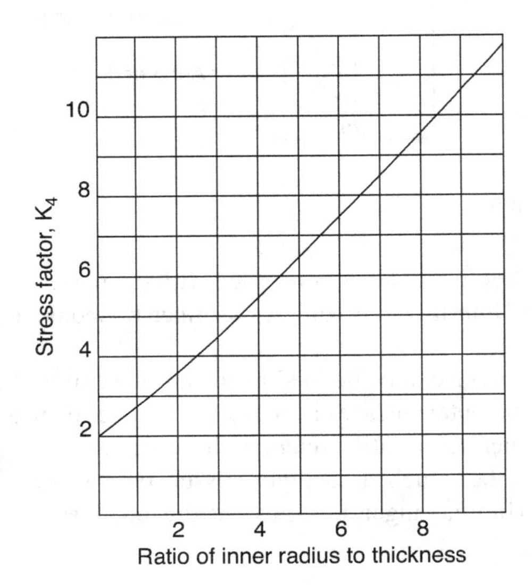

FIGURE 11.3 Stress factor for a thick cylinder under external pressure.

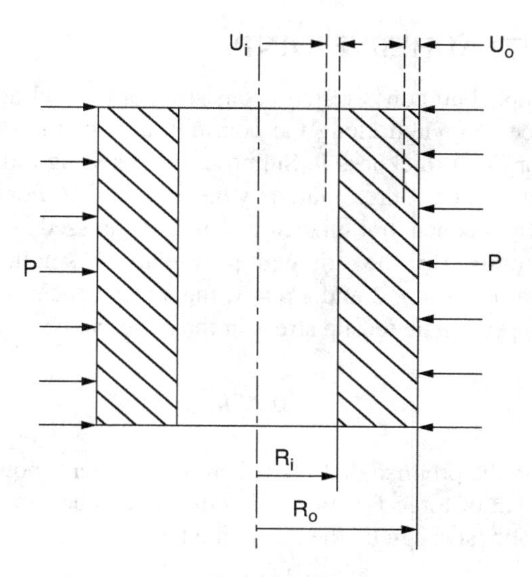

FIGURE 11.4 Notation for a thick cylinder under external pressure.

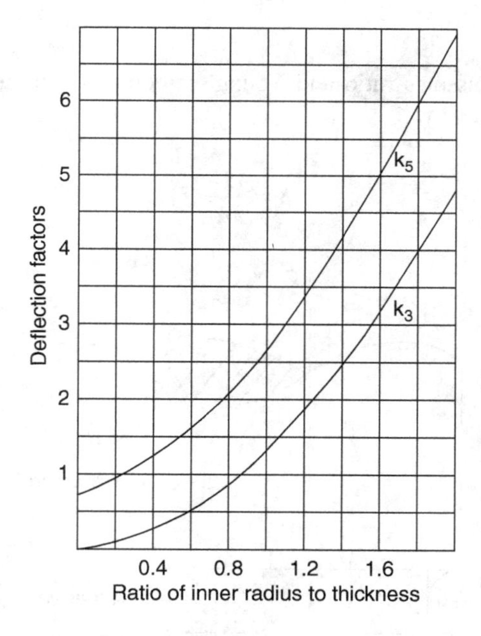

FIGURE 11.5 Deflection factors for thick cylinders under external pressure.

Note that from the first term in Eq. (11.13) comes Eq. (11.10), when h = $R_0 - R_i$ and $E = E_a$, defining the appropriate variables of the adhesives sleeve as illustrated in Figure 11.2. Similarly, the remaining term in Eq. (11.13) follows directly from Eq. (11.12) when $R_0 = a$. The parameter Δ is the radial displacement which is expected to be generated by the adhesive sleeve during the process of shrinkage.

Numerically, Δ consists of a radial contraction of the adherend with the simultaneous radial expansion of the adhesive, as show in Figure 11.2. These effects are individually described by Eqs. (11.11) and (11.10), respectively. While Eq. (11.13) gives some general insight into the mechanics of shrinkage in a tubular lap joint, it is offered here merely as a gross approximation to a complex problem where the behavior of the actual materials also involves the relaxation phenomena and a doubtful conformance to the Hookean response.

11.6 STRESSES IN BUTT ADHESIVE JOINTS

Figure 11.6 illustrates a simple butt adhesive joint consisting of two cylinders joined end-to-end.

An external tensile force F applied along the common axis of the two cylinders deforms the adhesive disk having the original thickness h. Suppose, the resulting radius of the groove, caused by the Poisson's ratio effect, can be approximated by the amount $h/2$. For the case of a round bar in tension, the conventional stress concentration factor k can be expressed as a function of e/ρ as shown in Figure 11.7 [5], on the premise that the ratio $\rho/(a-\rho)$ is relatively small.

For a radius of curvature $\rho = e = h/2$ and $e/\rho = 1$, the stress concentration chart in Figure 11.7 gives $k = 2.2$. Hence, the approximate tensile stress in the adhesive is:

$$S_a = 0.7F/a^2 \tag{11.14}$$

Equation (11.14) is based on the premise that the adherend cylinders shown in Figure 11.6 are relatively rigid and that the effect of force F is uniformly distributed across the butt joint. The change in radial dimension in the adhesive can be then described as:

$$U_0 = \frac{a\upsilon_a S}{E_a} \tag{11.15}$$

where υ_a and E_a denote Poisson's ratio and Young's modulus of the adhesive, respectively. The nominal tensile stress is given by:

$$S = \frac{F}{\pi a^2} \tag{11.16}$$

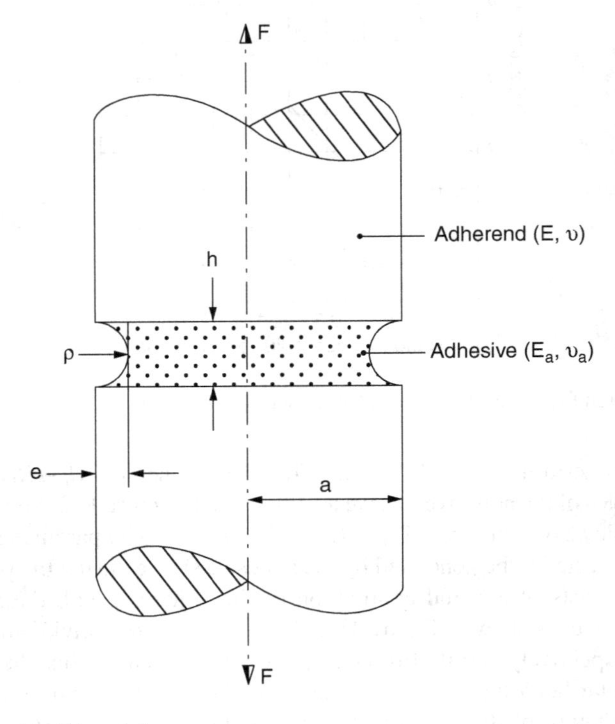

FIGURE 11.6 A cylindrical butt joint.

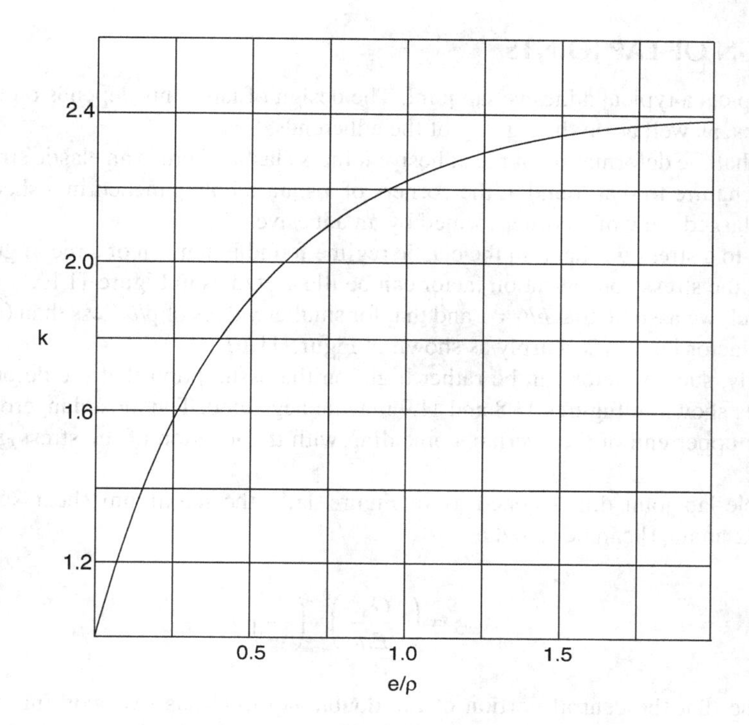

FIGURE 11.7 Stress concentration for a grooved bar in tension.

Based on the above model, involving Eqs. (11.12), (11.15), and (11.16), the radial stress P in the adhesive layer can be approximated by the expression:

$$P = 0.45\upsilon_a F/a^2 \tag{11.17}$$

The foregoing calculations, although crude, indicate that the radial stress in a disc-type adhesive for a typical butt joint can be a significant fraction of the nominal axial stress. The radial stress is then likely to have a detrimental effect when the cracks in the adhesive are oriented normally to the adherend-adhesive boundary. The existence of a relatively high radial stress in a butt-joint configuration has been verified experimentally [6]. The relevant experiment was conducted on a two-plate system held by a layer of a commercial quality of the epoxy adhesive.

For a disc-type adhesive layer, the externally applied stress increases as long as a/h ratio is increased. This type of behavior is tested by keeping the disc radius "a" constant and by gradually decreasing the disc thickness h. In these circumstances, the adherend may break before a plastic adhesive starts flowing.

The increase of the rupture stress with the decrease in thickness of the adhesive layer has, on occasions, been reported in the trade literature. In the majority of analyzed cases, the adherends were of a cylindrical form as in Figure 11.6. The adherends included such materials as copper, aluminum, steel, or glass to mention a few. The adhesives were either of a plastic or metallic nature. The failure modes may have been influenced by a number of factors such as strain ratios and the rates of setting of the adhesives.

The rates of curing and the dimensional parameters of the adhesive joints may precipitate the formation of residual stresses. The precise effect of these stresses on the ultimate strength of butt joints and other configurations is still difficult to predict reliably.

11.7 DESIGN OF LAP JOINTS

Figure 11.8 depicts a typical adhesive lap joint. The design of lap joints depends on the knowledge of the adhesives, as well as on the rigidity of the adherends.

Assuming that the deformation of the adhesive joint is elastic, there is an elastic stress concentration similar in nature to that found at the corners of a square hole punched in a sheet. Figure 11.9 presents an enlarged view of a corner formed by an adhesive.

By analogy to a stress gradient, in the elastic regime found in tension of a rectangular bar with a shoulder fillet, the stress concentration factor can be illustrated as in Figure 11.10.

In this model, we assume that $h/\rho = 1$ and that for smaller values of ρ/b (less than 0.05), the stress concentration factor increases sharply as shown in Figure 11.10.

Theoretically, such a factor can be rather high, on the assumption that the deformation of the adherend plates shown in Figures 11.8 and 11.9 may be neglected. The most dangerous shear stress is found at the upper end of the overlap, coinciding with the location of the stress gradient shown in Figure 11.8.

For a double-lap joint dimensioned as in Figure 11.8, the maximum shear stress originally derived by Bikerman [1] can be stated as:

$$\tau_{\max} = \left(\frac{G_a}{2Ehc}\right)^{0.5}\left(\frac{F}{w}\right) \tag{11.18}$$

On the premise that the central portion of the double-lap joint has two working surfaces of the adhesive in shear resisting the external force F, the stress concentration in shear can be expressed as:

$$k_s = \frac{\tau_{\max}}{\tau_{\text{ave}}} \tag{11.19}$$

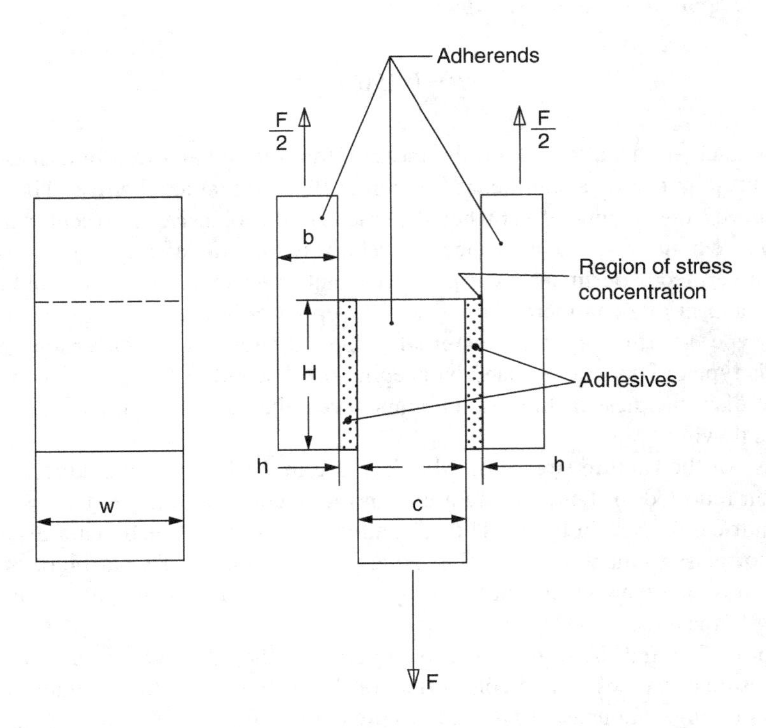

FIGURE 11.8 A typical double-lap joint.

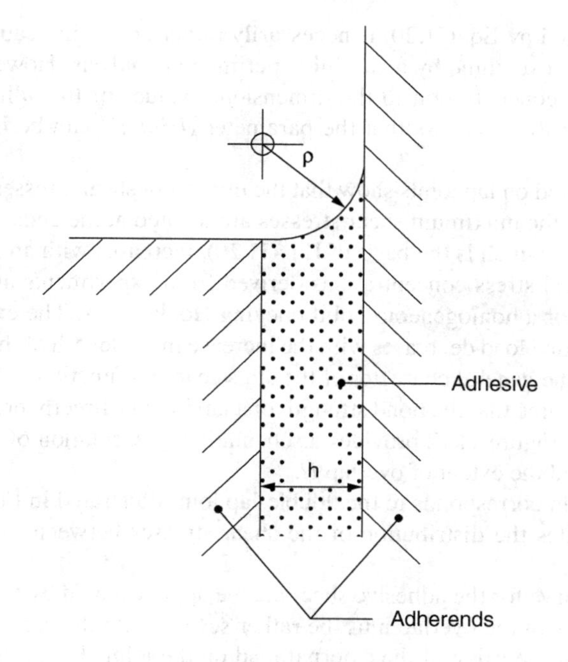

FIGURE 11.9 Detail of an adhesive boundary.

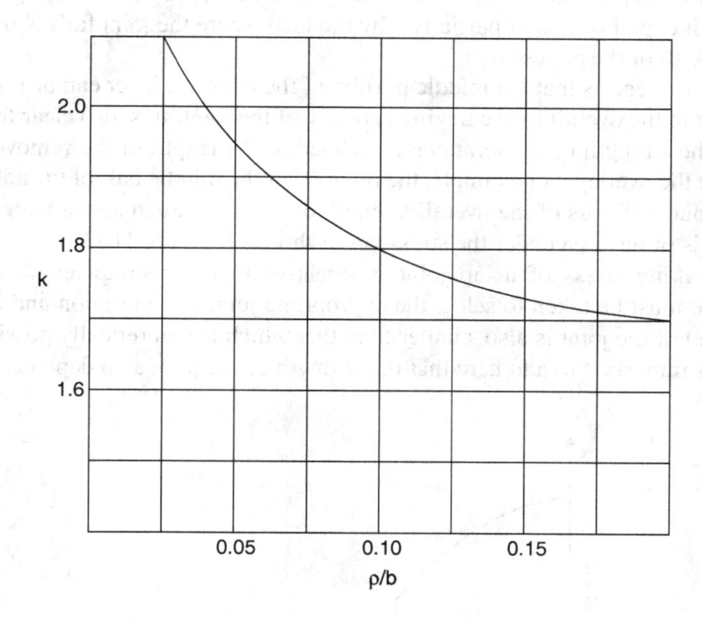

FIGURE 11.10 Approximate stress factor for the adhesive boundary.

and

$$k_s = H\left(\frac{2G_a}{Ehc}\right)^{0.5} \tag{11.20}$$

where G_a is the modulus of rigidity of the adhesive and E is the Young's modulus of the adherends. Figure 11.8 shows the width w of the adherends as well as other geometric parameters. τ_{ave} denotes the average shear stress.

The model represented by Eq. (11.20) is necessarily rather crude, indicating the various limitations which can only be overcome by a careful experimental analysis. However, the theory shows the sensitivity of stress concentration to the dimensional values of the adhesive layer and to the length of the overlap. It then appears that the parameter $(H^2/hc)^{0.5}$ can be important in design of adhesive lap joints.

Experiments conducted on lap joints show that the minimum shear stresses are found in the middle of an overlap, while the maximum shear stresses are located at the ends. The ratio of the maximum to average stresses, which is the basis of Eq. (11.20), increases with an increase in overlap.

Fortunately, the actual stress concentrations derived from experiments are not as damaging as they would have been for a homogeneous solid obeying Hooke's law. The experimental stress in a lap joint for a given applied load decreases with the increase in the length of the overlap. Figure 11.11 graphically represents the general character of the stress overlap function.

It is reasonable to assume that the bond strength of a lap joint is directly proportional to the width of the adhesive layer w. Figure 11.12 provides a schematic representation of the joint configuration showing the width w and the extent of overlap H.

The force equilibrium corresponds to the double-lap joint illustrated in Figure 11.8.

Figure 11.13 illustrates the distribution of the shear stresses between the edges of a complete lap joint.

If the distribution curve for the adhesive shear can be approximated by a parabolic function, the stress peaks at the edges of the overlap must be rather severe, with the central portion of the adhesive carrying only a small portion of the external load on the joint. The stress concentration factor k_s, approximated by Eq. (11.20), is then reasonably consistent with the stress gradient shown in Figure 11.13 and it explains at least partially why the load where the joint fails is likely to be below the nominal strength of the adhesive.

This model also suggests that the middle portion of the adhesive layer can be removed without a serious detriment to the overall load carrying capacity of the joint. It is also clear from Figure 11.13 that the loss of the strength of the joint can be related to the length of the removed adhesive film near the center of the overlap. For example, the removal of the middle half of the adhesive layer may result in only about 20% loss of the overall strength of the joint. Such an estimate can be approximated on the basis of the area under the stress curve shown in Figure 11.13.

Since the breaking stress of a lap joint is sensitive to the overlap length as illustrated in Figure 11.11, care must be taken to select the appropriate joint configuration and dimensions. The actual failure load of the joint is also smaller than that which is theoretically possible, as shown in Figure 11.14. It is important to add here that the strength of the joint also depends on the thickness

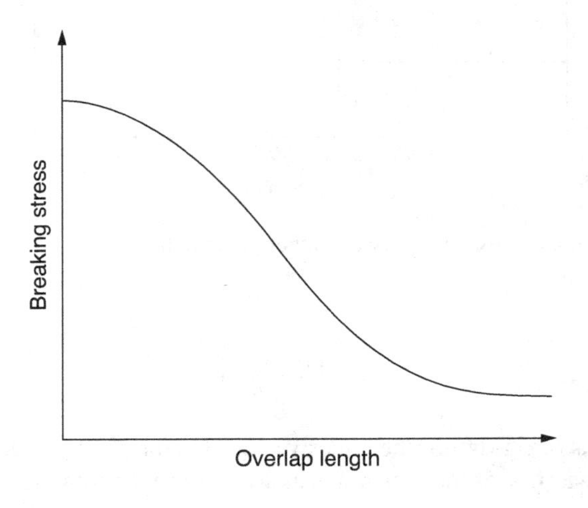

FIGURE 11.11 General trend in the stress-overlap relation.

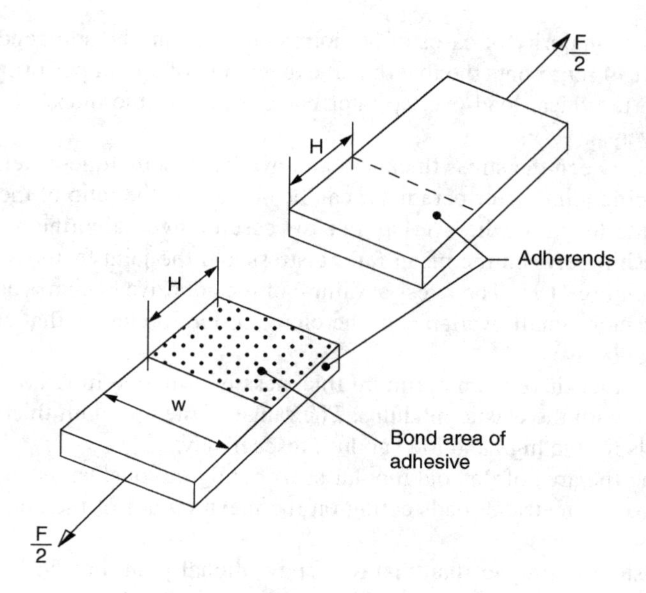

FIGURE 11.12 Elements of a lap joint.

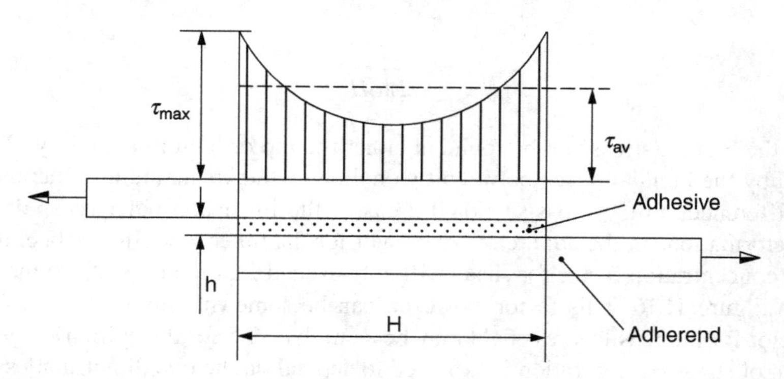

FIGURE 11.13 Shear distribution in a lap joint.

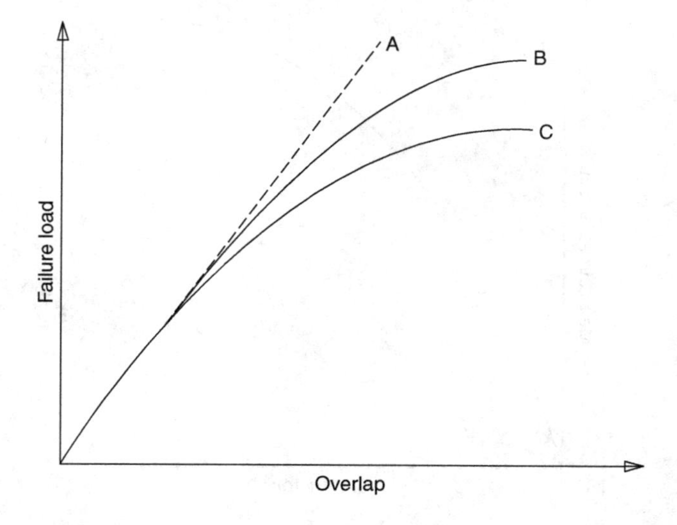

FIGURE 11.14 Effect of metal thickness and overlap.

of the adherend. In the majority of cases of lap joint construction, the adherend is made of metal. Curve A in Figure 11.14 represents the theoretical expectation of a joint performance, while curves B and C correspond to typical load-overlap interrelationships for the thicker and thinner metallic adherends, respectively.

The relationship between the stress thickness and overlap is sometimes referred to as the "joint factor" [7]. The specific joint factor parameter can be defined as the ratio of the square root of the metal thickness to the length of the overlap. For the case of steel, aluminum alloy and clad aluminum, the relationship between the mean failing stress and the joint factor is characterized by a smooth curve as in Figure 11.15. The stress at failure of the adhesive becomes asymptotic at higher values of the joint factor. Small overlap can, therefore, be interpreted as that compatible with the true strength of the adhesive.

Experience shows that there is an optimum thickness of adhesive in relation to bond strength. This optimum varies with the elastic modulus. The range of the optimum thickness appears to be between 2 and 6 mils for the high and low moduli, respectively.

Prior to reviewing the area of flat and tubular scarf configurations, let us consider a cylindrical lap joint subjected to symmetrical loads acting on the mean circles of the adherends as shown in Figure 11.16.

This adhesive system is simpler than that of a conventional joint, involving flat plate elements. Here, the mean shear stress can be expressed in elementary terms despite the effect of cylindrical geometry as:

$$\tau = \frac{F}{2\pi a H} \tag{11.21}$$

Theoretically, the highest stress can be found at either the top or bottom boundary of the adhesive sleeve, formed by the annular space between two hollow cylinders acting as adherends. However, the obvious differences in the cross-sectional areas of the inner and outer tubes dictate that the stresses and deformations in the adherend are greater for the inner tube. In practice, it is not likely that the stress concentration factor for shear will ever exceed 2 in the case of a typical cylindrical joint shown in Figure 11.16. This factor, however, can be somewhat higher for the flat lap joints. Analogous factor for the tensile stress field may be as high as 2.5 for the cylindrical joint. In either case, the extent of stress concentration is expected to depend on the two dimensionless parameters. In terms of the notation given in Figure 11.16, these parameters are: $b/2a$ and H/b.

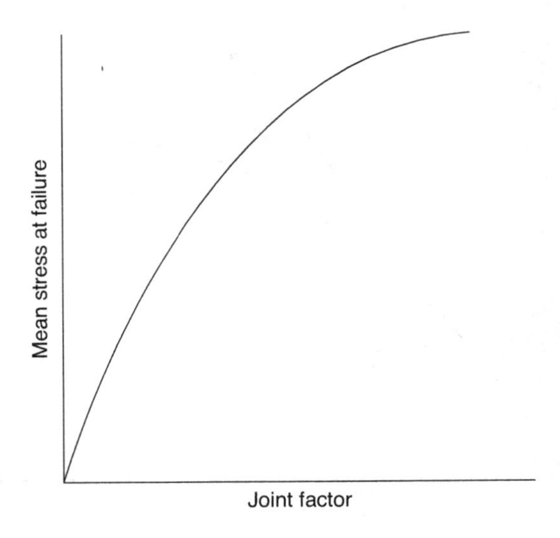

FIGURE 11.15 Failure stress as a function of the joint factor.

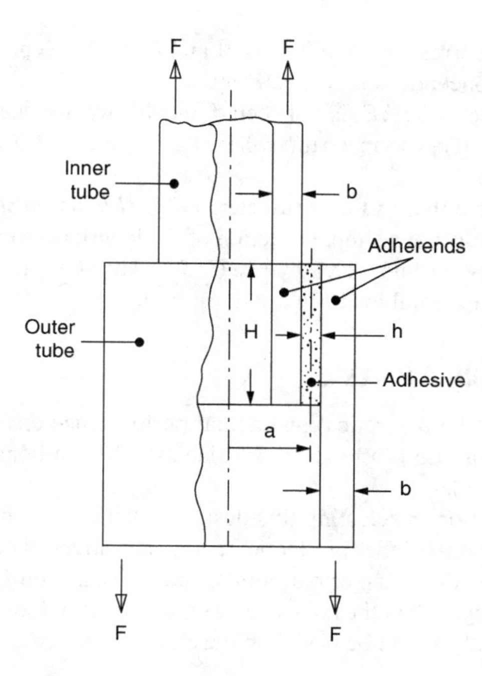

FIGURE 11.16 A cylindrical lap joint.

The stress concentration factor can be plotted as a function of the rigidity parameter defined as hE/bE_a, and where E and E_a denote the moduli of elasticity of adherend and adhesive, respectively. There is a considerable similarity between the stress concentration characteristics for the shear and tensile gradients found in a cylindrical lap joint. The maximum stress concentration factors for this type of an adhesive joint can be bounded by the two curves as shown in Figure 11.17. These

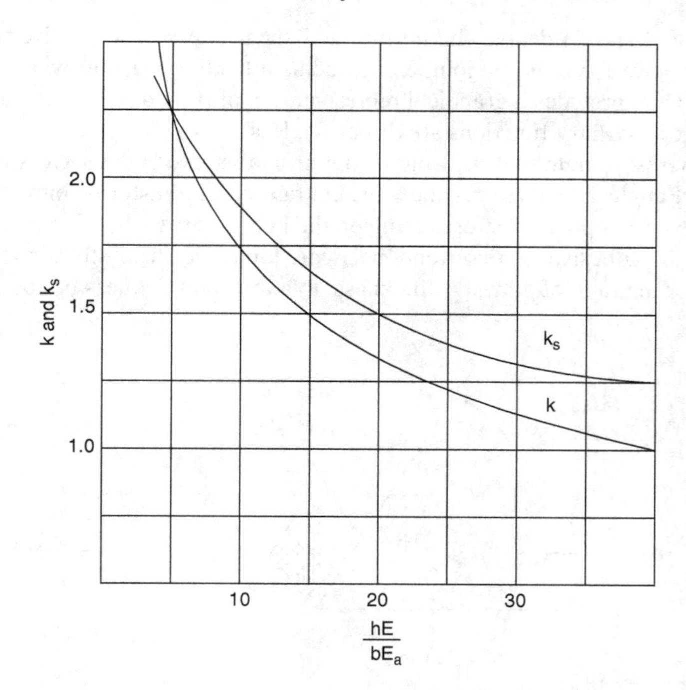

FIGURE 11.17 Example of stress concentration for a cylindrical lap joint [11].

particular boundary curves correspond to $b/2a = 0.01$ and $H/b = 10$. Symbols k and k_s refer to tensile and shear modes of stress concentration, respectively.

When the rigidity parameter hE/bE_a is between 5 and 10, we are dealing with a relatively stiff adhesive film. At the other end of the spectrum, where hE/bE_a is about 100, it is customary to regard the adhesive joint as flexible.

It is also of interest to note that as the parameters $b/2a$, H/b, and hE/bE_a vary, we find that the strength of the adhesive joint is not a simple function of the length of overlap H, as might have been deduced from the elementary formula given by Eq. (11.21). The strength is defined here as the ability of the joint to sustain an external load F without rupture.

11.8 DESIGN OF SCARF JOINTS

While instructive to review the theoretical and actual performance characteristics of butt and lap joints, the most efficient joint used in modern practice of adhesive bonding is the so-called scarf joint, as shown in Figure 11.18.

The only practical limitation in selecting this design is that the machining and tooling requirements in fabrication may lead to higher production costs. The stress distribution is generally better in a scarf joint than in conventional lap or butt joints, and the scarf configuration appears to be less sensitive to eccentric bonding. When the adhesive layer, shown as a dotted region in Figure 11.18, is stiff and thin, the design analysis can be based on the elementary stress equations:

$$S = \frac{FA_1}{b} \tag{11.22}$$

and

$$\tau = \frac{FA_2}{b} \tag{11.23}$$

Equations (11.22) and (11.23) denote the normal and shear stresses, respectively, where F is the tensile or a compressive force on the joint expressed as a loading per unit width of the adherend, and where Figure 11.19 provides a graphical representation of A_1 and $A_2 \cdot A_1$, A_2, and A_3 are design analysis functions or auxiliary functions are dimensionless.

It may be of interest to note that the ratio of the normal stress to the shear stress varies as the tangent of the scarf angle ϕ tends to $\pi/2$ the scarf configuration transforms into a butt joint, giving the nominal tensile or compressive stress acting on the bonded area.

At the ends of the adhesive in a conventional scarf joint, the shear stress concentration factor can be about 1.5 for the ratio of adhesive thickness, to adherend thickness of about 0.1. This stress

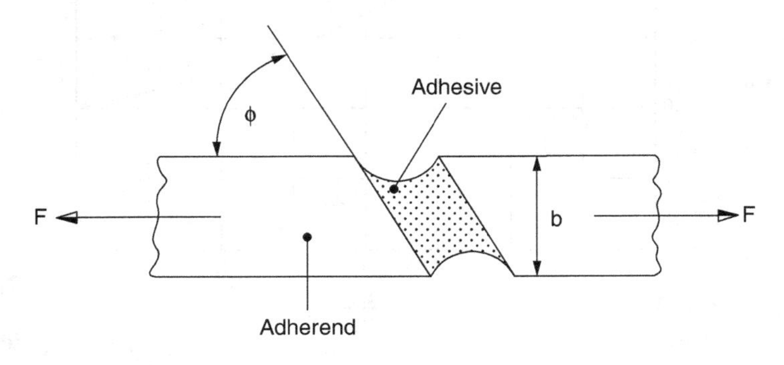

FIGURE 11.18 Typical scarf joint.

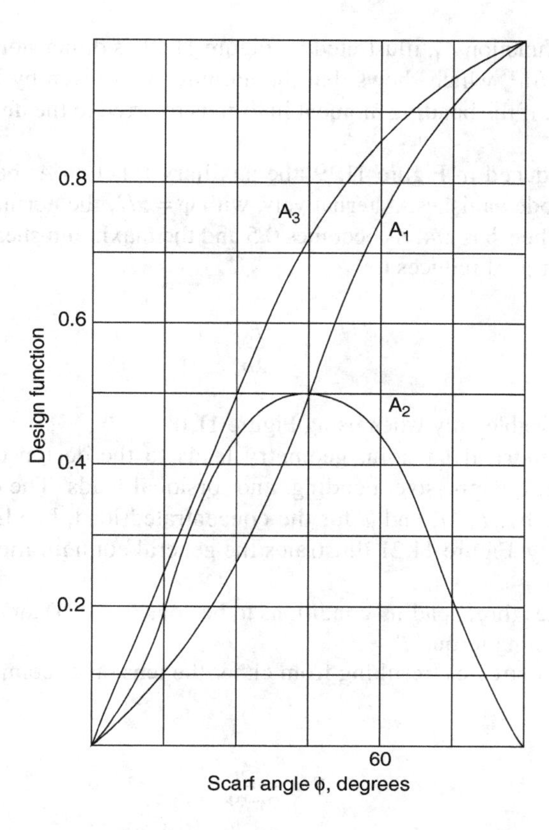

FIGURE 11.19 Auxiliary functions for scarf joints design.

concentration, however, decreases with the decrease in thickness ratio and it becomes negligible when the ratio approaches 0.01.

Suppose now that a flat scarf joint as in Figure 11.20 is subjected to a pure bending moment M, expressed per unit width of the adherend. Then, using the conventional theory of beam bending, the relevant normal stress becomes:

$$S = \frac{6MA_1}{b^2} \tag{11.24}$$

On the premise of the theory of pure bending, the shear force is obtained by taking the first derivative of the moment, with respect to distance over which the moment varies. In this case, the moment is assumed to be constant, making the derivative equal to zero.

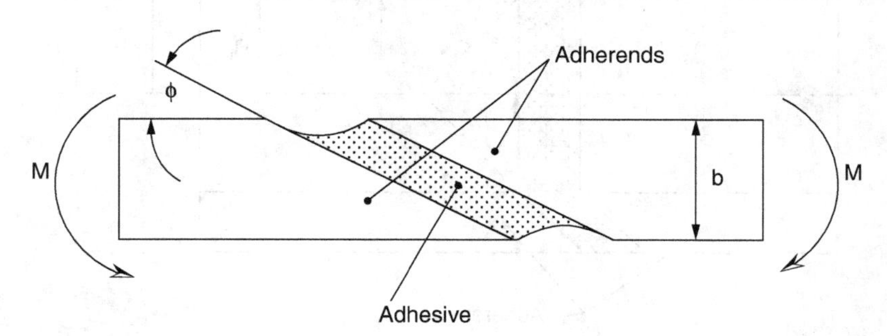

FIGURE 11.20 Scarf joint in pure bending.

The design analysis function A_1, illustrated by Figure 11.19, is dimensionless. A simple check of units gives lb–in/in × 1/in², which shows that the normal stress given by Eq. (11.24) has conventional dimensions of psi, if the bending moment in lb-in is referred to the unit width of the adherend in inches.

When $\phi = 0$ is introduced in Figure 11.19, the auxiliary function A_1 becomes zero so that the direct tension in this mode vanishes. Alternatively, with $\phi = \pi/2$, the normal stress attains a maximum value of $6M/b^2$. When ϕ is $\pi/4$, A_2 becomes 0.5 and the maximum shear stress τ for the tensile response of the flat scarf joint reduces to:

$$\tau = \frac{F}{2b} \tag{11.25}$$

Equation (11.25) is applicable only when using Figure 11.18.

Development of cylindrical lap joint geometry leads to the design of a tubular scarf joint which can carry tensile, compressive, bending, and torsional loads. The corresponding external forces are denoted here by: *F*, *M*, and *T* for the concentrated load, bending moment, and twisting moment, respectively. Figure 11.21 illustrates the general configuration of a tubular adhesive scarf joint.

Figure 11.22 illustrates three loading conditions to be considered: 1) an axial force; 2) a bending moment *M*; and 3) a twisting torque *T*.

The normal and shear stresses, resulting from either the tension or compression caused by force *F*, can be described as:

$$S = \frac{FA_1}{2\pi Rb} \tag{11.26}$$

and

$$\tau = \frac{FA_2}{2\pi Rb} \tag{11.27}$$

In Eqs. (11.26) and (11.27), $2R = R_0 + R_i$ and $b = R_0 - R_i$. This is done simply for convenience, particularly when the adherend tubes are relatively thin.

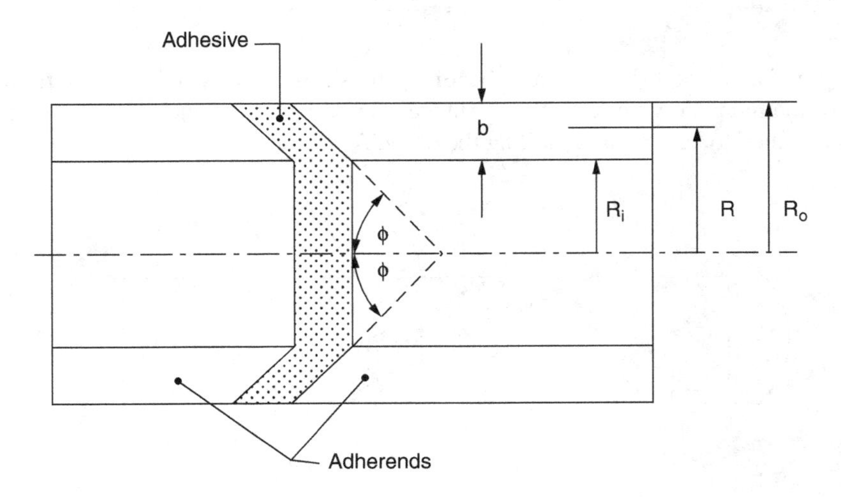

FIGURE 11.21 Tubular scarf joint.

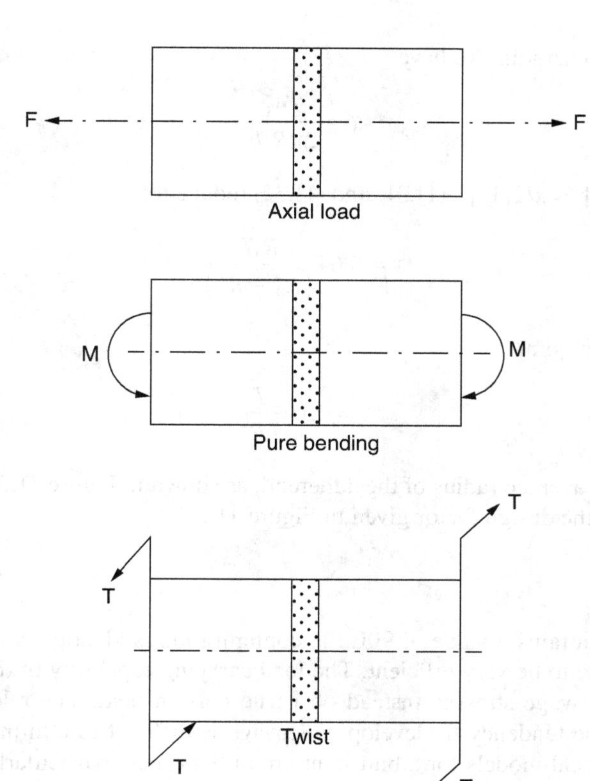

FIGURE 11.22 Types of loading on tubular scarf joint.

Again, when the scarf angle tends to a small value, both shear and normal stresses tend to vanish. For the case when $\phi = \pi/2$, the shear stress vanishes while the tensile (or compressive) stress attains a maximum value equal to $F/2\pi Rb$. The maximum shear is found for $\phi = \pi/4$. This yields:

$$\tau = \frac{F}{4\pi Rb} \tag{11.28}$$

When the tubular scarf joint is subjected to a pure bending as shown in Figure 11.22, the normal stress can be represented by the equation:

$$S = \frac{4MR_0A_1}{\pi\left(R_0^4 - R_i^4\right)} \tag{11.29}$$

For a very thin wall of the tubular joint, the above expression can be written as:

$$S = \frac{MA_1}{\pi R^2 b} \tag{11.30}$$

The case of pure torsion results in a zero normal stress, while the torsional stress becomes:

$$\tau = \frac{2R_0TA_3}{\pi\left(R_0^4 - R_i^4\right)} \tag{11.31}$$

For a thin wall of the scarf joint we have:

$$\tau = \frac{TA_3}{2\pi R^2 b} \tag{11.32}$$

When the scarf angle ϕ is $\pi/2$, Eqs. (11.31) and (11.32) reduce to:

$$\tau = 0.64\frac{R_0 T}{\left(R_0^4 - R_i^4\right)} \tag{11.33}$$

and for a thin wall scarf joint

$$\tau = 0.16\frac{T}{bR^2} \tag{11.34}$$

It is noted that R is the average radius of the adherend, as shown in Figure 11.21, and b denotes wall thickness, while A_3 is the design factor given in Figure 11.19.

11.9 PEELING

When the scarf angle attains a value of 90E the configuration is identical with that of a butt joint which is not considered to be very efficient. The load carrying capability of this joint is influenced by the presence of cleavage stresses instead of a true tension field. The relevant bonded area is relatively small and the tendency to develop a cleavage is difficult to eliminate. Also, as implied previously, the theoretical models for a butt joint are rather crude, particularly in the case of rigid adherends held together by a small layer of adhesive.

In terms of mechanics, cleavage may be viewed as a peel force. In general, adhesives exhibit poor resistance to peel forces because their effect is concentrated in a small area of the joint.

Consider a peeling test is conducted on a system involving a flexible ribbon, adhesive layer, and a rigid plate as shown in Figure 11.23.

In analyzing a typical peel joint and developing an approximate model, for a peel force oriented 90E to the rigid substrate, it can be postulated that the shear stresses in the adhesive are negligible. In addition, each fiber of the adhesive resisting a pull-out force F of the ribbon is assumed to act in the direction of this force.

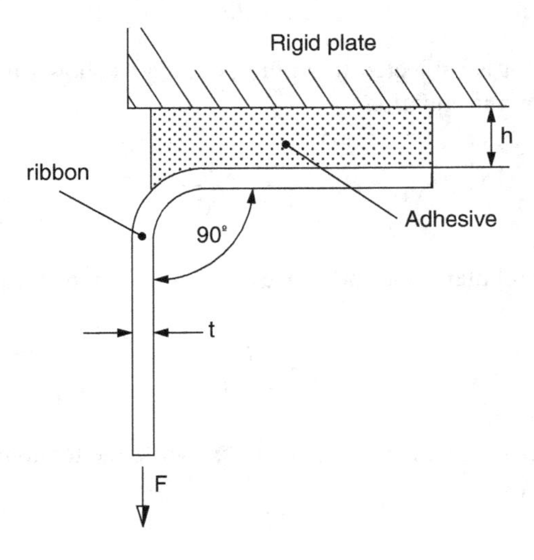

FIGURE 11.23 Test configuration in a peeling test.

Tearing action occurs when the extension of the fiber reaches its critical value and breaks the adhesive layer. That is, the notch tensile strength of the adhesive bond between the adherend and the adhesive is exceeded. Following the premise that the adhesive and the ribbon behave as Hookean solids, Bikerman [1] derives an expression for a peel force F inclined 90E to the rigid plate as shown in Figure 11.23. The relevant formula is:

$$F = 0.38 w S_a \left(\frac{E}{E_a} \right)^{0.25} t^{0.75} h^{0.25} \tag{11.35}$$

where w is the width of the adhesive layer (normal to the paper in Figure 11.23), S_a denotes breaking stress of the adhesive, while E and E_a stand for the moduli of elasticity of the ribbon and adhesive, respectively. Figure 11.23 shows the dimensions t and h.

With the product of the three terms involving linear dimensions in Eq. (11.35), the resulting peel force is given in lbs. Experience shows that quite often the term $0.38(E/E_a)^{0.25}$ is of the order of unity, simplifying Eq. (11.35) yields:

$$F = w S_a t^{0.75} h^{0.25} \tag{11.36}$$

When $t = h$ the formula given by Eq. (11.35) implies that the external force can be equal to the strength of the adhesive in tension. The fact that cleavage stresses are a part of the peeling process imply that the peeling force should be lower than the conventional tensile force required to break the adhesive.

It should be noted that the foregoing discussion of peel force parameter applies to a limiting case when the ribbon is very flexible and the adhesive layer is extremely strong. Unfortunately, this and other models of this type fail to predict peeling forces accurately because of the effects of local stress concentrations and the possibility of existence of a bending moment when the ribbon is not flexible.

While the formulas given by Eqs. (11.35) and (11.36) are invaluable as a research tool, the use of these formulas for direct design purposes must be supplemented with comprehensive tests. It appears, for example, that the degree of brittleness of the adhesive can play an unusually important role during the peeling process, in addition to other parameters such as the peeling angle and the effect of the beam on an elastic foundation. The latter is possible when the peeling force F, acting on the adhesive layer, creates a sinusoidal response of the adhesive with a diminishing amplitude of deformation and stresses. Such waves are rapidly damped out, in agreement with the theory of beams on an elastic foundation. Nevertheless, this phenomenon adds another layer of complexity to the already involved theoretical and experimental problem.

Lack of agreement between theory and experiment involving peel strength is not particularly surprising, if we assume that notch tensile strength of the adhesive relates directly to the tearing mechanism. However, this implies that an initial crack of a notch exists in the adhesive bond at the time the external peel force is applied. This leads directly to an indication that peel strength is the resistance of an adhesive bond to further failure [7].

Therefore, to predict peel strength, it is necessary to know notch tensile strength from an experiment on a butt joint with an artificially induced notch around all the edges of the adhesive bond. Such an experiment should yield the notch tensile strength, as well as the modulus of elasticity of the adhesive.

If, instead of using Eqs. (11.35) and (11.36), the peel strength of a bonded panel is treated with the aid of the theory of beams on an elastic foundation, the expression for the bending moment referred to unit width of the adhesive film can be expressed as:

$$M = (2EIh)^{1/2} \left(\frac{S_a^2}{2E_a} \right)^{1/2} \tag{11.37}$$

where S_a denotes the notch tensile strength of the adhesive bond between the metal adherend and the adhesive, while E_a is the modulus of elasticity of the adhesive. Note, that the resultant dimensions of in.-lb/in. are those of a force and that for a given geometry and material properties obeying Hooke's law, Eq. (11.37) may be restated as:

$$M = AU^{1/2} \tag{11.38}$$

where A represents a constant and U is the elastic strain energy absorbed per unit volume of the adhesive [7].

The values of peel strength are expected to be influenced by a number of variables in addition to peel angle or relative rigidity of the individual components. Process variables, including curing cycles and cleaning, loading rates and thickness of the adhesive should also be scrutinized. Thick adhesive layers tend to produce higher peel strength values but lower shear strength, obviously requiring a design trade-off. Insufficiently cured adhesive systems give initially high peel strength but are subject to strength decline during aging. In addition, thicker adhesive layers are less resistant to fatigue, although their shock properties may well be enhanced. It appears then that the design and fabrication aspects of peel joints are full of technical compromises.

11.10 SPECIAL APPLICATIONS

Experience in the aircraft industry has shown that use of adhesives has at times led to the replacement of such elements as rivets, bolts, nails, pegs, and clamps. In terms of fabrication processes, the adhesion often competes with welding, brazing, and soldering. The mechanism of adhesion helps to distribute and mitigate stress concentrations commonly found in bolted, riveted, and welded joints. Mitigation of stress gradients improves fatigue life in automobile, boat, and aircraft parts, as well as domestic equipment. For example, the average washing machine has at least 20 adhesively bonded components. Shock and impact characteristics are improved in tooling applications, while the greatest gain is found in joining dissimilar materials such as metals, plastics, wood, and ceramics. The adhesive joint allows sufficient mechanical compliance in parts subject to temperature warpage. Last, but not least, the adhesives can be contoured and formed in various fabrication processes, including those producing small electronic components, medical products, dental equipment, insulators, conductors, and seals of almost infinite variety.

In the perspective of the designer, there are certain limitations relative to the mechanical properties and choice of the adhesive products. Adhesives can be limited in resistance to temperature, hardness, surface effects, directionality, and quality control. Furthermore, the choice of properties is made more difficult by various claims of "cure-all" materials in trade publications.

Standard handbook material [7] indicates that modern technology has progressed to the point that successful applications of adhesives are possible in the field of machinery. While in the usual structures the adhesive is called upon to resist both tensile and shear loading, the machinery adhesives are typically utilized for their shear strength. One of the modern products known as anaerobic adhesive can be utilized up to about 350E F.

Because of proprietary restrictions, detailed adhesive design information may be difficult to obtain. It is known, however, that this material can be selected for machinery as a thread-locking adhesive. The recommended formula for the torque developed in the thread-locking process can be stated after Cagle [7] as follows:

$$T_p = B_1 \pi D_p^2 L_e \tag{11.39}$$

where B_1, D_p, and L_e are:

B_1: adhesion constant
D_p: thread pitch dia
L_e: engaged thread length

Table 11.1 provides values of the adhesive constant for several applications.

Similarly, the expected shearing torque needed to overcome the resistance of anaerobic adhesive in a threaded connection can be expressed in terms of the torsional strength constant B_2 of the thread metal. The shearing torque given by Cagle [7] is:

$$T_S = B_2 D_P^3 \tag{11.40}$$

Table 11.2 provides values of the torsional strength constant B_2 for commonly used metals.

The expressions given by Eqs. (11.39) and (11.40) suggest that when T_p is equal to or greater than T_S, it is expected that the screw thread will fail during the assembly prior to overcoming the thread-locking strength of the anaerobic adhesive. For the case of equal thread-locking and ultimate strength in torsion, Eqs. (11.39) and (11.40) give:

$$\frac{L_e}{D_p} = 0.318\frac{B_2}{B_1} \tag{11.41}$$

Thread-locking adhesives can be used in dynamic and vibratory load applications involving cap screws and adjustment screws, as well as locking devices for the shaft-hub assemblies. These assemblies are normally retained by press fits, keys, taper, splines, pins, brazing, or setscrews.

When a design system requires bonding of a cylindrical member, the maximum axial force F to overcome the strength of the adhesive is:

$$F = \pi DH\tau B_3 \tag{11.42}$$

where the term πDH represents the bonded area. The shear strength of the adhesive expressed psi is given by τ. The size factor B_3 derived experimentally can be obtained from the design curve illustrated in Figure 11.24.

The approximate formula for the axial strength can also be stated as:

$$F = (3.05 - 0.14DH)DH\tau \tag{11.43}$$

TABLE 11.1
Average Adhesion Constant

	Stud-locking Adhesive	Nut-locking Adhesive	Screw-locking Adhesive	Pipe-sealing Adhesive
B_1, psi	5,000	2,000	1,000	500

TABLE 11.2
Torsional Strength Constants for Common Metals

Material	Constant B_2, psi
Mild steel	8,700
Medium carbon steel	15,000
Yellow brass	5,800
Stainless steel	11,000
Soft aluminum	4,000
Hard aluminum	5,000

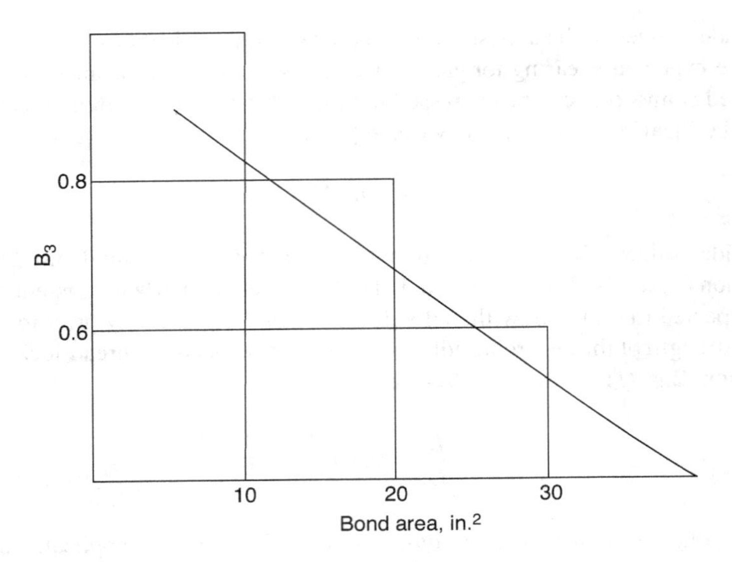

FIGURE 11.24 Size factor B_3 for cylindrical adhesion [7].

In this case, the bond area is allowed to vary between 5 and 40 in^2, and Eq. (11.43) represents an empirical relation based on Figure 11.24. The essential difference between Eqs. (11.21) and (11.42) lies in the size factor B_3.

The design of an adhesive joint can be further complicated by "built-in" stresses when the adherends are subject to a differential thermal expansion. The general thermoelastic relation applicable to this case is:

$$\theta_1\alpha_1E_1 = \theta_2\alpha_2E_2 \tag{11.44}$$

where θ_1 and θ_2 are the temperature differences between the adherends and the curing temperature; E_1 and E_2 are the elastic moduli, and α_1 and α_2 are the coefficients of thermal expansion.

In Eq. (11.44) it is assumed that there are no shear strains in the adhesive which could alter the "built-in" stresses.

When the adherends can be assumed to be relatively rigid, the maximum shear stress developed at the free end of the adhesive may be calculated from the expression:

$$\tau_{max} = \frac{(\theta_1\alpha_1 - \theta_2\alpha_2)G_aH}{2h} \tag{11.45}$$

where, in this handbook-type formula [7], the symbols H, h and G_a denote the length, the thickness, and the modulus of rigidity of the adhesive layer, respectively.

Equation (11.45) should be a reasonable model for design of lap-shear joints subjected to thermal differentials. However, this model does not account for all the environmental, materials, and geometrical parameters that can be encountered. In the limit when the thermal gradients become very small, the corresponding shear stress must vanish as indicated by Eq. (11.45).

11.11 MATERIALS

When an adhesive joint fails to perform as intended, the cause of weakness can be in the adherend, the adhesive, or the boundary layer. Because of interactions between the adhesive, adherend, and the environment, careful analysis must be made in defining the cause of joint failure. Often the remedy is to alter one of the three interacting components.

When the adhesive material is identified as the cause of joint failure, the problem may be due to the deterioration of the adhesive, stress concentration, shrinkage, or inherent flaws in the material. The selection of an optimum adhesive depends on a number of technical and economic constraints. Polyethylenes, for example, are not suitable as household cements because of the many impurities present in the common products. At the opposite extreme, certain adhesives can be designed for the joints fabricated under strict controls so that the formation of an adhesive joint weak zone can be avoided.

As a general rule one can state that the rheological behavior of the adhesive should be quite similar to that of the adherends. Of course, a compromise is in order when the two adherends are quite different. Bikerman suggests that the relevant adhesive should have the properties equal to arithmetic average of the moduli of elasticity and strains of the two adherends [1]. Selection of the best adhesive, however, will have to be based on the available handbook data [7,8]. The process of selection, as well as the choice of the products, are equally challenging.

It can generally be stated that epoxies are the best adhesives for mechanical purposes, since they can be used with the great majority of engineering materials. Their physical properties, such as wetting penetration ability and viscosity, are excellent for bond development. The epoxies have good internal strength in cohesion, surpassing that of thin gage metals, plastics, glass, concrete, and wood. However, when this adhesive is used in a joint involving heavy gage metallic adherends, the cohesion strength may be lower than the adhesive strength of the joint.

Shrinkage stresses of the epoxies are relatively low. This feature is particularly important in bonding dissimilar materials such as glass or titanium, where significant thermal stresses can be present due to the fabrication. In addition, low-temperature cure and creep, as well as insensitivity to moisture, are the specific advantages in selecting epoxy as an adhesive.

While the basic properties of epoxies are impressive, fillers are often used for improvement of the mechanical strength. Depending on the requirements and economy, the fillers will range between beach sand, aluminum oxide, silver flakes, and boron fibers. In general, the fillers have the effect of controlling the electrical properties and lowering the coefficient of thermal expansion of the adhesive. For the specific limitations on the choice of fillers, the design engineer is advised to consult trade publications.

Polyester resins are widely used in the commercial field for the various structural applications involving plastics. Polyester adhesives are far too numerous for inclusion in this brief chapter. They are often employed in laminated structures, provided working temperatures remain essentially below 250E F. For somewhat higher temperatures, heterocyclic polymers appear to do better although their chemistry and curing characteristics await further studies. Anaerobic adhesives, suggested for machinery in thread-locking applications described in this chapter, appear to be in the forefront of the adhesion technology.

In reviewing some of the physical properties of the adhesives, it may be instructive to make a brief comparison with some of the typical adherends, with special reference to the modulus of elasticity, thermal expansion, and density. Table 11.3 provides such a comparison.

TABLE 11.3
Properties of Typical Adherends and Adhesives

Material	Application	Modulus of Elasticity, psi	Thermal Expansion, in/in EF	Weight Density, lb/in^3
Stainless steel	Adherend	29×10^6	9×10^{-6}	0.280
Aluminum	Adherend	10.5×10^6	12.2×10^{-6}	0.097
Alumina	Adherend	50×10^6	4.4×10^{-6}	0.137
Epoxy (unfilled)	Adhesive	0.5×10^6	30.6×10^{-6}	0.043
Epoxy (filled)	Adhesive	2×10^6	17.8×10^{-6}	0.065
Nitrite rubber phenolic	Adhesive	0.04×10^6	27.8×10^{-6}	0.036

Correspondingly, Table 11.4 provides a listing of material properties for sealants—specifically, polymer and urethane components.

The values given in Table 11.4 are, at best, approximate and are used here as a general illustration of the order of magnitudes involved. It is noted that peel strength values are rather small when compared to the tensile and shear strengths. In general, the design preference for adhesion type of performance can be stated in the following descending order: shear, tension, cleavage, and peel.

11.12 GUIDE FOR DESIGNERS

Due to the empirical nature of the material properties and the majority of the design formulas, it is extremely important to select the appropriate methods of mechanical testing for a particular configuration of the adhesive joint. Figure 11.25 depicts a simple test for measuring the strength of a cylindrical lap joint, involving "pull-through" action.

In this test, a length of cord or rope is pulled through the adhesive layer of depth *H*. According to Bikerman [1], the force *F* needed to pull out the cord is proportional to tan h of the embedded length *H*. This function does not appear to be dependent on the type of the material of the cord. Cord materials tried in the experiments included cotton, nylon, and rayon, as well as brass-plated steel.

Similar tests, however, with glass-polyester interface, indicate a more linear relationship between the pull-out load and the embedded length. It is concluded that a weaker boundary layer exists along the glass-polyester bond.

TABLE 11.4
Typical Properties of Sealants

Property	Polymer Basis	Urethane Basis
Hardness, Shore A	45	80
Tensile strength, psi	400	3,000
Elongation in/in, %	600	600
Shrinkage in volume, %	2.5	
Peel strength lb/in of width	65	40
Lap shear, psi	250	1,800

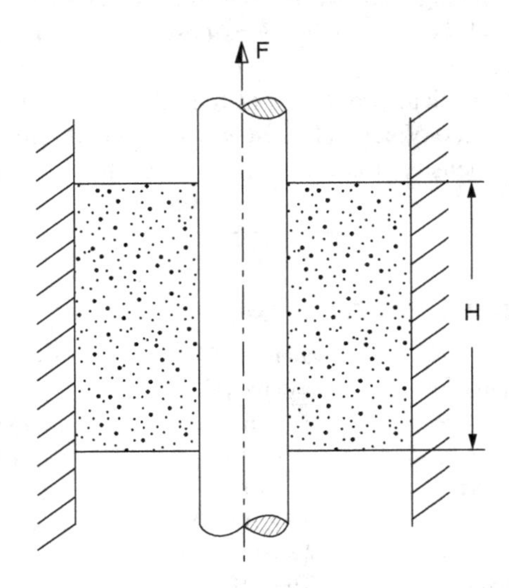

FIGURE 11.25 Concept of a "Pull through" test.

Figure 11.26 shows another test method where the coating of the adhesion is pressurized causing a bulge prior to rupture. The stress at failure appears to be a function of internal pressure P and the mechanical properties of the adhesive layer acting as a membrane. If the bulge can be treated as a portion of a spherical vessel, then the maximum tensile stress prior to burst is on the order of $Pr/2h$. However, from Figure 11.26 we see that the real mechanism of membrane failure can also be approximated on the basis of peeling.

The disadvantage of both types of tests is that they are conducted to destruction. Nondestructive tests are much more sophisticated because they are based on the resonance properties of the adhesive joints. The breaking stress is likely to be roughly proportional to the elastic modulus but the most practical approach is to use the resonance test for comparing the structural response of several joint configurations.

One of the main difficulties that a designer has to face is the selection of the adhesive for a particular joint. Skiest's handbook on adhesives [8] provides a broad and useful compilation of data for a variety of conditions.

In this section, we provide a summary of selection criteria applicable to a broad set of bonding problems. Federal, military, and trade specifications attempt to control the basic chemical compositions, although a few commercial designations of adhesives can be identified as being of the right composition for a particular design choice. The reason for this is that the selection process is highly subjective in assigning the degree of importance to the various factors controlling the design.

Perhaps the most crucial constraint of adhesive selection is the environment. The relevant influencing factors include temperature, pressure, humidity, and stress history, which are difficult to define in detail. When the adhesive performs an interim function, such as holding the adherends together for fabrication purposes, the environment may be of a relatively limited importance. However, when this function is extended to shipping and storage conditions prior to releasing the product for service, the environment can play a decisive role.

Polymer type of adhesives are sensitive to heat and exposure to moisture and oxygen. Added problems include the effects of static and cyclic stresses. When metal adherends are held together by structural adhesives the combination of moisture and stress is particularly undesirable when the preparation of the bonded surfaces is inadequate. Furthermore, the exposed surfaces of the adhesive joints are susceptible to chemical deterioration and initiation of fracture from a variety of sources.

The effect of temperature on curing, brittle behavior, deformation, and strength characteristics is highly involved, so that the determination of the optimum joint performance is particularly difficult to attain. There are, no doubt, many examples of hazardous situations which can be traced to thermal effects. For example, heat can alter viscoelastic properties of the adhesive in the case of bonded and sealed electronic equipment containing electromechanical devices. Other adhesive

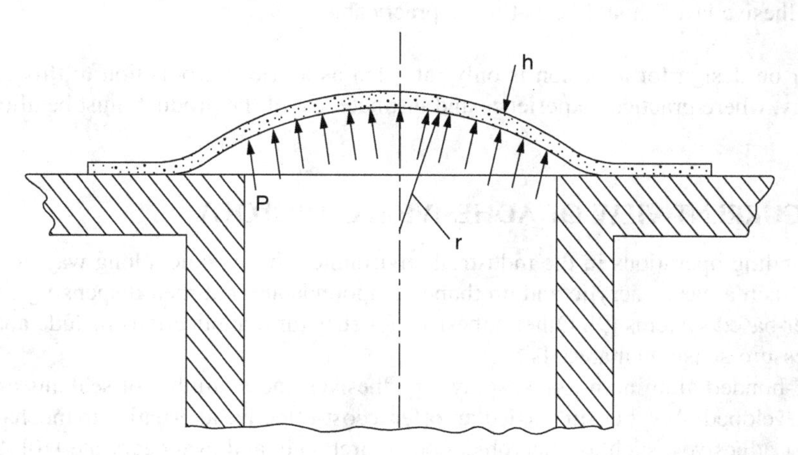

FIGURE 11.26 Burst test for adhesion.

formulations subjected to heating display significant outgassing. The resulting vapors can condense on electrical and mechanical components, causing corrosion.

While careful material selection and thorough evaluation of the products is highly desirable, joint design may ultimately dictate the method of fabrication and control of induced stresses. For example, design for shear, instead of peeling or tension, can help to mitigate stress concentrations. However, every shearing load transfer may not be a panacea where poorly mated adherend surfaces or improper distribution of the adhesives are involved.

Although the nature and configuration of adherends are determined by design, they play a special role in the selection of the adhesive. For example, brittle or highly compressible adherends can hardly be used with adhesives requiring high bonding pressures. Similarly, the thermoplastic adherends are not suitable for use with adhesives made for high curing temperatures. Also, when warpage of the adherends is controlled by bonding pressure, the relevant adhesive should be resistant to creep. At the same time, polymeric adherends can be subject to environmental stress cracking, even if other design considerations have been identified and understood. The list of constraints, special effects, and design parameters involved in the selection of adherends is long.

The corresponding list of adhesive concerns includes a number of characteristics of chemical, environmental, and stress analysis nature. The chemistry can be decisive in durability and compatibility with other materials. The working properties, such as consistency, tack range, and pot life, may determine the choice of the fabrication process. Finally, the conventional mechanical properties will define the extent of stress redistribution and strain sensitivity.

Unfortunately, many variables and design details cannot be spelled out in sufficient detail to assure optimal material selections. This difficulty of choice is compounded by the variations in material quality and the multitude of industrial products on the market. Specific data sheets prepared by a given supplier are probably the best, practical form of design information, provided the generic classes of the materials are well defined and understood.

Although no theory or materials knowledge is sufficient to account for all factors of geometry, properties, and environment involved in a particular adhesive joint, there are a number of practical rules which the designer should always keep in mind:

- Make the bonded area as large as possible.
- Utilize the bonded area in contributing to the joint strength.
- Load the adhesive in the direction of its maximum strength.
- Minimize stress in the weak direction.
- The design preference in a descending order should be: shear, tension, cleavage, and peel.
- Avoid irregular thicknesses of the adhesive.
- The adhesive layer should be as thin as practicable.

This chapter on design for adhesion is only intended as a brief introduction to this complex area of technology, where practical experience and ultimate test of the product must be allowed the last word.

11.13 A CURRENT VIEW OF ADHESIVE TECHNOLOGY

Adhesive bonding operations in the industrial environment have come a long way and include the advances in such areas as acrylic and urethane compounds and hot-melt dispensing equipment, as well as water-based systems [9]. Other adhesives for structural applications include anaerobic, silicon, and pressure-sensitive materials.

Adhesive-bonded aluminum joints, spraying adhesives and a number of sealants and modifiers are highly developed. Acrylics, in particular, offer a cost-effective alternative to mechanical fasteners and other adhesives, such as anaerobic, epoxy, urethane, and cyanoacrylate [10]. Such a comparison is based on bond, peel, and impact strength, in addition to the simplicity of curing.

Acrylic-based adhesives are adaptable to various production conditions, require little surface preparation, and cure at room temperature. Easy curing process eliminates costly ovens and 90% of full strength can be developed in not more than 2 h. Full mechanical properties can be assured after a 24 h cure. These adhesives can be applied by spraying, rolling, or brushing the accelerator onto one or both substrates. The pre-coated parts can be stored dry for up to 3 months or longer. Application of the adhesive resin to such parts and bringing them into contact starts the cure process immediately.

Modern developments in the application of acrylic adhesives now allow bonding the surfaces contaminated by oil or other lubricants. Until recently, such an operation was not possible without a time-consuming process of vapor-degreasing, sanding, or grit-blasting. All that is required involves a simple solvent wipe.

Acrylic adhesives assure structural integrity of the bond without expensive machinery and skilled labor. When these adhesives allow the development of mechanical joints without conventional fasteners, the problems of localized stresses and preload control can be avoided. Fastening of dissimilar materials in a thermal environment and developing welded joints can be eliminated with the use of acrylics which can maintain highly stressed bonds up to the point of metal separation. The only substrates for which acrylics are not recommended include bronze, titanium, terneplate, and some types of galvanized steel. In a nonmetallic family of materials, on which acrylics should not be used, we have polyolefins, Teflon, styrene foam, and cured rubber.

The mechanical properties of acrylics remain quite stable over a significant range of hostile environment. Some acrylic grades can withstand 400E F and show good resistance to ultraviolet light and moisture. The acrylic compounds can be made brittle or plastic and can also fulfill the sealant function while still assuring a strong bond. The combination of fast cure, low exothermic heat, and appreciable bond strength of acrylic adhesives can solve many fabrication problems in an economical manner. In a particular case involving plexiglass and laminated components, the use of conventional fasteners is likely to cause crazing and cracking due to high clamping pressures. The ability of acrylics to tolerate local strain concentrations while retaining good bond strength assures that this adhesive will remain popular with design and production engineers in many branches of industry.

It is well known that the wood processing industry makes extensive use of adhesives. The chemical spectrum of the materials for this purpose involves: amino resin, phenolic resin, tannin, lignin, diisocyanate, and polyvinyl acetate [11]. Research efforts in this field are concerned with the properties and performance of wood adhesives of interest to wood technologists, physical chemists, forest products researchers, and polymer scientists.

11.14 SUMMARY

The technology of adhesion is concerned with the analysis of forces existing between the adherends and the adhesives. The adherend is defined as a solid body held by a layer of adhesive.

The adhesives can be applied to fractured as well as newly fabricated surfaces. Roughness and waviness of the surface affect the value of the conventional friction factor and the forces of adhesion. Machining creates a preference in terms of the orientation of crystals. Other factors influencing the mechanics of adhesion include corrosion, impurities, the actual area of interface, and the phenomenon of adsorption.

Solid-to-solid adhesion results from the pressing together of the two surfaces in a vacuum. Elastic spring-back of a driven nail or rod wringing inside a cylinder is other examples of creating interface bond. Still other bonds, resulting from special coatings or hooking action of surface irregularities, create unusually strong adhesive joints. However, a mechanical or chemical process of surface cleaning is required to get the best results.

There are many methods of setting and controlling the flow of industrial adhesives. Some effects such as shrinking during setting can have an undesirable influence on the mechanical strength of the joint.

Failure of the joint along the adherend-adhesive boundary is difficult to postulate. Larger samples of adhesive bonds are generally weaker than the smaller test pieces. When extending this conclusion to the concept of the adhesive film, one finds that the breaking stress of the interface may be appreciable. This is not the case with household products. The nominal stress across the interface can be obtained by dividing the force by the appropriate area of cross-section. This represents the molecular cohesion which is a function of the composition and the atomic structure of the adhesive.

Conventional stress equations apply to the analysis of the adhesive layer under conditions of external forces, external constraints, and shrinkage. Similarly, the analysis of the adherend materials is based on the theory of elasticity. The foregoing sections contain a number of suitable equations for design of flat and cylindrical adhesive joints. The formulas are only approximate because of the possibility of relaxation and nonconformance of the materials to the Hookean response.

The strength of butt adhesive joints may be estimated with the aid of the conventional factors of stress concentration recommended for grooved bars in tension. Radial stresses are present in a disc-type adhesive layer in a typical butt joint. These stresses appear to be relatively high when compared with the nominal axial stress intensity. Furthermore, adhesives develop residual stresses affecting the performance of a butt joint. Typical adherends found in industry include steel, aluminum, copper, and glass.

The design of a lap joint depends on the rigidity of the adherends. Stress gradients can be appreciable and require special care in developing a model for the calculations. Stress concentration in shear is influenced by the geometry and the elastic constants of the joint components.

The actual stress in a lap joint for a given applied load decreases with the increase in the length of the overlap. The distribution of the adhesive shear follows a parabolic curve, with the stress peaks found at the edges of overlap. The overall strength of the joint also depends on the thickness of the adherend. Small overlap strength is truly representative of the adhesive performance. The flexibility of the adhesive film can be defined by a simple parameter involving material thicknesses and the elastic constants.

The most efficient adhesive bonding is found in the scarf joint. The ratio of the normal to shear stress varies as the tangent of scarf angle. The maximum stress concentration factor in shear is 1.5 compared with 2.0 for the lap joint. The scarf joint can carry tensile, compressive, bending, and torsional loads.

Adhesives generally exhibit limited resistance to peeling. The failure takes place when the notch tensile strength of the adhesive bond is exceeded. The peeling force is lower than the force sufficient to cause tensile failure. Accurate theoretical prediction of peeling resistance is not possible without experiments. The peel strength is affected by peel angle and rigidity of joint components, as well as process variables such as curing, cleaning, and aging.

The process of adhesion can, under right circumstances, compete with riveting, bolting, or welding. However, the adhesives are limited in resistance to temperature, hardness, surface effects, and directionality. Special grades of machinery adhesives can be used for shear applications and thread-locking functions. These adhesives perform well in dynamic and vibration environments and can be exposed to thermal differentials.

Analysis of interactions between the adhesive, adherend, and the environment is necessary for determining the weak zones of the joint. The selection of an optimum adhesive depends upon technical and economic constraints. A general rule of thumb is that the rheological behavior of the adhesive and the adherends should be similar. The epoxies are best for many applications. Fillers can be used for the enhancement of specific properties. Polyester resins are accepted throughout the industry in structural systems involving plastics below the temperature of 250E F.

The general design preference for the consideration of adhesion type of performance can be stated in the following descending order: shear, tension, cleavage, and peel.

The text includes a concise guide for designers struggling with the problems of material qualification for a particular job. While this is an important function, the method of design may ultimately dictate the fabrication constraints and the control of the induced stresses.

The choice is often difficult because of the multitude of industrial products on the market. Nevertheless, the best current source of information for design is available through the supplier.

11.15 DESIGN ANALYSES

EXAMPLE ANALYSIS 1

Figure a shows an overhead and a side view of an adhesive wooden joint. The joint consists of a thin slat of width b glued between two relatively sturdy, fixed wooden supports, as shown. If a force F of 5000 lb is exerted in tension on the slat, and if the adhesive shear strength τ_{max} is 100 psi, and if the adhesion is only 80% effective, determine the minimum length d of the adhesive needed to support the force when b is 6 in.

SOLUTION

From Figure a, the adhesive area is seen to be: $2bd$ or $12d$ in^2. (The adhesive is on *two* sides of the slat.) The shear stress τ on the adhesive is then:

$$\tau = \frac{F}{2bd} = \frac{5000}{12d} = \frac{416.7}{d} \tag{a}$$

If this value is set equal to 80% of the shear strength, we have:

$$\frac{416.7}{d} = (0.8)(100) = 80 \tag{b}$$

or

$$d = 5.21 \text{ in} \tag{c}$$

Overhead View

b

adhesive

F

d

Side View

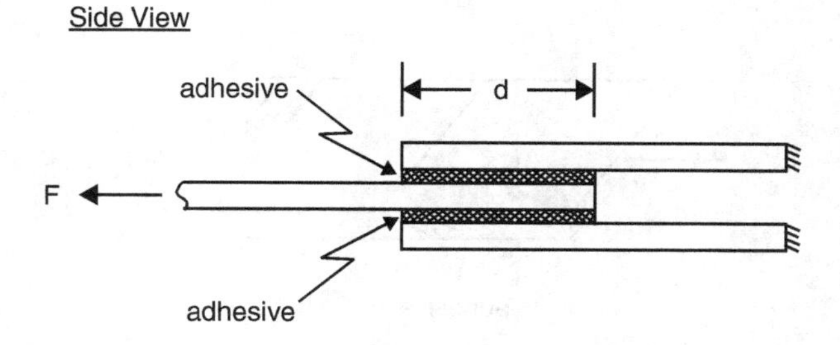

FIGURE A An adhesive wood joint.

EXAMPLE ANALYSIS 2

See Example Analysis 1. Repeat the analysis if $F = 25$ kN, $\tau_{max} = 750$ kPa, and $b = 0.15$ m.

Solution

With the problems being the same, except for the units and the magnitudes of the dimension and physical parameters, the solution procedure will also be the same. Therefore, in view of the solution to Example Analysis 1, we see that the shear stress τ on the adhesive is:

$$\tau = F/2bd = 25\times10^3/(2)(0.15)d = \left(83.33\times10^3/d\right)\text{Pa}$$

or

$$\tau = \left(\frac{83.33}{d}\right)\text{kPa} \tag{a}$$

where d is expressed in meters (m).

If this shear stress is set equal to 80% of the shear strength, we have:

$$\frac{83.33}{d} = (0.8)(750) = 600 \tag{b}$$

or

$$d = 83.33/600 = 0.138 \text{ m} \tag{c}$$

EXAMPLE ANALYSIS 3

Figure a contains a line drawing of a scarf joint. It consists of two plates with similar dimensions held together by adhesive at an inclined joint. Let the plate thickness be t and let the depth of the plates be d as shown. Let the inclination of the joint relative to the long dimension of the plates be θ and let the plates be held together by an adhesive.

Let the joined plates be placed in tension by an applied force F.

Find the normal and tangential stresses at the joint; that is, tensile and shear stresses on the adhesive.

Solution

Consider a free-body diagram of, say, the right end of the joint in Figure b, where F_N and F_T represent the resultants of the normal and tangential components of the forces acting on the adhesive surface of the joint, respectively.

By balancing forces in the normal and tangential directions, we obtain:

$$F_N = F\sin\theta \quad \text{and} \quad F_T = F\cos\theta \tag{a}$$

Next, observe from Figure c that the area A_a of the inclined adhesive region is simply:

$$A_a = d\ell \tag{b}$$

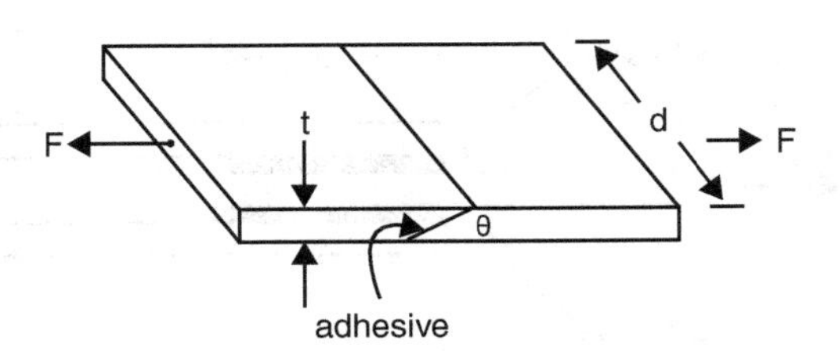

FIGURE A A scarf joint.

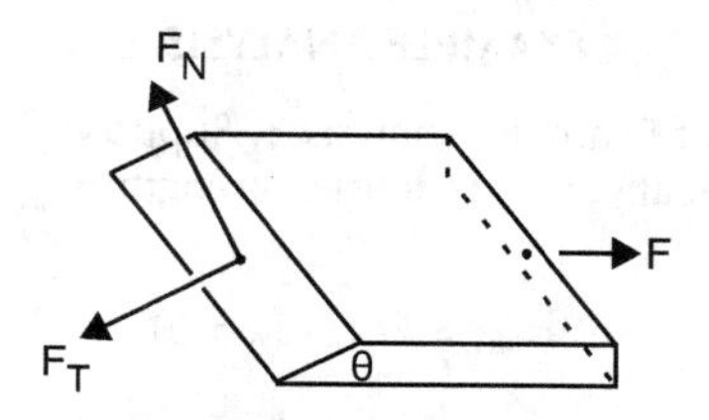

FIGURE B Free-body diagram of the right end of the scarf joint.

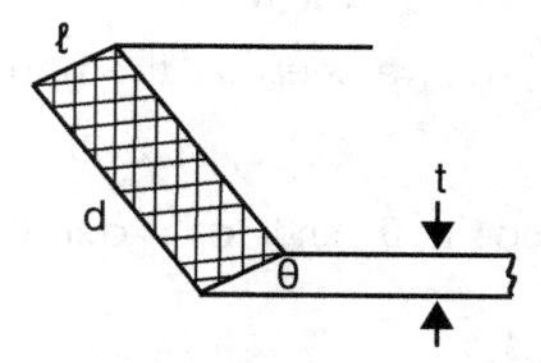

FIGURE C Adhesive surface.

where ℓ is the length of the inclined side of the adhesive section. By inspection of Figure c, we see that ℓ may be expressed as:

$$\ell = \frac{t}{\sin\theta} \tag{c}$$

(Note from elementary trigonometry that $\sin\theta = t/\ell$.)

By substituting from Eq. (c) into (b) we see that the adhesive area A_a may be expressed as:

$$A_a = \frac{dt}{\sin\theta} = \frac{A}{\sin\theta} \tag{d}$$

where A is simply the cross-section area of the plate numbers of the joint.

Finally, the normal and tangential stresses on the adhesive are:

$$\sigma_N = \frac{F_N}{A_a} = \left(\frac{F}{A}\right)\sin^2\theta \tag{e}$$

and

$$\sigma_T = \frac{F_T}{A_a} = \left(\frac{F}{A}\right)\sin\theta\cos\theta \tag{f}$$

Observe that F/A is simply the tensile stress σ in the plates. Therefore, σ_N and σ_T may be written as:

$$\sigma_N = \sigma\sin^2\theta \text{ and } \sigma_T = \sigma\sin\theta\cos\theta \tag{g}$$

Comment

Observe that with sines and cosine appearing in Eq. (g), five values of θ would seem to be significant: θ = 0°, 30°, 45°, 60°, and 90°. If θ is 0° we no longer have a scarf joint, but instead parallel plates. Similarly, θ = 90° is a butt joint, rather than a scarf joint. When θ = 45°, the shear stress and normal stresses are the same, with the shear stress having a maximum value. For θ being either 30° or 60°, the shear stress σ_T is $\left(\sqrt{3}/4\right)\sigma$, whereas the normal stress σ_N is larger for 60° at (3/4)σ.

EXAMPLE ANALYSIS 4

Consider again the scarf joint of Example Analysis 3. Suppose the adhesive strengths in tension and shear are different. Specifically, let the tensile strength σ_{max} and the shear strength τ_{max} be related as:

$$\sigma_{max} = k\tau_{max} \quad (k > 0) \tag{a}$$

Determine the scarf joint inclination θ so that the adhesive will fail simultaneously in tension and in shear.

Solution

From Eq. (g) of Example Analysis (31), we see that the normal (tension) and tangential (shear) stresses are:

$$\sigma_N = \sigma \sin^2\theta \quad \text{and} \quad \sigma_T = \sigma \sin\theta\cos\theta \tag{b}$$

where σ is simply the axial stress: *F*/*A*.

By letting σ_N become σ_{max} and σ_T become τ_{max}, Eq. (a) gives:

$$\sigma \sin^2\theta = k\sigma \sin\theta\cos\theta$$

or

$$\sin\theta = k\cos\theta \tag{c}$$

Therefore, the requested angle is:

$$\theta = \tan^{-1} k \tag{d}$$

Comment

When the plates of this scarf joint are loaded in tension as expected in a structural application, the inclination angle θ of Eq. (d) provides the design for the *strongest* scarf joint for the given strength properties of the adhesive.

REFERENCES

1. Bikerman, J. J. 1968. *The Science of Adhesive Joints*, 3rd Ed. New York and London: Academic Press, Inc.
2. ASME Handbook. 1965. *Metals Engineering Design.* New York: McGraw-Hill.
3. Peterson, R. E. 1953. *Stress Concentration Design Factors.* New York: John Wiley.
4. Savin, G. N. 1961. *Stress Concentration Around Holes*, New York: Pergamon Press.
5. Huston, R.L. and H. Jospehs. 2009. *Practical Stress Analysis in Engineering Design*, 3rd Ed. Boca Raton, FL: CRC Press.
6. Mylonas, C. 1955. *Proceedings of the Society of Experimental Stress Analysis*, 12:129.
7. Cagle, C. V. 1973. *Handbook of Adhesive Bonding.* New York: McGraw-Hill.
8. Skeist, I. 1990. *Handbook of Adhesives*, 3rd Ed. New York: Van Nostrand Reinhold Co.
9. Schneberger, G. L. (Editor). 1983. *Adhesives in Manufacturing.* New York and Basel: Marcel Dekker, Inc.
10. Gordon, S. 1983. Acrylic structural adhesives cut bonding costs. *Mechanical Engineering*, 105:60–65.
11. Pizzi, A. 1983–1989. *Wood Adhesives Chemistry and Technology*, (2 volumes). New York and Basel: Marcel Dekker, Inc.

SYMBOLS

A	Parameter in peel strength equation
A_1, A_2, A_3	Design factors
a	Radius of cylinder

B_1	Adhesive constant
B_2	Torsional constant for metal
b	Width of adherend plate
c	Thickness of adherend
D	Diameter of bonded cylinder
D_p	Thread pitch diameter
E, E_1, E_2	Young's modulus of adhesive
e	Depth of groove
F	External force
G_a	Modulus of rigidity of adhesive
H	Plate thickness; length of adhesive
h	General symbol for thickness of adhesive
h_0	Thickness of adhesive after shrinkage
h_1	Thickness of free adhesive
I	Second moment of area
I_a	Second moment of area for adhesive layer
K_3, K_5	Deflection factors
K_4	Stress factor
k	General symbol for stress concentration
k_S	Stress concentration in shear
L	Linear dimension
L_e	Length of engaged thread
M	Bending moment
P	Interface pressure
R	Mean radius of cylinder
R_o	Outer radius of cylinder
R_i	Inner radius of cylinder
r	Radius of curvature
S	General symbol for stress
S_a	Stress in adhesive
S_m	Stress in molecular cohesion
S_t	Tensile stress
S_θ	Tangential stress
T	Twisting moment
T_p	Thread-locking torque
T_s	Surface shearing torque
t	Thickness of ribbon in peel test
U	Energy per unit volume
u_i	Displacement at inner radius
u_o	Displacement at outer radius
w	Width of adherend
α_1, α_2	Coefficients of thermal expansion
θ_1, θ_2	Temperatures
Δ	Radial displacement
ΔL	Element of length
ϕ	Scarf angle
v_a	Poisson's ratio of adhesive
ρ	Groove radius
τ	General symbol for shear stress
τ_{ave}	Average shear stress
τ_{max}	Maximum shear stress

GLOSSARY

Accuracy: The nearness to the true value for a measurement.
Adhesive shrinkage: Volume decrease of an adhesive as it cures and solidifies.
Adhesive residual stress: The stress arising in an adhesive joint when shrinkage during curing is restricted.
Arc welding: Welding (heating) via an electrical short.
Adhesive: A glue or glue-like fastener placed between structural components providing a permanent joint.
Adhesive joint: A joint connected by any adhesive (glue-like) material (see Chapter 11).
Alpha error: Rejecting the null hypothesis when it is indeed valid—also known as: "error of the first kind" and as "producer error."
Alternative hypothesis: A hypothesis different than the null hypothesis.
ASTM: American Society for Testing Materials
Auto-frettage: Planned yielding with residual stress making a joint secure.
Beam: A long slender member. Nominally the length is at least 10 times the thickness or depth.
Bending: The deformation of a slender member (i.e., a beam) subjected to end moments.
Bending stress: Normal (tension or compression) stress occurring in a beam cross-section due to bending moments—also known as "flexural stress."
Beta error: Accepting the null hypothesis when it is instead false—also known as "error of the second kind" and as "consumer error."
Blind hole: An incompletely drilled hole.
Bolt: A commonly used and familiar cylindrical fastener with one end being threaded (for a mating nut) and the other end having a hexagonal head.
Bonding: The maturing and securing of an adhesive fastener or weld.
Braze: To solder or connect with a metal that melts at a lower temperature than the components being joined.
Brittle: A hard, but easily fractured characteristic.
Buckling: Extreme and often sudden large displacement or deformation of a structural component (see Section 3.9).
Butt joint: A joining of similar components end-to-end. (see Figure 1.2).
Butt weld: Weld joint of adjacent inline components (see Figure 9.3).
Button joint: A joint whose connector is a finger pushed through a cylindrical component.
Cantilever beams: A beam supported at an end by a rigid or built-in support.
C-Clamp: A clamp in the shape of a capital letter C.
Central limit theorem: For large populations, and for large sample size, the mean and variance will be approximately normal—also known as "distribution of the mean" (see Section 4.3).
Centroid: A "middle-type" point, with locating coordinates defined by Eqs. (5.9) and (5.10).
Chain link: The long component of a chain or of a chain drive mechanism.
Charpy V-notch: A test procedure for measuring toughness of a material—also known as CVN (see Section 3.19).
Chi-square distribution: See Section 4.12, Eq. (4.16)
Circular membrane: See Section 10.3.
Clamp: A detachable mechanism forming a joint which can be adjusted for tightness. An external component or components forming a joint by compressing internal components.
Clevis joint: A joint formed by two parallel cantilever beams pressed upon a component between the beams (see Figure 7.7).

Coarse threads: Threads with relatively large separation between corresponding points on the threads. The opposite of "fine threads."

Coefficient of friction: For structural components sliding on one another, the coefficient of friction is the ratio of the force causing the sliding to the force perpendicular to the sliding component surfaces. The coefficient of friction is dimensionless with values between zero and one.

Column: A beam or slender member in compression.

Compact flange: A shortened flange (see Figures 6.20 and 6.21).

Confidence interval: A measure of the number of values expected within a bracket centered about the mean (see Section 4.7).

Connection: A synonym for "joint."

Consumer error: Accepting the null hypothesis when instead it is false—also known as "error of the second kind" and as "beta error."

Contact stress: The stress arising from materials in contact and being pushed toward one another.

Control charts: A means of determining whether a process is in control or not (see Section 4.15).

Control limits: Boundaries of a control chart.

Corner weld: Weld joint of components forming a corner (see Figure 9.3).

Corrosion: An oxidation or other similar material degradation occurring between structural components so that the components are connected as a permanent joint. A wearing away.

Cotter pin: A generally circular component for insertion into a similarly shaped, but slightly larger, cavity for the purpose of securing a joint (see Section 8.2).

Coupling: A detachable mechanism forming a joint. A connector of two components of a joint (see Section 8.5).

Crack: A narrow separation of material.

Creep: A slow deformation or displacement of components relative to each other.

CVN: Acronym for Charpy V-notch test.

Cyclic load: An oscillating load.

Degrees of freedom: A measurement of the variability of a set of data.

Diaphragm stress: Stress on a membrane joint due to internal or external pressures.

Dishing effect: The transferring of load or stress from one structural component to another.

Dispersion: Either the range (largest to smallest) or standard deviation of a set of data.

Distribution of the mean: For large population, and for large sample size, the mean and variance will be approximately normal—also known as "Central Limit Theorem" (see Section 4.3).

Double shear rivet test: See Figure 2.1.

DT: Acronym for dynamic tear.

Ductile: A relatively pliable characteristic.

Dynamic: A changing or time-dependent condition (opposite of "static").

Eccentric shear: A condition occurring when a multiple-riveted joint has unequal shear loading among the rivets. A joint experiencing off-center and/or twisting loading (see Figure 5.19).

Edge weld: Weld joint of parallel adjacent components at their edge (see Figure 9.3).

Effective eccentricity: An estimate of the eccentricity of a joint loading.

Engineering tolerance limit: Outer limits of acceptability of a given characteristic.

Error of the first kind: Rejecting the null hypothesis when it is indeed valid—also known as: "alpha error" and as "producer error."

Error of the second kind: Accepting the null hypothesis when it is instead false—also known as: "beta error" and as "consumer error."

Eyebar joint: A flexible joint connected via a pin—also known as a "knuckle joint."

***F*-Distribution:** See Section 4.13, Eq. (4.17).

Factor of safety: A measure of overdesign to keep a joint from slipping or failing due to unexpected stress concentration.

Failure: Collapse or separation of a material usually due to excessive loading and/or fatigue.
Failure in cohesion: An adhesive failure.
Fastener: The connecting component or substance for a joint.
Fatigue: A condition of material failure due to excessive use and/or cycling forces.
Flexible joint: A joint with a relatively soft gasket or with flexible fasteners.
Fracture: A material failure resulting in separation often in the form of a crack.
Fracture control: Means of preventing or arresting crack growth.
Fracture energy: The energy required to break (or "fracture") a material—also known as "toughness" and "work of fracture."
Free-body diagram: A drawing or sketch of a structural element showing a representation of all the forces acting on the structure.
Fillet: A rounded surface at the intersection of two planes so as to prevent a sharp corner at the intersection.
Fillet weld: A welding at the corner intersection of perpendicular components (see Figure 9.10).
Fine threads: Threads with a relatively short separation between corresponding points on the threads. The opposite of "coarse threads."
Finger-tight: Referring to threaded fasteners, just enough tightening (or torque) to keep the joint from being or becoming loose. Synonym: "snug."
Flange: An external or internal rim or ridge used for strengthening a structural component (see Chapter 6).
Flexural stress: See "bending stress."
Force: Commonly regarded as a "push" or a "pull."
Frayed-out wire joint: A joint with metal wire flayed into a softer material (see Figure 1.5).
Frozen adhesive joint: An adhesive joint whose stresses are unchanged by the curing of the adhesive.
Gasket: A thin material between parallel surfaces of jointed structural components.
Gaussian distribution: A normal distribution (see Chapter 4).
Griffith theory of fracture: See Section 3.15. In essence, a structure will not break as long as a crack length is less than a critical length.
Groove weld: Weld joint where the jointed component's geometry is designed to accept filler material (see Figure 9.4).
Gusset: A plate or rib for strengthening a joint (see Figure 6.13).
Hard joint: A perfectly rigid joint.
HAZ: In welding, the "heat affected zone."
Heat treating: Raising a material's temperature to improve its physical characteristics.
Hookean solid: A material behaving Hooke's law—that is, the strain is proportional to the stress.
Hoop stress: Tangential stress in a ring.
Hub: The central region of a circular component.
Hypothesis testing: See Section 4.7.
Impact riveting: The forcing of a rivet into its space by sudden impulsive loading ("hammering").
Interface: The interior contacting surfaces of a joint.
Interference fit: A permanent joint formed by placing a larger structural component within a smaller component. This is usually accomplished by temperature differences via heating and cooling.
Joint factor: In adhesive joints, it is the relation between the component stress and component overlap. Specifically, the ratio of the square root of component thickness to the overlap length (see Figure 11.15).
K: Symbol representing "torque coefficient" or "nut factor" (see Eq. (2.20)).
Key: A slender element designed to be placed in a slot (or "keyway") forming a joint between a shaft and a structural component (see Figures 8.5 and 8.6).
Keyway: A slot intended to accommodate a slender element (a "key") to form a joint between a shaft and a structural component.

Knuckle joint: A flexible joint connected via a pin—also known as an "eyebar joint."
Lamé stresses: Stresses arising in adhesive joints due to adhesive shrinkage during curing.
Lap joint: A joining of components by placing one on top of the other (see Figure 1.3).
Lap weld: Weld joint of overlapping components (see Figure 9.3).
Level of significance: The probability of making an error of the first kind (or an alpha error)—also known as "significant level."
Mean: Average
Membrane joint: See Chapter 10.
Modulus of elasticity: The ratio between normal stress and strain.
Modulus of rigidity: See "shear modulus."
Mohr's circle: A procedure for evaluating maximum and minimum stresses in a plane. (see Eqs. (3.13) and (3.14)).
Moment of inertia: A geometric property of a beam cross-section which measures the flexibility of the beam. It is usually denoted by the symbol: I and it has the units of l^4. Technically it is the "second moment of area." See strength of materials text for a precise definition.
Multipass procedure: Selective tightening of bolts with a gasketed joint so as to obtain a uniform compression between the jointed parallel surfaces.
Mylar barrier: See Section 10.4.
NDT: Acronym for nil-ductility-transition—the temperature where steel increases in fracture toughness.
Neutral axis: When a beam is bent, the stress created at a cross-section and directed normal (perpendicular) to the cross-section varies linearly from tension to compression across the cross-section, with the largest values occurring at the perimeter (upper and lower boundaries) of the cross-section. With this linear variation of the stress, there is a point near the center of the cross-section where the stress is *zero*—a neutral (or null) stress point. The neutral axis is the locus of these neutral points along the beam axis.
Nil-ductility-transition: The temperature where steel increases in fracture toughness and becomes more ductile.
Normal distribution: See Section 4.3.
Normal stress: See "stress".
Null hypothesis: Baseline of anticipated data.
Nut: A commonly used and familiar fastener component having an annular form with interior threads and a hexagonal boundary. The threads are designed to mate with those of a bolt.
Nut factor: A parameter K used to incorporate bearing friction and thread friction in bolt tightening—also known as "torque coefficient" (see Eq. (2.20)).
OC: Operating Characteristic Curve.
Operating characteristic curve: The probability of making an error of the second kind (erroneously accepting the null hypothesis)—also known as "OC."
Outgassing: The leaking of a gas or liquid through an adhesive or welded joint.
Peel strength: In adhesive joints, the resistance of the adhesive to being pulled apart.
Peeling: In adhesive joints the pulling apart of the adhesion.
Pin: A slender element, usually with cylindrical shape, to be inserted into a cavity between structural components, forming a detachable joint.
Pitch: The distance between two corresponding points of adjacent identical components.
Pitch diameter: For circular components the mean diameter where the pitch is measured.
Plastic deformation: A deformation occurring at a constant force, or at only a slight force increase, and with the deformation remaining after a force is removed.
Poisson ratio: The ratio of contraction to elongation for a deformed structural component, also known as the "transverse contraction ratio."
Population: A set or collection of objects being studied or examined.

Precision: The closeness of a series of measurements to one another—not necessarily accurate measurements.

Preload: The amount of load on a bolt or stud obtained by tightening before a structural load is placed on the joint. A preload is intended to keep the combined loading from becoming zero or undesirably small.

Press fit: A connection formed by forcing components together.

Producer error: Rejecting the null hypothesis when it is indeed valid—also known as: "alpha error" and as "error of the first kind."

Proof load: Load which stresses a bolt, screw, or stud up to a maximum value without a permanent set.

Prying: See Section 5.19.

Pull through: For cylindrical adhesive joints the tendency for one cylinder component to slide relative to the other.

Quality: A measure of goodness, reliability, robustness, and precision of a joint or structural component.

Quenching: Sudden cooling of a hot material by dipping into a cold liquid.

Random measurement errors: Errors occurring aside from systemic measurement errors except for blunders.

Random sample: A selection of items or objects without regard to any characteristic of an individual item or object. That is, each has an equal chance of being selected.

Relaxation: A reduction of load and/or stress.

Retaining ring: A ring for fitting into a circular slot or circular space to form a fastener on a shaft.

Ribbed flange: A stiffened flange via a rib perpendicular to the flange plane.

Rivet: A commonly used fastener having a cylindrical shape with rounded ends overlapping the jointed components.

Rivet shear: See Figure 2.2.

Rolling diaphragm: See Figure 10.11.

Root: The base, or innermost, point of a thread.

Root diameter: The diameter (or twice the radius) measured at the root of a thread.

Rule of thumb: A commonly accepted procedure for making an approximation in a design.

Rupture stress: The stress where fracture and then failure in a component occurs.

SAE: Society of Automotive Engineers

Sample: A subset of a population.

Scarf joint: A joining of components at an angle to their overall dimensions (see Figure 1.1). A joint where the jointed components have their interface geometry cut to fit together—usually by tapering or chamfering (see Figure 10.18).

Screw: A threaded fastener, which may be straight or slightly tapered.

Sealing: Sufficient snugness of a fastener so that neither liquids nor gasses can pass through the fastener.

Second moment of area: See "moment of Inertia."

Section modulus: The moment of inertia I of a beam cross section divided by the distance C from the neutral axis to the extreme outer perimeter of the cross-section. (I/C)

Set screw: A screw through a circular component projecting into another circular component thus forming a detachable fastener.

Shank: 1. The unthreaded portion of a bolt. 2. A straight narrow component of a joint.

Shear: A sliding or tendency to slide, of parallel surfaces relative to one another. It can refer to either force or deformation.

Either "shear force," "shear stress," "shear deformation," or "shear strain." See the corresponding glossary listings.

Shear deformation: A distortion caused by parallel, but oppositely directed forces.

Shear force: A force tangent to a surface.

Shear modulus: The ratio between shear stress and twice the shear strain. Also known as "modulus of rigidity."

Shear stress: See "stress".

Shear lag: The non-uniform distribution of shear stress among a series of fasteners (see Section 5.16).

Shelf life: The time for which a used fastener is expected to be reliable when placed into service.

Shrink: A means of creating a cylindrical or conical connection by reducing the size of the outer component usually by cooling a previously heated state.

Significant level: The probability of making an error of the first kind (or an alpha error)—also known as "significant level."

Simple supported beam: A beam supported at one end by a pin (or hinge) and at the other end by a roller (or small cylinder) allowing lengthwise movement.

Slip: A usually unintended sliding or movement of joined components relative to one another.

Snap fasteners: A fastener that becomes secure by a sudden spring through movement.

Snug: See "finger-tight."

Solder: A relatively low temperature melting material which when melted to liquid and placed between structural components and then cooled becomes an adhesive forming a permanent joint.

Spherical cap barrier: See Section 10.7.

Splice: A joining of two components by a similar component.

Split hub joint: 1. A joint formed by clamping a component about a circular shaft (see Figure 1.8).
2. A cylindrical joint whose outer member has a gap which when closed forms a tight connection (see Section 7.3).

Spin riveting: Insertion of a rivet into its space by rapid rotation of a structural component.

Spline: A type of key attached to one of two connecting components and which slides into an opening of the other component.

Spring: As a fastener, a spring is deformable but its fastening force increases linearly with the deformation.

Squeeze riveting: Compressing a rivet into its space.

Standard deviation: The square root of the variance.

Static: A stationary or time-independent condition (opposite of "dynamic").

Statistical inference: A conclusion based on a statistical analysis.

Statistical tolerance limit: A measurable characteristic of a population expected to lie within a stated interval.

Statically indeterminate system: A structure where a free-body diagram produces fewer equations than there are unknowns.

Stiffener: A strengthening structure.

Strain: A dimensionless measurement of deformation per unit length.

Strength: The character or force value of a material when the material is loaded to a point of failure.

Stress: A force per unit area. When the force is directed perpendicular or "normal" to a surface area it is called a "normal stress." If the force pushes against the surface the corresponding stress is called "compressive stress." If the force pulls away from the surface the corresponding stress is called "tension," "tensile stress," and/or "normal stress."

Stress concentration: A high localized value of stress.

Stress concentration factor: A number used in design to account for stress concentration.

Stress factor: A number used in design to account for complex geometry.

Stress gradient: A local, and generally significant, change in stress.

Stud: A rod or bar used as a fastener, as similar to a bolt or rivet.

Student's statistic: See Section 4.6—also known as "student t-test."

Student t-test: See Section 4.6—also known as "student's statistic."

Systemic measurement errors: Errors due to inaccuracies in a measurement device.

Tapped hole: A circular hole or circular cavity with internal threads.
Tee weld: Weld joint of perpendicular components (see Figure 9.3).
Tempering: Softening or toughening by heating at a relatively low temperature.
Tensile: A synonym for "tension."
Tensile stress: See "stress".
Tension: See "stress".
Threaded fasteners: A fastener that becomes secure by a sudden spring through movement.
Tightening torque: The torque or moment used in tightening a nut—also known as "wrenching torque."
Tolerance interval: A term often used to refer to either: "confidence interval," "engineering tolerance limit," or "statistical tolerance limit"—also known as "tolerance limit."
Tolerance limit: A term often used to refer to either: "confidence interval," "engineering tolerance limit," or "statistical tolerance limit"—also known as "tolerance interval."
Tongue and groove: A linear (lengthwise) joint with one component fitting inside the other.
Torque coefficient: A parameter K used to incorporate bearing friction and thread friction in bolt tightening—also known as "nut factor" (see Eq. (2.20)).
Torsion: Twisting of a beam or shaft (see Section 3.7).
Toughness: The energy required to break (or "fracture") a material—also known as "fracture energy" and "work of fracture."
Transverse contraction ratio: See Poisson ratio.
Transverse stress: Shear stress in a bent beam cross-section.
t-Statistic: See Section 4.6.
Tubular joint: A joint formed by intersecting tubes as in Figure 7.16.
Turn-of-nut technique: A complete turn of a nut *after* it has initially been tightened to a preload.
Two population distribution: See Sections 4.9 and 4.10.
Variance: The average of the square of the deviation from the mean.
von Mises stress: The square root of one half of the sum of the squares of differences in principal stresses.
Washer: An annular component placed between a nut or a bolt head and a connected component.
Weld: As a verb, it is the joining of two metal components via heating the components, with a filler, until they all melt together; as a noun, a weld is the connected joint.
Welded joint: The connection of components by melting their materials together and including a filler material.
Window joint: A membrane or diaphragm covering an opening in a structural component.
Woodruff key: A rounded key as in Figure 8.7.
Work of fracture: The energy required to break (or "fracture") a material—also known as "toughness" and "fracture energy."
Working load: The load on a joint in service.
Wrenching torque: The torque or moment used in tightening a nut—also known as "tightening torque."
Yield: The condition in a material when deformation begins to increase nonlinearly with an imposed loading.
Yield strength: The stress at the point of yielding (see "yield").
Yoke clamp: A clamp in the form of a draft animal's yoke as in Figure 7.13.
z-score: See Eq. (4.4).

Index

J

K

L

M

N

O

P

Q

R

S